OSHA

Technical Manual

Fifth Edition

U.S.Department of Labor
Occupational Safety and Health Administration

Government Institutes
a Division of ABS Group Inc.
Rockville, MD

Government Institutes, a Division of ABS Group Inc.
4 Research Place, Rockville, Maryland 20850, USA
Phone: (301) 921-2300
Fax: (301) 921-0373
Email: giinfo@govinst.com
Internet: http://www.govinst.com

03 02 01 00 99 5 4 3 2 1

This publication was prepared by the U.S. Department of Labor, Occupational Safety and Health Administration, for use within OSHA. Government Institutes determined that it contained information of interest to those outside OSHA so we are reproducing this material in order to serve those interested.

Production: Mark McIntosh
Cover Design: Tom Ehart
Cover Production: Gene Baldwin

ISBN: 0-86587-674-6

Printed in the United States of America

TABLE OF CONTENTS

EXTENSIVE TABLE OF CONTENTS

SECTION I: GENERAL INFORMATION

CHAPTER 1. INTRODUCTION TO OSHA TECHNICAL MANUAL (OTM)

TABLES AND FIGURES

EQUATIONS

SECTION I: CHAPTER 1 **[This section is a new addition to the OTM]**

INTRODUCTION TO OSHA TECHNICAL MANUAL (OTM)

Contents:

I. PURPOSE.

A. The OTM provides technical information and guidance on occupational safety and health topics. The content is based on currently available research publications, OSHA standards, and consensus standards. The purpose of the manual is to assist OSHA Compliance Safety and Health Officers in hazard recognition and to provide guidance in accident prevention. It also serves as a source of advice for CSHO's on safety and health issues.

B. The manual is not a substitute for OSHA standards and is not used for establishing Agency compliance policies. In the unlikely event of any inconsistencies between material in the manual and the requirements of the Occupational Safety and Health Act and the standards and regulations promulgated thereunder, the latter are controlling. The manual has ten sections and each section is divided into specific chapters. (See reference section below for a complete list of sections, chapters, and updates.) The information herein is valuable in establishing sound safety and health programs.

II. SCOPE.

This instruction applies OSHA-wide.

III. CANCELLATION.

OSHA Instruction TED 1.15 is cancelled.

IV. REFERENCES.

This directive may be used as a resource reference for compliance officers, in addition to OSHA Instruction CPL 2.103, Field Inspection Reference Manual (FIRM), which serves as a primary resource.

V. ACTION INFORMATION.

A. RESPONSIBLE OFFICE. Office of Science and Technology Assessment

B. ACTION OFFICES. None

C. INFORMATION OFFICES. All National Office Directorates and Regions, State and Consultation Designees

VI. ACTIONS REQUIRED.

OSHA's compliance safety and health officers should use this as one of their resources for technical information on occupational safety and health matters.

VII. SIGNIFICANT CHANGES.

The following are major changes that are made in this directive:

A. This Introduction (Section I) was added to the OTM.

B. In Section II, Chapter 1, "Personal Sampling for Air Contaminants," the word "additive" was replaced by "synergistic" to reflect synergistic exposures and their measurement. All sections were revised, and a new Appendix (II:1-7) was added. Chapter 2, "Sampling for Surface Contamination," was extensively revised. In Chapter 3, "Technical Equipment," the section on batteries was augmented; private delivery services were added to the mailing information; and brand names were removed from the Appendixes. In Chapter 4, "Sample Shipping and Handling," mailing information was revised; and under hazardous materials "biological samples" was added.

C. In Section III, Chapter 5, "Noise Measurement," the wording in the text describing the procedure for measuring attenuation for dual hearing protection was modified to be consistent with the wording in Figure III 5-1, "Calculating Hearing Protection Attenuation." A new section, "New Developments in Hearing Protection Labeling," and an Appendix on Sample NHCA Labels and Recommendations of the NHCA Task Force on Hearing Protector Effectiveness were added. A table showing Threshold Limit Values for Ultrasound was replaced with one adhering to 1998 standards published by ACGIH, Figure titled "Calculating Hearing Protector Attenuation" was changed to "Case Scenarios," which consists of clearer examples of when hearing protectors are required. In Chapter 7, "Legionnaires' Disease," the micron specification of a recommended filter was changed from 1 to 0.2.

D. In Section IV, Chapter 2, "Petroleum Refining Process,"15 tables were redesigned for a clearer representation of the process flows.

E. In Section VII, Chapter 1, "Back Disorders and Injuries," Appendix VII:1-1 was replaced to reflect NIOSH's revised equation for the design and evaluation of manual lifting tasks. The remaining Appendixes were replaced by the current Appendixes VII:1-2, VII:1-3, VII:1-4.

F. In Section VIII, a new chapter (Chapter 2, "Respiratory Protection") was added.

VIII. CONTENT OF THE MANUAL.

A. The OTM is divided into the following sections and chapters:

1. General Information

   Introduction to OSHA Technical Manual (OTM)

2. Sampling and measurement methods

   Personal sampling for air contaminants Sampling for surface contamination
   Technical equipment Sample shipping and handling

3. Health hazards

   Polymer matrix materials
   Indoor air quality investigation
   Ventilation investigation
   Heat stress

Noise measurement
Laser hazards
Legionnaires' disease

4. Safety hazards

   Oil well derrick stability
   Petroleum refining processes
   Pressure vessel guidelines
   Industrial robots

5. Construction operations

   Demolition
   Excavations
   Controlling lead exposures in the construction industry

6. Health care facilities

   Hospital investigations
   Controlling occupational exposure to hazardous drugs

7. Ergonomics

   Back disorders and injuries

8. Personal protective equipment

   Chemical protective clothing
   Respiratory protection

9. Safety and health management

10. Miscellaneous issues

    Metric system conversion

B. OTM chapters are current as of the date of the original OTM publication, September 22, 1995, except for a) the chapters listed in the abstract under the headings "Recent Updates," and b) two chapters that were published on May 24, 1996 (Section III, Chapter 7, "Legionnaires' Disease" and Section IV, Chapter 2, "Petroleum Refining Processes") and are current as of that date.

C. Section VIII, Chapter 2 ("Respiratory Protection") is a new chapter and current as of May 1998.

D. The bibliography at the end of each chapter provides a complete reference list.

SECTION II: CHAPTER 1 **[New additions and extensively revised]**

PERSONAL SAMPLING FOR AIR CONTAMINANTS

Contents:

I. INTRODUCTION.

A. Effective and efficient sampling strategies require planning and foresight to ensure the most productive and thorough evaluation of air contaminants in the workplace. Air sampling should be conducted when necessary and laboratory resources should be used wisely to avoid delays in reporting results of necessary sampling. When possible, evaluate the potential for employee overexposure by observing work practices and screening samples before conducting any partial or full-shift air sampling.

B. Screening with portable monitors, gravimetric sampling, or detector tubes should be used to evaluate the following:

1. Exposures to substances with exceptionally high permissible exposure limits (PEL's) in relatively dust-free atmospheres, e.g., ferric oxide and aluminum oxide;

2. Intermittent processes involving substances without short-term exposure limits (STEL's);

3. Engineering controls, work practices, or isolation of process; and the need for CSHO protection;

4. Substances that have Ceiling exposure limits. (There are validated direct-reading sampling devices available specifically for these substances.)

C. Take a sufficient number of samples to obtain a representative estimate of exposure. Contaminant concentrations vary seasonally, with weather, with production levels, and in a single location or job class. The number of samples taken depends on the error of measurement and differences in results. Consult the NIOSH Occupational Exposure Sampling Strategy Manual for further information.

D. If the employer has conducted air sampling and monitoring in the past, review the records.

E. Bulk samples are often required to assist the Salt Lake Technical Center (SLTC) in the proper analysis of field samples. (See Section II, Chapter 4, Sample Shipping and Handling.) Some contaminants in these categories are:

- silica
- Portland cement
- asbestos
- mineral oil and oil mist
- chlorodiphenyl
- hydrogenated terphenyls
- chlorinated camphene
- fugitive grain dust
- explosibility testing

F. Bulk samples can also be taken and analyzed to support any Hazard Communication inspections (i.e., Material Safety Data Sheet determinations).

II. GENERAL SAMPLING PROCEDURES.

NOTE: Radio frequency electromagnetic fields can interfere with the proper operation of industrial hygiene instruments. This interference is called electromagnetic susceptibility (EMS). Determine if there is a potential for such interference. Likely sources of radio frequency interference are walkie-talkies, vehicles equipped with mobile radio transmitters, RF heat sealers, etc. If there is a potential for such interference, select sampling instruments that are properly rated for EMS to avoid faulty data or malfunction.

A. SAMPLING PREPARATIONS. Screen the sampling area with detector tubes, if appropriate. Determine the appropriate sampling technique [see the Chemical Sampling Information section of the Technical Information link found via the Web (http://www.osha-slc.gov/SLTC/index.html) or the Technical Information section of OSHA CD-ROM]. prepare and calibrate the equipment and prepare the filter media (if sample weights are necessary, see Section VI., Filter weighing, below.). Figure II:1-1a shows the appropriate assembly for the three piece sampling cassette used to collect specific particulates. Other sampling media usage and preparation can be found within this chapter, via Technical Links (Chemical Sampling Information or specific analytical methods), or the OSHA CD.

B. SELECT THE EMPLOYEE to be sampled and discuss the purpose of the sampling. Inform the employee when and where the equipment will be removed. Stress the importance of not removing or tampering with the sampling equipment. Instruct the employee to notify the supervisor or the CSHO if the sampler requires temporary removal.

C. STARTING SAMPLING.

1. Place the sampling equipment on the employee so that it does not interfere with work performance. Attach the collection device (filter cassette, charcoal tube, etc.) to the shirt collar or as close as practical to the nose and mouth of the employee, i.e., in a hemisphere forward of the shoulders with a radius of approximately six to nine inches. The inlet should always be in a downward vertical position to avoid gross contamination. Position the excess tubing so that it does not interfere with the work of the employee.

2. Turn on the pump and record the starting time.

3. Observe the pump operation for a short time after starting to make sure it is operating correctly.

4. Record the information required by the Air Sampling Data Form (OSHA 91A).

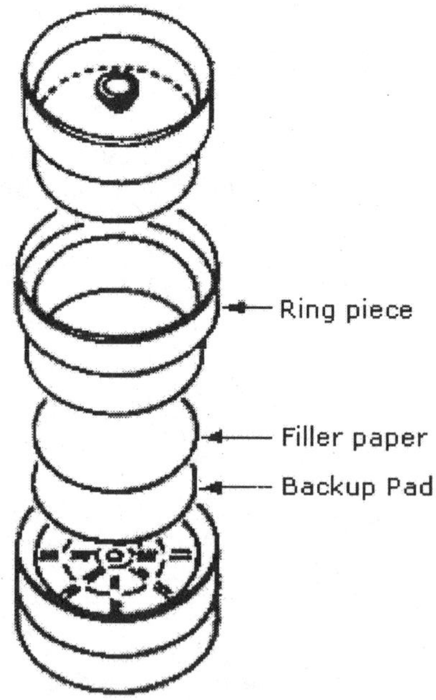

FIGURE II:1-1a. EXPLODED VIEW OF THREE-PIECE CASSETTE SHOWS PLACEMENT OF BACK-UP PAD.

D. MONITORING.

1. Check pump every two hours. More frequent checks may be necessary when heavy filter loading is possible. Ensure that the sampler is still assembled properly and that the hose has not become pinched or detached from the cassette or the pump. For filters, observe for symmetrical deposition of particulate on the filter, unexpected large particles, or other evidence of sample tampering with the sample or pump. Record the flow rate and any relevant observations.

2. Periodically monitor the employee throughout the workday to ensure that sample integrity is maintained and cyclical activities and work practices are identified. Turn off or remove sampling pumps immediately prior to an employee leaving a potentially contaminated area (such as when he/she goes to lunch or on a break in a clean area). If these areas also appear contaminated and are considered part of the workplace, continue sampling and assess the need for surface contamination measurements (see also Section II, Chapter 2, Sampling for Surface Contamination).

3. Take photographs (as appropriate) and detailed notes concerning visible airborne contaminants, work practices, potential interferences, movements, and other conditions to assist in determining appropriate engineering controls.

4. Prepare blank(s) during the sample period for each type of sample collected. (Also see Section II, Chapter 4, Sample Shipping and Handling.) One blank will suffice for up to 20 samples for any given analysis/sampling period except asbestos, which requires a minimum of two field blanks. The blanks should be opened but not used to take samples (charcoal tubes, filters etc.). They should be handled in the same manner as any sampling media used in sampling air contaminants, with the exception that no air is drawn through them.

E. FINAL PROCEDURES.

1. Before removing the pump at the end of the sample period, if there is a pump rotameter, check the flow rate to ensure that the rotameter ball is still at the calibrated mark. If the ball is no longer at the mark, record the pump rotameter reading.

2. Turn off the pump and record the ending time.

3. Remove the collection device from the pump and seal it with an OSHA-21 form as soon as possible. The seal should be attached across sample inlet and outlet so that tampering is not possible. (See Figures II:1-1b and II:1-1c.)

4. Prepare the samples for mailing to the Salt Lake Technical Center (SLTC) for analysis. (Also see Section II, Chapter 4.) Mail bulk samples and air samples separately to avoid cross-contamination. If any sample materials could be considered hazardous, always consult and follow appropriate shipping regulations to assure safe handling during shipment. Pack the samples securely to avoid any rattle or shock damage (do not use expanded polystyrene packing). Use bubble sheeting as packing. Put identifying paperwork in every package. Do not send samples in plastic bags or in envelopes. Use OSHA Form 91A. Print legibly on all forms.

5. Recalibrate pumps after each day of sampling (before charging). Take proper care of battery packs (see Section II, Chapter 3 - III. Batteries).

6. When calibrating conditions are significantly different from sampling site conditions such as large temperature and pressure differences, call the Salt Lake City Technical Center.

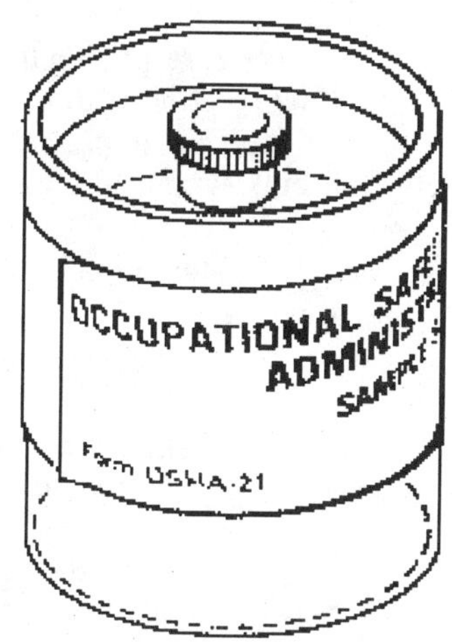

FIGURE II: 1-1b. IMPROPERLY SEALED CASSETTE ALLOWS ACCESS TO INLET AND OUTLET AFTER SAMPLE HAS BEEN TAKEN

III. SAMPLING MEDIA.

A. DETECTOR TUBES. Each pump should be leak-tested before use. Calibrate the detector tube pump for proper volume at least quarterly or after 100 tubes. (See Appendix II:1-1.)

B. TOTAL DUST AND METAL FUME - AEROSOL SAMPLERS.

1. Collect total dust on a pre-weighed, low-ash polyvinyl chloride (PVC) filter (See Section VI, Filter Weighing below) at a flow rate of about 2 liters per minute (L/min), depending on the rate required to prevent overloading.

2. Collect metal fumes on a 0.8-micron mixed cellulose ester (MCE) filter at a flow rate of approximately 1.5 L/min, not to exceed 2.0 L/min. When the gravimetric weight needs to be determined for welding fumes, use the sampling device described in Section VI, Filter Weighing, for gravimetric determinations.

3. Take care to avoid overloading thc filtcr, as cvidenced by any loose particulate.

4. Calibrate personal sampling pumps before and after each day of sampling, using a bubble meter method (electronic or mechanical) or the precision rotameter method (that has been calibrated against a bubble meter) , as described in Section V, part E-2, below. Assure the calibration equipment bubble meter, etc.) is within its prescribed calibration interval, and record the serial number of the calibration bubble meter (or rotameter) in your data.

   NOTE: Using precision rotameters for pump calibration should be avoided if at all possible. Tests have indicated significant measurement errors due to pump pulsations.

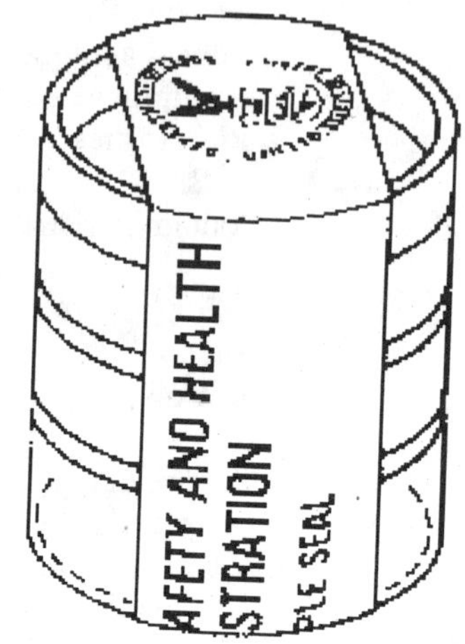

FIGURE II: 1-1c. PROPERLY SEALED CASSETTE WITH OSHA-21 FORM COVERING INLET AND OUTLET PORTS PROVIDES SECURITY.

5. PVC filters used for gravimetric analysis need to be pre- and post-weighed. See Section VI, Filter Weighing for further details.

C. RESPIRABLE DUST SAMPLERS.

Collect respirable dust using a clean cyclone equipped with a pre-weighed low-ash polyvinyl chloride filter at a flow rate of 1.5 to 2.0 L/min. (See Figure II:1-2.). Pre-weighed filters are available from SLTC or CTC (see Section VI, Filter Weighing, below, for further information). Collect silica only as a respirable dust, and, if possible, submit a bulk sample also to SLTC.

1. Calibration Procedures

   a. Perform the calibration at the pressure and temperature where the sampling is to be conducted.

   b. For respirable dust sampling using a cyclone, or for total dust sampling using an open-face filter cassette, set up the calibration apparatus as shown in Figure II:1-9.

   c. Place the open-face filter cassette, or cyclone assembly in a 1-liter jar. The jar is provided with a special cover.

   d. Connect the tubing from the electronic bubble meter to the inlet of the jar.

   e. Connect the tubing from the outlet of the cyclone holder assembly or from the filter cassette to the outlet of the jar and then to the sampling pump.

f. Calibrate the pump with a light and heavy load, both pre- and post-sampling. All readings must be within five percent of each other.

2. Alternative Calibration Method - "Jarless Cyclone Calibration."

The following "Jarless Cyclone Calibration" is an alternative method to calibrating the cyclone with a 1-liter jar. This method will be included in the NIOSH Manual of Analytical Methods (NMAM), Fourth Edition.

```
                Pressure Gauge
                      |
air flow              |                          air flow
  ⇦      Pump ----- TEE ----- Load ----- Bubble Meter   ⇦
```

a. Connect the air sampling pump to a TEE fitting, a pressure gauge (0" to 50" H2O vacuum full scale) or water manometer, a light load (typically a 5-µm, 37mm filter), and a calibrated electronic bubble meter (or standard bubble tube).

b. Adjust the air sampling pump to 1.7 L/min as indicated on the bubble meter, and adjust the loading, if necessary, to produce a 2" to 5" H2O indication on the pressure (vacuum)gauge.

c. Increase the loading (typically five to six 0.8-µm, 37-mm filters) until the pressure (vacuum) gauge indicates between 25" and 35" H2O, and check the flow rate again. The flow rate should remain at 1.7 L/min ±5%.

d. Finally, replace the load and bubble meter with the cyclone having a clean filter Installed, and verify that the loading caused by the cyclone assembly is between 2" and 5" H2O.

```
                Pressure Gauge
                      |
air flow              |                    air flow
  ⇦      Pump ----- TEE ----- Cyclone Assembly   ⇦
```

e. This calibration method actually performs a dynamic test of the pump under load.

3. Cyclone Cleaning.

a. Unscrew the grit pot from the cyclone. Empty the grit pot by turning it upside down and tapping it gently on a solid surface.

b. Clean the cyclone thoroughly and gently after each use in warm soapy water or, preferably, wash in an ultrasonic bath. Rinse thoroughly in clean water, shake off excess water, and set aside to dry before reassembly. Never insert anything into the cyclone during cleaning. See Figure II:1-2.

c. Inspect the cyclone parts for signs of wear or damage such as scoring, rifling, or a loose coupler. Replace the units or parts if they appear damaged.

d. Leak test the cyclone before use unless it has been leak tested within the past month. A cyclone leak Test Kit and Cyclone Leak Test Procedure are provided in each office for this purpose.

e. Detailed instructions on leak testing are available from the Directorate of Technical Support, Cincinnati Technical Center (CTC) and are available through Technical Links on our Web Page for silica.

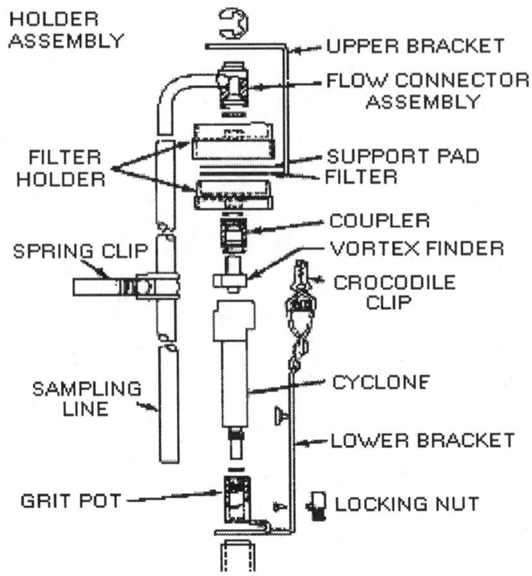

FIGURE II: 1-2. CYCLONE ASSEMBLY

D. SOLID SORBENT TUBES.

1. Organic vapors and gases may be collected on activated charcoal, silica gel, or other adsorption tubes using low-flow pumps (see Figure II:1-3).

2. Immediately before sampling, break off the ends of the flame-sealed tube so as to provide an opening approximately half the internal diameter of the tube. Wear eye protection when breaking ends. Use tube holders, if available, to minimize the hazards of broken glass. Do not use the charging inlet or the exhaust outlet of the pump to break the ends of the tubes.

3. Use the smaller section of the tube as a back-up and position it near the sampling pump. The tube shall be held or attached in an approximately vertical position with the inlet either up or down during sampling.

4. Draw the air to be sampled directly into the inlet of the tube. This air is not to be passed through any hose or tubing before entering the tube.

5. Cap the tube with the supplied plastic caps immediately after sampling and seal with an OSHA-21 form as soon as possible. (See Figures II:1-4a and b.) Do not ship tubes with bulk material.

6. Tubes may be furnished by SLTC with either caps or flame-sealed glass ends. If using the capped version, simply uncap during the sampling period and recap at the end of the sampling period.

7. For organic vapors and gases, low-flow pumps are required. Refer to the Chemical Sampling Information for flow rates recommended for specific chemicals.

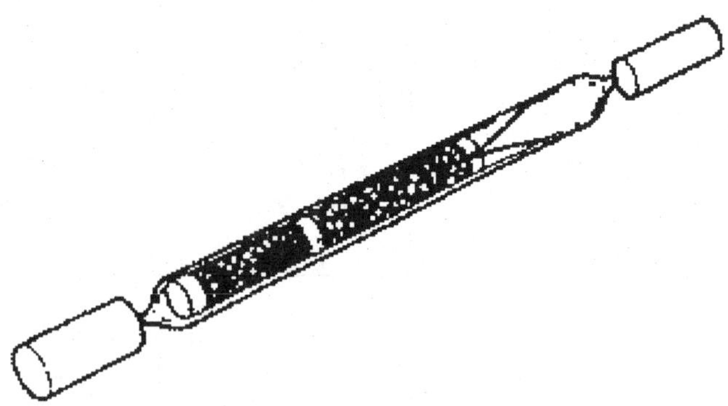

FIGURE II: 1-3. THE CHARCOAL OR "C" TUBE WITH GLASS SEALED END AND NIOSH-APPROVED CAPS BEFORE SAMPLING.

8. With sorbent tubes, flow rates may have to be lowered or smaller air volumes (half the maximum) used when there is high humidity (above 90%) in the sampling area or relatively high concentrations of other organic vapors are present.

9. Set up the calibration apparatus as shown in Figure II:1-8 replacing the cassette with the solid sorbent tube to be used in the sampling (e.g., charcoal, silica gel, etc.). If a sampling protocol requires the use of two charcoal tubes, the calibration train must include two charcoal tubes. The air flow must be in the direction of the arrow on the tube. Calibrate the pump.

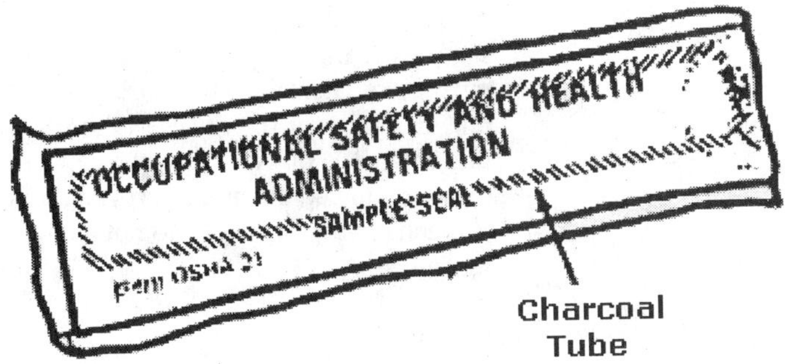

FIGURE II: 1-4a. CORRECTLY SEALED C-TUBE. SAMPLE IS COMPLETELY ENCLOSED IN THE SEAL, AND NO TAMPERING IS POSSIBLE

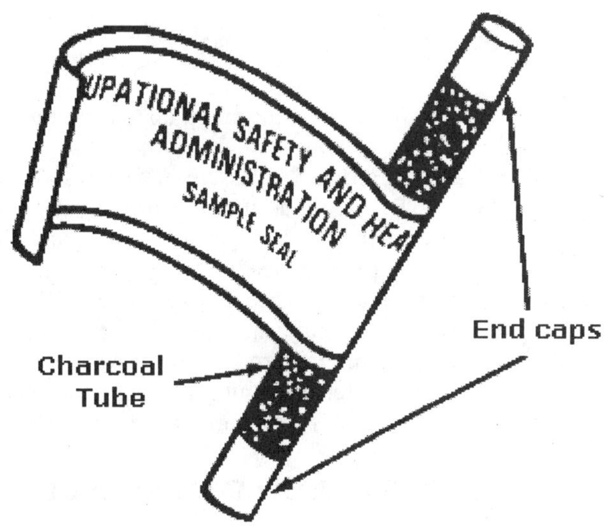

INCORRECTLY SEALED C-TUBE. END CAPS CAN BE REMOVED AND SAMPLE INTEGRITY JEOPARDIZED WITHOUT DISTURBING THE SEAL.

E. MIDGET IMPINGERS AND BUBBLERS.

1. Method.

a. Take care in preparing bubblers and impingers to see that frits or tips are not damaged and that joints can be securely tightened.

b. Rinse the impinger or bubbler, Figure II:1-5, with the appropriate reagent (see Technical Links, Chemical Sampling Information via the Web or the OSHA-CD and Appendix II:1-4). Then add the specified amount of this reagent to the bubbler or impinger flask either in the office or at the sampling location. If flasks containing the reagent are transported, caps must be placed on the bubbler or impinger stem and side arm.

c. To prevent overflow, do not add over 10 ml of liquid to the midget impingers or bubblers.

d. Collect contaminants in an impinger or bubbler at a maximum flow rate of 1.0 L/min. Because bubblers tend to offer better collection efficiency than impingers, they are the preferred method over impingers for gas and vapor collection. Impingers are used, if absolutely necessary, for particle counting. Contact the SLTC prior to collecting any samples for particle (dust) counting.

e. The impinger or bubbler may either be hand-held by the industrial hygienist or attached to the employee's clothing using an impinger or bubbler holster. In either case, it is very important that the impinger or bubbler does not tilt and cause the reagent to flow down the side arm to the hose and into the pump.

NOTE: Attach a trap in line to the pump, if possible.

f. In some instances, it will be necessary to add additional reagent during the sampling period to prevent the amount of reagent from dropping below one half of the original amount.

g. After sampling, remove the glass stopper and stem from the impinger or bubbler flask.

h. Rinse the absorbing solution adhering to the outside and inside of the stem directly into the impinger or bubbler flask with a small amount (1-2 ml) of the sampling reagent. Pour the contents of the flask into a 20-ml glass bottle (preferably a scintillation vial with inert caps and liners). Avoid using metal cap liners or other materials that may react with the samples. Teflon cap liners with polypropylene caps are normally inert to most materials. Rinse the flask with a small amount (1-2 ml) of the reagent and pour the rinse solution into the bottle or vial. Tape the cap shut by wrapping the tape in the direction of cap closure to prevent it from coming loose due to vibration. If electrical tape is used, do not stretch tape since it will contract and loosen cap.

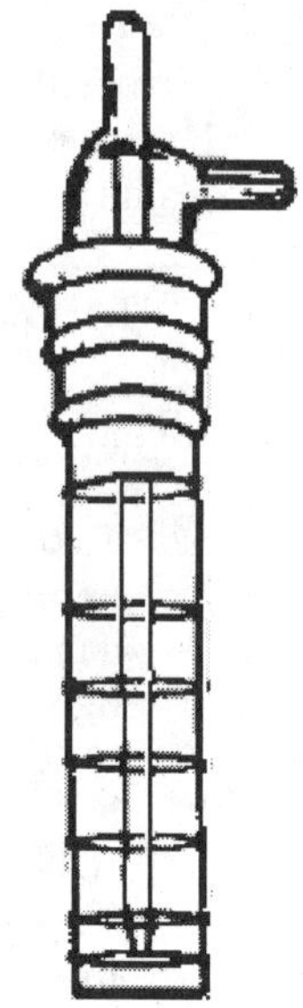

FIGURE II: 1-5. A TYPICAL GLASS BUBBLER

2. Calibration.

a. Set up the calibration apparatus as shown in Figure II:1-8 and replace the cassette with the impinger or bubbler filled with the amount of liquid reagent specified in the sampling method. (Refer to Chemical Sampling Information via Technical Links or the OSHA CD.)

b. Connect the tubing from the electronic bubble meter to the inlet of the impinger or bubbler.

c. Connect the outlet of the impinger or bubbler to the tubing to the pump.

d. Calibrate the pump up to a maximum flow rate of 1.0 L/min.

F. VAPOR BADGES. Passive-diffusion sorbent badges (also known as monitors), Figure II:1-6, are useful for screening and compliance monitoring for certain chemical exposures, especially vapors and gases. The major advantage is that no sampling pump is needed to take the samples. Some badges have been validated for use in compliance. Badges are available from the SLTC to detect mercury, nitrous oxide, ethylene oxide, formaldehyde, and other organic substances. Specific sampling procedures for each type of badge are also supplied with the monitors. Interfering substances should be noted during sampling and recorded. Contact the SLTC for further information regarding badge availability and use.

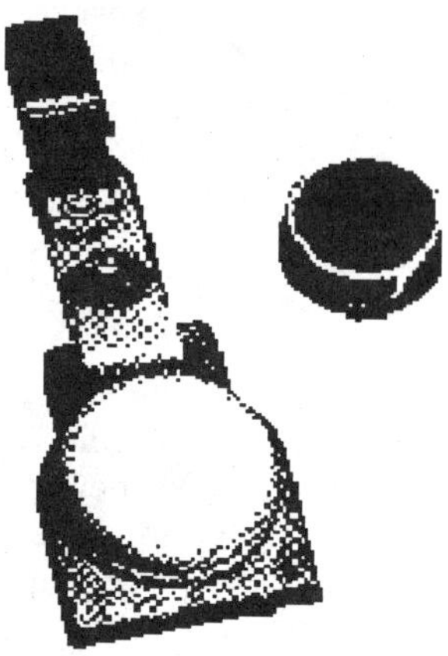

FIGURE II:1-6. VAPOR BADGE WITH CLOTHING CLIP.

IV. SPECIAL SAMPLING PROCEDURES

A. ASBESTOS.

1. Collect asbestos on a special 0.8 micrometer pore size, 25-mm diameter mixed cellulose ester filter with a back-up pad. Use a fully conductive cassette with conductive extension cowl, Figure II:1-7.

2. Sample open face (the filter/cowl is open to the sampling environment; the extension cowl is in place but the cowl end piece is removed from the extension) in the worker's breathing zone.

3. Ensure that the bottom joint (between the extension and the conical black piece) of the assette is sealed tightly with a shrink band or electrical tape. Point the open face of the cassette down to minimize contamination.

4. Use a flow rate in the range of 0.5 to 5 L/min. One liter per minute is suggested for general sampling. Office environments allow flow rates of up to 5 L/min. Calibrate pump before and after sampling. Calibration may be done either as in Figure II:1-8 or Figure II:1-9. Do not use nylon or stainless-steel adaptors if in-line (Figure II:1-8) calibration is done.
5. Sample for as long a time as possible without overloading (obscuring) the filter.

6. Instruct the employee to avoid knocking the cassette and to avoid using a compressed-air source that might dislodge the sample while sampling.

7. Submit 10% blanks, with a minimum of two blanks in all cases.

8. Where possible, collect and submit to the SLTC a bulk sample of the material suspected to be in the air.

9. Mail bulk sample and air samples separately to avoid cross-contamination. Pack the samples securely to avoid any rattle or shock damage (do not use expanded polystyrene packing). Use bubble sheeting as packing. Put identifying paperwork in every package. Do not send samples in plastic bags or envelopes. Use OSHA Form 91A. Print legibly on all forms.

10. For exceptional sampling conditions or high flow rates, contact the SLTC. More detailed instructions can be obtained from SLTC.

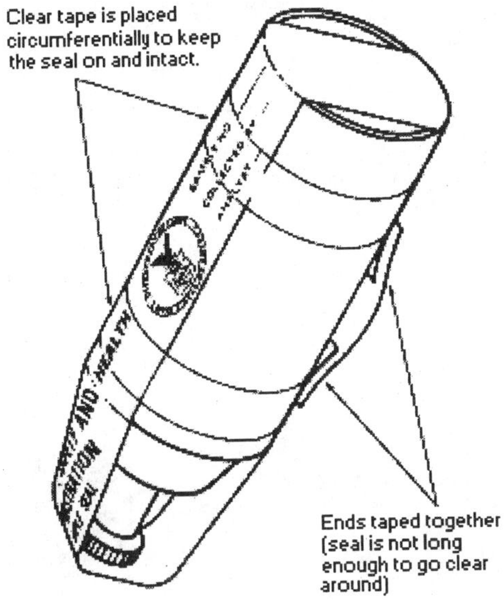

FIGURE II: 7. A STANDARD ASBESTOS CASSETTE (25mm) SEALED WITH AN OSHA 21 FORM.

B. SAMPLING FOR WELDING FUMES.

1. When sampling for welding fumes, the filter cassette must be placed inside the welding helmet to obtain an accurate measurement of the employee's exposure.

   NOTE: The policy of placing the inlet of the sampling device inside of personal protective equipment for the face or eyes applies only to welding hoods. The inlet of the sampling device shall be placed outside other face or eye personal protective equipment, such as face shields for projectiles or chemical splashes.

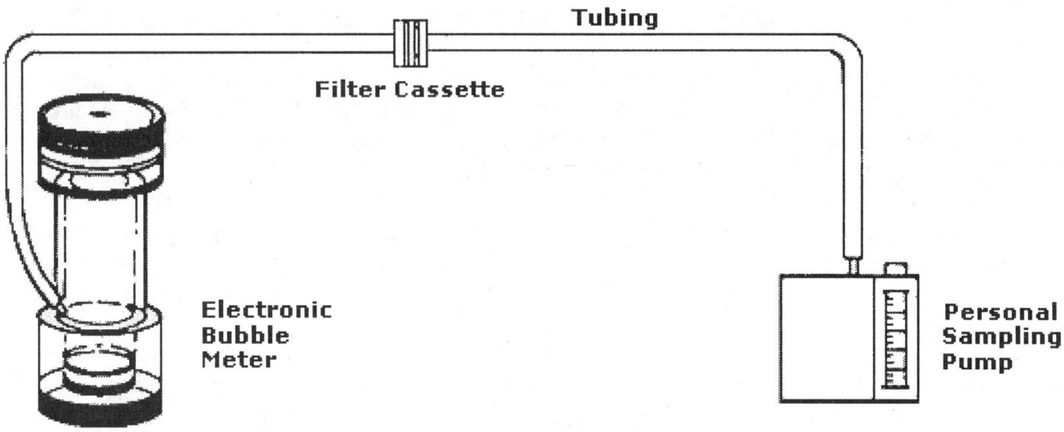

FIGURE II: 1-8. FOR CALIBRATION, THE CASSETTE IS ATTACHED TO AN ELECTRONIC BUBBLE METER.

2. Welding fume samples are normally taken using 37-mm filters and cassettes; however, if these cassettes will not fit inside the helmet, 25-mm filters and cassettes can be used. Care must be taken not to overload the 25-mm cassette when sampling.

3. The Assistant Regional Administrator for Technical Support should be consulted in the case of technical difficulties.

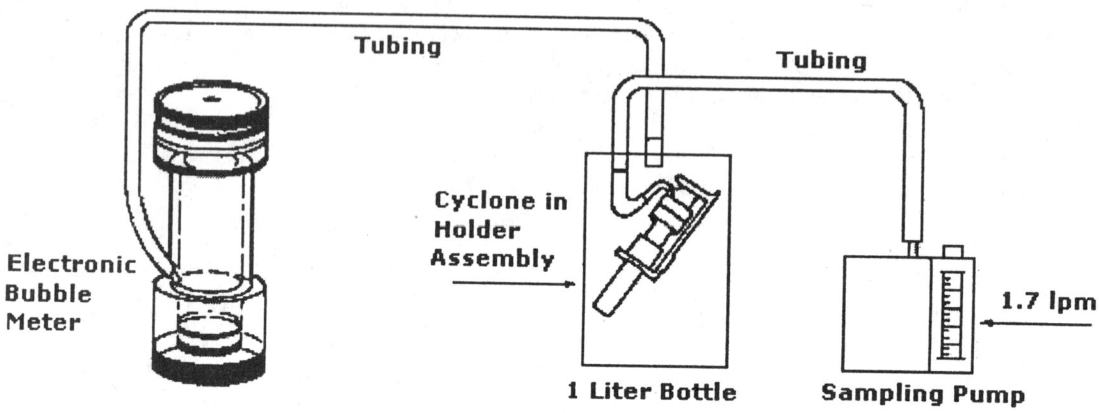

FIGURE II: 1-9. THE CYCLONE IS CALIBRATED BY PLACING THE CYCLONE IN A 1 LITER VESSEL ATTACHED TO AN ELECTRONIC BUBBLE METER.

V. EQUIPMENT PREPARATION AND CALIBRATION.

A. ALKALINE BATTERIES. Replace alkaline batteries as needed. Keep fresh replacement batteries with the equipment.

B .RECHARGEABLE NI-CAD BATTERIES. Check the rechargeable Ni-Cad batteries in older pumps under load (e.g., turn pump on and check voltage at charging jack) before use. (See Section II, Chapter 3, Part III. - Batteries)

C. TIME OF CALIBRATION. Calibrate personal sampling pumps before and after each day of sampling, using either the electronic bubble-meter method or the precision-rotameter method calibrated against a bubble meter. Make sure that the calibration equipment (bubble meter, etc.) is within its prescribed calibration interval and record the serial number of the calibration instrument in your data.

D. ELECTRONIC FLOW CALIBRATORS. These units are high-accuracy electronic bubble flow meters that provide instantaneous air-flow readings and cumulative averaging of multiple samples. These calibrators measure the flow rate of gases and present the results as volume per unit of time and should be used to calibrate all air-sampling pumps. Appendix II:1-1 provides more information on this piece of equipment.

E. CALIBRATION. When a sampling train requires an unusual combination of sampling media (e.g., glass fiber filter preceding impinger), the same media and devices should be in line during calibration.

1. Electronic Bubble Meter Method.

a. Allow the pump to run five minutes prior to voltage check and calibration.

b. Assemble the cassette filter holder, using the appropriate filter for the sampling method. Compress cassette by using a mechanical press or other means of applying pressure. Use shrink tape around cassette to cover joints and prevent leakage. If a cassette adapter is used, care should be taken to ensure that it does not come in contact with the back-up pad.

NOTE: When calibrating with a bubble meter, cassette adaptors can cause moderate to severe pressure drop at high flow rates in the sampling train and affect the calibration result. If adaptors are used for sampling, they should also be used when calibrating.

CAUTION: Nylon adapters can restrict air flow due to plugging. Stainless-steel adapters are preferred.

c. Connect the collection device, tubing, pump, and calibration apparatus as shown in Figure II:1-8 for the cassette sampler and Figure II:1-9 for the cyclone sampler.

d. Visually inspect all Tygon tubing connections.

e. Wet the inside of the electronic flow cell with the soap solution supplied by pushing on the button several times.

f. Turn on the pump and adjust the pump rotameter, if available, to the appropriate flow rate.

g. Press the button on the electronic bubble meter. Visually capture a single bubble and electronically time the bubble. The accompanying printer will automatically record the calibration reading in liters per minute.

h. Repeat the step until two readings are within 2%.

i. If necessary, adjust the pump while it is still running.

j. Repeat the procedures described above for all pumps to be used for sampling. The same cassette and filter may be used for calibrations involving the same sampling method.

2. Precision Rotameter Method.

The precision rotameter, Figure II:1-10, is a secondary calibration device. If it is to be used in place of a primary device such as a bubble meter, take care to ensure that any error introduced will be minimal and noted.

NOTE: Using precision rotameters for pump calibration should be avoided if possible. Tests have indicated significant measurement errors due to pump pulsations. These pulsations are typically not observable by the user.

a. Replacing the Bubble Meter. The precision rotameter may be used for calibrating the personal sampling pump in lieu of a bubble meter, provided it is:

- Calibrated regularly, at least monthly, with an electronic bubble meter or a bubble meter, as described in Appendix II:1-3.
- Disassembled, cleaned as necessary, and recalibrated. (It should be used with care to avoid dirt and dust contamination, which may affect the flow.)
- Not used at substantially different temperature and/or pressure from conditions present when the rotameter was calibrated against the primary source.
- Used in such a way that the pressure drop across it is minimized.

b. Unusual Conditions. If altitude or temperature at the sampling site are substantially different from those at the calibration site, it is necessary to calibrate the precision rotameter at the sampling site.

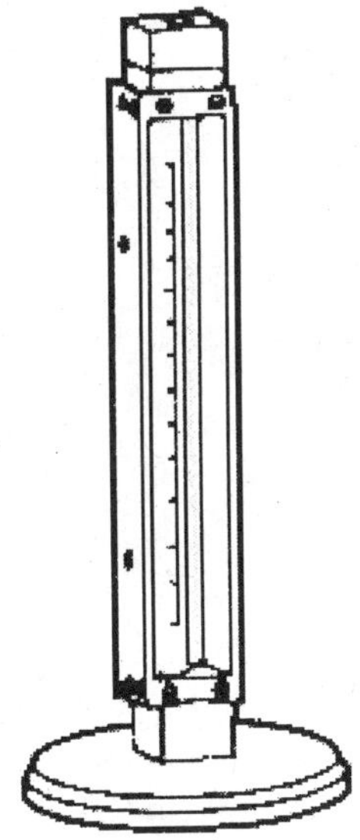

FIGURE II: 1-10. A SINGLE COLUMN PRECISION ROTAMETER CAN BE USED AS A SECONDARY CALIBRATION DEVICE

3. Manual Buret Bubble Meter Method. See Appendix II:1-3.

VI. FILTER WEIGHING.

The SLTC is providing pre-weighed filters for gravimetric analysis. These filters should reduce sample preparation time by CSHO's in the field because the filters are weighed at SLTC and shipped to the field assembled and ready for use in inspections. The filters will then be returned to SLTC for gravimetric and any other additional analyses.

A. THE FILTER MEDIA is 37-mm diameter, low-ash polyvinyl chloride. These filters should be used for silica (quartz) analysis, aluminum, and any other appropriate substance having a high PEL or requiring gravimetric analysis. The filters may be used without the cyclone attached for total dust analyses. Please indicate on the 91A form all analytes of interest. If gravimetric analysis yields a result less than the PEL for the requested substance(s), no further analysis will be provided unless specifically requested. The filter/cassette unit, shown in Figure II:1-11a, is expensive; please only use it for samples requiring crystalline silica gravimetric analysis or other gravimetric analysis. Appendix II:1-7 is a partial listing of suggested substances that can be sampled and gravimetrically analyzed using the pre-weighed cassettes.

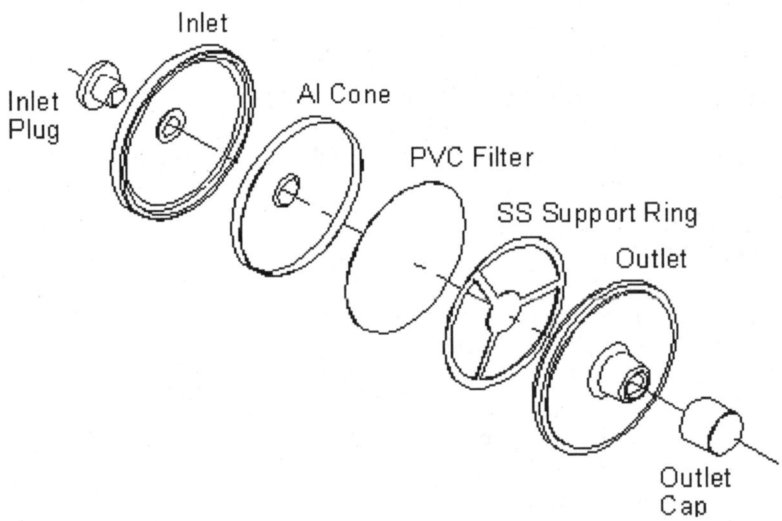

FIGURE II: 1-11a. THE FILTER/ CASSETTE UNIT

B. AVAILABILITY. These filters are shipped pre-weighed and assembled in the cassettes. They are available directly from the SLTC or through the CTC "Expendables Supplies" menu in the "OSHA Property Management Inventory System (OPMIS)." (Listed as "Cassette, Aerosol, 37MM, Preassembled, **Pre-weighed at SLTC**, #FES0000161).

C. INSTRUCTIONS FOR CALIBRATION, SAMPLING AND SUBMISSION OF SAMPLES. Be sure to follow all appropriate protocols for calibration, sampling and submission of samples. A blank should be included with every set of samples. Due to the slightly smaller size of the sampler, check frequently to prevent filter overloading. This can be accomplished by viewing the inlet sampling port of the cassette. Visual observation of the airborne dust concentration around the worker may assist in determining how frequently to check the filter for overloading.

1. Two versions of the pre-weighed cassettes exist. New cassettes have a stainless steel filter support (shown in Figure II:1-11b), and older cassettes have a Tyvek® backup pad. The stainless steel support seems to be less affected by humidity (providing a more stable blank weight) than the old pad. Both styles may even be used in a

single inspection; however, it is very important that the blanks submitted with a set of samples have the same backup pad as the samples.

2. As shown in Figure II:1-11c, the inlet side of the cassette is marked on the polystyrene cassette. This is the side of the filter cassette with the metallic antistatic shield. The Tyvek® backup pad or the stainless steel support is visible from the outlet side of the assembly. Each of the filter assemblies is bar coded for weighing purposes. To aid in the tracking of the filters, please use the barcode number for the sample submission number when filling out OSHA form 91A. For your convenience, the barcode number is printed on a label attached to the outlet side of the cassette.

FIGURE II: 1-11b. STAINLESS STEEL FILTER SUPPORT

3. The filter/cassette assembly can be used with both nylon cyclone and holder assemblies currently in field use; however, the standard MSA coupler (used with a standard 2- or 3-piece cassette) will not fit these cassettes. Another coupler available from MSA (part #457391), which is plastic instead of stainless steel, can be obtained from CTC.

4. If a compliance officer wishes to employ field balance weighing techniques instead of the SLTC weighing program, please consult a prior version of the Technical Manual (Sampling Chapter, Section VI) or contact the SLTC for further instructions.

FIGURE II: 1-11C. SIDES OF THE FILTER CASSETTE

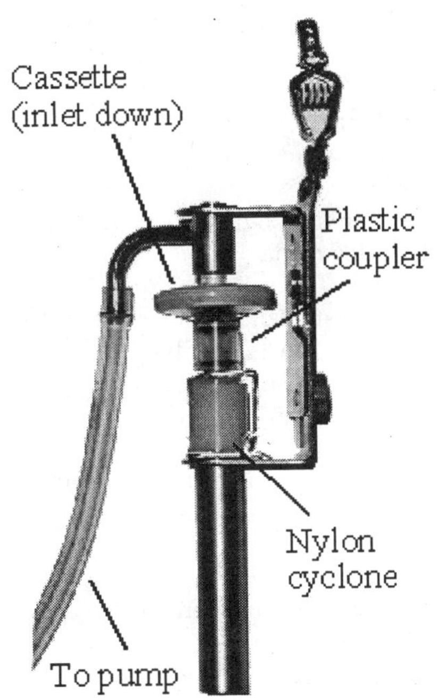

FIGURE II: 1-11d. MSA CYCLONE HOLDER WITH CASSETTE

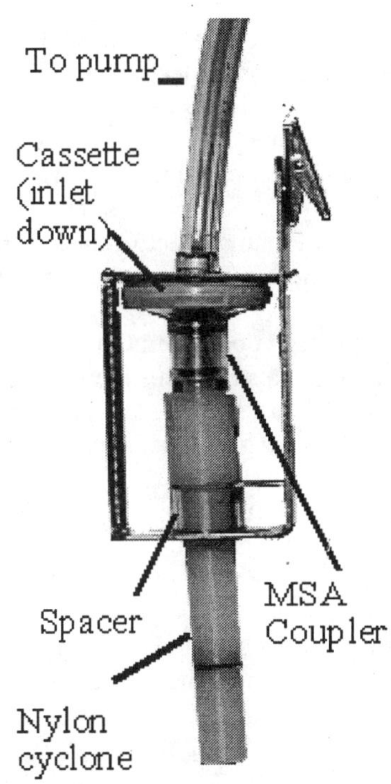

FIGURE II: 1-11e. BENDIX HOLDER WITH CASSETTE

VII. BIBLIOGRAPHY.

American Industrial Hygiene Association (AIHA). 1987. *Fundamentals of Analytical Procedures in Industrial Hygiene.* AIHA: Akron, OH.

Hesketh, H.E. 1986. *Fine Particles in Gaseous Media.* Lewis Publishers, Inc.: Chelsea, MA.

Lioy, P.J. 1989. *Air Sampling Instruments for Evaluation of Atmospheric Contaminants.* American Conference of Governmental Industrial Hygienists: Cincinnati.

Lodge, J.P., Jr. (Ed.) 1988. *Methods of Air Sampling and Analysis.* Lewis Publishers, Inc.: Chelsea, MA.

National Institute for Occupational Safety and Health (NIOSH). 1977. *Occupational Exposure Sampling Strategy Manual.* DHEW (NIOSH) Publication No. 77-173. U. S. Government Printing Office, Washington, D.C.

Occupational Safety and Health Administration, U.S. Dept. of Labor. 1995. OSHA Computerized Information System (OCIS) Chemical Sampling Information. U.S. Government Printing Office: Washington, D.C.

APPENDIX II:1-1. DETECTOR TUBES AND PUMPS

PRINCIPLE AND DESCRIPTION.

Detector tube pumps are portable equipment which, when used with a variety of commercially available detector tubes, are capable of measuring the concentrations of a wide variety of compounds in industrial atmospheres.

Operation consists of using the pump to draw a known volume of air through a detector tube designed to measure the concentration of the substance of interest. The concentration is determined by a colorimetric change of an indicator which is present in the tube contents. Most detector tubes can be obtained locally.

APPLICATIONS AND LIMITATIONS.

Detector tubes and pumps are screening instruments which may be used to measure more than 200 organic and inorganic gases and vapors or for leak detection. Some aerosols can also be measured.

Detector tubes of a given brand are to be used only with a pump of the same brand. The tubes are calibrated specifically for the same brand of pump and may give erroneous results if used with a pump of another brand.

A limitation of many detector tubes is the lack of specificity. Many indicators are not highly selective and can cross-react with other compounds. Manufacturers' manuals describe the effects of interfering contaminants.

Another important consideration is sampling time. Detector tubes give only an instantaneous interpretation of environmental hazards. This may be beneficial in potentially dangerous situations or when ceiling exposure determinations are sufficient. When long-term assessment of occupational environments is necessary, short-term detector-tube measurements may not reflect time-weighted average levels of the hazardous substances present.

Detector tubes normally have a shelf life at 25 °C of one to two years. Refrigeration during storage lengthens the shelf life. Outdated detector tubes (i.e., beyond the printed expiration date) should never be used. The OSHA Training Institute can sometimes use these outdated tubes for training purposes.

Several different types and brands of detector tubes have been evaluated for screening and compliance use by SLTC. Information regarding these evaluations can be obtained by contacting SLTC.

PERFORMANCE DATA.

Specific manufacturers' models of detector tubes are listed in the Chemical Sampling Information. The specific tubes listed are designed to cover a concentration range that is near the PEL. Concentration ranges are tube-dependent and can be anywhere from one-hundredth to several thousand ppm. The limits of detection depend on the particular detector tube. Accuracy ranges vary with each detector tube.

The pump may be hand-held during operation (weight: 8-11 ounces), or it may be an automatic type (weight: about 4 pounds) that collects a sample using a preset number of pump strokes. A full pump stroke for either type of short-term pump has a volume of about 100 ml.

In most cases where only one pump stroke is required, sampling time is about one to two minutes. Determinations for which more pump strokes are required take proportionately longer.

Contact the CTC for information regarding long-term maintenance.

LEAKAGE TEST.

Each day prior to use, perform a pump leakage test by inserting an unopened detector tube into the pump and attempt to draw in 100 ml of air. After a few minutes, check for pump leakage by examining pump compression for bellows-type pumps or return to resting position for piston-type pumps. Automatic pumps should be tested according to the manufacturer's instructions.

In the event of leakage that cannot be repaired in the field, send the pump to the CTC for repair. Record that the leakage test was made on the Direct-Reading Data Form (OSHA-93).

CALIBRATION TEST.

Calibrate the detector tube pump for proper volume measurement at least quarterly. Simply connect the pump directly to the bubble meter with a detector tube in-line. Use a detector tube and pump from the same manufacturer.

Wet the inside of the 100 ml bubble meter with soap solution. For volume calibration, experiment to get the soap bubble even with the zero (0) ml mark of the buret. For piston-type pumps, pull the pump handle all the way out (full pump stroke) and note where the soap bubble stops; for bellows-type pumps, compress the bellows fully; for automatic pumps, program the pump to take a full pump stroke. For either type pump, the bubble should stop between the 95 ml and 105 ml marks. Allow 4 minutes for the pump to draw the full amount of air (This time interval varies with the type of detector tube being used in-line with the calibration setup).

Also check the volume for 50 ml (one-half pump stroke) and 25 ml (one-quarter pump stroke) if pertinent. As in Section 1 above, ± 5% error is permissible. If the error is greater than ± 5%, send the pump to CTC for repair and recalibration. Record the calibration information required on the Calibration Log (OSHA-93). It may be necessary to clean or replace the rubber bung or tube holder if a large number of tubes have been taken with any pump.

ADDITIONAL INFORMATION.

DRAEGER, MODEL 31 (BELLOWS). When checking the pump for leaks with an unopened tube, the bellows should not be completely expanded after 10 minutes. For the DRAEGER ACCURO PUMP (BELLOWS), a 15-minute period is used and the end-of-stroke indicator should not be noticeable after this period.

DRAEGER, QUANTIMETER 1000, MODEL 1 (AUTOMATIC). A battery pack is an integral part of this pump. The pack must be charged prior to initial use. One charge is good for 1000 pump strokes. During heavy use, it should be recharged daily. If a "U" (undervoltage) message is continuously displayed in the readout window of this pump, the battery pack should be immediately recharged. A leak test is performed by turning the system on, setting the pump stroke indicator to "2" or greater, inserting an unopened tube into the holder and pressing the start/stop key. When the second stroke has not started after 30 minutes, the device is considered sufficiently gas-tight.

MATHESON-KITAGAWA, MODEL 8014-400A (PISTON). When checking the pump for leaks with an unopened tube, the pump handle should be pulled back to the 100-ml mark and locked. After 2 minutes, the handle should be released carefully. It should return to zero or resting position. After taking 100-200 samples, the pump should be cleaned and relubricated. This involves removing the piston from the cylinder, removing the inlet and pressure-relief valve from the front end of the pump, cleaning, and relubricating.

MINE SAFETY APPLIANCES, SAMPLAIR PUMP, MODEL A, PART NO. 46399 (PISTON). The pump contains a flow-rate control orifice protected by a plastic filter which periodically needs to be cleaned or replaced. To check the flow rate, the pump is connected to a buret and the piston is withdrawn to the 100-ml position with no tube in the tube holder. After 24-26 seconds, 80 ml of air should be admitted to the pump. Every 6 months the piston should be relubricated with the oil provided.

MINE SAFETY APPLIANCES KWIK DRAW™ SAMPLING PUMP, PART NO. 487500 (BELLOWS). The pump contains a filter disk that needs periodic cleaning or replacement. The bellows shaft can be cleaned and lubricated with automotive wax if operation becomes jerky. This pump is tested for leakage by inserting an unopened tube into the holder, deflating the pump fully and releasing. After 10 minutes the distance of the bellows to the frame is ½ inch or greater.

SENSIDYNE-GASTEC, MODEL 800, PART NO. 7010657-1 (PISTON). This pump can be checked for leaks as mentioned for the Kitagawa pump; however, the handle should be released after 1 minute, and should return to within 6-mm or less of resting or fully closed position. Periodic relubrication of the pump head, the piston gasket, and the piston check valve is needed and is use-dependent.

SPECIAL CONSIDERATIONS.

Detector tubes should be refrigerated when not in use to prolong shelf life. Detector tubes should not be used when cold. They should be kept at room temperature or in a shirt pocket for one hour prior to use. Lubrication of the piston pump may be required if volume error is greater than 5%.

APPENDIX II:1-2. ELECTRONIC FLOW CALIBRATORS

DESCRIPTION.

These units are high-accuracy electronic bubble flow meters that provide instantaneous airflow readings and a cumulative averaging of multiple samples. These calibrators measure the flow rate of gases and report volume per unit of time.

The timer is capable of detecting a soap film at 80-microsecond intervals. This speed allows under steady flow conditions an accuracy of ± 0.5% of any display reading. Repeatability is ± 0.5% of any display.

The range with different cells is from 1 ml/min to 30 L/min. Battery power will last 8 hours with continuous use. Charge for 16 hours. Can be operated from A/C charger.

MAINTENANCE OF CALIBRATOR.

1. Cleaning Before Use. Remove the flow cell and gently flush with tap water. The acrylic flow cell can be easily scratched. Wipe with cloth only. Do not allow center tube, where sensors detect soap film to be scratched or get dirty. Never clean with acetone, alcohol or other cleaning solutions. Use only soap and warm water. When cleaning prior to storage, allow flow cell to air dry. If stubborn residue persists, it is possible to remove the bottom plate. Squirt a few drops of soap into the slot between base and flow cell to ease removal.

2. Leak Testing. The system should be leak checked at 6" H2O by connecting a manometer to the outlet boss and evacuating the inlet to 6" H2O. No leakage should be observed.

3. Verification of Calibration. The calibrator is factory calibrated using a standard traceable to National Institute of Standards and Technology, formerly called the National Bureau of Standards, (NBS). Attempts to verify calibrator against a glass one-liter burette should be conducted at 1000 ml/min for maximum accuracy. The calibrator is linear throughout the entire range.

SHIPPING AND HANDLING.

When transporting, especially by air, it is important that one side of the seal tube which connects the inlet and outlet boss, be removed for equalizing internal pressure within the calibrator. Do not transport unit with soap solution or storage tubing in place.

PRECAUTIONS AND WARNINGS.

1. Avoid the use of chemical solvents on flow cell, calibrator case and faceplate. Generally, soap and water will remove any dirt.
2. Never pressurize the flow cell at any time with more than 25 inches of water pressure.
3. Do not charge batteries for longer than 16 hours.
4. Do not leave A/C adapter plugged into calibrator when not in use, as this could damage the battery supply.
5. Black close fitting covers help to reduce evaporation of soap in the flow cell when not is use.
6. Do not store flow cell for a period of one week or longer with soap. Clean and store dry.
7. The Calibrator Soap is a precisely concentrated and sterilized solution formulated to provide a clean, frictionless soap film bubble over the wide, dynamic range of the calibrator. The sterile nature of the soap is important in the prevention of residue build-up in the flow cell center tube, which could cause inaccurate readings. The use of any other soap is not recommended. Proper soap solution is available from CTC's expendable supplies program (AESP).

APPENDIX II:1-3. MANUAL BURET BUBBLE METER TECHNIQUE

When a sampling train requires an unusual combination of sampling media (e.g., glass fiber filter preceding impinger), the same media/devices should be in line during calibration. Calibrate personal sampling pumps before and after each day of sampling.

BUBBLE METER METHOD

1. Allow the pump to run 5 minutes prior to voltage check and calibration.
2. Assemble the cassette filter holder using the appropriate filter for the sampling method. If a cassette adapter is used, care should be taken to ensure that it does not come in contact with the back-up pad.

   NOTE: When calibrating with a bubble meter, the use of cassette adaptors can cause moderate to severe pressure drop in the sampling train, which will affect the calibration result. If adaptors are used for sampling, then they should be used when calibrating.
3. Connect the collection device, tubing, pump and calibration apparatus as shown in Figures II:1-12 and II:1-13.
4. A visual inspection should be made of all Tygon tubing connections.
5. Wet the inside of a 1-liter buret with a soap solution.
6. Turn on the pump and adjust the pump rotameter to the appropriate flow rate setting.
7. Momentarily submerge the opening of the buret in order to capture a film of soap.
8. Draw two or three bubbles up the buret in order to ensure that at least one bubble will complete its run.
9. Visually capture a single bubble and time the bubble from 0 to 1000 ml for high flow pumps or 0 to 100 ml for low flow pumps.

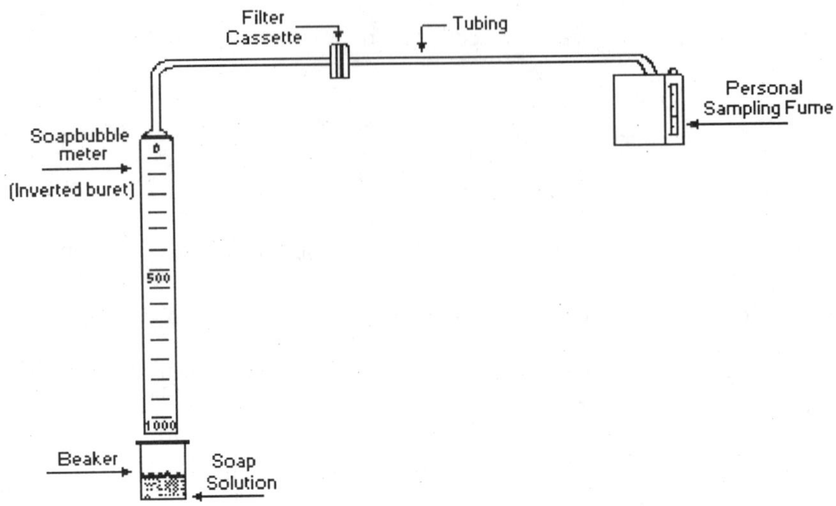

FIGURE II: 1-12. CALIBRATION SET-UP FOR PERSONAL SAMPLING WITH FILTER CASSETTE.

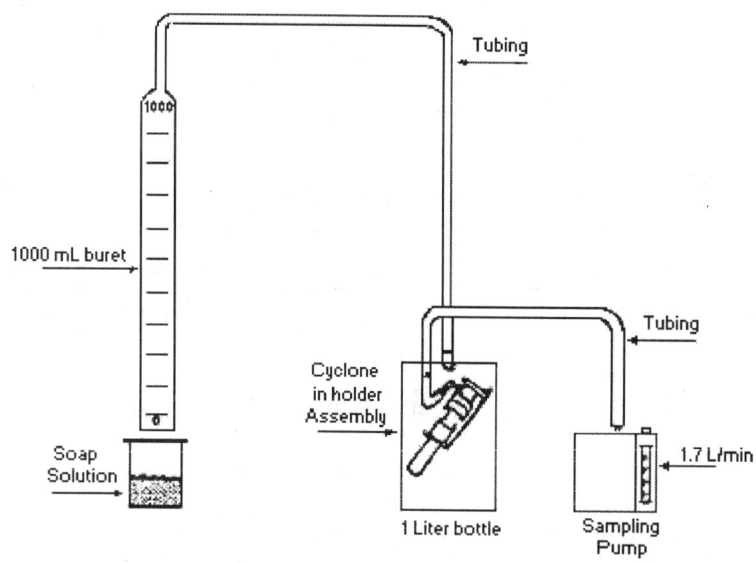

FIGURE II: 1-13 CALIBRATION OF CYCLONE RESPIRABLE DUST SAMPLER USING A BUBBLE METER.

10. The timing accuracy must be within ±1 second of the time corresponding to the desired flow rate. If the time is not within the range of accuracy, adjust the flow rate and repeat steps 9 and 10 until the correct flow rate is achieved. Perform steps 9 and 10 at least twice.

11. While the pump is still running, mark the pump or record on the OSHA-91 the position of the center of the float in the pump rotameter as a reference.

12. Repeat the procedures described above for all pumps to be used for sampling. The same cassette and filter may be used for all calibrations involving the same sampling method.

APPENDIX II:1-4. SHELF LIFE OF SAMPLING MEDIA PROVIDED BY SLTC

Sampling medium	***Shelf life***	***Comments***
Sodium hydroxide (all normalities)	6 months	
Hydrochloric acid Sulfuric acid Methanol in water	1 year	Same for all concentrations of all solutions.
Solution for bis-chloromethyl ether (BCME) and chloromethyl methyl ether (CMME)	2 months	Prepared on request*
Hydroxylammonium chloride solutions (for ketene collection)	2 weeks	Prepared on request*
Hydroxylammonium chloride-sodium hydroxide mixed solutions (for ketene collection)	Stable only 2 hours	Must be prepared fresh just prior to use.
Folin's reagent	5 days	Prepared on request*
Passive monitors		Must be used before the expiration date (if given) printed on the monitor package.
Nitrogen oxides collection tubes		Should be stored in a refrigerator.
Sampler for ozone (Nitrite-treated filter collection device)	28 days	Prepared on request*
Coated filter sampler for di-isocyanates (MDI, HDI, TDI, etc.)	1 year	Prepared on request*
Treated filter sampler for collection of anhydrides	30 days	Prepared on request*

* Please notify SLTC of need two days in advance to allow for preparation time.

APPENDIX II:1-5. SAMPLING FOR SPECIAL ANALYSES

CRYSTALLINE SILICA SAMPLES ANALYZED BY X-RAY DIFFRACTION (XRD).

Air Samples.

Respirable dust samples for quartz, cristobalite, and tridymite are analyzed by X-ray diffraction (XRD). XRD is the preferred analytical method due to its sensitivity, minimum requirements for sample preparation and ability to identify polymorphs (different crystalline forms) of free silica.

The analysis of crystalline free silica by XRD requires that the particle size distribution of the samples be matched as closely as possible to the standards. This is best accomplished by collecting a respirable sample.

Respirable dust samples are collected on a tared low ash PVC filter using a 10-mm nylon cyclone at a flow rate of 1.5 to 2.0 L/min. A sample not collected in this manner is considered a total dust (or nonrespirable) sample. Because of the difficulty in matching sample particle size with analytical standards, CSHO's are discouraged from submitting total dust samples. If the sample collected is nonrespirable, the laboratory must be advised on the OSHA-91 form.

Samples are analyzed for cristobalite or tridymite only upon request.

Quartz (also cristobalite and tridymite) is initially identified by its major (primary) X-ray diffraction peak. A few substances also have peaks near the same location, and it is necessary to confirm quartz (also cristobalite or tridymite) using secondary and/or tertiary peaks. To assist the analyst in identifying interference, the CSHO should provide information concerning potential presence of other substances in the workplace. The following substances should be noted:

- Aluminum phosphate
- Feldspars (microcline, orthoclase, plagioclase)
- Graphite
- Iron carbide
- Lead sulfate
- Micas (biotite, muscovite)
- Montmorillonite
- Potash
- Sillimanite
- Silver chloride
- Talc
- Zircon (Zirconium silicate).

Total air volume shall accompany all filter samples. If a field balance is used to provide pre- and post weights instead of SLTC- or CTC-supplied pre-weighed filters, sample weights shall also accompany the filter samples. Sample weights of 0.5 mg to 3.0 mg are preferred. For those samples that are still weighed in the field, do not submit a sample(s) unless its weight or the combined weights of all particulate filters representing an individual exposure exceed 0.04 mg. If heavy sample loading is noted during the sampling period, it is recommended that the filter cassette be changed to avoid collecting a sample with a weight greater than 5.0 mg.

Laboratory results for air samples are usually reported under one of four categories:

- Percent Quartz (or Cristobalite). Applicable for a respirable sample in which the amount of quartz (or cristobalite) in the sample was confirmed.
- Less Than or Equal to Value in Units of Percent. Less than or equal to values are used when the adjusted 8-hour exposure is found to be less than the PEL, based on the sample's primary diffraction peak. The value reported represents the maximum amount of quartz (or cristobalite) that could be present. However, the presence of quartz (or cristobalite) was not confirmed using secondary and/or tertiary peaks in the sample since the sample could not be in violation of the PEL.

- Approximate Values in Units of Percent. The particle size distribution in a total dust sample is unknown and error in the XRD analysis may be greater than for respirable samples. Therefore, for total dust samples, an approximate result is given.
- Nondetected. A sample reported as nondetected indicates that the quantity of quartz (or cristobalite) present in the sample is not greater than the detection limit of the instrument. The detection limit is usually 10 µg for quartz and 30 µg for cristobalite. If less than a full-shift sample was collected, the CSHO should evaluate a nondetected result to determine whether adequate sampling was performed. If the presence of quartz (or cristobalite) is suspected in this case, the Industrial Hygienist may want to sample for a longer period of time to increase the sample weights.

Bulk Samples.

Bulk samples should be submitted for all silica analyses, if possible. They have the following purposes:

- To confirm the presence of quartz or cristobalite in respirable samples, or to assess the presence of other substances that may interfere in the analysis of respirable samples.
- To determine the approximate percentage of quartz (or cristobalite) in the bulk sample.
- To support Hazard Communication inspections.

A bulk sample must be representative of the airborne free silica content of the work environment sampled; otherwise, it will be of no value.

The laboratory's order of preference for bulk samples for an evaluation of personal exposure is:

- A representative settled-dust (rafter) sample.
- A bulk sample of the raw material used in the manufacturing process (most practical if used for Hazard Communication inspections).

The type of bulk sample submitted to the laboratory should be stated on the OSHA-91 form and cross-referenced to the appropriate air samples.

Normally, any reported results for bulk sample analysis for quartz (also cristobalite or tridymite) will be approximate because of the difficulty in matrix and particle size matching of the bulk material with the analytical standards used during analysis.

SAMPLE CALCULATIONS FOR CRYSTALLINE SILICA EXPOSURES.

Where the employee is exposed to combinations of silica dust (i.e., quartz, cristobalite, and tridymite), the additive effects of the mixture will be considered.

For the PEL calculation specified in 29 CFR 1910.1000, Table Z-3, the percent silica will be determined by doubling the percentage of cristobalite and/or tridymite and adding it to the percentage of quartz, according to the following formula. The PEL mixture pertains to the respirable fraction. (Refer to Figure II:1-14.)

Two consecutive samples from the same employee taken for a combined exposure to silica dusts have the following results, shown below in Figure II:1-14, Sample Calculation for Silica Exposure.

Sample	Sampling period (min)	Total volume (L)	Respirable weight (mg)	Respirable concentration (mg/m^3)	Laboratory results (%)
A	238	405	0.855	2.1	5.2 quartz 2.3 cristobalite ND tridymite
B	192	326	0.619	1.9	4.8 quartz 1.7 cristobalite ND tridymite
Total	430	731	1.474		

Key: ND = Not detectable.

FIGURE II: 1-14. SAMPLE CALCULATION FOR SILICA EXPOSURE

Calculation of the TWA from the sampling and analytical data:

Step 1. Calculate the percentage of quartz, cristobalite, and tridymite in the respirable particulate collected.

Quartz:
5.2 (0.855/1.474) + 4.8 (0.619/1.474) = 3.0 + 2.0 = 5.0%
Cristobalite: 2.3 (0.855/1.474) + 1.7 (0.619/1.474) = 1.3 + 0.7 = 2.0%

Step 2. Calculate the PEL for the mixture.

$$\text{PELmixture} = \frac{10\ \text{mg/m}^3}{\%\ \text{quartz} + 2(\%\ \text{cristobalite}) + 2(\%\ \text{tridymite}) + 2}$$

= 10/[5.0 + 2(2.0) + 2(0) +2] = 10/11.0 = 0.91 mg/m3

Step 3. Calculate the employee's exposure.

Exposure = (Sample wt. A + Sample wt. B)/Total volume = (0.855 + 0.619)/0.731 = 2.0 mg/m3

Step 4. Adjust (where necessary) for less than 8-hour sampling period.

TWA = (2.0 mg/m3)[(430 min)/(480 min)] = 1.8mg/m3

Step 5. Calculate the severity of the exposure.

(1.8 mg/m3)/(0.91 mg/m3) = 2.0

SAMPLES ANALYZED BY INDUCTIVELY COUPLED PLASMA (ICP)

Metals.

Where two or more of the following analytes are requested on the same filter, an ICP analysis may be conducted. However, the compliance officer should specify the metals of interest in the event samples cannot be analyzed by the ICP method. A computer printout of the following 13 analytes may be reported:

- Antimony

- Beryllium
- Cadmium*
- Chromium
- Cobalt
- Copper
- Iron
- Lead
- Manganese
- Molybdenum
- Nickel
- Vanadium
- Zinc

* Cadmium can be analyzed if air volumes are greater than 200 liters.

If requested, the laboratory can analyze for "solder-type" elements, such as:

- Antimony
- Beryllium
- Cadmium
- Copper
- Lead
- Silver
- Tin
- Zinc

Samples taken during abrasive blasting operations are no longer analyzed by ICP because of difficulties with heavy loadings. These samples can be analyzed by atomic absorption spectrometry (AAS) for specific metals (i.e, Pb, Cd, Cr, Fe).

SAMPLES ANALYZED BY X-RAY FLUORESCENCE (XRF).

Filter, wipe, and bulk samples can be qualitatively analyzed by XRF. Requests for XRF analyses should be preceded by a phone call to SLTC to determine the extent and value of the analysis. Packaging and shipping of such samples should be done in a manner consistent with directions previously given in this chapter.

APPENDIX II:1-6. SAMPLING AND ANALYTICAL ERRORS (SAE's)

DEFINITION OF SAE's.

When an employee is sampled and the results analyzed, the measured exposure will rarely be the same as the true exposure. This variation is due to sampling and analytical errors, or SAE's. The total error depends on the combined effects of the contributing errors inherent in sampling, analysis, and pump flow.

DEFINITION OF CONFIDENCE LIMITS.

Error factors determined by statistical methods shall be incorporated into the sample results to obtain the lowest value that the true exposure could be (with a given degree of confidence) and also the highest value the true exposure could be (also with some degree of confidence).

The lower value is called the lower confidence limit (LCL), and the upper value is the upper confidence limit (UCL). These confidence limits are termed one-sided since the only concern is with being confident that the true exposure is on one side of the PEL.

DETERMINING SAE's.

SAE's that provide a 95% confidence limit have been developed and are listed on each OSHA-91B report form (most current SAE's). If there is no SAE listed in the OSHA-91B for a specific substance, call the SLTC. If using detector tubes or direct-reading instruments, use the SAE's provided by the manufacturer.

ENVIRONMENTAL VARIABLES.

Environmental variables generally far exceed sampling and analytical errors. Samples taken on a given day are used by OSHA to determine compliance with PEL's. However, where the employer has previously monitored the work area, the CSHO should review the long-term pattern and compare it with the results. When OSHA's samples fit the long-term pattern, it helps to support the compliance determination. When OSHA's results differ substantially from the historical pattern, the CSHO should investigate the cause of this difference and perhaps conduct additional sampling.

CONFIDENCE LIMITS.

One-sided confidence limits can be used to classify the measured exposure into one of three categories:

1. If the measured results do not exceed the standard and the UCL also does not exceed the standard, we can be 95% confident that the employer is in compliance. (See Equation II:1-6E.)

2. If the measured exposure exceeds the PEL and the LCL of that exposure also exceeds the PEL, we can be 95% confident that the employer is in noncompliance, and a violation is established. (See Equation II:1-6F.)

3. If the measured exposure does not exceed the PEL, but the UCL of that exposure does exceed the PEL, we cannot be 95% confident that the employer is in compliance. (See Equation II:1-6E.) Likewise, if the measured exposure exceeds the PEL, but the LCL of that exposure is below the PEL, we cannot be 95% confident that the employer is in noncompliance. (See Equation II:1-6F.) In both of these cases, the measured exposure can be termed a "possible overexposure."

A violation is not established if the measured exposure is in the "possible overexposure" region. It should be noted that the closer the LCL comes to exceeding the PEL, the more probable it becomes that the employer is in noncompliance.

If measured results are in this region, the CSHO should consider further sampling, taking into consideration the seriousness of the hazard, pending citations, and how close the LCL is to exceeding the PEL.

If further sampling is not conducted, or if additional measured exposures still fall into the "possible overexposure" region, the CSHO should carefully explain to the employer and employee representative in the closing conference that the exposed employee(s) may be overexposed but that there was insufficient data to document noncompliance. The employer should be encouraged to voluntarily reduce the exposure and/or to conduct further sampling to assure that exposures are not in excess of the standard.

SAMPLING METHODS.

The LCL and UCL are calculated differently depending upon the type of sampling method used. Sampling methods can be classified into one of three categories:

1. Full-period, Continuous Single Sampling. Full-period, continuous single sampling is defined as sampling over the entire sample period with only one sample. The sampling may be for a full-shift sample or for a short period ceiling determination.

2. Full-period, Consecutive Sampling. Full-period, consecutive sampling is defined as sampling using multiple consecutive samples of equal or unequal time duration which, if combined, equal the total duration of the sample period. An example would be taking four 2-hour charcoal tube samples. There are several advantages to this type of sampling:

    - If a single sample is lost during the sampling period due to pump failure, gross contamination, etc., at least some data will have been collected to evaluate the exposure.

- The use of multiple samples will result in slightly lower sampling and analytical errors.
- Collection of several samples allows conclusions to be reached concerning the manner in which differing segments of the work day affect overall exposure.

3. Grab Sampling. Grab sampling is defined as collecting a number of short-term samples at various times during the sample period which, when combined, provide an estimate of exposure over the total period. Common examples include the use of detector tubes or direct-reading instrumentation (with intermittent readings).

CALCULATIONS.

If the initial and final calibration flow rates are different, a volume calculated using the highest flow rate should be reported to the laboratory. If compliance is not established using the lowest flow rate, further sampling should be considered.

Generally, sampling is conducted at approximately the same temperature and pressure as calibration, in which case no correction for temperature and pressure is required and the sample volume reported to the laboratory is the volume actually measured. Where sampling is conducted at a substantially different temperature or pressure than calibration, an adjustment to the measured air volume may be required depending on sampling pump used, in order to obtain the actual air volume sampled. The actual volume of air sampled at the sampling site is reported, and used in all calculations.

For particulates, the laboratory reports mg/m3 of contaminant using the actual volume of air collected at the sampling site. The value in mg/m3 can be compared directly to OSHA Toxic and Hazardous Substances Standards (e.g., 29 CFR 1910.1000).

The SLTC normally does not measure concentrations of gases and vapors directly in parts per million (ppm). Rather, most analytical techniques determine the total weight of contaminant in collection medium. Using the air volume provided by the CSHO, the lab calculates concentration in mg/m3 and converts this to ppm at 25°C and 760 mm Hg using Equation II:1-6A. This result is to be compared with the PEL without adjustment for temperature and pressure at the sampling site.

$$ppm(NTP) = mg/m^3(24.45)/Mwt$$

Where: 24.45 = molar volume at 25°C (298°K) and 760 mm Hg
Mwt = molecular weight
NTP = Normal Temperature and Pressure, 25°C and 760 mm Hg.

EQUATION II: 1-6 A

If it is necessary to know the actual concentration in ppm at the sampling site, it can be derived from the laboratory results reported in ppm at NTP by using the following equation:

$$ppm(PT) = ppm(NTP)[(760)/(P)][(T)/(298)]$$

where:
P = sampling site pressure (mm of Hg)
T = sampling site temperature (°K)
298 = temperature in degrees Kelvin (273°K + 25°)

EQUATION II: 1-6B

$$\text{Since ppm(NTP)} = \text{mg/m}^3(24.45)/(\text{Mwt})$$
$$\text{ppm(PT)} = (\text{mg/m}^3)(24.45/\text{Mwt})(760/P)(T/298)$$

EQUATION II: 1-6 C

NOTE: When a laboratory result is reported as mg/m3 contaminant, concentrations expressed as ppm (PT) cannot be compared directly to the standards table without converting to NTP.

NOTE: Barometric pressure can be obtained by calling the local weather station or airport, request the unadjusted barometric pressure. If these sources are not available, then a rule of thumb is: for every 1,000 feet of elevation, the barometric pressure decreases by 1 in. Hg.

CALCULATION METHOD FOR A FULL-PERIOD, CONTINUOUS SINGLE SAMPLE.

Obtain the full-period sampling result (value X), the PEL and the SAE. The SAE can be obtained from the OSHA 91B or by contacting SLTC.

Divide X by the PEL to determine Y, the exposure severity. That is:

$$Y = X/PEL$$

EQUATION II: 1-6 D

Compute the UCL95% as follows:

$$UCL_{95\%} = Y + SAE$$

EQUATION II: 1-6 E

Compute the LCL95% as follows:

$$LCL_{95\%} = Y - SAE$$

EQUATION II: 1-6 F

Classify the exposure according to the following classification system:

- If the UCL < 1, a violation does not exist.
- If LCL < 1 and the UCL > 1, classify as possible overexposure.
- If LCL > 1, a violation exists.

SAMPLE CALCULATION FOR FULL-PERIOD, CONTINUOUS SINGLE SAMPLE.

A single fiberglass filter and personal pump were used to sample for carbaryl for a 7-hour period. The CSHO was able to document that the exposure during the remaining unsampled one-half hour of the 8-hour shift would equal the exposure measured during the 7-hour period. The laboratory reported 6.07 mg/m3. The SAE for this method is 0.23. The PEL is 5.0 mg/m3.

Step 1. Calculate the exposure severity.
Y = 6.07/5.0 = 1.21

Step 2. Calculate confidence limits.

LCL = 1.21 - 0.23 = 0.98

Since the LCL does not exceed 1.0, noncompliance is not established. The UCL is calculated:

UCL = 1.21 + 0.23 = 1.44

Step 3. Classify the exposure.

Since the LCL < 1.0 and the UCL > 1.0, classify as possible overexposure.

CALCULATION METHOD FOR FULL-PERIOD CONSECUTIVE SAMPLING.

The use of multiple consecutive samples will result in slightly lower sampling and analytical errors than the use of one continuous sample since the inherent errors tend to partially cancel each other. The mathematical calculations, however, are somewhat more complicated. If preferred, the CSHO may first determine if compliance or noncompliance can be established using the calculation method noted for a full-period, continuous, single-sample measurement. If results fall into the "possible overexposure" region using this method, a more exact calculation should be performed using equation II:1-6G below.

Classify the exposure according to the following classification system:

- If UCL < 1, a violation does not exist.
- If LCL < 1, and the UCL > 1, classify as possible overexposure.
- If LCL > 1, a violation exists.

- Obtain X_1, X_2 ... X_n, the n consecutive concentrations n one workshift and their time durations, T_1, T_2 ... T_n.
- Also obtain the SAE in listed in the OSHA-91B sample report form.
- Compute the TWA exposure.
- Divide the TWA exposure by the PEL to find Y, the standardized average (TWA/PEL).
- Compute the $UCL_{95\%}$ as follows:

  $UCL_{95\%}$ = Y + SAE (Equation II:1-6E)
- Compute the $LCL_{95\%}$ as follows:

  $LCL_{95\%}$ = Y - SAE (Equation II:1-6F)

EQUATION II: 1-6 G

When the LCL < 1.0 and UCL > 1.0, the results are in the "possible overexposure" region and the CSHO must analyze the data using the more exact calculation for full-period consecutive sampling, as follows:

$$LCL=Y-\frac{SAE\sqrt{T_1^2X_1^2+T_2^2X_2^2+\ldots+T_n^2X_n^2}}{PEL(T_1+T_2+\ldots+T_n)}$$

EQUATION II: 1-6 H

SAMPLE CALCULATION FOR FULL-PERIOD CONSECUTIVE SAMPLING.

If two consecutive samples had been taken for carbaryl instead of one continuous sample, and the following results were obtained:

	-- Samples -- A	B
Sampling rate (L/min)	2.0	2.0
Time (min)	240	210
Volume (L)	480	420
Weight (mg)	3.005	2.457
Concentration (mg/m3)	6.26	5.85

The SAE for carbaryl is 0.23

Step 1. Calculate the UCL and the LCL from the sampling and analytical results:

TWA = [(6.26 mg/m3) 240 min + (5.85 mg/m3) 210 min]/450 min = 6.07 mg/m3
Y = 6.07 mg/m3/PEL = 6.07/5.0 = 1.21
Assuming a continuous sample: LCL = 1.21 - 0.23 = 0.98
UCL = 1.21 + 0.23 = 1.44

Step 2. Since the LCL < 1.0 and UCL > 1.0, the results are in the possible overexposure region, and the CSHO must analyze the data using the more exact calculation for full-period consecutive sampling as follows:

$$LCL = 1.21 - \frac{0.23\sqrt{(240\,\text{min})^2(6.26\,mg/m^3)^2 + (210\,\text{min})^2(5.85\,mg/m^3)^2}}{5.0\,mg/m^3(240+210\,\text{min})}$$

$$= 1.21 - 0.20 = 1.01$$

Since the LCL > 1.0, a violation is established.

GRAB SAMPLING.

If a series of grab samples (e.g., detector tubes) is used to determine compliance with either an 8-hour TWA limit or a ceiling limit, consult with the ARA for Technical Support regarding sampling strategy and the necessary statistical treatment of the results obtained.

SAE's FOR EXPOSURE TO CHEMICAL MIXTURES.

Often an employee is simultaneously exposed to a variety of chemical substances in the workplace. Additive toxic effects on a target organ is common for such exposures in many construction and manufacturing processes. This type of exposure can also occur when impurities are present in single chemical operations. Permissible exposure limits [or Threshold Limit Values (TLV's) such as the welding fume TLV of 5 mg/m3] for mixtures address the complex problem of additive exposures and their health effects. In addition, 29 CFR 1910.1000 contains a computational approach to assess exposure to a mixture. This calculation should be used when components in the mixture pose an additive threat to worker health. While these calculations can be used for synergistic exposures, a separate consideration must be performed. For synergistic mixtures, an individualized assessment of toxicity using the most current scientific data is conducted to consider the total physiologic burden. [See the OSHA Field Inspection Reference Manual (FIRM) for further details.] The SAE can still be calculated for synergistic mixtures; however, Equation II:1-6I is adjusted to reflect any synergistic effects.

Whether using a single standard or the mixture calculation, the sampling and analytical error (SAE) of the individual constituents must be considered before arriving at a final compliance decision. These SAE's can be pooled and weighted to give a control limit for the additive mixture. To illustrate this control limit, the following example using the mixture calculation is expressed in the following equation.

$$E_m = C_1/L_1 + C_2/L_2 + \ldots + C_n/L_n$$

Where:

E_m = equivalent exposure for a mixture
(E_m should be < 1 for compliance)
C = concentration of a particular substance
L = PEL

EQUATION II: 1-6 I

For example, to calculate exposure to three different but additive substances:

Material	***8-hr. exposure***	***8-hr TWA PEL (ppm)***	***SAE***
Substance 1	500	1000	0.089
Substance 2	80	200	0.11
Substance 3	70	200	0.18

Using Equation II:1-6I: Em = 500/1000 + 80/200 + 70/200 = 1.25

Since Em > 1, an overexposure appears to have occurred; however, the SAE for each substance also needs to be considered:

Exposure ratio (for each substance) Yn = Cn/Ln
Ratio to total exposure R1 = Y1/Em ... Rn = Yn/Em

The SAE's (95% confidence) of the substance comprising the mixture can be pooled by:

(Rst)2 = [(R1)2 (SAE1)2+(R 2)2 (SAE2)2+ ... (Rn)2 (SAEn)2]

The mixture Control Limit (CL) is equivalent to: 1 + RSt

If Em < CL, then an overexposure has not been established at the 95% confidence level; further sampling may be necessary.

If Em > 1 and Em > CL, then an overexposure has occurred (95% confidence).

Using the mixture data above:

Therefore Em > CL and an overexposure has occurred within 95% confidence limits. This calculation is also used when considering a standard such as the one for total welding fumes. A computer program that will calculate a control limit for any additive mixture is available for personal computers. The program will run on any IBM-compatible computer in DOS and is available on the OSHA CD under the heading "software."

APPENDIX II:1-7. PARTIAL LIST OF SUBSTANCES FOR AUTOWEIGHING SUBMISSION

Substance	Code	Limit
alpha-Alumina..........		
Total dust...........	0160	15
Respirable fraction..	A201	5
Aluminum Metal (as Al).		
Total dust...........	A100	15
Respirable fraction..	A110	5
Ammonium sulfamate.....		
Total dust...........	0185	15
Respirable fraction..	A111	5
Barium sulfate.........		
Total dust...........	B101	15
Respirable fraction..	B104	5
Bismuth telluride Undoped.............		
Total dust...........	0370	15
Respirable fraction..	B110	5
Boron oxide............		
Total dust...........	0380	15
Calcium Carbonate......		
Total dust...........	0505	15
Respirable fraction..	C130	5
Calcium hydroxide......		
Total dust...........		15
Respirable fraction..		5
Calcium oxide..........	0520	5
Calcium silicate.......		
Total dust...........	C112	15
Respirable fraction..	C122	5
Calcium sulfate........		
Total dust...........	C104	15
Respirable fraction..	C123	5
Carbon black...........	0527	3.5
Cellulose.............		
Total dust...........	0575	15
Respirable fraction..	C124	5
Coal dust (less than 5% SiO(2)), respirable fraction..	9040	
Oil Mist (Mineral)	5010	5
Particulates not otherwise regulated (PNOR)............		
Total dust...........	9135	15
Respirable fraction..	9130	5
Pentaerythritol........		
Total dust...........	1987	15
Respirable fraction..	P157	5
Perlite................		
Total dust...........	2035	15
Respirable fraction..	P101	5
Plaster of paris.......		
Total dust...........	2127	15
Respirable fraction..	P102	5
Portland cement........		
Total dust...........	0557	15
Respirable fraction..	P104	5
Rouge..................		
Total dust...........	2229	15
Respirable fraction..	R102	5
Silica, amorphous, precipitated and gel.	9050	
Silica, amorpous, diatomaceous earth, containing less than 1% crystalline silica	S112	
Silica, crystalline cristobalite, respirable dust......	9015	
Silica, crystalline quartz, respirable dust.........	9010	
Silica, crystalline tripoli (as quartz), respirable dust......	S114	
Silica, crystalline tridymite, respirable dust......	9017	
Silica, fused,		

Substance	Code	Value
Coal dust (greater than or equal to 5% SiO(2)), respirable fraction............	C120	
Cotton Dust (The SLTC will supply special pre weighed cotton dust filters and cassettes on request.)		
Dicyclopentadienyl iron		
Total dust..........	0904	15
Respirable fraction..	D100	5
Emery.................		
Total dust..........	1016	15
Respirable fraction..	E102	5
Grain dust (oat, wheat, barley)........	G109	10
Glycerin (mist)........		
Total dust..........	1363	15
Total dust..........	1363	15
Respirable fraction..	G115	5
Graphite, synthetic....		
Total dust..........	1366	15
Respirable Fraction..	G100	5
Gypsum................		
Total dust..........	1367	15
Respirable fraction..	G101	5
Kaolin................		
Total dust..........	1568	15
Respirable fraction..	K100	5
Limestone.............		
Total dust..........	1593	15
Respirable fraction..	L100	5
Magnesite.............		
Total dust..........	M113	15
Respirable fraction..	1615	5
Magnesium oxide fume...		
Total Particulate....	1610	15
Marble................		
Total dust..........	1626	15
Respirable fraction..	M114	5
respirable dust......	9013	
Silicates (less than 1% crystalline silica)		
Mica (respirable dust)........		
Soapstone, total dust	9025	
Soapstone, respirable dust................	S121	
Talc (containing no asbestos), respirable dust......	9030	
Silicon...............		
Total dust..........	2235	15
Respirable fraction..	S120	5
Silicon carbide........		
Total dust..........	2236	15
Respirable fraction..	S123	5
Starch................		
Total dust..........	2263	15
Respirable fraction..	S124	5
Sucrose...............		
Total dust..........	2285	15
Respirable fraction..	S130	5
Tantalum, metal and oxide dust..........	2325	5
Titanium dioxide......		
Total dust..........	2440	15
Vegetable oil mist.....		
Total dust..........	V126	15
Respirable fraction..	V127	5
Zinc oxide fume........	2610	5
Zinc oxide............		
Total dust..........	Z102	15
Respirable fraction..	Z103	5
Zinc stearate..........		
Total dust..........	2616	15
Respirable fraction..	Z104	5
Zirconium compounds (as Zr).............	2620	5

SECTION II: CHAPTER 2 **[Extensively Revised and Featuring New Additions]**

SAMPLING FOR SURFACE CONTAMINATION

Contents:

I. INTRODUCTION.

A. Worksite analysis (i.e., hazard assessment) is a basic component of an effective safety and health program. A complete worksite analysis requires the assessment of surface contamination since workers may be exposed to these contaminants directly through dermal and ingestive routes (e.g., isocyanates, pesticides), and indirectly through inhalation of contaminants that become re-entrained in the air (e.g., asbestos, lead).

B. Dermal and ingestive routes of entry are much more significant than inhalation for a large number of chemicals. For example, a fifteen-minute exposure of the hands and forearms to liquid glycol ethers [2-methoxy-ethanol (ME) and 2-ethoxy-ethanol (EE)] will result in a dose to the body well in excess of the eight-hour inhalation dose at their recommended air exposure limits. (Biological monitoring for the urinary metabolites methoxyacetic acid and ethoxyacetic acid was used to estimate the absorption via skin and lung.) Unfortunately, many industrial hygienists are only familiar with air sampling and fail to evaluate significant exposures caused by surface contamination.

C. Wipe sampling is an important tool of worksite analysis for both identifying hazardous conditions, and in evaluating the effectiveness of personal protective equipment, housekeeping, and decontamination programs. As described below, wipe sampling is an important tool for assessing compliance with certain OSHA requirements even though there are few specific criteria for acceptable surface contamination amounts.

D. The terms wipe sampling, swipe sampling, and smear sampling are synonyms that describe the techniques used to assess surface contamination on the skin, work surfaces, and PPE surfaces (e.g., gloves, respirators, aprons, etc.) The term "wipe sampling" will be used in this document.

II. THE USE OF SURFACE CONTAMINATION SAMPLING IN EVALUATING SAFETY AND HEALTH PROGRAMS.

29 CFR 1910.132 requires employers to "assess the workplace to determine if hazards are present, or are likely to be present, which necessitate the use of personal protective equipment (PPE)." To this purpose, wipe sampling can be useful in categorizing work areas for certain types of controls, such as PPE and/or special cleaning and decontamination. It is also useful in assessing the effectiveness of these controls, including proper work practices. Examples are provided below for three generalized work areas: controlled areas that require the use of PPE, controlled areas that require the use of special cleaning and/or decontamination, and non-controlled work areas that require neither PPE or special cleaning.

A. CONTROLLED WORK AREAS REQUIRING PPE.

1. These are areas where it has been determined (e.g., from an employer's hazard assessment) that PPE is necessary to prevent dermal exposures to a surface contaminant in spite of an aggressive, yet feasible cleaning regimen. Many production areas and specific job tasks fall into this category.

2. Wipe sampling can be used in assessing the effectiveness of the PPE program. Many elements of PPE programs are intended to prevent contamination to certain locations, such as the use of gloves to prevent contamination to the skin of the hands. Surface contamination found in those "protected" locations usually indicates a problem with the program. For example, the presence of surface contamination inside a glove is normally the result of either PPE failure (e.g., the contaminate soaked through the glove material or a tear in the glove), and/or an improper work practice for using the PPE, such as the worker inserting a contaminated hand inside the glove. Additional sampling and observation can be used to determine the specific source of the program failure and possible abatement (e.g., changing gloves more often, checking for tears before donning, cleaning hands before donning, etc.). Sampling after abatement measures are implemented can be used to show the effectiveness of the abatement.

3. It is important to recognize that this sampling is not attempting to assess the health risk resulting from the contamination inside the glove. Rather, it is to identify failures in the PPE program. Therefore, the criteria for concluding that contamination exists does not need to be quantitative. Criteria and reproducible procedures should be selected that provide confidence that contamination has not been adequately controlled (i.e., contaminant levels are above background). The use of wipe pads that change color upon contact with the contaminant is ideal both in locating contamination and as a visual tool in training workers on the consequences of poor work practices.

B. CONTROLLED WORK AREAS REQUIRING SPECIAL CLEANING.

1. Wipe sampling in these areas can show that a feasible and practical regimen of special cleaning and/or decontamination precludes the need for PPE or additional cleaning. The cleaning of lunch room tables, and the decontamination of equipment before being removed from a restricted area are examples of this category. Other examples include cleaning surfaces to reduce accumulation of toxic materials (e.g., asbestos, lead, beryllium) that may become re-suspended in air and thus contribute to airborne exposures.

2. Wipe sampling is used in these areas as a quality control test of the specialized cleaning (or decontamination) regimen. Therefore, samples are taken to assess contamination levels of those surfaces for which the special cleaning is required. Samples found in excess of an acceptable, task-specific, surface contamination limit (see below) indicate a failure in the cleaning or decontamination program. More aggressive training and supervision of the cleaning procedures and/or scheduling may need to be implemented.

3. Again, it is important to recognize that this sampling is not attempting to assess the health risk resulting from the contamination. Rather, it is to ensure that the cleaning and decontamination regimen is being effectively implemented. Establishing an acceptable contamination limit will depend on the purpose of the cleaning, and what is feasible for the procedures utilized. For example, periodic vacuuming of floor surfaces in a lead production area may be used to reduce the amount of lead dust available for re-entrainment, but significant lead contamination of the floor would still be expected. An acceptable surface contamination limit for this type of cleaning would be set much higher than a limit used to evaluate cleaning of tables in the break room.

4. A few surface contamination concentration guidelines have been published, but typically concentration limits must be established by an employer for a specific task. The limits should be based on sufficient initial sampling to determine a "normal" range of contamination that can be expected after utilizing prescribed cleaning procedures. It would be appropriate to include documentation for the limits and their purpose, in the worksite Safety and Health Program.

C. NON-CONTROLLED WORK AREAS.

1. These are work areas for which no special cleaning or PPE are required by the Safety and Health Program. Examples of this category are office areas that are physically separated from the production areas. These areas are often "assumed" to have no significant contamination. Wipe sampling is useful in demonstrating the lack of contamination. If samples do show contamination, further investigation would be needed to determine the cause. Consistent positive results would require a re-assessment of whether the area requires controls.

2. As with sampling to evaluate PPE programs described above, procedures and criteria for sampling non-controlled areas need to provide confidence that contamination has not occurred (i.e., surface concentrations are

not above background). Again, the use of wipe pads which change color upon contact with the contaminant is ideal. The "direct reading" capability makes it possible to quickly screen an entire work area (and a single pad may be used for multiple locations within the area).

3. Sample those locations within the non-controlled area that accumulate dust (e.g., tops of filing cabinets), and surfaces that have potential for contamination from production areas (e.g., paper work brought in from the production areas).

4. Additional surfaces to consider for sampling include those that may come into contact with food and other materials that are ingested or placed in the mouth (e.g., chewing tobacco, gum, cigarettes).

   Contaminated smoking materials may allow toxic materials or their combustion products (e.g., lead, mercury) to enter the body via the lungs. Wiping of surfaces that smoking materials may touch, including the hands, may be useful in evaluating this possible route of exposure.

D. EVALUATION OF SAMPLING RESULTS.

1. The investigator must use professional judgment on a case-by-case basis when evaluating the significance of wipe-sampling results. As described above, acceptable surface contamination amounts will vary widely for the same toxic agent depending on the purpose and location of the sample. Any concentration above background is sufficient to identify a problem with the PPE program for some sample locations.

2. When evaluating results, consider the toxicity and the contribution of skin absorption and/or gastrointestinal absorption to the total dose. Additional factors are the ambient-air concentrations, skin irritation, etc.

3. The OSHA Technical Links Internet site includes Chemical Sampling Information which lists substances that have a potential for ingestion toxicity, skin absorption, and/or a hazardous effect on skin. This information may be found under the "Health" notation. Additional toxicological information concerning chronic skin absorption, dermatitis, etc. should be used in determining if the resulting exposure presents a potential employee hazard (see Bibliography and other references in Technical Links).

III. MEDIA AND TECHNIQUES FOR WIPE SAMPLING.

A. SAMPLING SURFACES FOR CONTAMINATION.

1. Techniques and media for collection of wipe samples from surfaces vary with the agent and purpose of the sample. It is recommended that the Technical Links Chemical Sampling Information, and/or the Salt Lake Technical Center's Applied Industrial Hygiene and Chemical Procedures Group be consulted when selecting a sampling procedure for a specific chemical or contaminant.

2. Classic wipe sampling techniques involve wiping a surface with a filter, which is then submitted to the SLTC for chemical analysis.

   a. Glass fiber filters (GFF), 37 mm in diameter as used in air sampling, are recommended for many chemicals that are analyzed by gas chromatography (GC), or high-performance liquid chromatography (HPLC). These may be used dry, or wetted with an appropriate solvent as specified in the Technical Links Chemical Sampling Information file.

   b. Paper filters are generally used for collection of metals. Mixed cellulose ester filter discs (AA filters) or smear tabs, or their equivalent, are most often recommended.

   c. Polyvinyl chloride filters are available for substances which are unstable on paper-type filters.

   d. Squares of a gauze material, available from the Salt Lake Technical Center upon request, may be used for many organic substances, and have the advantage of being more durable than filter media, especially when wiping rough surfaces. They may be used dry, or wetted with water or solvent to enhance collection efficiency.

e. Charcoal-impregnated pads may be useful for collection of volatile solvents from surfaces. They work by trapping the solvent on activated charcoal, similar to air sampling charcoal tubes.

f. In certain specialized cases, such as isocyanates and aromatic amines, highly reactive and unstable compounds must be collected on a filter medium that has been treated with a derivatizing reagent. These are available from the SLTC quality control department.

3. For a limited number of chemicals, direct-reading colorimetric wipe sampling procedures are available for qualitative or semi-quantitative detection of surface contaminants. These can be used for acids and bases, isocyanates, aromatic amines, organic solvents (not solvent specific), lead, platinum salts, explosives and hydrazine. Contact the Salt Lake Technical Center for more information.

4. For a variety of pesticides and certain other toxic chemicals, immunoassay kits can provide qualitative or semi-quantitative information on-site, and within about an hour. Some wet chemistry is required. Contact the Salt Lake Technical Center for more information.

B. SAMPLING SKIN FOR CONTAMINATION.

1. Techniques and media for wipe sampling of skin contamination vary with the agent and purpose of the sample. It is recommended that the Technical Links Chemical Sampling Information, and/or the Salt Lake Technical Center's Applied Industrial Hygiene and Chemical Procedures Group be consulted when selecting a sampling procedure for a specific chemical or contaminant.

2. There are concerns related to direct wipe sampling of the skin, including the possibility of promoting skin absorption with the use of certain solvents. Contact the Salt Lake Technical Center prior to taking wipe samples directly on the skin to receive agent specific procedures and precautions. Where feasible, biological monitoring is often the most effective means of assessing overall absorption of a contaminant, including through the skin.

3. Before any skin wipe is taken, explain why you want the sample and ask the employee about possible skin allergies to the chemicals in the sampling medium or wetting solution. Employees may elect not to allow sampling of their skin.

4. As an alternative to direct skin sampling, an indirect measurement of skin contamination (as well as PPE failure) can be assessed by wipe sampling surfaces that workers can touch (e.g., table tops, handles, control knobs, inside surfaces of protective equipment).

5. Classic wipe sampling techniques as described earlier, employing glassfiber filters, mixed cellulose ester filters or smear tabs, or gauze squares, charcoal impregnated pads, may be used for sampling contaminants on the skin. If it is deemed desirable to moisten the collecting medium to improve collection efficiency, procedures will normally utilize distilled or de-ionized water, or a 50% solution of isopropyl alcohol in water.

6. Hand washes may be appropriate in some cases. Twenty ml of distilled or de-ionized water, or a dilute solution of mild soap may be added to a zipper-style sandwich bag. The hand to be sampled is inserted, and the bag held tightly closed around the wrist. After a few seconds of agitation, the hand is carefully removed, and the wash solution is poured back into a scintillation vial for shipment to the laboratory.

7. For a limited number of chemicals, direct-reading colorimetric wipe sampling procedures are available for qualitative or semi-quantitative detection of contaminants. These can be used for acids and bases, isocyanates, aromatic amines, organic solvents (not solvent-specific), platinum salts, and hydrazine. The technique differs from that used for surface wipes. Contact the Salt Lake Technical Center for more information.

8. The same technology employed in the colorimetric wipe sampling procedures described above has been applied to a band-aid-type format. These can be applied to the hands inside gloves to demonstrate glove permeability or breakthrough. They can serve as an excellent tool in employee training.

C. PROCEDURES FOR COLLECTING WIPE SAMPLES.

1. Preloading a group of vials with appropriate filters is a convenient method to carry the sample media to the worksite. (The smear tabs should be inserted with the tab end out.) Clean plastic gloves should be worn when handling the filters. The gloves should not be powdered.

2. The following are general procedures for taking wipe samples:

   a. If multiple samples are to be taken at the worksite, prepare a rough sketch of the area to be wipe sampled.

   b. A new set of clean, impervious gloves should be used for each sample to avoid contamination of the filter by previous samples (and the possibility of false positives) and to prevent contact with the substance.

   c. Withdraw the filter from the vial with your fingers or clean tweezers. If a damp wipe sample is desired, moisten the filter with distilled water or other solvent as recommended.

   d. Depending on the purpose of the sample, it may be useful to determine the concentration of contamination (e.g., in micrograms of agent per area). For these samples, it is necessary to record the area of the surface wiped (e.g., 100 cm2). This would normally not be necessary for samples taken to simply show the presence of the contaminant.

   e. Firm pressure should be applied when wiping.

   f. Start at the outside edge and progress toward the center of the surface area by wiping in concentric squares of decreasing size.

   g. Without allowing the filter to come into contact with any other surface, fold the filter with the exposed side in, then fold it over again. Place the filter in a sample vial, cap and number it, and note the number at the sample location on the sketch. Include notes with the sketch giving any further description of the sample (e.g., "Fred Employee's respirator, inside"; "Lunch table").

   h. At least one blank filter treated in the same fashion, but without wiping, should be submitted for each sampled area.

   i. Some substances should have solvent added to the vial as soon as the wipe sample is placed in the vial (e.g., benzidine). These substances are indicated with an "X" next to the solvent notation in the Technical Links Chemical Sampling Information File.

   j. Submit the samples to the Salt Lake Technical Center (SLTC) with an OSHA 91 form.

IV. BIBLIOGRAPHY.

Adams, R.M. 1983. *Occupational Skin Disease*. New York: Grune and Stratton.

Benezra, C. et al. 1982. *Occupational Contact Dermatitis*. Clinical and Chemical Aspect. Philadelphia; Saunders. 1st ed.

Chaiyuth, C. and L. Levin. "A laboratory evaluation of wipe testing base on lead oxide surface contamination." *Am. Ind. Hyg. Assoc*. J. 45:311-317, 1984.

Clayton, G.D. and F.E. Clayton. 1981. *Patty's Industrial Hygiene and Toxicology.* New York: John Wiley and Sons. Vol. II.

Fisher, A.A. 1986. *Contact Dermatitis*. Philadelphia: Lea and Febriger. 3rd ed.

Gellin, G. and H.I. Malbach. 1982. *Occupational and Industrial Dermatology*. Chicago: Year Book Medical Publisher.

Lees, P.S.J. et al. "Evidence for dermal absorption as the major route of body entry during exposure of transformer maintenance and repairman to PCBs." *Am. Ind. Hyg Assoc. J.* 48:257-264, 1987.

Occupational Safety and Health Administration (OSHA), U.S. Dept. of Labor. 1995. "Chemical Sampling Information." Washington, D.C. Available on the OSHA Internet site.

Kezic, Sanja; et al.; "Dermal absorption of vaporous and liquid 2-methoxyethanol and 2-ethoxyethanol in volunteers," *Occup. Environ. Med.*, 54, 1997, pp. 38-43.

APPENDIX II:2-1. SUBSTANCES LISTED WITH A SKIN NOTATION BY THE ACGIH TLV's AND/OR THE OSHA PEL's

NAME:	CAS # :
B) Acrylamide	79-06-1
T) Acrylic acid	79-10-7
T) Acrylonitrile	107-13-1
B) Aldrin	309-00-2
B) Allyl Alcohol	107-18-6
T) 4-Aminodiphenyl	92-67-1
B) Aniline	62-53-3
B) Anisidine (o,p-Isomers)	29191-52-4
B) Azinphos-Methyl	86-50-0
T) Benzidine	92-87-5
B) Bromoform	75-25-2
B) 2-Butoxyethanol	111-76-2
T) n-Butyl Alcohol	71-36-3
B) Butylamine	109-73-9
B) tert-Butyl Chromate (as CrO_3)	1189-85-1
T) o-sec-Butylphenol	89-72-5
T) Captafol	2425-06-1
T) Carbon Disulfide	75-15-0
T) Carbon Tetrachloride	56-23-5
T) Catechol	120-80-9
B) Chlordane	57-74-9
B) Chlorinated Camphene	8001-35-2
T) Chloroacetyl Chloride	79-04-9
T) o-Chlorobenzylidene	2698-41-1
T) Malononitrile	
B) Chlorodiphenyl (42% Cl)	53469-21-9
B) Chlorodiphenyl (54% Cl)	11097-69-1
B) Chloroprene	126-99-8
T) Chlorpyrifos	2921-88-2
B) Cresol (All Isomers)	1319-77-3
B) Cumene	98-82-8
B) Cyanide (as Cn)	57-12-5
T) Cyclohexanol	108-93-0
T) Cyclohexanone	108-94-1
T) Cyclonite	121-82-4
P) DDT	50-29-3
B) Decaborane	17702-41-9
B) Demeton	8065-48-3
T) Diazinon	333-41-5
T) 2-n-Dibutylaminoethanol	102-81-8
T) 3,3'-Dichlorobenzidine	91-94-1
B) Dichloroethyl Ether	111-44-4

NAME:	CAS # :
T) Hexachloroethane	67-72-1
B) Hexachloronaphthalene	1335-87-1
T) Hexafluoroacetone	684-16-2
T) Hexamethyl phosphoramide	680-31-9
T) 2-Hexanone	591-78-6
B) Hydrazine	302-01-2
B) Hydrogen Cyanide	74-90-8
B) Lindane	58-89-9
B) Malathion (Total Dust)	121-75-5
T) Manganese Cyclopentadienyl Tricarbonyl (as Mn)	12079-65-1
T) Mercury (organo) Alkyl Compounds (as Hg)	
T) Mercury (Vapor) (as Hg)	7439-97-6
B) Methyl Acrylate	96-33-3
T) Methyl Acrylonitrile	126-98-7
T) Methyl Alcohol	67-56-1
B) Methyl Bromide	74-83-9
B) Methyl Cellosolve	109-86-4
B) Methyl Cellosolve Acetate	110-49-6
T) Methyl Chloride	74-87-3
T) o-Methylcyclohexanone	583-60-8
T) Methylcyclopentadienyl	12108-13-3
T) Methyl Demeton	8022-00-2
T) 4,4'-Methylene-bis (2-Chloroaniline)	101-14-4
T) 4,4'-Methylenedianiline	101-77-9
B) Methyl Iodide	74-88-4
B) Methyl Isobutyl Carbinol	108-11-2
B) Methyl Isocyanate	624-83-9
T) Methyl Parathion	298-00-0
T) Monocrotophos	6923-22-4
T) Monomethyl Aniline	100-61-8
B) Monomethyl Hydrazine	60-34-4
B) Morpholine	110-91-8
B) Nicotine	54-11-5
B) p-Nitroaniline	100-01-6
B) Nitrobenzene	98-95-3
T) 4-Nitrodiphenyl	92-93-3
B) p-Nitrochlorobenzene	100-00-5
B) Nitroglycerin	55-63-0
T) N-Nitrosodimethylamine	62-75-9

T) 1,3-Dichloropropene	542-75-6
B) Dichlorvos (DDVP)	62-73-7
T) Dicrotophos	141-66-2
B) Dieldrin	60-57-1
T) Diethanolamine	111-42-2
T) Diethylamine	109-89-7
B) Diethylamino ethanol	100-37-8
T) Diethylenetriamine	111-40-0
B) Diisopropylamine	108-18-9
B) Dimethyl Acetamide	127-19-5
B) Dimethylaniline (N,N-Dimethylaniline)	121-69-7
T) Dimethyl-1,2-dibromo-2, 2-dichloroethyl phosphate	300-76-5
B) Dimethylformamide	68-12-2
B) 1,l-Dimethylhydrazine	57-14-7
B) Dimethyl Sulfate	77-78-1
B) Dinitrobenzene (All isomers)	25154-54-5
B) Dinitro-o-cresol	534-52-1
B) Dinitrotoluene	25321-14-6
B) Dioxane	123-91-1
T) Dioxathion	78-34-2
B) Dipropylene Glycol Methyl Ether	34590-94-8
T) Diquat	85-00-7
T) Disulfoton	298-04-4
B) EPN	2104-64-5
T) Endosulfan	115-29-7
B) Endrin	72-20-8
B) Epichlorohydrin	106-89-8
T) Ethion	563-12-2
B) 2-Ethoxyethanol	110-80-5
B) 2-Ethoxyethyl Acetate	111-15-9
P) Ethyl Acrylate	140-88-5
T) Ethylamine	75-04-7
T) Ethyl Bromide	74-96-4
B) Ethylene Chlorohydrin	107-07-3
T) Ethylenediamine	107-15-3
T) Ethylene Dibromide	106-93-4
B) Ethylene Glycol Dinitrate	628-96-6
B) Ethyleneimine	151-56-4
B) N-Ethylmorpholine	100-74-3
T) Fenamiphos	22224-92-6
T) Fenthion	55-38-9
T) Fonofos	944-22-9
T) Formamide	75-12-7
T) Heptachlor	76-44-8
T) Hexachlorobenzene	118-74-1
T) Hexachlorobutadiene	87-68-3

B) Nitrotoluene	99-08-1
B) Octachloronaphthalene	2234-13-1
P) Paraquat, respirable dust	1910-42-5
B) Parathion	56-38-2
T) Pentachloronaphthalene	1321-64-8
B) Pentachlorophenol	87-86-5
B) Phenol	108-95-2
T) Phenothiazine	92-84-2
P) p-Phenylene Diamine	106-50-3
T) Phenyl Glycidyl Ether	122-60-1
B) Phenylhydrazine	100-63-0
T) Phorate	298-02-2
B) Phosdrin	7786-34-7
P) Picric Acid	88-89-1
T) Propargyl Alcohol	107-19-7
T) Propyl Alcohol	71-23-8
T) 1,2-Propylene Glycol Dinitrate	6423-43-4
B) Propylene Imine	75-55-8
B) Sodium Fluoroacetate	62-74-8
T) Styrene	100-42-5
B) TEDP	3689-24-5
B) TEPP	107-49-3
T) 1,1,2,2-Tetrachloroethane	79-34-5
P) Tetrachloronaphthalene	1335-88-2
B) Tetraethyl Lead (as Pb)	78-00-2
B) Tetramethyl Lead (as Pb)	75-74-1
B) Tetramethyl Succinonitrile	3333-52-6
P) Tetryl	479-45-8
B) Thallium, soluble	7440-28-0
T) Thioglycolic Acid	68-11-1
T) Tin, organic compounds	1983-10-4
T) o-Tolidine	119-93-7
T) Toluene	108-88-3
T) m-Toluidine	108-44-1
B) o-Toluidine	95-53-4
T) p-Toluidine	106-49-0
T) 1,1,2-Trichloroethane	79-00-5
B) Trichloronaphthalene	1321-65-9
T) 1,2,3-Trichloropropane	96-18-4
T) Triethylamine	121-44-8
B) 2,4,6-Trinitrotoluene	118-96-7
T) Triorthocresyl Phosphate	78-30-8
T) Vinyl Cyclohexene Dioxide	106-87-6
T) m-Xylene-alpha,alpha'-diamine	1477-55-0
T) Xylidine	1300-73-8

Note: P = Skin notation in the PEL
T = Skin notation in the TLV
B = Skin notation in both references

APPENDIX II:2-2. EXAMPLE PROCEDURES FOR ISOCYANATES

This example is provided for information only and should not be taken as the basis for OSHA policy.

Prepared by: Richard Lawrence, Chemist Date: August 21, 1997

Aromatic Isocyanate Surface Contamination Sampling and Evaluation Techniques

The OSHA Salt Lake Technical Center received a request for assistance at a facility that uses methylene bisphenyl isocyanate (MDI, CAS no. 101-68-8) in their production process. It was decided that an inspection would include screening and sampling for possible MDI surface contamination (as well as air sampling for MDI). The following surface monitoring techniques were evaluated in preparation for that inspection.

A route of workplace exposure to chemicals with low vapor pressures, such as aromatic isocyanates, may be through skin contact with contaminated solvents or surfaces (1,2). Aromatic isocyanates present a respiratory sensitization hazard (1,2). Laboratory studies with animals have indicated that respiratory sensitization to both TDI (3,4) and MDI (5) can be induced by dermal contact alone. The ability to determine surface contamination may be useful in evaluating the effectiveness of housekeeping, decontamination and chemical protective equipment.

Direct reading indicators, such as commercially prepared pads that change color when they come in contact with specific chemicals (or classes of chemicals), are available for aromatic isocyanates. These types of indicators may be used as a screening tool, when assessing the extent of surface contamination, because they are inexpensive and the results are immediate. If an indicator wipe yields a positive result, a wipe sample can then be taken and sent to a laboratory for confirmation. The mention of the commercial products Swypes® and Permea-Tec® does not constitute an endorsement but is inteded solely as an identification of the type of product deemed suitable for the use intended.

The use of direct reading instruments and indicators can be effective in helping employers to comply with the OSHA personal protective equipment standard, 29 CFR 1910.132 (d)(1)(I) and (f)(1)(iii) and (iv).

Wipe Sampling for Screening

Materials:
At the time of the inspection, the OSHA Salt Lake Technical Center had an aromatic isocyanate indicator wipe kit in stock called a Swype® kit (6). This kit consists of indicator wipes called Swype®, a spray bottle of a developer solution that contains a chemical which activates the Swype®, and indicator patches that are worn under PPE, such as gloves, to test the effectiveness. The kit also contains a spray bottle of decontamination solution and a skin cleaner. The effectiveness of the skin cleaner and decontamination solution was not investigated.

1. Gloves and other personal protective equipment must be worn during testing, as appropriate. "Best" style 727 nitrile gloves should provide protection to the hygienist's hands for the time required to perform the screening.

2. Gloves must be changed after a positive result to avoid cross contamination of any subsequent tests.

3. Spray the area to be sampled lightly with the Developer Solution. Use as little as needed to ensure that the surface is wet. Excess solution will dilute contamination, possibly below the detection limit. When testing a vertical surface or knob, some of the solution may begin to "run-off" or drip. This "run-off" should be captured onto the pad to ensure that any contaminant present has not been lost. The Developer Solution contains a proprietary component that activates the Swype® pad.

4. Wait approximately 30 seconds for any aromatic isocyanate to dissolve, then wipe the surface with a surface Swype® pad.

5. Allow 2 to 3 minutes for the color to develop. A pastel red-orange or pink color indicates aromatic isocyanate contamination. The color varies depending on the type of isocyanate present. The surface Swype® detection limit is approximately 3-5 µg.

6. Record appropriate information as needed.

7. If the surface Swype® tests indicate a positive for contamination, the hygienist may want to take corresponding surface wipe samples for laboratory quantitation and confirmation in key samples.

Wipe Sampling for Laboratory Analysis

Materials:
Glass fiber filters, scintillation vials and a derivatizing solution are required for this sampling procedure.

1. Gloves and other personal protective equipment must be worn during sampling, as appropriate. "Best" style 878 butyl gloves provide protection to the hygienist's hands for the time required to perform the sampling.

2. Gloves must be changed after obtaining each sample to avoid cross contamination of any subsequent samples.

3. A solution of 10.0 mg 1-(2-pyridyl) piperazine per milliliter acetonitrile is prepared. The 1-(2-pyridyl) piperazine is a derivatizing agent that stabilizes the isocyanate. The derivatives also allow a greater sensitivity during the analysis.

4. Estimate the number of samples that will be taken. An overestimate might be better. Prepare vials by adding 1.0 milliliter of the 1-(2-pyridyl) piperazine solution to each vial. It is recommended that the solution be pipetted into the vials in a controlled environment, before sampling, to eliminate any chance for contamination.

5. Untreated glass fiber filters are the appropriate wipe media.

6. Reagent grade acetonitrile is used to wet the filter. The acetonitrile acts as a solvent to dissolve and extract any contaminant present.

7. Using a dropper, wet the filter until almost dripping.

8. Select an area immediately adjacent to the area where the Swype® yielded a positive test result.

9. Using the filter, wipe an area about 100 cm2, rubbing the entire area side to side, then up and down. In many cases (such as knobs and levers) it may not be possible to wipe 100 cm2.

10. Place the filter in a scintillation vial containing the derivatizing solution, label the vial, and record appropriate information.

11. The samples are ready to be analyzed by a laboratory. The OSHA Salt Lake Technical Center uses OSHA method 47 for analysis of MDI, and OSHA method 42 for analysis of 2,6-TDI and 2,4-TDI.

Monitoring Inside Protective Equipment (gloves, suits).

Studies have shown that solvents containing chemicals may act as a vehicle allowing the chemicals to permeate gloves and protective clothing (7). The Permea-Tec® aromatic isocyanates detection system may be used for worksite evaluation of chemical protective PPE. For example, a negative result (no color change) of the Permea-Tec®, after being worn under a glove for a time period, in a work environment known to have contaminants present, demonstrates that the glove protection was effective for that time period, in those working conditions. The Permea-Tec® is an indicating pad that is attached to a band-aid-like adhesive strip.

1. Place one or more Permea-Tec® patches (pad side out) on the fingers, palms, wrist, lower arm (near cuff of glove), wherever there is likely to be permeation or contamination.

2. Workers should then don their PPE and work for a time period as they normally would. (If the workers normally change gloves every two hours, for example, then the time period is two hours.)

3. After the shift, allow the workers to doff the gloves as they normally would, then collect, identify, and note the color of the pads.

4. In most cases the Permea-Tec® pad should not need any further treatment. If solvent (containing isocyanates) permeation has occurred, then this solvent should be sufficient to activate the pad.

5. If permeation or penetration of the PPE by the solvent containing aromatic isocyanates has occurred, a reaction turns the pad a pastel red-orange to pink. It is a reliable indicator to a detection limit of 3-5 μg aromatic isocyanates.

Evaluating PPE for Dry Chemicals

1. There may be situations where the Industrial Hygienist may want to use Permea-Tec® pads for dry chemicals. (For example: unprotected handling of components that are assumed to be totally cured.) In this case, after the pads have been collected, 3 drops of solvent (methanol works well) needs to be placed directly onto the pad. Again gloves must be worn during sampling and solvent dispensing.

2. The methanol (or other solvent) wicks into the pad and enables a reaction that turns the pad a pastel red-orange to pink if aromatic isocyanate contamination is present. It is a reliable indicator to a detection limit of 3-5 μg.

References

1. "Preventing Asthma and Death from Diisocyanate Exposure", Department of Health, Education and Welfare, NIOSH, ALERT, Cincinnati, OH, March, 1996, DHHS (NIOSH) Publication No. 96-111, 2-3.

2. Material Safety Data Sheet for MDI, DOW Chemical Company, Midland, MI, 1995, MSD002334, 1.

3. Bickis, U. Investigation of dermally induced airway hyperreactivity to toluene diisocyanate in guinea pigs. Ph.D. thesis, Department of Pharmacology and Toxicology, Queen's University, Kingston, Canada. November, 1994.

4. Karol, M.H., Hauth, B.A., Riley, E.J., and Magreni, C.M. Dermal contact with toluene diisocyanate (TDI) produced respiratory tract hypersensitivity in guinea pigs. Toxicol. Appl. Pharmacol, 1981, 58, 221-230.

5. Rattray, N.J., Botham, P.A., Hext, P.M., Woodcock, D.R., Fielding, I., Dearman, R.J., and Kimber, I. Induction of respiratory hypersensitivity to diphenylmethane-4,4'-diisocyanate (MDI) in guinea pigs. Influence of route of exposure. Toxicology, 1994, 88, 15-30.

6. CLI, Colormetric Laboratories, Inc. 1261A Rand Road, Des Plaines, IL, 60016-3402, Telephone: (847) 696-3036.

7. Gunderson, E. C., Kingsley, B. A., Witham,C. L. and Bromberg, D.C. A Practical Study in Laboratory and Workplace Permeation Testing. Appl. Ind. Hyg., 1989, Vol. 4, 12, 324-329.

APPENDIX II:2-3. EXAMPLE PROCEDURES FOR AROMATIC AMINES

This example is provided for information only and should not be taken as a basis for OSHA policy.

Prepared by: Richard Lawrence, Chemist Date: August 21, 1997

Aromatic Amine Surface Contamination Sampling and Evaluation Techniques

Air sampling by itself may be an inadequate technique to evaluate potential worker exposure to compounds with a low vapor pressure, such as aromatic amines (e.g., MOCA, MDA). A major route of workplace exposure to these chemicals is through skin contact with contaminated surfaces. Aromatic amines are persistent chemicals and once released into the work environment they may remain on surfaces for months and even years. When contaminated surfaces are contacted by unprotected skin, significant exposures are possible. This screening and sampling technique can be useful in evaluating the effectiveness of chemical protective equipment, housekeeping, and decontamination.

Direct reading indicators, such as commercially prepared pads that change color when they come in contact with specific chemicals (or classes of chemicals), are available for aromatic amines. These types of indicators may be used as a screening tool, when assessing the extent of surface contamination, because they are inexpensive and the results are immediate. If an indicator wipe yields a positive result, a wipe sample can then be taken and sent to a laboratory for confirmation.

The mention of the commercial products Swypes® and Permea-Tec® does not constitute an endorsement but is intended solely as an identification of the type of product deemed suitable for the use intended.

The use of direct reading instruments and indicators can be effective in helping employers to comply with the OSHA personal protective equipment standard, 29 CFR 1910.132 (d)(1)(I) and (f)(1)(iii) and (iv).

Using the surface contamination Swype®*.

1. Gloves and other personal protective equipment must be worn during testing, as appropriate.

2. Gloves must be changed after a positive result to avoid cross contamination of any subsequent tests.

3. Spray the area to be sampled lightly with the Cleaning Solution. Use as little as needed to ensure that the surface is wet. Excess solution may dilute any contamination, possibly to less than the detection limit. When testing a vertical surface or knob, some of the solution may begin to "run-off" or drip. This "run-off" should be captured onto the pad to ensure that any contaminant present has not been lost. The Cleaning Solution contains a proprietary component which activates the Swype® pad.

4. Wait approximately 30 seconds for any aromatic amine to dissolve, than wipe the surface with a surface Swype® pad.

5. Allow 2 to 3 minutes for the color to develop. A red color indicates aromatic amine contamination. The surface Swype® detection limit is approximately 3-5 µg.

6. Record appropriate information as needed.

7. If the surface Swype® tests indicate a positive for contamination, the hygienist may want to take corresponding surface wipe samples for laboratory quantitation. The laboratory may also confirm the presence of the contaminant in key samples .

Taking a surface wipe sample.

1. Gloves and other personal protective equipment must be worn during sampling, as appropriate.

2. Gloves must be changed after obtaining each sample to avoid cross contamination of any subsequent samples.

3. Glass fiber filters that have been prepared for air sampling for MOCA or MDA are the appropriate wipe media. These filters have been prepared by soaking each filter with 0.5 ml of 0.26 N sulfuric acid and drying them in an oven. The sulfuric acid converts the amine to a more stable amine salt.

4. A solution of 60/40 v/v methanol/water is used to wet the filter. The methanol acts as a solvent to dissolve and extract any contaminant present. The water activates the sulfuric acid already in the filter.

5. Using a dropper, wet the filter until it is almost dripping.

6. Select an area immediately adjacent to the one where the Swype® yielded a positive test result.

7. Using the filter, wipe an area about 100 cm2, rubbing the entire area from side to side, then up and down. In many cases, such as knobs and levers, it may not be possible to wipe 100 cm2.

8. Place the filter in a scintillation vial, label the vial, and record appropriate information.

9. The samples are ready to be analyzed by a laboratory. The OSHA Salt Lake Technical Center uses OSHA method 71 for analysis of MOCA, and OSHA method 57 for analysis of MDA.

Monitoring Inside Protective Equipment (gloves, suits).

Field and laboratory studies have shown that aromatic amines may permeate gloves and protective clothing. The Permea-Tec® aromatic amine detection system may be used for work site evaluation of chemical protective PPE. For example, a negative result (no color change) of the Permea-Tec®, after being worn under a glove for a time period, in a work environment known to have contaminants present, demonstrates that the glove protection was effective for that time period, in those working conditions. The Permea-Tec® is an indicating pad similar to the Swype® attached to a band-aid like adhesive strip.

1. Have the selected workers wear one or more Permea-Tec® patches (pad side out) on the fingers, palms, wrist, lower arm (near cuff of glove), wherever there is likely to be permeation or contamination.

2. After the shift, allow the workers to doff the gloves as they normally would, then collect, ID, and develop the pads.

3. To develop, ten drops of water are applied slowly to the white strip. The white strip contains a component that activates the Permea-Tec® pad.

4. The water wicks up into the pad and the pad turns red if aromatic amine contamination is present. It is a reliable indicator to a detection limit of 3-5 µg

The following compounds are among the suspected agents that can be detected through this procedure:

- methylene dianiline (MDA)
- 4,4'-methylene bis(2-chloroaniline)
- benzidine
- a-napthylamine
- b-napthylamine
- 4-aminobiphenyl
- o-toluidine
- aniline
- 2,4-toluenediamine
- 1,3-phenylenediamine
- napthylenediamine
- 2,4-xylidine
- o-chloroaniline
- 3,4-dichloroaniline
- r-nitroaniline

* Mention of the Swypes® and Permea-Tec® brand names does not constitute an endorsement of those products.

SECTION II: CHAPTER 3

TECHNICAL EQUIPMENT

Contents:

I. INTRODUCTION.

The Directorate of Technical Support provides the overall management of technical equipment, including calibration and maintenance. Each person to whom technical equipment is assigned is responsible for ensuring that the equipment is properly maintained and calibrated according to Agency schedules.

This chapter discusses the types, calibration, maintenance, and operation of equipment commonly used by OSHA compliance personnel in the field. It is not a comprehensive discussion of all available equipment, nor a review of technical equipment. Mention of a specific product is not meant to imply approval or promotion by OSHA, but merely indicates past procurement policy. Some technical equipment can be connected to a computer for calculations and printouts. Consult the manufacturer's manual or call the OSHA Cincinnati Technical Center (CTC).

II. CALIBRATION.

OSHA's Cincinnati Technical Center (CTC) calibrates and repairs equipment and instruments, and it serves as a source of technical information on instruments and measurement technology.

A. SHIPPING INSTRUCTIONS. Equipment shall be packed and sent to the CTC when repairs are necessary, calibration is due, or when a valid reason exists for requesting a more frequent calibration or check of the equipment. Send all parts of the instrument, not just those needing repairs. If the instrument needs repairs or any special attention, attach a note to the instrument stating the associated problem as clearly as possible. Call the CTC, if there are any questions.

1. Cahn balances shall be repaired by the manufacturer or by local skilled repair companies.

2. Equipment needing regular calibration by the CTC is listed in Appendix II:3-1.

3. Send the Maintenance and Calibration Instrument Service Request (OSHA-79 Form, Rev. May 1981) in each box of equipment, and list the equipment in that box.

4. Use one line per item.

5. Enter the instrument's description, the manufacturer's code, model code, and serial number.

6. Enter the return address.

7. List repairs needed and special instructions on a tag or note attached to the instrument. Include the name and telephone number of a contact person.

8. Place the equipment in a clean plastic bag. Packing material should be polystyrene foam, polyurethane foam, or crumpled newspaper. Do not use vermiculite, woodchips, or other fibrous or powdery material that may create fine dust, and clog the instrument(s).

9. Do not send in equipment which has been contaminated. All contaminated equipment should be decontaminated on-site after use. If equipment must be sent in to the CTC after it has been decontaminated, indicate such, including what the equipment was contaminated with. This information should be clearly shown on the OSHA-79, as well as the individual piece of equipment being tagged and identified.

B. POSTAL REGULATIONS. Packages to be shipped by the postal service cannot exceed 100 inches in length plus girth or 40 pounds in weight. All markings (old registration, certification, addresses, etc.) must be removed from reused shipping containers or covered so that only new markings are visible.

C. OTHER SHIPPING MODES. There are other shipping alternatives (UPS, Fed Ex, etc.) to the Postal Service. The regulations of any private shipping concern will need to be followed as well as Department of transportation regulations.

D. SPECIAL INSTRUCTIONS.

1. Instruments requiring repair or special instructions must be tagged with the symptoms of the malfunction and/or the special instruction written on the tag, including the contact person. The special instructions in Appendix II:3-2 may apply. All hazardous materials must be marked and the carrier informed.

2. The CTC has specialized equipment in addition to that described here. This equipment is available on loan. Contact the CTC directly for all loans.

3. The Health Response Team (HRT) and the Salt Lake Technical Center (SLTC) may sometimes have specialized equipment such as ozone meters, portable gas chromatographs, and radon and bioaerosol monitors available for field use. Contact HRT and SLTC for further information.

III. BATTERIES.

A. ALKALINE BATTERIES. Replace frequently before they become depleted, or carry fresh replacements. When replacing a battery, never mix types (alkaline, carbon zinc, etc.) or capacity and age. Doing so can have negative affects on all the batteries. Remove batteries if equipment will not be used for an extended period of time.

B. RECHARGEABLE NI-CAD BATTERIES.

1. Check the batteries under load (e.g., turn pump on and check voltage at charging jack, if one is available and this can be done safely) before use. See manufacturer's instructions for locations to check voltage. Use 1.2 volts per Ni-Cad cell for an estimate of the fully charged voltage of a rechargeable battery pack.

2. It is undesirable to discharge a multicell Ni-Cad battery pack to voltage levels that are below 1.0 volts per cell; doing this will drive a reverse current through some of the cells and can permanently damage them.

3. Rechargeable Ni-Cad batteries should be charged only in accordance with manufacturer's instructions. Chargers are generally designed to charge batteries in approximately 8 to 16 hours at a high charge rate. A battery can be overcharged and ruined when a high charge rate is applied for too long a time. However, Ni-Cad batteries may be left on a proper trickle charge indefinitely to maintain them at peak capacity. In this case, discharging for a period equal to the longest effective field service time may be necessary, because of short-term memory imprinting. However, do not let the battery run down overnight or longer. Turn the instrument OFF when the battery reaches the proper discharge level.

   NOTE: Battery care is important in assuring uninterrupted sampling. A pump battery pack, for example, should be discharged to the recommended level before charging, at least every their use. If the pump is allowed to run down until the battery reaches the low battery Fault condition, the pump should be turned OFF soon after the Fault condition stops the pump. Leaving some pumps (such as the GilAir) ON for a long time after this Fault condition can damage the battery pack. Also, avoid overcharging the battery pack.

C. OTHER RECHARGEABLE BATTERIES. Other types of rechargeable batteries are being used in equipment such as lead-acid, nickel-metal hydride, etc. Make sure the manufacturer's instructions are followed concerning the handling and recharging of these types of batteries.

IV. ADVERSE CONDITIONS.

A. ADVERSE TEMPERATURE EFFECTS. High ambient temperature, above 100°F and/or radiant heat (e.g., from nearby molten metal) can cause flow faults in air sampling pumps. If these conditions are likely, use the pump with a higher operating temperature range, as opposed to a pump with a lower operating temperature range. Temperature can also affect the accuracy of instrument readings or operation. Check the operating manual for the proper operating temperature range.

B. EXPLOSIVE ATMOSPHERES.

1. Instruments shall not be used in atmospheres where the potential for explosion exists (see 29 CFR 1910.307) unless the instrument is listed by a Nationally Recognized Testing Laboratory (NRTL; see 29 CFR 1910.7) for use in the type of atmosphere present. Check the class and division ratings.

2. When batteries are being replaced, use only the type of battery specified on the safety approval label.

3. Do not assume that an instrument is intrinsically safe. Verify by contacting the instrument's maker or the CTC, if uncertain.

C. ATMOSPHERES CONTAINING CARCINOGENS. A plastic bag should be used to cover equipment when carcinogens are present. Decontamination procedures for special environments are available through the Directorate of Technical Support and should be followed after using equipment in carcinogenic environments. If at all possible, decontaminate the equipment after use on-site.

V. DIRECT-READING INSTRUMENTS.

A. MERCURY ANALYZER-GOLD FILM ANALYZER.

1. Description and Application. This instrument is available from the CTC's loan program (ALEP). A gold-film analyzer draws a precise volume of air over a gold-film sensor. A microprocessor computes the concentration of mercury in milligrams per cubic meter and displays the results on the digital meter. The meter is selective for mercury and eliminates interference from water vapor, sulfur dioxide, aromatic hydrocarbons, and particulates. However, hydrogen sulfide is an interferant.

2. Calibration. Calibration should be performed by the manufacturer or a laboratory with the special facilities to generate known concentrations of mercury vapor. Instruments should be returned to the manufacturer or a calibration laboratory on a scheduled basis.

3. Special Considerations. In high concentrations of mercury vapor the gold film saturates quickly. Check operating manual for more specific information.

4. Maintenance. Mercury vapor instruments generally contain rechargeable battery packs, filter medium, pumps, and valves which require periodic maintenance. Except for routine charging of the battery pack, most periodic maintenance will be performed during the scheduled annual calibrations. However, depending on usage, routine maintenance should also include burning mercury deposits off of the gold-film and changing the zero filter when necessary. See operating manual for specific instructions.

B. OZONE METER.

1. Description and Application. This instrument is available from the SLTC. The detector uses a thin-film semiconductor sensor. A thin-film platinum heater is formed on one side of an alumina substrate. A thin-film platinum electrode is formed on the other side, and a thin-film semiconductor is formed over the platinum electrode by vapor deposition. The semiconductor film, when kept at a high temperature by the heater, will vary in resistance due to the absorption and decomposition of ozone. The change in resistance is converted to a change of voltage by the constant-current circuit.

   The measuring range of the instrument is 0.01 ppm to 9.5 ppm ozone in air. The readings are displayed on a liquid crystal display that reads ozone concentrations directly. The temperature range is 0°-40° C, and the relative humidity range is 10%-80% RH.

2. Calibration. Calibrate instrument before and after each use. Be sure to use a well-ventilated area since ozone levels may exceed the PEL for short periods. Calibration requires a source of ozone. Controlled ozone concentrations are difficult to generate in the field, and this calibration is normally performed at SLTC. Gas that is either specially desiccated or humidified must not be used for preparing calibration standards, as readings will be inaccurate.

3. Special Considerations.

   a. The instrument is not intrinsically safe.

   b. The instrument must not be exposed to water, rain, high humidity, high temperature, or extreme temperature fluctuation.

   c. The instrument must not be used or stored in an atmosphere containing silicon compounds, or the sensor will be poisoned.

   d. The instrument is not to be used for detecting gases other than ozone. Measurements must not be performed when the presence of organic solvents, reducing gases (such as nitrogen monoxide, etc.), or smoke is suspected; readings may be low.

4. Maintenance. The intake-filter unit-Teflon sampling tube should be clean and connected firmly. These should be checked before each operation. Check pump aspiration and sensitivity before each operation.

C. TOXIC GAS METERS.

1. Description and Application. This analyzer uses an electrochemical voltametric sensor or polarographic cell to provide continuous analyses and electronic recording. In operation, sample gas is drawn through the sensor and absorbed on an electrocatalytic sensing electrode, after passing through a diffusion medium. An electrochemical reaction generates an electric current directly proportional to the gas concentration. The sample concentration is displayed directly in parts per million. Since the method of analysis is not absolute, prior calibration against a known standard is required. Exhaustive tests have shown the method to be linear; thus, calibration at a single concentration, along with checking the zero point, is sufficient.

Types: sulfur dioxide, hydrogen cyanide, hydrogen chloride, hydrazine, carbon monoxide, hydrogen sulfide, nitrogen oxides, chlorine, and ethylene oxide. Can be combined with combustible gas and oxygen meters.

2. Calibration. Calibrate the direct-reading gas monitor before and after each use in accordance with the manufacturer's instructions and with the appropriate calibration gases.

3. Special Considerations.

   a. Interference from other gases can be a problem. See manufacturer literature.

   b. When calibrating under external pressure, the pump must be disconnected from the sensor to avoid sensor damage. If the span gas is directly fed into the instrument from a regulated pressurized cylinder, the flow rate should be set to match the normal sampling rate.

   c. Due to the high reaction rate of the gas in the sensor, substantially lower flow rates result in lower readings. This high reaction rate makes rapid fall time possible simply by shutting off the pump. Calibration from a sample bag connected to the instrument is the preferred method.

D. PHOTOIONIZATION METERS.

1. Description and Applications. Ionization is based upon making a gas conductive by the creation of electrically charged atoms, molecules, or electrons and the collection of these charged particles under the influence of an applied electric field. The photoionization analyzer is a screening instrument used to measure a wide variety of organic and some inorganic compounds. It is also useful as a leak detector.

   The limit of detection for most contaminants is approximately 0.1 ppm.

2. Calibration. The procedure for calibration involves applying the calibration gas (typically 100 ppm isobutylene) to the instrument and checking the reading.

3. Special Considerations. The specificity of the instrument depends on the sensitivity of the detector to the substance being measured, the number of interfering compounds present, and the concentration of the substance being measured relative to any interferences. Many models now have built-in correction or correlation factors. After calibrating the unit on isobutylene, select the gas to be measured. The instrument will automatically correct for the relative sensitivity of the gas selected. Some instruments are listed by an NRTL for hazardous locations. Check the operating manual for specific conditions.

4. Maintenance. Keeping these instruments in top operating shape means charging the battery, cleaning the ultraviolet lamp window and light source, and replacing the dust filter. The exterior of the instrument can be wiped clean with a damp cloth and mild detergent if necessary. Keep the cloth away from the sample inlet, however, and do not attempt to clean while the instrument is connected to line power.

E. INFRARED ANALYZERS.

1. Description and Applications. The infrared analyzer has been used within OSHA as a screening tool for a number of gases and vapors (contact the SLTC or the CTC) and is presently the recommended screening method for substances with no feasible sampling and analytical method. See Chemical Sampling Information in OCIS for specific substances. These analyzers are often factory-programmed to measure many gases and are also user-programmable to measure other gases.

   A microprocessor automatically controls the spectrometer, averages the measurement signal, and calculates absorbance values. Analysis results can be displayed either in parts per million (ppm) or absorbance units (AU). The variable path-length gas cell gives the analyzer the capability of measuring concentration levels from below 1 ppm up to percent levels.

   Some typical screening applications are:

- Carbon monoxide and carbon dioxide, especially useful for indoor air assessments;
- Anesthetic gases including, e.g., nitrous oxide, halothane, enflurane, penthrane, and isoflurane;
- Ethylene oxide; and
- Fumigants including e.g. ethylene dibromide, chloropicrin, and methyl bromide.

2. Calibration. The analyzer and any strip-chart recorder should be calibrated before and after each use in accordance with the manufacturer's instructions.

3. Special Considerations. The infrared analyzer may be only semispecific for sampling some gases and vapors because of interference by other chemicals with similar absorption wavelengths.

4. Maintenance. No field maintenance of this device should be attempted except items specifically detailed in the instruction book such as filter replacements and battery charging.

F. DIRECT-READING PARTICLE MONITORS.

1. Condensation Nuclei.

a. Description and Applications. Condensation-nuclei counters are based upon a miniature, continuous-flow condensation nucleus counter (CNC) that takes particles too small to be easily detected, enlarges them to a detectable size, and counts them. Submicrometer particles are grown to supermicrometer alcohol droplets by first saturating the particles with alcohol vapor as they pass through a heated saturator lined with alcohol soaked felt, and then condensing the alcohol on the particles in a cooled condenser. Optics focus laser light into a sensing volume.

As the droplets pass through the sensing volume, the particles scatter the light. The light is directed onto a photodiode which generates an electrical pulse from each droplet. The concentration of particles is counted by determining the number of pulses generated. Applications include the testing of respirators and dust monitors.

The counter counts individual airborne particles from sources such as smoke, dust, and exhaust fumes. Models typically operate in one of three possible modes, each with a particular application. In the "count" mode, the counter measures the concentration of these airborne particles. In the "test" (or fit test) mode, measurements are taken inside and outside a respirator and a fit factor is calculated. In the "sequential" mode, the instrument measures the concentration on either side of a filter and calculates filter penetration.

This instrument is sensitive to particles as small as 0.02 micrometers. However, it is insensitive to variations in size, shape, composition, and refractive index.

b. Calibration. Check the counter before and after each use in accordance with the manufacturer's instructions. This usually involves checking the zero of the instrument. Annual calibration is handled through the CTC.

c. Special Considerations. Reagent-grade isopropyl alcohol for use in these types of instruments is available from the CTC's expendable supply program (AESP).

- Dry the saturator felt by installing a freshly charged battery pack without adding alcohol. Allow the instrument to run until the LO message (low battery) or the E-E message (low particle count) appears. Some instruments allow you to remove the alcohol cartridge for storage purposes.
- Remove the battery pack.
- Install the tube plugs into the ends of the twin-tube assembly.

d. Maintenance. Isopropyl alcohol must be added to the unit every 5-6 hours of operation, per the manufacturer's instructions. Take care not to overfill the unit. Under normal conditions, a fully charged battery pack will last for about 5 hours of operation. Low battery packs should be charged for at least 6 hours, and battery packs should not be stored in a discharged condition.

2. Photodetection.

   a. Description and Applications. Photodetectors operate on the principle of the detection of scattered electromagnetic radiation in the near infrared. Photodetectors can be used to monitor total and respirable particulates. The device measures the concentration of airborne particulates and aerosols including dust, fumes, smoke, fog, mist, etc.

   b. Calibration. Factory calibration is adequate.

   c. Special Considerations. Certain instruments have been designed to satisfy the requirements for intrinsically safe operation in methane-air mixtures.

   d. Maintenance. When the photodetector is not being operated, it should be placed in its plastic bag, which should then be closed. This will minimize the amount of particle contamination of the inner surfaces of the sensing chamber.

      After prolonged operation in or exposure to particulate-laden air, the interior walls and the two glass windows of the sensing chamber may have become contaminated with particles. Although repeated updating of the zero reference following the manufacturer's procedure will correct errors resulting from such particle accumulations, this contamination could affect the accuracy of the measurements as a result of excessive spurious scattering and significant attenuation to the radiation passing through the glass windows of the sensing chamber.

G. COMBUSTIBLE GAS METERS.

1. Description and Applications. These meters use elements which are made of various materials such as platinum or palladium as an oxidizing catalyst. The element is one leg of a Wheatstone bridge circuit. These meters measure gas concentration as a percentage of the lower explosive limit of the calibrated gas.

   The oxygen meter displays the concentration of oxygen in percent by volume measured with a galvanic cell. Other electrochemical sensors are available to measure carbon monoxide, hydrogen sulfide, and other toxic gases. Some units have an audible alarm that warns of low oxygen levels or malfunction.

2. Calibration. Before using the monitor each day, calibrate the instrument to a known concentration of combustible gas (usually methane) equivalent to 25%-50% LEL full-scale concentration.

   The monitor must be calibrated to the altitude at which it will be used. Changes in total atmospheric pressure from changes in altitude will influence the instrument's measurement of the air's oxygen content. The unit's instruction manual provides additional details on calibration of sensors.

3. Special Considerations.

   a. Silicone compound vapors, leaded gasoline, and sulfur compounds will cause desensitization of the combustible sensor and produce erroneous (low) readings.

   b. High relative humidity (90%-100%) causes hydroxylation, which reduces sensitivity and causes erratic behavior including inability to calibrate.

   c. Oxygen deficiency or enrichment such as in steam or inert atmospheres will cause erroneous readings for combustible gases.

   d. In drying ovens or unusually hot locations, solvent vapors with high boiling points may condense in the sampling lines and produce erroneous (low) readings.

e. High concentrations of chlorinated hydrocarbons such as trichloroethylene or acid gases such as sulfur dioxide will depress the meter reading in the presence of a high concentration of combustible gas.

f. High-molecular-weight alcohols can burn out the meters filaments.

g. If the flash point is greater than the ambient temperature, an erroneous (low) concentration will be indicated. If the closed vessel is then heated by welding or cutting, the vapors will increase and the atmosphere may become explosive.

h. For gases and vapors other than those for which a device was calibrated, users should consult the manufacturer's instructions and correction curves.

4. Maintenance. The instrument requires no short-term maintenance other than regular calibration and recharging of batteries. Use a soft cloth to wipe dirt, oil, moisture, or foreign material from the instrument. Check the bridge sensors periodically, at least every six months, for proper functioning. A thermal combustion-oxygen sensor uses electrochemical cells to measure combustible gases and oxygen. It is not widely used in the area offices.

H. OXYGEN METERS. These oxygen-measuring devices can include coulometric and fluorescence measurement, paramagnetic analysis, and polarographic methods. The output of most electrochemical oxygen sensors is dependent on the partial pressure of oxygen in the atmosphere. They do not actually measure concentration directly. An instrument calibrated at sea level and used at higher elevations, such as mountains, will indicate a value lower than the actual concentration.

VI. BIOAEROSOL MONITORS.

A. DESCRIPTION AND APPLICATIONS. A bioaerosol meter, usually a two-stage sampler, is also a multiorifice cascade impactor. This unit is used when size distribution is not required and only respirable-nonrespirable segregation or total counts are needed.

Ninety-five to 100 percent of viable particles above 0.8 microns in an aerosol can be collected on a variety of bacteriological agar. Trypticase soy agar is normally used to collect bacteria, and malt extract agar is normally use to collect fungi. They can be used in assessing sick- (or tight-) building syndrome and mass psychogenic illness.

These samplers are also capable of collecting virus particles. However, there is no convenient, practical method for cultivation and enumeration of these particles.

B. CALIBRATION. Bioaerosol meters must be calibrated before use. This can be done using an electronic calibration system with a high-flow cell, available through the HRT.

C. SPECIAL CONSIDERATIONS. Prior to sampling, determine the type of collection media required and an analytical laboratory. The HRT can provide this information. This specialized equipment is available from the HRT with accompanying instructions.

D. MAINTENANCE. The sampler should be decontaminated prior to use by sterilizer or chemical decontamination with isopropanol.

VII. RADIATION MONITORS AND METERS.

A. LIGHT.

1. Description and Applications. The light meter is a portable unit designed to measure visible, UV, and near-UV light in the workplace. (The CTC has a UV light meter available through its loan program, ALEP.)

The light meter is capable of reading any optical unit of energy or power level if the appropriate detector has been calibrated with the meter. The spectral range of the instrument is limited only by the choice of detector.

Steady-state measurements can be made from a steady-state source using the "normal operation" mode. Average measurements can be obtained from a flickering or modulated light source with the meter set in the "fast function" position. Flash measurements can be measured using the "integrate" function. (OSHA currently does not own any light meters with integrating capacity.)

2. Calibration. No field calibration is available. These instruments are generally very stable and require only periodic calibration. Units should be sent to the CTC for calibration.

3. Special Considerations. Exposure of the photomultiplier to bright illumination when the power is applied can damage the sensitive cathode or conduct excessive current.
4. Maintenance. Little maintenance is required unless the unit is subjected to extreme conditions of corrosion or temperature. Clean the optical unit with lens paper to avoid scratching.

B. IONIZING RADIATION.

1. Description and Applications. The ionizing radiation survey meter is useful for measuring radon decay products from air samples collected on filters. Wipe samples collected on a filter can also be counted with this detector, and general area sampling can be done. Several types of ionizing radiation meters are available from the CTC's loan program (ALEP).

   The survey meter with the scintillation detector can be used to measure the presence of radon-decay products in a dust sample. The barometric pressure should be noted for ionizing radiation chambers.

2. Calibration. No field calibration is available. Periodic calibration by a laboratory is essential and should be handled by the CTC.

C. NONIONIZING RADIATION.

1. Description and Applications. Various nonionizing radiation survey meters are available through the CTC's ALEP for measuring electromagnetic fields. The frequency ranges covered by OSHA's instruments are: 10 Hz to 300 kHz, 0.5 MHz to 6000 MHz, 6 GHz to 40 GHz, and the 2.45 GHz microwave oven frequency. These instruments are capable of measuring the electric field strength (E-field), magnetic field strength (H-field), or both depending on the instrument.

   Depending on the instrument, electromagnetic field strengths from power lines, transformers, video display terminals, RF induction heaters, RF heat sealers, radio & television transmitters, microwave ovens and other sources can be measured.

2. Calibration. No field calibration is available. Periodic calibration by a laboratory is essential and should be handled by the CTC.

3. Special Considerations.

   a. Some of the instruments have an automatic instrument zeroing, other instruments may require "zeroing" the instrument in a "zero-field" condition. Check the manual for guidance.

   b. Some units have a peak memory-hold circuit that retains the highest reading in memory.

   c. Some units operate with either electric (E) or magnetic (H) field probes based on diode-dipole antenna design. Total field strength is measured at the meter regardless of the field orientation or probe receiving angle. The diode-dipole antenna design of the probe is much more resistant to burnout from overload than the thermocouple design of probes used with other meters.

4. Maintenance. No field maintenance is required other than replacing the alkaline batteries when needed.

VIII. AIR VELOCITY MONITORS AND METERS.

A. FLOW HOODS.

1. Description and Applications. These instruments measure air velocities at air supply or exhaust outlets.

2. Calibration. No field calibration is available. Periodic calibration by a laboratory is essential.

3. Maintenance. These typically require little field maintenance other than battery-pack servicing and zero balancing of analog scales. (Check manufacturer's manual.)

B. THERMOANEMOMETER.

1. Description and Applications. These instruments monitor the effectiveness of ventilation by measuring air velocities.

2. Calibration. No field calibration is available. Periodic calibration by a laboratory is essential.

3. Maintenance. These typically require little field maintenance other than battery-pack servicing and zero balancing of analog scales. (Check manufacturer's manual.)

C. OTHER AIR VELOCITY METERS. Other air velocity meters include rotating-vane and swinging-vane velometers.

NOTE: Barometric pressure and air temperature should be noted when using air velocity meters.

IX. NOISE MONITORS AND METERS.

A. SOUND LEVEL METERS.

1. Description and Applications. The sound level meter is a lightweight instrument for the measurement of sound pressure level (SPL) in decibels. All ANSI-approved meters meet minimum requirements that include an A-weighted network, a slow-response meter characteristic, and a fully graduated scale with measurements ranging from 80 to 130 dBA. The Type II meter is most frequently used. Applications are in worker exposure and noise evaluations.

2. Octave Band Analyzer. Some sound level meters may have an octave or one-third octave band filter attached or integrated into the instrument. The filters are used to analyze the frequency content of noise. They are also valuable for the calibration of audiometers and to determine the suitability of various types of noise control.

3. Calibration. In normal operation, calibration of the instrument usually requires only checking. Prior to and immediately after taking measurements, it is a good practice to check, using a calibrator, the ability of the sound level instrument to correctly measure sound levels. As long as the sound level readout is within 0.2 dB of the known source, it is suggested that no adjustments to the calibration pot be made. If large fluctuations in the level occur (more than 1 dB), then either the calibrator or the instrument may have a problem.

4. Special Considerations.

    a. Always check the batteries prior to use. Use the microphone windscreen to protect the microphone when the wearer will be outdoors or in dusty or dirty areas. (The windscreen will not protect the microphone from rain or extreme humidity.)

    b. Never use any other type of covering over the microphone (e.g., plastic bag or plastic wrap) to protect it from moisture. These materials will distort the noise pickup, and the readings will be invalid.

c. Never try to clean a microphone, particularly with compressed air, since damage is likely to result. Although dirt and exposure will damage microphones, regular use of an acoustical calibrator will detect such damage so that the microphones can be replaced.

d. Remove the batteries from any meter that will be stored for more than 5 days. Protect meters from extreme heat and humidity.

5. Maintenance. No field maintenance is required other than replacement of batteries.

B. PERSONAL DOSIMETERS.

1. Description and Applications. These meters can be worn by personnel to obtain individual readings of noise exposure. Typical dosimeters consist of a pocket-sized monitor with remote microphone and an indicator for readout of exposure data. Some have a preset threshold; others have a selector switch that may be preset. It is also possible to select the threshold, criterion level, and exchange rate on many dosimeters.

2. Calibration. Field calibrate at the measurement site according to the manufacturer's instructions both before and after each use. Use an acoustical calibrator that was designed to be used with the particular model noise dosimeter being used.

3. Special Considerations.

a. Always check the batteries prior to use. Be very careful with the microphone cable. Never kink, stretch, pinch, or otherwise damage the cable.

b. Use the microphone windscreen to protect the microphone when the wearer will be outdoors or in dusty or dirty areas. (The windscreen will not protect the microphone from rain or extreme humidity.)

c. Never use any type of covering over the microphone (e.g., plastic bag or plastic wrap) to protect it from moisture. Such materials will distort the noise pickup, and the readings will be invalid.

d. Never try to clean a microphone, particularly with compressed air, since damage is likely to result. Although dirt and exposure to industrial environments will damage the microphones, regular use of an acoustical calibrator will detect such damage so that microphones can be replaced.

e. Remove the batteries when the dosimeter will be stored for more than 5 days. Protect dosimeters from extreme heat and humidity.

4. Maintenance. No field maintenance is required other than replacement of batteries.

X. ELECTRONIC TESTING METERS.

A. DESCRIPTION AND APPLICATIONS. Electrical testing meters include multimeters, clip-on current meters, megohmmeters, battery testers, ground-wire impedance testers, 120-V AC receptacle testers, ground fault interrupt testers, electrostatic meters, and AC voltage detectors.

Multimeters measure AC or DC voltage or current and resistance. They can check for AC leakage, proper line voltage, batteries, continuity, ground connection, integrity of shielded connections, fuses, etc. Other specialized equipment is described in Appendix II:3-3.

B. CALIBRATION. Few, if any, field calibrations are available. Check manufacturer's manual. Periodic calibration is to be handled by the CTC.

C. MAINTENANCE. No field maintenance is required other than battery-pack servicing.

XI. HEAT STRESS INSTRUMENTS.

A. DESCRIPTION AND APPLICATIONS. Heat-stress monitors are portable instruments used to measure environmental factors that may cause heat-related injuries. Personal heat-stress monitors measure body temperature and sometimes heartbeat through sensor belts around the chest or ear-canal sensors.

B. CALIBRATION. Most calibration is done by a laboratory.

C. MAINTENANCE. Some field servicing is required (check manufacturer's manual).

XII. BIBLIOGRAPHY.

A.M. Best Co. (AMB). 1990. *Bests Safety Directory*. AMB: Odwick, NJ.

Hering, S.V., Ed. 1989. *Air Sampling Instruments for Evaluation of Atmospheric Contaminants*. American Conference of Governmental Industrial Hygienists:
Cincinnati, Ohio.

APPENDIX II:3-1. CALIBRATION INTERVALS.

The information below is for general reference only. Calibration or servicing due dates are shown on the equipment. Contact the CTC for specific information.

TABLE II: 3-1. CALIBRATION INTERVALS

Instrument	*Interval (years)*
Air Sampling Pumps	*
Air Sampling Pump Calibrators	1
Air Velocity Meters	2
Battery Chargers	*
Distance/Range Finders	*
Electronic Distance Meters	1
Dust Monitors	1
Electrical Testers	1
GFCI Testers	2
Multimeters	2
Tic Tracers	*
Ergonomics	
Force Gauge Meter	2
Gas Monitors	1
National Draeger Model 190 & PAC III	2
BioSystems Toxi Plus & Ultra, CO H_2S	2
Heat Stress Monitors	1
Ionizing radiation	1
Light Meters	2
Nonionizing radiation meters	1
Pressure gauges	2
Respirators	
Powered Air Purifying	1
Fit Testers	1
Sound Instruments	1
General Purpose (Type 2)	1*
Stop Time Meters	1
Tachometers	1
Vibration Meters	1

* = Field Inspect/Send to the CTC for repair only

APPENDIX II: 3-2. GENERAL PROCEDURES FOR RETURNING INSTRUMENTS TO CTC.

Note: Never use a carrying case like a shipping case. Carrying cases should be carefully stuffed to avoid any instrument movement during shipping and securely packed in a cardboard box.

Simpson 260 Multimeter: Pack extra carefully, as they are extremely susceptible to damage during shipping.

All Hot Wire Anemometers: Return with their probes. Instrument and probe serial numbers are usually the same.

Heat Stress Monitors: Send all associated probes and accessories.

Air Sampling Pumps: Return with battery packs. Several types of battery packs can be repaired. If instrument is shipped with batteries or battery pack inside, turn off all switches.

Remove all batteries from sound measuring instruments (do not send batteries with the instrument).

GenRad 1982 Sound Level Meter and Analyzer: Send all associated attenuators and pre-amplifiers with the instruments.

Any instrument with gauges, meters, glass or plastic parts exposed should have special protection over or around these parts before final packing for shipment. If a case has been furnished with the instrument, the case should be used whenever the instrument is not actually being operated. The case provides necessary protection. Styrofoam packing, bubbled polyethylene film, or crumpled newspaper may be used for packing.

Those instruments not specifically listed should be shipped using the customary precautions. Contact CTC if you have questions about specific instruments.

APPENDIX II:3-3. INSTRUMENT CHART.

The information shown in the table below is for reference only. Not every compliance officer or field office will have every type of instrument. Many of the instruments can be found in and are available through the CTC's loan program (ALEP) or through the SLTC.

TABLE II:3-2. INSTRUMENT USE

Type of instrument	***Measured substance***	***Application***
PHYSICAL MEASUREMENTS		
Electrobalances	Any weighable	filter weighing
Stop time meter	Time	calibration
Tachometers	Mechanical speed	flywheels, belts, cylinders, lathes, etc.
Electrical testers	Electricity	circuits
Multimeters	Electricity	circuits
Noise dosimeters	Noise	noisy locations
SLM kits type 1	Noise	noisy locations
Omnicals	Noise meter calibration	noise meters
Vibration meters	Excessive vibration	bearings, gear trains, housings, walls
Thermoanemometer	Air movement	ventilation
Hand pumps	Detector tubes	screening
Pressure gauges	Air pressure	compressor air lines
Pumps, low	Air volume	sampling with charcoal tubes
Pumps, medium flow	Air volume	sampling
Gilibrators	Air pump calibration	pump calibration
Fibrous aerosol monitors	Fibers in air	asbestos
Dust monitors	Respirable dust	mines, sandblasting, dusty operations
	Respirable dust, particles	indoor air, QNFT
Soil test kit	Soil quality	trenching, excavating

TABLE II: 3-3. GAS AND VAPOR METERS

Type of instrument	*Measured substance*	*Application*
Double-range meters	Combustible gas, O_2	confined spaces
Triple range meters	Toxic, O_2, combustible gas	confined spaces
Quad range meters	2 toxics, O_2, combustible gas	
CO dosimeter	CO	garages, indoor air quality
Carbon dioxide meter	CO_2	indoor air quality
Infrared analyzers	CO, CO_2, organic substances	traces indoor air, leaks, spills
Hydrogen cyanide monitors	Hydrogen cyanide	plants
Hydrogen sulfide meters	Hydrogen sulfide	farms, sewers
Mercury vapor meters	Mercury	mercury plants, spills
NO and NO_2 meters	NO and NO_2 combustion	
Ozone analyzers	O_3	water or air purification, IAQ

TABLE II: 3-4. RADIATION METERS

Type of instrument	*Measured substance*	*Application*
Heat stress meters	Ambient (environmental) heat	foundries, furnaces, and ovens
Photoionization	Ionizable substances	indoor air, leaks, spills
Light meters	Light	indoor lighting, UV exposure
Nonionizing radiation meters	Nonionizing radiation	communications, microwaves, heaters
Ionizing radiation meters	Ionizing radiation	nuclear waste or plants
Electrostatic field tester	Static electric fields	hazardous locations
RF instruments	Electromagnetic fields	RF heat sealers, VDT'S, induction motors

TABLE II: 3-5. BIOLOGICAL MONITORS

Type of instrument	*Measured substance*	*Application*
Microbial sampler	Microbes	indoor air quality

SECTION II: CHAPTER 4

SAMPLE SHIPPING AND HANDLING

Contents:

I. INTRODUCTION.

This chapter contains sample handling, packaging, and mailing instructions for industrial hygiene samples to be shipped to the Salt Lake Technical Center (SLTC) or another accredited facility. Certain Department of Transportation (DOT) Regulations (49 CFR) may apply to shipment of materials. The Technical Center has staff who have received training in the shipment of these materials. OSHA staff may call for guidance on these matters.

A. SAMPLE COLLECTION.

1. Chemical Information File. Collect all samples following the procedures outlined for the specific chemical or agent in the Chemical Information File. The current version of the Chemical Information File is available on OCIS and Internet from Technical links (http://www.osha-slc.gov/SLTC/)

2. Single Component Analysis. Particular attention should be given to substances that must be submitted for analysis of a single component of a mixture.

3. Interference.

   a. Lab notification. Laboratory analysis methods and their results may be susceptible to interference by compounds that are sometimes present in the sample. For this reason, the Technical Center must be notified if a suspected interfering substance may be present in the sample.

   b. Interfering substances. The following substances should be noted, if suspected or known to be present, on the OSHA-91A form:

      - Solvents. Solvents with the same boiling point and polarity as the substance being tested may cause interference, although mass spectral identification will usually resolve any conflict.
      - Free silica. The following chemicals should be noted on the OSHA-91A if they are considered to be present in the work environment or as part of the sample: aluminum phosphate; feldspars (microcline, orthoclase, plagioclase); graphite; iron carbide; lead sulfate; micas (biotite, muscovite); montmorillonite; potash; sillimanite; silver chloride; talc; and zircon (zirconium silicate).
      - Asbestos. All fibrous materials and high nonfibrous dust levels.
      - Metals. High concentrations of other metals and inorganic dust.

B. BULK SAMPLES.

1. Bulk samples should be submitted to the Technical Center only in the following circumstances:

   a. When an analysis is required to support a potential violation (e.g. 1% silica in sand-blasting operations).

   b. As an analytical reference, or to assess solvent or interference.

c. When the chemical composition of the material is incomplete or unknown. Discuss composition of an undefined sample with the supervisory CSHO and the manufacturer of the material.

d. The analysis of bulk samples will generally be semi-quantitative and cannot be evaluated on the basis of the sampling analytical errors (SAE'S) stated in the Chemical Information File.

2. Bulk samples are required for analyses of the following:

- asbestos
- mineral oil and oil mist
- chlorinated camphene
- chlorodiphenyl(PCB)
- silica (quartz, cristobalite)
- hydrogenated terphenyls
- Portland cement
- chlorinated diphenyl oxide
- fugitive grain dust
- explosibility testing

3. Determine labeling and packaging requirements of the material according to DOT regulations before shipping bulk samples. Staff at the Technical Center can help OSHA staff with these regulations.

II. MAILING INSTRUCTIONS.

A. SAMPLE IDENTIFICATION. Samples sent to the Technical Center shall be packaged with a copy of the Air Sampling Data Form (OSHA-91A). All samples shall be sent to the Technical Center by certified mail or its equivalent in order to maintain a chain of legal accountability or custody. If a sample cannot be shipped to the Technical Center by certified mail, an alternate mode of transportation (Federal Express, Emery Air Express, United Parcel Service, etc.) can be used. A chain of custody for all samples must be obtained regardless of the mode of transportation.

B. FILTER CASSETTES.

1. Container Parts. The container used by OSHA for mailing filters has four parts:

- A reusable exterior brown cardboard mailing carton;
- Two brown cardboard flapped inserts;
- A white box to hold the filter cassettes. This box is shipped inside the brown exterior carton; and
- An interior white die-cut cardboard tray to hold the cassettes in place.

NOTE: The U.S. Postal Service offers mailing boxes for Priority packages at no cost. The only charge is the actual cost of the Priority Mail when the box is sent. Priority mail charges are the same as first class mail for widths over 11 ounces. Your Local post office will deliver a pre-requested number of boxes per a designated time period as part of a regular subscription service; for example, 10 boxes every month. You can also just pick up a few extra boxes at your local post office if you need them. This is an alternative that area offices may wish to consider if they are sending in samples on a regular basis. These boxes come in several sizes such as:

8.25 × 5.25 × 1.5
12.25 × 15.5 × 3
11.25 × 14 × 2.25

2. Reusable Container. All of the container parts are reusable. Do not discard the brown exterior mailing carton or the flapped inserts.

3. Use of Container. The containers should be used as follows:

a. Place the cassettes in the circular slots of the die-cut white cardboard tray. The samples should be sealed to ensure sample stability and legal accountability.

   NOTE: For asbestos, do not use the cut-out cardboard tray. Securely fasten the cassettes to bubble sheet packing material so that they will not be damaged by outside shocks or by striking against each other.

b. Remove or obliterate all old mailing labels.

c. Include a copy of Form OSHA-91A in the box.

d. Do not use or place Styrofoam bits or any static-producing materials in containers with asbestos or other fibrous air samples.

e. Do not ship bulk samples in the same mailing package as air samples.

C. SOLID SORBENT TUBES. These tubes may be mailed to SLTC in cylindrical containers or other sturdy containers. Tubes should be prepared for mailing to prevent individual tubes from becoming mixed with packing material; tubes may be held together with a rubber band, for example. Include a copy of the Form OSHA-91A with the samples.

D. MIDGET IMPINGER OR FRITTED GLASS BUBBLER SAMPLES.

1. Transfer each solution from the impinger or bubbler to a 20-milliliter (mL) vial. Ensure that the cap contains the proper liner. Tighten the cap snugly and wrap an elastic tape counterclockwise around the cap so that as the tape shrinks it will tighten the lid. Also, verify that the liner is not warped due to over-tightening.

2. Label the vial with the appropriate sample number for your identification. Affix an OSHA-20 label to the vial in the proper manner.

3. If the sampling solution was provided from the Technical Center and is a hazardous material by DOT regulations, an instruction sheet on the necessary labeling and packaging for return will have been provided inside the original package in which the solution was mailed to the OSHA office. If you have any questions on shipping the material, call the SLTC for assistance.

4. Include a copy of the Form OSHA-91A in the container.

5. Ship by certified mail if the samples are not hazardous to the SLTC. If the samples are hazardous, follow the instructions in the original package from the SLTC for proper labeling and packaging. If there are questions, OSHA staff may call SLTC for assistance.

E. WIPE SAMPLES. Wipe samples/filters should be mailed in 20-mL vials. If a hazardous solvent was used to wet the filters, proper labeling and packaging may be required. Wipe samples must be identified as such on the accompanying Form OSHA-91A.

F. BULK SAMPLES.

1. Bulk solvent samples should never be mailed to the Technical Center in the same package with any other type of air sample. They should be shipped in 20-mL glass vials with Teflon™-lined caps. Tighten the cap and wrap elastic tape counterclockwise so that as the tape shrinks it will tighten the lid. Label the vial with the appropriate sample number for your identification. Affix an OSHA-20 label to the vial in the proper manner. If the material is hazardous according to DOT regulations, it should be properly labeled and packaged. OSHA staff may call the SLTC for guidance.

2. A copy of the Form OSHA-91A must accompany each bulk sample. The sheet must identify the shipped material as a bulk sample and must list the air sample numbers corresponding to the bulk sample. The air

sample's Form OSHA-91A should also indicate that an associated bulk sample is being shipped and the mode of shipment to the Technical Center. If available, include a copy of the material safety data sheet for the bulk sample.

G. SOIL SAMPLES.

1. Bagging. Soil should be placed in a heavy-duty plastic bag that will not tear, and secured and sealed air-tight with tape. Place the first plastic bag in a second heavy-duty bag for additional protection. The Technical Center provides heavy-duty cotton bags upon request.

2. Size of Samples. Samples should vary from one pint for very fine-grained samples to two quarts for coarse gravel. A typical sample should be approximately one quart and weigh three pounds.

3. Sample Identification. Each plastic bag should be sealed for identification with an official Form 21 seal containing a field number, sampling date, and the sampler's name. A laboratory number will be assigned to each sample at the Technical Center.

4. Sample Shipping. The heavy-duty bags containing soil samples should be tied at the top and placed in a box for shipment to the Technical Center. The Form OSHA 91A worksheets should not be in contact with the soil.

   A Form OSHA 91A must be submitted for each soil sample. All of the information in boxes 1 through 9 must be correctly entered, or the sample cannot be analyzed. The field number should be entered in Box 21, and the soil specification or other specific requests should be entered in Box 30. The Technical Center may perform additional tests on occasion.

III. FEDERAL MAILING REGULATIONS.

A. JURISDICTION. When shipping hazardous materials to the Technical Center, Federal interstate transportation regulations must be followed. Such regulations may prohibit the use of the United States Postal Service (USPS).

B. RESPONSIBILITY.

1. The shipper is responsible for compliance with applicable transportation or postal laws and regulations governing acceptability to the carrier and additional packaging requirements.

2. All items that are acceptable for mailing are subject to provisions of Part 124, USPS Manual and Publication 52 of the USPS, Acceptance of Hazardous or Perishable Articles.

3. The Transportation Safety Act of 1974 extended the Department of Transportation's (DOT) authority over transportation of hazardous or restricted materials. The full text of the hazardous materials regulations is contained in Title 49, Code of Federal Regulations, Part 100-199. It is the shipper's responsibility to comply with all applicable DOT regulations. The SLTC has staff with training in the shipment of hazardous material. OSHA staff may call for guidance on these matters.

C. HAZARDOUS MATERIALS.

1. The main categories of hazardous materials sent to the Technical Center are:

   - poisons
   - flammable liquids
   - oxidizers
   - flammable solids
   - corrosive materials (acids and alkalies)
   - irritating materials
   - biological samples (Legionella, etc)

2. Publication 49 CFR Table 172.101 is the key to understanding current DOT regulations for domestic shipment of hazardous materials. If hazardous materials are to be shipped internationally, then either the International Civil Aviation Organization (ICAO) technical instructions or the International Air Transport Association (IATA) instructions are to be used. To ensure that current regulations are followed, it is important to use only the most recent edition of 49 CFR, ICAO, or IATA.

3. The USPS and private carriers base their shipping procedures for hazardous materials on the DOT 49 CFR regulations. These regulations are the minimum acceptable for hazardous materials. In some cases, the carriers have chosen to be more restrictive than DOT regulations. In using these procedures, it is the shipper's responsibility to determine if the carrier they plan to use is more restrictive than DOT. The shipper must comply with the carrier's requirements.

4. For all modes of transportation, the carrier must be clearly informed that hazardous material is being tendered, as specified below.

Notification must be given. Any person who violates a provision of Title 49 in shipping a hazardous material shall be subject to a civil penalty of not more than $10,000 per violation, and if any such violation is a continuing one, each day of the violation constitutes a separate offense. A person who willfully violates a provision of this title and is convicted of a criminal offense is subject to a fine of not more than $25,000, imprisonment for a term not to exceed 5 years, or both.

5. The great variety of chemicals precludes the listing of each item that may be mailed. Publications available from the United States Postal Service give an indication of what can be mailed.

   Certain chemicals are not mailable as bulk samples. Special handling or analysis may be needed in these cases. Call the Technical Center for guidance.

   Most solid sorbent tubes, silica-gel tubes, filters, and wipe samples will not be classified as hazardous materials and can be shipped as regular certified mail through the USPS.

   When a restricted article is tendered for shipment, the customer is required properly to identify, classify, package, mark, label, and certify all articles as specified in Title 49.

   Since all samples are subject to possible litigation, there has to be a chain and/or proof of custody of the samples from the field to the Technical Center. The preferred form is the certified mail receipt. Samples shipped by certified mail go first class (airmail).

   Detailed instructions on sample shipping according to DOT regulations are available directly from the SLTC.

SECTION III: CHAPTER 1

POLYMER MATRIX MATERIALS: ADVANCED COMPOSITES

Contents:

I. INTRODUCTION.

The composites industry in the United States includes three manufacturing areas: Polymers, metals, and ceramics.

A. Composites are classified according to their matrix phase. There are polymer matrix composites (PMC's), ceramic matrix composites (CMC's), and metal matrix composites (MMC's). Materials within these categories are often called "advanced" if they combine the properties of high strength and high stiffness, low weight, corrosion resistance, and in some cases special electrical properties. This combination of properties makes advanced composites very attractive for aircraft and aerospace structural parts.

B. This chapter deals with a segment of the polymer composite industry known as advanced polymer matrix composites, or advanced composites. Since the reinforced plastics, or polymer matrix composite industry is much larger than the subject of this chapter, the term "advanced composites" is used here to define this special segment of the industry. Information on this industry has been developed for use by OSHA field personnel to help them understand this new and growing technology.

C. Advanced composites have been identified as an important growth sector in U.S. manufacturing. This identification has led to more use of these materials in existing facilities as well as an increase in the number of advanced composites manufacturing locations. Field staff may expect to encounter composites more frequently in the course of their assignments. At the same time, much of the technology is new and not presented formally in secondary or undergraduate education.

D. Information is presented here on the technology as practiced in current operations. The technology of advanced composites manufacture is continually evolving, and field personnel will learn here what to expect in these processing facilities in the way of materials handled, manufacturing methods, machinery, potential worker exposures, and other relevant health and safety information.

E. The information presented necessarily makes reference to industrial hygiene and safe work practices, but this manual is not intended to provide comprehensive guidelines for assessing compliance with regulations. Much of the terminology used in this manual is peculiar to the composites industry, and for this reason a glossary of terms has been provided in Appendix III:1-1.

II. OVERVIEW OF THE INDUSTRY.

Polymer-matrix composites manufacturing is a multibillion dollar industry in the U.S. and one of the few in which the U.S. is conceded to be slightly more advanced than competitors abroad. Composite products range from skateboards to components of the space shuttle. The industry can be generally divided into two basic segments, industrial composites and advanced composites. Several of the composites manufacturing processes are common to both segments. The two basic segments are described below.

A. INDUSTRIAL COMPOSITES. The industrial composites industry has been in place for over 40 years in the U.S. This large industry utilizes various resin systems including polyester, epoxy, and other specialty resins. These materials, along with a catalyst or curing agent and some type of fiber reinforcement (typically glass fibers) are used in the production of a wide spectrum of industrial components and consumer goods: boats, piping, auto bodies, and a variety of other parts and components.

B. ADVANCED COMPOSITES.

1. This sector of the composites industry is characterized by the use of expensive, high-performance resin systems and high-strength, high-stiffness fiber reinforcement. The aerospace industry, including military and commercial aircraft of all types, is the major customer for advanced composites. These materials have also been adopted for use by the sporting goods suppliers who sell high-performance equipment to the golf, tennis, fishing, and archery markets.

2. While aerospace is the predominant market for advanced composites today, the industrial and automotive markets will increasingly see the use of advanced composites toward the year 2000. At present, both manual and automated processes are employed in making advanced-composite parts. As automated processes become more predominant, the costs of advanced composites are expected to decline to the point at which these materials will be used widely in electronic, machinery, and surface transportation equipment.

3. Suppliers of advanced composite materials tend to be larger companies capable of doing the research and development necessary to provide the high-performance resin systems used in this segment of the industry. End-users also tend to be large, and many are in the aircraft and aerospace businesses.

4. Advanced composite systems are divided into two basic types, thermosets and thermoplastics. Thermosets are by far the predominant type in use today. Thermosets are subdivided into several resin systems including epoxies, phenolics, polyurethanes, and polyimides. Of these, epoxy systems currently dominate the advanced composite industry. Both thermoset and thermoplastic systems will be discussed in more detail in Section IV of this chapter.

III. THE MANUFACTURING PROCESS.

A. ELEMENTS. The feature common to all composite processes is the combining of a resin, a curing agent, some type of reinforcing fiber, and in some cases a solvent. Typically, heat and pressure are used to shape and "cure" the mixture into a finished part. In composites, the resin acts to hold the fibers together and protect them, and to transfer the load to the fibers in the fabricated composite part. The curing agent, also known as hardener, acts as a catalyst and helps in curing the resin to a hard plastic. The reinforcing fiber imparts strength and other required properties to the composite. Solvents may serve three purposes:

- as part of the resin mixture;
- as part of the process; and
- as a cleaning agent for removing residue from the process equipment.

B. MAJOR PROCESSES. Diagrams of the major processes used in the advanced composites industry are provided in Section A of this chapter. The processes vary widely in type of equipment and potential worker exposure. Several of the processes are automated; however, some are manual and require worker contact with the part during manufacture. The basic process types are described below.

1. Formulation is the process where the resin, curing agent, and any other component required are mixed together. This process may involve adding the components manually into a small mixing vessel or, in the case of larger processes, the components may be pumped into a mixing vessel. The potential hazards involve skin, eye, and respiratory contact with the ingredients or final formulation.

2. Prepregging is the process where the resin and curing agent mixture are impregnated into the reinforcing fiber. These impregnated reinforcements (also known as prepregs) take three main forms: woven fabrics, roving, and unidirectional tape. Fabrics and tapes are provided as continuous rolls in widths up to 72 inches and lengths up to several hundred feet. The fabric or tape thickness constitutes one ply in the construction of a multi-ply lay-up. Impregnated roving is wound onto cores or bobbins and is used for filament winding. Once the resin mixture has been impregnated onto the fibers, the prepreg must be stored in a refrigerator or freezer until ready for use in the manufacturing process. This cold storage prevents the chemical reaction from occurring prematurely. Prepreg materials are used widely in the advanced composite industry, particularly in aircraft and aerospace. Potential exposure is generally from handling of the fiber or resin.

3. Open Molding processes are those where the part being manufactured is exposed to the atmosphere. The worker typically handles the part manually, and there is a higher potential for exposure. The resin mixture may be a liquid being formed onto a reinforcing material or it may be in the form of a prepreg material being formed for final cure.

4. Closed Molding processes are those in which all or part of the manufacture takes place in a closed vessel or chamber. The liquid resin mixture or prepreg material may be handled or formed manually into the container for the curing step. In the case of liquid resin mixtures, these may be pumped into the container, usually a mold of some type, for the curing step. These processes usually have less worker exposure potential, particularly if the entire process is closed.

5. Sequential or batch processes involve manufacture of a single part at a time, in sequence. This type of process is usually required where the part being made is small and complex in shape, when the curing phase is critical, when finishing work must be minimized, or where a small number of parts is involved.

6. Continuous processes are typically automated to some degree and are used to produce larger numbers of identical parts relatively quickly. These processes are typified by pumping of the resin mixture into the mold, followed by closed curing.

IV. POLYMER MATRIX COMPOSITE (PMC) RESIN SYSTEMS.

The advanced composite processes are discussed in more detail in Section V of this chapter. Seven manufacturing processes are covered, along with two preliminary processes and two finishing processes. The number and variety of processes should give some indication of the wide spectrum of workplaces likely to be encountered by field personnel. Potential worker exposure obviously will also vary widely, depending on the size and type of process being used. Since the advanced composite industry is relatively new and still developing, other processes may be developing or changing to meet new performance requirements. Advanced composites exhibit desirable physical and chemical properties that include light weight coupled with high stiffness and strength along the direction of the reinforcing fiber, dimensional stability, temperature and chemical resistance, flex performance, and relatively easy processing. Advanced composites are replacing metal components in many uses, particularly in the aerospace industry.

A. RESINS. The resin systems used to manufacture advanced composites are of two basic types: thermosetting and thermoplastic. Thermosetting resins predominate today, while thermoplastics have only a minor role in advanced composites manufacture.

B. THERMOSETS.

1. Thermoset resins require addition of a curing agent or hardener and impregnation onto a reinforcing material, followed by a curing step to produce a cured or finished part. Once cured, the part cannot be changed or reformed, except for finishing. Some of the more common thermosets include:

- epoxies
- polyurethanes
- phenolic and amino resins
- bismaleimides (BMI, polyimides)

- polyamides

2. Of these, epoxies are the most commonly used in today's PMC industry. Epoxy resins have been in use in U.S. industry for over 40 years. The basic epoxy compounds most commonly used in industry are the reaction product of epichlorohydrin and bisphenol-A. Epoxy compounds are also referred to as glycidyl compounds. There are several types of epoxy compounds including glycidyl ethers (or diglycidyl ethers), glycidyl esters, and glycidyl amines. Several of these compounds are reactive diluents and are sometimes added to the basic resin to modify performance characteristics. The epoxy molecule can also be expanded or cross-linked with other molecules to form a wide variety of resin products, each with distinct performance characteristics. These resins range from low-viscosity liquids to high-molecular weight solids. Typically they are high-viscosity liquids.

3. Since epoxies are relatively high molecular-weight compounds, the potential for respiratory exposure is fairly low. The potential for respiratory exposure is increased when the resin mixture is applied by spraying or when curing temperatures are high enough to volatilize the resin mixture. The potential for dermal exposure is typically much greater than respiratory exposure when working with epoxies. Several advanced composite processes involve some worker contact with the resin mixture. These and the other processes are discussed in more detail in Section V of this chapter.

4. The second of the essential ingredients of an advanced composite system is the curing agent or hardener. These compounds are very important because they control the reaction rate and determine the performance characteristics of the finished part. Since these compounds act as catalysts for the reaction, they must contain active sites on their molecules.

5. Some of the most commonly used curing agents in the advanced composite industry are the aromatic amines. Two of the most common are 4,4'-methylene-dianiline (MDA) and 4,4'-sulfonyldianiline (DDS). Like the epoxies, these compounds have a very low vapor pressure and usually do not present an airborne hazard unless in a mixture that is sprayed or cured at high temperatures. However, potential for dermal exposure is frequently high. The aromatic amines may permeate many of the commonly used protective gloves and thus may be particularly difficult to protect against.

6. Several other types of curing agents are also used in the advanced composite industry. These include aliphatic and cycloaliphatic amines, polyaminoamides, amides, and anhydrides. Again, the choice of curing agent depends on the cure and performance characteristics desired for the finished part.

7. Polyurethanes are another group of resins used in advanced composite processes. These compounds are formed by reacting the polyol component with an isocyanate compound, typically toluene diisocyanate (TDI); methylene diisocyanate (MDI) and hexamethylene diisocyanate (HDI) are also widely used. While the polyols are relatively innocuous, the isocyanates can represent a significant respiratory hazard as well as a dermal hazard.

8. Phenolic and amino resins are another group of PMC resins. With respect to the phenol-formaldehyde resins, the well-known hazards of both phenol and formaldehyde must be protected against. In addition to traces of free formaldehyde, they may also contain free phenol, and contact with these resins in the uncured state is to be avoided. The urea- and melamine-formaldehyde resins present similar hazards. Free formaldehyde, which is present in trace amounts and may be liberated when their resins are processed, can irritate the mucous membranes.

9. The bismaleimides and polyamides are relative newcomers to the advanced composite industry and have not been studied to the extent of the other resins.

C. THERMOPLASTICS. Thermoplastics currently represent a relatively small part of the PMC industry. They are typically supplied as nonreactive solids (no chemical reaction occurs during processing) and require only heat and pressure to form the finished part. Unlike the thermosets, the thermoplastics can usually be reheated and reformed into another shape, if desired.

D. FIBER REINFORCEMENTS.

1. Fiber reinforcement materials are added to the resin system to provide strength to the finished part. The selection of reinforcement material is based on the properties desired in the finished product. These materials do not react with the resin but are an integral part of the advanced composite system.

2. Potential worker exposure is typically higher in facilities that manufacture the fibers or use them to produce prepreg material. Most of the fibers in use are considered to be in the nonrespirable range. However, they do have the potential to cause eye, skin, and upper respiratory tract irritation as a result of the mechanical properties of the fibers.

3. The three basic types of fiber reinforcement materials in use in the advanced composite industry are:

    - carbon/graphite
    - aramid
    - glass fibers
    -

   Fibers used in advanced composite manufacture come in various forms, including:

    - yarns
    - rovings
    - chopped strands
    - woven fabric
    - mats

   Each of these has its own special application. When prepreg materials are used in parts manufacture, woven fabric or mats are required. In processes such as filament wet winding or pultrusion, yarns and rovings are used. 4.The most commonly used reinforcement materials are carbon/graphite fibers. (The terms graphite and carbon are often used interchangeably.) This is due to the fact that many of the desired performance characteristics require the use of carbon/graphite fibers. Currently, these fibers are produced from three types of materials known as precursor fibers:

    - polyacrylonitrile (PAN)
    - rayon
    - petroleum pitch

   The carbon/graphite fibers are produced by the controlled burning off of the oxygen, nitrogen, and other non-carbon parts of the precursor fiber, leaving only carbon in the fiber. Following this burning off (or oxidizing) step, the fibers are run through a furnace to produce either carbon or graphite fibers. Carbon fibers are produced at furnace temperatures of 1,000-2,000° C, while graphite fibers require temperatures of 2,000-3,000° C. At these temperatures the carbon atoms in the fibers are rearranged to impart the required characteristics to the finished fiber. The PAN-based fiber is the more commonly used precursor in the advanced composite industry today.

5. Aramid fibers are another human-made product. These fibers are produced by manufacturing the basic polymer, then spinning it into either a paper-like configuration or into fiber. Aramid fibers have several useful characteristics:

    - high strength and modulus;
    - temperature stability;
    - flex performance;
    - dimensional stability;
    - chemical resistance; and
    - textile processibility.

6. Textile (continuous filament) glass fibers are the type used in composite reinforcement. These fibers differ from the wool type in that they are die-drawn rather than spun.

7. A number of solvents are used in the advanced composites industry. These may be introduced into the workplace in three basic ways:

    - as part of the resin or curing agent;
    - during the manufacturing process; or
    - as part of the cleanup process.

    Most of the solvents used may be introduced in any or all of the three ways above. For this reason it would be difficult, if not impossible, to separate the solvents into the categories of use. The solvents discussed in this section are grouped by chemical class:

    - ketones
    - alcohols
    - chlorinated hydrocarbons
    - others

    Several solvents may be used in any one composite process. One or more may be introduced as part of the resin or curing agent, while another may be a part of the manufacturing process. Still another may be used for cleanup. Thus the hazard information for all products used in the process must be considered when evaluating potential exposures. The supplier's Material Safety Data Sheet (MSDS) should be consulted for more specific hazard information.

    Composite residues are often difficult to clean from operation equipment and molds. Various solvents have been used for cleaning, with varying degrees of success. Solvents in the workplace may be found in several areas:

    - in small containers near process equipment;
    - in larger containers (drums or vats) for soaking and
    - cleaning; and
    - in process equipment containers (tanks, reactors, molds, etc.).

V. DESCRIPTION OF PROCESSES.

A brief description of each process is given, followed by a basic diagram. Details on health hazard information and workplace controls are provided in Sections VI and VII of this chapter.

A. RESIN FORMULATION. Resin formulation consists of mixing epoxy or other resins with other ingredients to achieve desired performance parameters. These ingredients may be curing agents, accelerators, reactive diluents, pigments, etc.

B. PREPREGGING. Prepregging involves the application of formulated resin products, in solution or molten form, to a reinforcement such as carbon, fiberglass or aramid fiber or cloth. The reinforcement is saturated by dipping through the liquid resin (solution form, see Figure III:1-1) or by being

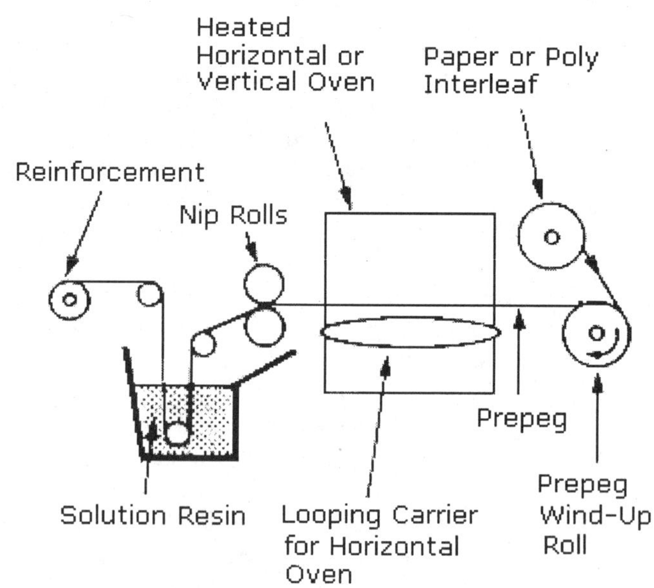

FIGURE III: 1-1. SOLUTION PREPEGGING.

impregnated through heat and pressure (hot melt form, see Figure III:1-2).

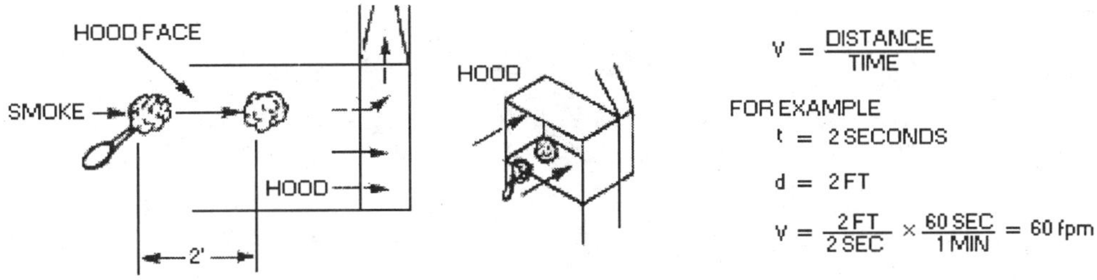

FIGURE III: 1-2. HOT MELT PREPEGGING

C. WET FILAMENT WINDING. In the filament wet winding process, continuous fiber reinforcement materials are drawn through a container of resin mixture (Figure III:1-3) and formed onto a rotating mandrel to achieve the desired shape. After winding, the part is cured in an oven.

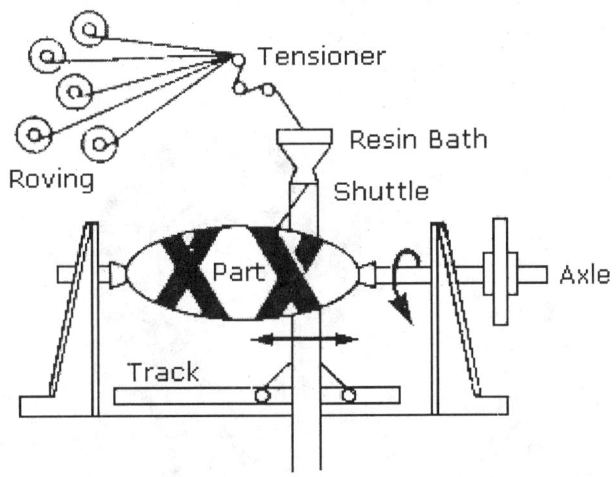

FIGURE III: 1-3. WET FILAMENT WINDING

D. HAND LAY-UP OF PREPREG. A prepreg product is laid down and formed to the desired shape (Figure III:1-4). Several layers may be required. After forming, the lay-up assembly is moved to an autoclave for cure under heat, vacuum and pressure.

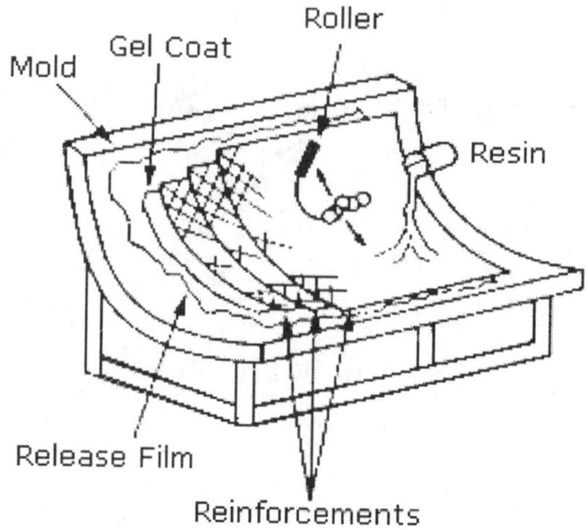

FIGURE III: 1-4. HAND LAY-UP OF PREPREG.

E. AUTOMATED TAPE LAY-UP. In this process, the prepreg tape material is fed through an automated tape application machine (robot). The tape is applied across the surface of a mold in multiple layers by the preprogrammed robot (Figure III:1-5).

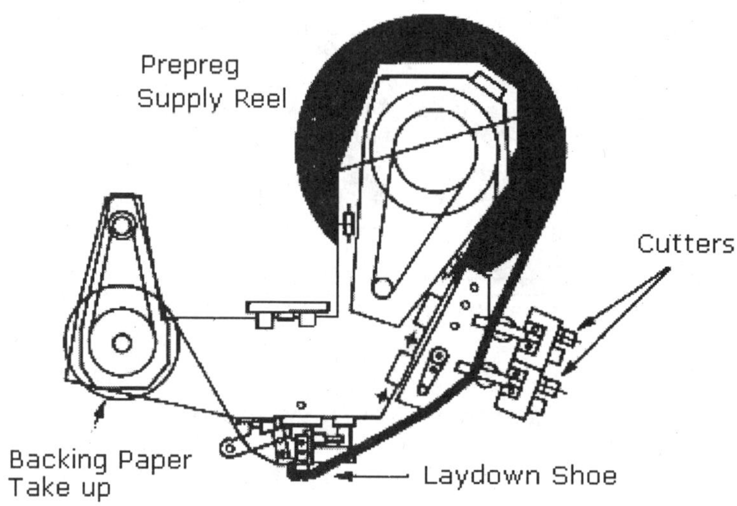

FIGURE III: 1-5. AUTOMATED LAY-UP.

F. RESIN TRANSFER MOLDING. Resin transfer molding is used when parts with two smooth surfaces are required or when a low-pressure molding process is advantageous. Fiber reinforcement fabric or mat is laid by hand into a mold and resin mixture is poured or injected into the mold cavity. The part is then cured under heat and pressure (Figure III:1-6).

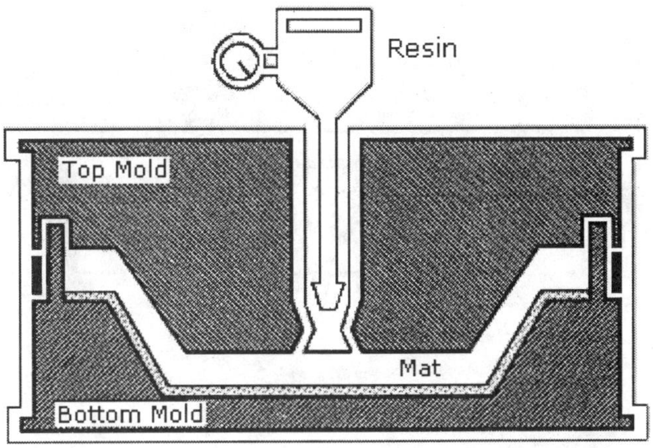

FIGURE III: 1-6. RESIN TRANSFER MOLDING

G. PULTRUSION. In the pultrusion process, continuous roving strands are pulled from a creel through a strand-tensioning device into a resin bath. The coated strands are then passed through a heated die where curing occurs. The continuous cured part, usually a rod or similar shape, is then cut to the desired length (Figure III:1-7).

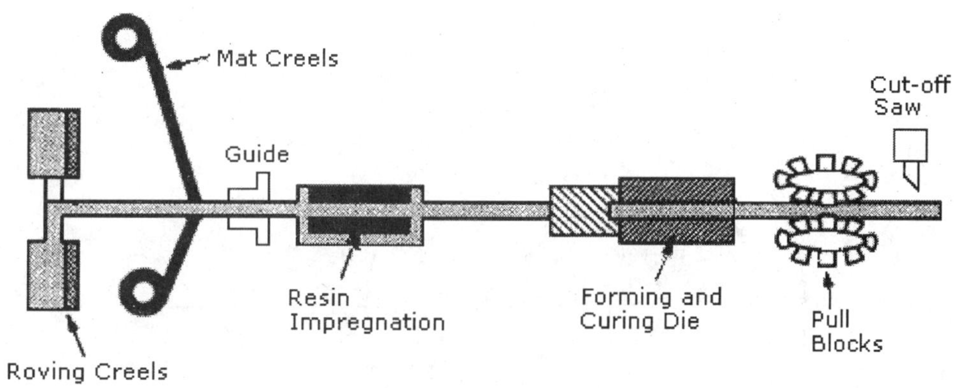

FIGURE III: 1-7. PULTRUSION

H. INJECTION MOLDING. One of the older plastics processes, injection molding is also the most closed process. It is not normally used in PMC processes due to fiber damage in the plasticating barrel. Thermoplastic granules are fed via a hopper into a screw-like plasticating barrel where melting occurs (Figure III:1-8). The melted plastic is injected into a heated mold where the part is formed. This process is often fully automated.

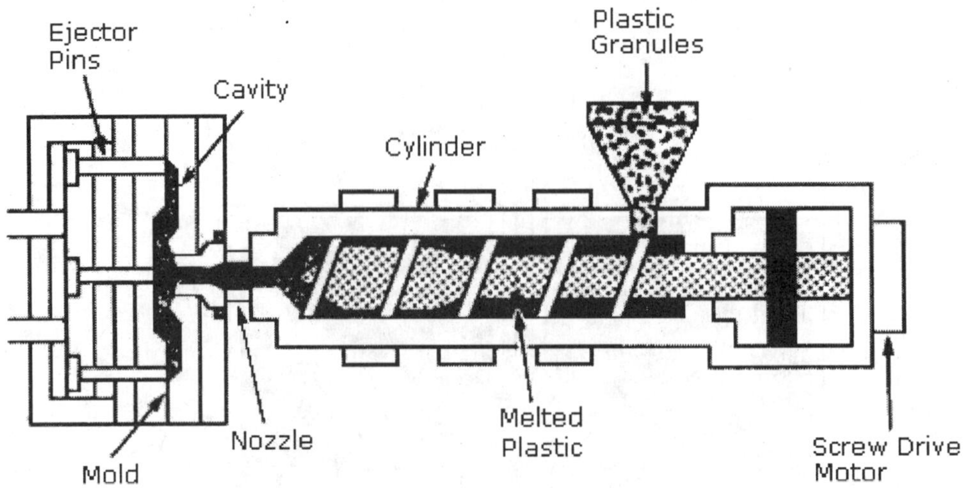

FIGURE III: 1-8. INJECTION MOLDING

I. VACUUM BAGGING & AUTOCLAVE CURING. Most parts made by hand lay-up or automated tape lay-up must be cured by a combination of heat, pressure, vacuum, and inert atmosphere. To achieve proper cure, the part is placed into a plastic bag inside an autoclave (Figure III:1-9). A vacuum is applied to the bag to remove air and volatile products. Heat and pressure are applied for curing. Usually an inert atmosphere is provided inside the autoclave through the introduction of nitrogen or carbon dioxide. Exotherms may occur if the curing step is not done properly.

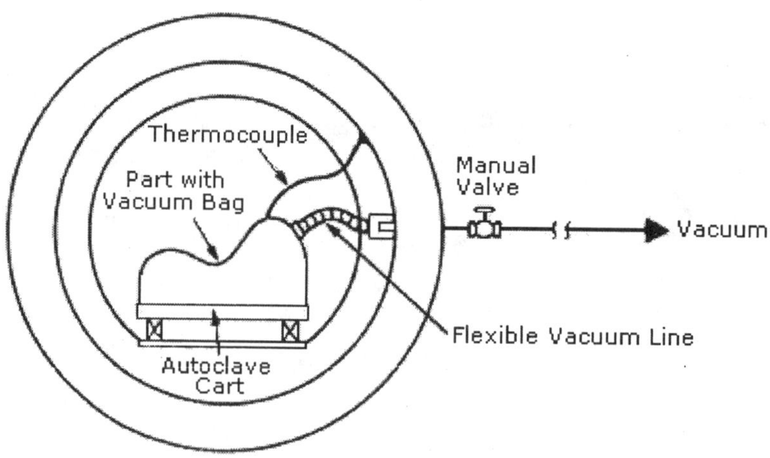

FIGURE III: 1-9. VACUUM BAGGING AND AUTOCLAVE

J. MACHINING AND FINISHING. Many of the parts made in PMC processes require some machining and/or finishing work. This may involve drilling, sanding, grinding, or other manual touch-up work. These processes vary widely, depending on the size of the finished part and the amount of finishing work required.

K. FIELD REPAIR. Repair of damaged PMC parts is frequently required. The process may consist of several steps including cutting out of the damaged material, depainting of the surface to be repaired, patching and sanding of the damaged area, and repainting of the repaired area.

VI. HEALTH HAZARDS.

Potential health hazards associated with the use of advanced composites (Table III:1-1) can be controlled through the implementation of an effective industrial hygiene program. Use of safe work practices, engineering controls, and proper personal protective equipment depends upon an appreciation of health hazard information for a safe work environment.

A. RESINS. The resins used in advanced composite processes have high molecular weights (MW > 10,000) with low vapor pressures. High molecular weight is generally associated with decreased volatility. In an epoxy system, the resin components have very low vapor pressures and they are not present as a volatilized airborne hazard.

1. As discussed earlier, epoxy resins are currently the most commonly used resins in the advanced composite industry. The basic epoxy molecule is a reaction product of epichlorohydrin (ECH) and bisphenol-A (BPA). Some epoxies contain trace amounts of residual ECH typically in the range of <1 to 10 ppm (by weight). Industrial hygiene air monitoring for ECH has been done in a number of workplaces, involving a variety of epoxy resin end-uses. Most of the monitoring has shown no detectable levels of ECH in the air. Uncured epoxy resins can present a significant dermal exposure hazard. In many workplaces, manual processing results in potential skin exposure. This can result in skin irritation, rashes and, subsequently, dermatitis if contact is prolonged. Sensitization to the resins can also develop and may require a change of work assignment.

2. Polyurethane resins are reaction products of polyols and isocyanates. The significant hazard associated with these resin systems is the presence of isocyanates. Exposure to highly toxic isocyanates can have adverse health effects. exposure to the vapor may cause irritation of the eyes, respiratory tract and skin. Irritation may be severe enough to produce bronchitis and pulmonary edema. Polyurethane resins contacting the eyes may cause severe irritation, and if polyurethane resins are allowed to remain in contact with the skin, they may produce redness, swelling, and blistering of the skin. Respiratory sensitization (an allergic, asthmatic-type reaction) may occur. Among the isocyanates, there is also evidence of cross-sensitization, in which a worker is sensitized to one isocyanate but reacts to others as well.

3. The phenol-formaldehyde resins must be handled with adequate ventilation. Traces of free formaldehyde and phenol may be present. Contact with these resins should be avoided because of the toxicity of these components and the skin-absorption potential of phenol. These components may also be given off during the curing process.

4. The acute toxicity of urea-formaldehyde resins is very similar to the phenol-formaldehyde resins. Free formaldehyde, which is present in trace amounts and may be liberated when the resins are processed, can have an irritating effect on mucous membranes. Skin sensitization to formaldehyde has been observed.

5. The health effects of bismaleimide resin systems have not been extensively studied. Manufacturers of these materials indicate that prolonged or repeated contact may cause skin irritation or sensitization. Dust or vapor from heated products may cause irritation of the eyes, nose, and throat.

6. Polyether and polyester polyols present no particular health hazard in industrial processing.

7. Thermoplastic resins in general are not considered harmful to workers' health. These resins appear harmless when ingested, and no skin irritation has been reported. No toxic effects are known to be associated with the inhalation of thermoplastic-resin dust. Treating it as nuisance or inert dust seems appropriate, although the presence of unreacted monomers may be of concern. These materials present a thermal hazard when handling. Molding operations may give off vapors which are irritating to the eyes and cause cold-like symptoms. Some thermoplastics are styrene-based, and presence of this monomer may be of concern.

TABLE III: 1-1. ORGAN SYSTEM TARGET.

Composite component	Organ system target (possible target)	Known (possible) health effect
Resins		
Epoxy resins	Skin, lungs, eyes	Contact and allergic dermatitis, conjunctivitis
Polyurethane resins	Lungs, skin, eyes	Respiratory sensitization, contact dermatitis, conjunctivitis
Phenol formaldehyde	Skin, lungs, eyes	As above (potential carcinogen)
Bismaleimides (BMI)	Skin, lungs, eyes	As above (potential carcinogen)
Polyamides	Skin, lungs, eyes	As above (potential carcinogen)
Reinforcing materials		
Aramid fibers	Skin (lungs)	Skin and respiratory irritation, contact dermatitis (chronic interstitial lung disease)
Carbon/graphite fibers	Skin (lungs)	As noted for aramid fibers
Glass fibers (continuous filament)	Skin (lungs)	As noted above
Hardeners and curing agents		
Diaminodiphenylsulfone	--	No known effects with workplace exposure
Methylenedianiline	Liver, skin	Hepatotoxicity, suspect human carcinogen
Other aromatic amines		
Meta-phenylenediamine (MPDA)	Liver, skin (kidney, bladder)	Hepatitis, contact dermatitis (kidney and bladder cancer)
Aliphatic and cyclo-aliphatic amines	Eyes, skin	Severe irritation, contact dermatitis
Polyaminoamide	Eyes, skin	Irritation (sensitization)
Anhydride	Eyes, lungs, skin	Severe eye and skin irritation, respiratory sensitization, contact dermatitis

B. CURING AGENTS.

1. Curing agents, or hardeners, used with the epoxy resins are mostly amines, amides, or anhydrides. Two of the most widely used are the aromatic amines, MDA (4,4'-methylenedianiline) and DDS (4,4'-diaminodiphenylsulfone).

2. Analysis and review of epidemiologic data and human and animal toxicity data indicates that occupational exposure to MDA may result in reversible liver toxicity (hepatotoxicity). The retina of the eye might be damaged not only by direct contact but also from MDA absorbed through ingestion. MDA is an animal carcinogen and a suspect human carcinogen by any exposure route: ingestion, inhalation, or dermal.

3. Frequently, curing agents containing mixtures of these amines can cause skin staining in processes requiring dermal contact, even when protective gloves are used. Brown and orange stains on walls and ceilings have also

been reported. The skin staining has been attributed to MDA; dermal absorption is approximately 2% per hour. Soap and water, rather than any organic solvent, should be used for skin clean-up to avoid any solvent increase of transdermal absorption.

4. The OSHA permissible exposure limits (PEL'S) for MDA are 10 ppb (parts per billion) expressed as an 8-hour time-weighted average, and a short-term exposure limit (STEL) of 100 ppb averaged over any 15-minute period for either general industry or construction uses of MDA. The FR 57(154): 35630 (August 10, 1992) issue published the Final Rule for 29 CFR Parts 1910 and 1926: Occupational Exposure to 4,4'Methylenedianiline (MDA).

5. Another of the amines, DDS, has a significant amount of toxicological data as its pharmaceutical grade, Dapsone™, has been used for years to treat leprosy and certain types of chronic dermal inflammation. However, at low airborne concentrations, there are no known effects from workplace exposure.

6. Other aromatic amines used in the advanced composites industry include m-phenylene diamine and the various isomers of toluenediamine. These aromatic amines are considered to be only slightly irritating to the skin.

7. Aliphatic and cycloaliphatic amines are strong bases and are considered to be severe eye and skin irritants. Inhalation of these amines can cause irritation of the nose and throat, and lung irritation with respiratory distress. Some of these amines are also skin and respiratory-tract sensitizers. Vapors of the volatile amines may cause conjunctivitis and visual disturbances.

8. Polyaminoamide hardeners have a less irritating effect on the skin and mucous membranes than the aliphatic and cycloaliphatic amine hardeners, but may cause sensitization.

9. Amide hardeners generally have only a slight irritant effect. Should the handling of these hardeners generate dust, measures should be taken to prevent inhalation.

10. The dusts of high-melting solids like most anhydride curing agents are severe eye and skin irritants. Some hydrophthalic anhydrides have high vapor pressures at the usual processing and curing temperatures and the vapors evolved during use of these curing agents can have an irritating effect on the skin, eyes, and respiratory tract. Exposure to the high-melting solids like trimellitic anhydride and tetraphthalic anhydride can cause respiratory sensitization.

C. REINFORCEMENT FIBERS. Most of the reinforcing materials used in the industry have the potential to cause eye, skin, and upper respiratory tract irritation as a result of the mechanical-irritant properties of the fibers. The potential synergism has not been clearly defined. The chemical irritation caused by resins can compound the mechanical irritation caused by the fibers.

1. Carbon/graphite fibers dominate the advanced composites industry and may be made from any of three precursors, as discussed in Section C. However, the PAN-based carbon fibers are the predominant form in use today. It is important to ascertain which type of carbon-fiber precursor is used in order to evaluate the hazards.

   Pitch-based carbon fibers may be associated with an increased risk of skin cancer, although the evidence is weak. PAN-based carbon fibers did not cause tumors when the same test was conducted. Standard mutagenicity tests conducted on PAN-based carbon fibers were negative.

   The principal hazards of carbon-fiber handling are mechanical irritation and abrasion similar to that of glass fibers. Skin rashes are common and reportedly more severe than from glass fibers. Carbon fibers commonly in use are also greater than six micrometers in diameter, making them unlikely to be respirable. An ongoing survey of workers in a carbon-fiber production plant shows no pulmonary function abnormalities and no evidence of dust-related disease.

   Carbon fibers may be coated with a material to improve handling, known generically as sizing. The sizing materials are typically epoxy resins. They may be biologically active and cause irritation or sensitization.

2. Aramid fibers are made from a polymer, poly(p-phenylenediamine terephthalate). Animal and human skin tests of Kevlar™ aramid fibers show no potential for skin sensitization and low potential for irritation. While Kevlar™ fibers are too large to be inhaled (12-15 mm), they may be fractured into respirable fibrils in some composite manufacturing processes. Industrial process monitoring shows that airborne respirable fibril levels are low in typical operations. Measured exposure levels from composite machining are typically below 0.2 fibrils per milliliter of air (0.2 f/ml), as an 8-hour, time-weighted average (TWA), while continuous filament handling generates less than 0.1 f/ml. The physical structure of aramid fibers makes it extremely difficult to generate airborne concentrations.

3. Glass fibers, used as reinforcement in PMC processes, are a continuous-filament form and not the glass-wool (random) type. Practically all glass fibers for composite reinforcement are greater than six microns in diameter. Airborne fiber of this diameter does not reach the alveoli and is nonrespirable. Glass fibers break only into shorter fragments of the same diameter. Their diameter cannot be reduced by machining, milling, or other mechanical processes.

   Mechanical irritation of skin, eyes, nose, and throat are common hazards associated with glass-fiber exposure. Continuous-filament glass fiber is not considered fibrogenic. Lung clearance mechanisms are effective for glass fibers. In June 1987, the International Agency for Research on Cancer (IARC) categorized continuous-filament glass fibers as not classifiable with respect to human carcinogenicity. The evidence from human and animal studies was evaluated by IARC as insufficient to classify continuous-filament glass fibers a possible, probable, or confirmed cancer-causing material.

   Like carbon fibers, glass fibers may also be coated with a sizing material to improve handling. Sizing materials may be epoxy resins, polyvinyl acetate-chrome chloride, polyvinyl acetate-silane, polyester-silane, or epoxy-silane compounds. These materials may be biologically active and cause irritation or sensitization.

D. DUSTS.

1. Dusts may be generated in several ways in advanced composite processes. The most common dust-generating processes are machining and finishing of cured parts and in repair of damaged parts. Much of the dust generated in these processes can be very fine and should be considered respirable. Studies of some graphite-epoxy finishing operations found respirable fractions ranging from 25% to 100%.

2. More dust is usually generated in finishing and repair processes since large surface areas are involved. Grinding, routing and sanding are frequently used methods in both processes. The repair process may require the use of abrasive blasting as well as sanding to remove existing paint or coatings. Typically, a synthetic blasting agent, e.g., plastic media blast, is used. Ingredients of the paint or coating being removed, such as lead or chromates, may also be of concern. The repair process may also require cutting or sawing to remove the damaged part area, and both may generate significant amounts of airborne dust.

3. In general, studies on composite dusts indicate that:

   - The dusts are particulate in nature and usually contain few fibers;
   - The dusts are thermally stable up to 250 °C and exhibit a high degree of cure;
   - Toxicology studies indicate the dusts should probably be controlled at levels below the PEL for inert dust, but not approaching the PEL for crystalline quartz.

E. SOLVENTS.

1. Many of the solvents used in advanced composite processes are volatile and flammable. Most are skin and eye irritants, and some may be readily absorbed through the skin. Precautions must be taken when using organic solvents because they can facilitate the entry of toxic materials into the skin and organ systems. They may also enhance skin sensitization caused by the resin systems. Some (such as methyl alcohol) are poisonous, and all are capable of extracting fat from skin. Harmful effects from industrial exposures come principally from skin contact and inhalation.

2. Selection of the proper glove for protection is important. Permeation data are available for many industrial chemicals, especially solvents. However, in the case of resins and curing agents, not much data are available. This also is true for mixtures of solvents, as little or no testing has been done. Often the glove selection process is one of trial and error. If a skin rash or dermatitis is observed there are several possible causes:

    - the wrong gloves may have been selected;
    - improper work practices are being followed;
    - the employee is deficient in personal hygiene practices; or
    - adequate washing facilities are absent.

3. Several of the solvent classes most commonly found in the PMC workplace are listed below, along with general hazard information.

4. Several ketones are frequently found in PMC manufacture. These include:

    - acetone (DMK)
    - methyl ethyl ketone (MEK)
    - methyl isobutyl ketone (MIBK)

    These solvents may cause eye, nose, and throat irritation, and prolonged contact with the liquid may result in defatting of the skin and resultant dermatitis. In high concentrations, narcosis is produced with symptoms of headache, nausea, light-headedness, vomiting, dizziness, incoordination, and unconsciousness. Ketones are volatile and flammable. Acetone is a popular solvent used for cleanup and may be found around the workplace in containers for this purpose.

5. Some of the lower-boiling alcohols are sometimes used in composites manufacture. These include:

    - methanol (methyl alcohol)
    - ethanol (ethyl alcohol)
    - isopropanol (isopropyl alcohol)

    These alcohols do not usually present serious hazards in the industrial setting. Toxicity is usually related to irritation of the conjunctivae and the mucous membranes of the upper airway. Contact with the liquid may cause defatting of the skin and dermatitis. These alcohols are volatile and flammable.

6. Three chlorinated hydrocarbon compounds in particular are found in the composites workplace:

    - methylene chloride (dichloromethane)
    - 1,1,1-trichloroethane (methyl chloroform)
    - trichloroethylene

    Health effects typical of the group include irritation of the eyes and upper respiratory tract, dizziness, confusion, drowsiness, nausea, vomiting, and occasionally abdominal pain. Visual disturbances may also occur. Due to the solvents' defatting properties, repeated or prolonged skin contact with these liquids may cause dermatitis. Ability to depress the central nervous system is a characteristic property of all members of this group.

    These solvents are not particularly flammable. Many manufacturers have replaced the ketones with the above hydrocarbon solvents to reduce the risk of flammability.

7. Other solvents that may occasionally be used are:

    - toluene
    - xylene
    - tetrahydrofuran (THF)
    - dimethylsulfoxide (DMSO)

- dimethylformamide (DMF)
- gamma-butyrolactone (BLO)
- n-methyl pyrrolidone (NMP)
- n-butyl acetate
- glycol ethers

Technical literature including MSDS's from the solvent supplier should be consulted about t hese or any chemicals used with advanced composites.

VII. WORKPLACE CONTROLS.

Good workplace controls are essential in controlling exposure to process materials. Many of the materials, particularly the resins, curing agents, and fibers, present a potential dermal-exposure hazard. Many of the solvents and some of the curing agents present a potential inhalation hazard. Some materials present both a dermal and inhalation hazard. Ingestion may be a potential exposure hazard, but usually involves poor personal hygiene or contamination of eating facilities. The various types of workplace controls described below may typically be found in the advanced composite workplace.

A. ENGINEERING CONTROLS. Isolation (e.g., isolated storage, separate process areas, enclosures, closed systems) and local exhaust ventilation are the primary engineering controls found in advanced composites processes. These controls can be found in:

- Resin mixing areas;
- Heated curing areas including autoclaves;
- Finishing and repair areas; and
- Controlling off-gases from exotherms

B. WORK PRACTICE CONTROLS. Work practices, as distinguished from engineering controls, involve the way a task is performed. Some fundamental and easily implemented work practices that can be used to minimize exposures when working with advanced composites are:

- good employee training and education;
- following the proper procedures for production, process and control equipment;
- proper use, maintenance, and cleaning of personal protective equipment;
- good personal hygiene program;
- housekeeping;
- periodic inspection and maintenance of production, process and control equipment; and
- good supervision.

C. PERSONAL PROTECTIVE EQUIPMENT.

1. Gloves, protective clothing, and eye protection may frequently be required, especially when working with resins, curing agents, and solvents. Selection of the proper protective materials should be based on permeation data, if available. This type of data are often available for the solvents used, but very little data are available for the resins and curing agents.

2. In many advanced composites processes several chemicals or mixtures are involved. There are essentially no permeation data available for chemical mixtures. This means that, in many cases, glove and clothing selection must be a trial and error process.

3. Generally, the resins are of a larger molecular size and so are less likely to permeate protective materials than the curing agents and solvents. The aromatic amine curing agents are particularly difficult to protect against. In some advanced composites processes, close hand work and contact is required, and a glove must provide good tactility. Often this type of glove provides the least protection against the resin and curing agent.

4. Eye protection can be provided by standard safety glasses with side shields, goggles, or a face shield, as needed.

5. Respiratory protection is not required in many advanced composites processes, due to the low vapor pressure of the materials involved. However, respirators may be required where:

   - Airborne solvent levels are high;
   - Dust levels are high (resin mixing, finishing, repair);
   - Large surface areas and significant hand work are involved; and
   - Exotherms are experienced.

D. ADMINISTRATIVE CONTROLS. Employee exposures also can be controlled by scheduling operations with the highest exposures at a time when the fewest employees are present.

VIII. BIBLIOGRAPHY.

Advanced Composites Glossary. Blaise Technoire, San Marcos, Calif.

Doyle, E.J. *Suggested Strategies in Screening for Health Effects in Personnel who Work with Composites*. Applied Industrial Hygiene Special Issue: December 1989. Pp 64-67.

Engineered Materials Handbook, Vol. 1: Composites. ASM International, Metals Park, Ohio: 1987.

Lubin, G., Ed. *Handbook of Composites*. Van Nostrand Reinhold Co., New York: 1982.

Occupational Safety and Health Administration, Department of Labor, Final Rule: Occupational Exposure to 4,4'Methylenedianiline (MDA) for 29 CFR Parts 1910 and 1926. Federal Register 57(154): 35630 (August 10, 1992)

APPENDIX III:1-1. GLOSSARY.

[Adapted from *Advanced Composites Magazine*, 1992 Bluebook, Edgell Plastics Publications]

A-stage An early stage of polymerization of thermosetting resins in which the material is still soluble in certain liquids and is fusible. Also called resole. See also B-stage, C-stage.

Ablative A material that absorbs heat through a decomposition process called pyrolysis at or near the exposed surface.

Addition A polymerization reaction in which no by-products are formed.

Additives Ingredients mixed into resin to improve properties.

Adhesive failure A rupture of adhesive bond that appears to be a separation at the adhesive-adherend interface.

Amorphous Polymers that have no order to their molecules, thus no crystalline component.

Anisotropic The tendency of a material to exhibit different properties in response to stresses applied along axes in different directions.

Aramid Aromatic polyamide fibers characterized by excellent high-temperature, flame-resistance, and electrical properties. Aramid fibers are used to achieve high-strength, high-modulus reinforcement in plastic composites.

Areal weight The weight of fiber per unit area (width times length) of tape or fabric.

Aspect ratio The ratio of length to diameter of a fiber.

Autoclave A closed vessel that permits application of pressure and heat used for curing composites.

Autoclave molding A molding technique in which an entire assembly (lay-up) is placed into an autoclave at 50 to 100 psi, in order to consolidate layers of the part by removing entrapped air and volatiles. Usually includes application of heat to trigger curing of the resin.

Axial winding A type of filament winding in which the filaments are parallel to the axis.

B-stage Intermediate stage in the polymerization reaction of thermosets, following which material will soften with heat and is plastic and fusible. Also called resistal. The resin of an uncured prepreg or premix is usually in B-stage. See A-stage, C-stage.

Bag molding A technique in which the composite material is placed in a rigid mold and covered with a flexible bag, with pressure applied by vacuum, autoclave, press, or by inflating the bag.

Balanced laminate All lamina except those at 0/90 are placed in plus-minus pairs (not necessarily adjacent) symmetrically about the lay-up centerline.

Bearing strength The maximum bearing stress that will not cause a composite to fail when applied through a cylindrical fastener surface.

Bearing stress Applied load divided by bearing area (hole diameter times thickness).

Bias fabric A fabric in which warp and fill fibers are at an angle to the length.

Biaxial winding A type of filament winding in which the helical band is laid in sequence, side by side, with no crossover of the fibers.

Bidirectional laminate A reinforced plastic laminate in which the fibers are oriented in more than one direction in the plane of the laminate.

Bismaleimide A type of polyimide that cures by an addition reaction, avoiding formation of volatiles, and has temperature capabilities between those of epoxy and polyimide.

Bleeder cloth A layer of woven or nonwoven material, not a part of the composite, that allows excess gas and resin to escape during cure.

Bleedout The excess liquid resin appearing at the surface, primarily during filament winding.

Bond strength The amount of adhesion between bonded surfaces. As measured by load/bond area, the stress required to separate a layer of material from that to which it is bonded.

Boron fiber A fiber usually of a tungsten-filament core with elemental boron vapor deposited on it to impart strength and stiffness.

Braiding Weaving fibers into a tubular shape.

Breather A loosely woven material that does not come in contact with the resin but serves as a continuous vacuum path over a part in production.

Broadgoods Fibers woven into fabrics that may or may not be impregnated with resin, usually furnished in rolls.

Buckling (composite) A failure usually characterized by fiber deflection rather than breaking because of compressive action.

Butt joint A joint in which parts are joined with no overlap.

C-stage The final step in the cure of a thermoset resin; results in essentially irreversible hardening and insolubility.

Carbon/carbon A composite of carbon fiber in a carbon matrix.

Carbon fiber An important reinforcing fiber known for its light weight, high strength, and high stiffness that is produced by pyrolysis of an organic precursor fiber in an inert atmosphere at temperatures above 1,800° F. The material may also be graphitized by heat treating above 3,000° F.

Catalyst A substance used in small quantities to promote or control the curing of a compound without being consumed in the reaction.

Caul sheet Plate or sheet the same size and shape used in contact with a composite lay-up to transmit normal pressure and temperature during cure.

Ceramic-matrix composites Materials consisting of a ceramic or carbon fiber surrounded by a ceramic matrix, usually SiC (silicon carbide).

Chemical vapor deposition (CVD) A process in which desired reinforcement material is deposited from vapor phase onto a continuous core; boron on tungsten, for example.

Circuit One complete traverse of the fiber feed mechanism of a filament-winding machine.

Circumferential winding A type of filament winding in which the filaments are perpendicular to the axis.

Co-cured Cured and simultaneously bonded to another prepared surface.

Coefficient of thermal expansion A material's fractional change in length for a given unit change of temperature.

Commingled yarn A hybrid yarn made with two types of materials intermingled in a single yarn; for example, thermoplastic filaments intermingled with carbon filaments to form a single yarn.

Composite A material created from a fiber (or reinforcement) and an appropriate matrix material in order to maximize specific performance properties. The constituents do not dissolve or merge completely but retain their identities as they act in concert.

Compression molding A technique for molding thermoset plastics in which a part is shaped by placing the fiber and resin into an open mold cavity, closing the mold, and applying heat and pressure until the material has cured or achieved its final form.

Compressive strength A material's ability to resist a force that tends to crush or buckle; maximum compressive load a specimen sustains divided by the specimen's original cross-sectional area.

Condensation A polymerization reaction in which simple by-products (for example, water) are formed.

Consolidation A processing step that compresses fiber and matrix to reduce voids and achieve a desired density.

Contact molding A technique in which reinforcement and resin are placed in a mold, with cure taking place at room temperature with a catalyst/promoter system, or in a heated oven. No additional pressure is used.

Continuous filament An individual, small-diameter reinforcement that is flexible and indefinite in length.

Continuous-filament yarn Yarn that is formed by twisting two or more continuous filaments into a single continuous strand.

Continuous roving Parallel filaments coated with sizing, gathered together into single or multiple strands, and wound into a cylindrical package. It may be used to provide continuous reinforcement in woven roving, filament winding, pultrusion, prepregs, or high-strength molding compounds, or it may be used chopped.

Core The central component of a sandwich construction to which the sandwich faces or skins are attached; also, part of a complex mold that forms undercut parts.

Co-woven fabric A reinforcement fabric woven with two different types of fibers in individual yarns; for example, thermoplastic fibers woven side by side with carbon fibers.

Creep The dimensional change in a material under physical load over time beyond instantaneous elastic deformation.

Crimp A fiber's waviness, which determines the capacity of the fiber to cohere.

Critical length The minimum length of a fiber necessary for matrix shear loading to develop fiber ultimate strength by a matrix.

Cross laminated Material laminated so that some of the layers are oriented at various angles to the other layers with respect to the laminate grain. A cross-ply laminate usually has plies oriented only at 0°/90°.

Crystallinity The quality of having a molecular structure with atoms arranged in an orderly, three-dimensional pattern.

Cure To change the physical properties of a material irreversibly by chemical reaction via heat and catalysts, alone or in combination, with or without pressure.

Cure temperature The temperature at which a material attains final cure.

Curing agent A catalytic or reactive agent that brings about polymerization when it is added to a resin.

Damage tolerance A measure of the ability of structures to retain load-carrying capability after exposure to sudden loads (for example, ballistic impact).

Damping Diminishing the intensity of vibrations.

Debond An unplanned nonadhered or unbonded region in an assembly.

Delamination The separation of a laminated plastic material along the plane of its layers.

Denier A numbering system for yarn and filament in which yarn number is equal to weight in grams of 9000 meters of yarn.

Design allowable A limiting value for a material property that can be used to design a structural or mechanical system to a specified level of success with 95% statistical confidence. B-basis allowable: material property exceeds the design allowable 90 times out 100. A-basis allowable: material property exceeds the design allowable 99 times out of 100.

Doubler Localized area of extra layers of reinforcement, usually to provide stiffness or strength for fastening or other abrupt load transfers.

Draft angle A mandrel's taper or angle for ease of part removal.

Drape The ability of prepreg to conform to the shape of a contoured surface.

Dry winding A type of filament winding in which preimpregnated roving is used.

E-glass "Electrical glass"; the borosilicate glass most often used for the glass fibers in conventional reinforced plastics.

Fabric, nonwoven A material formed from fibers or yarns without interlacing.

Fabric, woven A material constructed of interlaced yarns, fibers, or filaments.

Fatigue The failure of a material's mechanical properties as a result of repeated stress.

Fatigue strength Maximum cyclical stress withstood for a given number of cycles before a material fails.

Fiber orientation The fiber alignment in a nonwoven or a mat laminate in which most of the fibers are in the same direction, thereby affording higher strength in that direction.

Fiber placement A continuous process for fabricating composite shapes with complex contours and/or cutouts by means of a device that lays preimpregnated fibers (in tow form) onto a nonuniform mandrel or tool. It differs from filament winding (below) in several ways: there is no limit on fiber angles; compaction takes place online via heat, pressure, or both; and fibers can be added and dropped as necessary. The process produces more complex shapes and permits a faster putdown rate than filament winding.

Filament winding A process for fabricating composites in which continuous reinforcing fibers, either preimpregnated with resin or drawn through a resin bath, are wound around a rotating, removable mandrel.

Filaments Individual fibers of indefinite length used in tows, yarns, or roving.

Film adhesive An adhesive in the form of a thin, dry, resin film with or without a carrier, commonly used for adhesion between layers of laminates.

Finish Material applied to fibers, after sizing is removed, to improve matrix-to-fiber coupling.

Fracture A rupture of the surface of a laminate because of external or internal forces, with or without complete separation.

Fracture toughness A measure of the damage tolerance of a material containing initial flaws or cracks.

Glass transition The reversible change in an amorphous polymer between a viscous or rubbery condition and a hard, relatively brittle one.

Glass-transition temperature (Tg) The approximate temperature at which increased molecular mobility results in significant changes in properties of a cured resin. The measured value of Tg can vary depending upon the test method.

Hand lay-up A fabrication method in which reinforcement layers, preimpregnated or coated afterwards, are placed in a mold by hand, then cured to the formed shape.

Hardener A substance used to promote or control curing action by taking part in it; as opposed to catalyst.

Heat-distortion temperature Temperature at which a test bar deflects a certain amount under specified temperature and a stated load.

Honeycomb Resin-impregnated material manufactured in, usually, hexagonal cells that serves as a core material in sandwich constructions. Honeycomb may also be metallic or polymer materials in a rigid, open-cell structure.

Hoop stress Circumferential stress in a cylindrically shaped part as a result of internal or external pressure.

Hybrid composite A composite with two or more reinforcing fibers.

Impact strength A material's ability to withstand shock loading as measured by the work done in fracturing a specimen.

Impregnate To saturate the voids and interstices of a reinforcement with a resin.

Impregnated fabric See Prepreg.

Interface The surface between two different materials: in fibers, the area at which the glass and sizing meet; in a laminate, the area at which the reinforcement and the laminating resin meet.

Interlaminar Existing or occurring between two or more adjacent laminae.

Interlaminar shear The shearing force tending to produce displacement between two laminae along the plane of their interface; usually the weakest element of a composite.

Isotropic Having uniform properties in all directions independent of the direction of load application.

Laminate ply One layer of a laminated product.

Lap joint A joint made by bonding overlapped portions of two adherends.

Lay-up The placement of layers of reinforcement in a mold.

Liquid-crystal polymers A newer type of thermoplastic, melt processible, with high orientation in molding, improved tensile strength, and high-temperature capability.

Mandrel. The form around which resin-impregnated fiber or tape is wound to form structural shapes or tubes.

Mat A fibrous reinforcing material comprised of chopped filaments (for chopped-strand mat) or swirled filaments (for continuous-strand mat) with a binder to maintain form; available in blankets of various widths, weights, and lengths.

Matrix A material in which the fiber of a composite is imbedded; it can be plastic, metal, ceramic, or glass.

Metal-matrix composites Materials in which continuous carbon, silicon carbide, or ceramic fibers are embedded in a metallic matrix material.

Modulus A measure of the ratio of load (stress) applied to the resultant deformation of a material, such as elasticity or shear.

Multifilament A yarn consisting of many continuous filaments.

Nondestructive inspection (NDI) A process or procedure for determining material or part characteristics without permanently altering the test subject. Nondestructive testing (NDT) is broadly considered synonymous with NDI.

Nonwoven roving A reinforcement composed of continuous rovings loosely gathered together.

Oriented materials Composites whose constituents are aligned in a particular way.

Out-life The period of time a prepreg material remains in a handleable form and with properties intact outside of the specified storage environment; for example, out of the freezer in the case of thermoset prepregs.

PAN See Polyacrylonitrile.

Peel ply Layer of material applied to a prepreg lay-up surface that is removed from the cured laminate prior to bonding operations and leaves a clean resin-rich surface ready for bonding.

Peel strength Strength of an adhesive bond obtained by stress applied "in a peeling mode."

Pitch A residual petroleum product used in the manufacture of certain carbon fibers.

Planar winding A type of filament winding in which the filament path lies on a plane that intersects the winding surface.

Ply The number of single yarns twisted together to form a plied yarn; one of the layers that make up a stack or laminate.

Polar winding A type of filament winding in which the filament path passes tangent to the polar opening at one end of the chamber and tangent to the opposite side of the polar opening at the other end of the chamber.

Polyacrylonitrile (PAN) A product used as a base material in the manufacture of certain carbon fibers.

Polymer A very large molecule formed by combining a large number of smaller molecules, called monomers, in a regular pattern.

Polymerization A chemical reaction in which the molecules of monomers are linked together to form polymers.

Postcure An additional elevated-temperature exposure that is performed often without tooling or pressure to improve elevated-temperature mechanical properties, for example.

Pot life The length of time a catalyzed thermosetting resin system retains a viscosity low enough for it to be suitable for processing.

Precure The full or partial setting of a resin or adhesive before the clamping operation is complete or before pressure is applied.

Precursor For carbon fibers, the rayon, PAN, or pitch fibers from which carbon fibers are made.

Preform A fibrous reinforcement preshaped to approximate contour and thickness desired in the finished part.

Prepreg Resin-impregnated cloth, mat, or filaments in flat form that can be stored for later use. The resin is often partially cured to a tack-free state called "B-staging." Catalysts, inhibitors, flame retardants, and other additives may be included to obtain specific end-use properties and improve processing, storage, and handling characteristics.

Pressure-bag molding A molding technique in which a flexible bag is placed over the contact lay-up in the mold, sealed, and clamped in place, and pressure applied by compressed air, which forces the bag against the part while the part cures.

Pultrusion A continuous process for manufacturing composites in rods, tubes, and structural shapes having a constant cross-section. After the reinforcement is passed through the resin-impregnation bath, it is drawn through a shaping die to form the desired cross-section; curing takes place before the laminate can depart from that cross-section.

Quasi-isotropic Approximating isotropy by orientation of plies in several directions.

Ramping A gradual, programmed increase or decrease in temperature or pressure to control the cure or cooling of composite parts.

Reinforcement A material added to the matrix to provide the required properties; ranges from short fibers through complex textile forms.

Release agents Materials that are used to prevent cured matrix material from bonding to tooling.

Release film An impermeable film layer that does not bond to the composite during cure.

Resin A material, generally a polymer, that has an indefinite and often high molecular weight and a softening or melting range and exhibits a tendency to flow when it is subjected to stress. Resins are used as the matrices to bind together the reinforcement material in composites.

Resin rich Localized area filled with resin but lacking reinforcement fiber.

Resin starved Localized area lacking sufficient resin for wetout of the fibers.

Resin-transfer molding (RTM) A molding process in which catalyzed resin is transferred into an enclosed mold into which the fiber reinforcement has been placed; cure normally is accomplished without external heat. RTM combines relatively low tooling and equipment costs with the ability to mold large structural parts.

Roving A collection of bundles of continuous filaments either as untwisted strands or as twisted yarns.

S-glass Structural glass; a magnesia/alumina/silicate glass reinforcement designed to provide very high tensile strength.

Sandwich construction A composite composed of a lightweight core material (usually honeycomb or foamed plastic) to which two relatively thin, dense, high-strength, functional, or decorative skins (also called faces) are adhered.

Scarf joint A bonded joint in which similar segments of adherends are cut away, with cut areas overlapped and bonded.

Selvage The narrow edge of woven fabric that runs parallel to the warp. It is made with stronger yarns in a tighter construction than the body of the fabric to prevent raveling.

Shelf life The length of time a material can be stored and continue to meet specification requirements and remain suitable for its intended use.

Silicon carbide fiber A reinforcing fiber with high strength and modulus; density is equal to that of aluminum. It is used in organic metal-matrix composites.

Sizing A compound that binds together and stiffens warp yarn to provide resistance to abrasion during weaving; normally removed and replaced with finish before matrix application.

Skin A layer of relatively dense material used in a sandwich construction on the surface of the core.

Specific gravity The density (mass per unit volume) of a material divided by that of water at a standard temperature.

Starved joint A joint that does not have the proper amount of adhesive because of insufficient spread or excessive pressure.

Stiffness The relationship of load to deformation for a particular material.

Storage life The amount of time a material can be stored and remain suitable for use.

Strain The elastic deformation of a material as a result of stress.

Stress The internal force that resists change in size or shape, expressed in force per unit area.

Structural adhesive An adhesive used for transferring loads between adherends.

Structural bond A bond joining load-bearing components of an assembly.

Tack The stickiness of a prepreg.

Tape A unidirectional woven prepreg, in widths up to 12 inches for carbon fiber, for example.

Tape laying A fabrication process in which prepreg tape is laid side by side or overlapped to form a structure.

Tensile strength The maximum tensile stress sustained by a plastic specimen before it fails in a tension test.

Thermal conductivity The ability of a material to conduct heat.

Thermoplastic A plastic material that is capable of being repeatedly softened by application of heat and repeatedly hardened by cooling.

Thermoset A plastic material that is capable of being cured by heat or catalyst into an infusible and insoluble material. Once cured, a thermoset cannot be returned to the uncured state.

Tooling resins Plastic resins, chiefly epoxy and silicone, that are used as tooling aids.

Toughness Tendency of a material to absorb work.

Tow An untwisted bundle of continuous filaments, usually designated by a number followed by "K," indicating multiplication by 1,000; for example, 12K tow has 12,000 filaments.

Unbond Area of a bonded surface in which bonding of adherends has failed to occur, or where two prepreg layers of a composite fail to adhere to each other; also denotes areas where bonding is deliberately prevented to simulate a defective bond.

Unidirectional Refers to fibers that are oriented in the same direction, such as unidirectional fabric, tape, or laminate, often called UD.

Vacuum bag molding A molding technique in which the part is cured inside a layer of film, from which entrapped air is removed by vacuum.

Viscosity The tendency of a material to resist flow.

Voids Pockets of entrapped gas that have been cured into a laminate.

Volatiles Materials in a sizing or a resin formulation that can be vaporized at room or slightly elevated temperature.

Warp The yarns running lengthwise and parallel to the selvage in a woven fabric.

Water jet A high-pressure stream of water used for cutting organic composites.

Weave The pattern by which a fabric is formed from interlacing yarns. In plain weave, the warp and fill fibers alternate to make both fabric faces identical; in satin weave, the pattern produces a satin appearance, with the warp tow over several fill tows and under the next one (for example, eight-harness satin would have warp tow over seven fill tows and under the eighth).

Weft The yarns running perpendicular to the warp in a woven fabric; also called "woof."

Wet lay-up The application of resin to dry reinforcement in the mold.

Wet winding A type of filament winding in which the fiber strand is impregnated with resin immediately before it contacts the mandrel.

Wetout The saturation of all voids between strands and filaments of porous materials with resin.

Wetting agent A surface-active agent that promotes wetting by decreasing the cohesion within a liquid.

Winding pattern The regularly recurring pattern of the filament path in a filament winding after a certain number of mandrel revolutions.

Woven roving A heavy, coarse fabric produced by the weaving of continuous roving bundles.

Wrinkle An imperfection in the surface of a laminate that looks like a crease or fold in one of the outer layers; it occurs in vacuum bag molding due to improper placement of the bag.

X axis The axis in the plane of the laminate used as 0 degree reference; the Y axis is the axis in the plane of the laminate perpendicular to the X axis; the Z axis is the reference axis normal to the laminate plane in composite laminates.

Yarn Continuously twisted fibers or strands suitable for use in weaving into fabrics.

Young's modulus The ratio of normal stress to the corresponding strain for tensile or compressive stresses less than the proportional limit of the material.

SECTION III: CHAPTER 2

INDOOR AIR QUALITY INVESTIGATION

Contents:

I. INTRODUCTION.

A. CAUSAL FACTORS. Modern office buildings are generally considered safe and healthful working environments. However, energy conservation measures instituted during the early 1970's have minimized the infiltration of outside air and contributed to the buildup of indoor air contaminants. Investigations of indoor air quality (IAQ) often fail to identify any harmful levels of specific toxic substances. Often employee complaints result from items such as cigarette smoke, odors, low-level contaminants, poor air circulation, thermal gradients, humidity, job pressures, lighting, work-station design, or noise. Appendix III:2-1 presents a brief discussion of these items.

B. INCIDENCE.

1. The range of investigations of indoor air quality problems encompasses complaints from one or two employees to episodes where entire facilities are shut down and evacuated until the events are investigated and problems corrected.

2. Complaints are often of a subjective, nonspecific nature and are associated with periods of occupancy. These symptoms often disappear when the employee leaves the workplace. They include headache, dizziness, nausea, tiredness, lack of concentration, and eye, nose, and throat irritation.

3. In approximately 500 indoor air quality investigations in the last decade, the National Institute for Occupational Safety and Health (NIOSH) found that the primary sources of indoor air quality problems are:

- Inadequate ventilation 52%
- Contamination from inside building 16%
- Contamination from outside building 10%
- Microbial contamination 5%
- Contamination from building fabric 4%
- Unknown sources 13%

C. RECOMMENDED VENTILATION RATES.

1. The American Society of Heating, Refrigerating, and Air-Conditioning Engineers (ASHRAE) established recommended ventilation rates for indoor environments in 1973.[2]

2. ASHRAE amended this standard in 1975 to specify the minimum value of 5 cubic feet per minute (CFM) of outdoor air per person be used in building design. This standard has been incorporated into the building codes of many cities and states. [3]

3. The 62-1989 standard recommends a minimum of 15 CFM of outdoor air per person for offices (reception areas) and 20 CFM per person for general office space with a moderate amount of smoking. Sixty cubic feet per minute per person is recommended for smoking lounges with local mechanical exhaust ventilation and no air recirculation. [4]

II. ACUTE HEALTH EFFECTS OF MAJOR INDOOR AIR CONTAMINANTS.

A. TYPES OF BUILDING PROBLEMS. Employee complaints can be due to two types of building problems: sick or tight building syndrome and building related illnesses.

1. Sick building syndrome is a condition associated with complaints of discomfort including headache; nausea; dizziness; dermatitis; eye, nose, throat, and respiratory irritation; coughing; difficulty concentrating; sensitivity to odors; muscle pain; and fatigue. The specific causes of the symptoms are often not known but sometimes are attributed to the effects of a combination of substances or individual susceptibility to low concentrations of contaminants. The symptoms are associated with periods of occupancy and often disappear after the worker leaves the worksite.

2. Building-related illnesses are those for which there is a clinically defined illness of known etiology and include infections such as legionellosis and allergic reactions such as hypersensitivity diseases and are often documented by physical signs and laboratory findings. A more thorough description of these illnesses can be found in the American conference of Governmental Industrial Hygienists (ACGIH) guidelines on evaluating bioaerosols.[5]

B. MAJOR INDOOR AIR CONTAMINANTS.

General. Although asbestos and radon have been listed below, acute health effects are not associated with these contaminants. These have been included due to recent concerns about their health effects. The investigator should be aware that there may be other health effects in addition to those listed.

1. Acetic Acid.

Sources: X-ray development equipment, silicone caulking compounds.

Acute health effects: Eye, respiratory and mucous membrane irritation.

2. Carbon Dioxide.
Sources: Unvented gas and kerosene appliances, improperly vented devices, processes or operations which produce combustion products, human respiration.

Acute health effects: Difficulty concentrating, drowsiness, increased respiration rate.

3. Carbon Monoxide.
Sources: Tobacco smoke, fossil-fuel engine exhausts, improperly vented fossil-fuel appliances.

Acute health effects: Dizziness, headache, nausea, cyanosis, cardiovascular effects, and death.

4. Formaldehyde.
Sources: Off-gassing from urea formaldehyde foam insulation, plywood, particle board, and paneling; carpeting and fabric; glues and adhesives; and combustion products including tobacco smoke.

Acute health effects: Hypersensitive or allergic reactions; skin rashes; eye, respiratory and mucous membrane irritation; odor annoyance.

5. Nitrogen Oxides.
   Sources: Combustion products from gas furnaces and appliances; tobacco smoke, welding, and gas- and diesel-engine exhausts.

   Acute health effects: Eye, respiratory and mucous membrane irritation.

6. Ozone.
   Sources: Copy machines, electrostatic air cleaners, electrical arcing, smog.

   Acute health effects: Eye, respiratory tract, mucous membrane irritation; aggravation of chronic respiratory diseases.

7. Radon.
   Sources: Ground beneath buildings, building materials, and groundwater.

   Acute health effects: No acute health effects are known but chronic exposure may lead to increased risk of lung cancer from alpha radiation.

8. Volatile Organic Compounds (VOC's). Volatile organic compounds include trichloroethylene, benzene, toluene, methyl ethyl ketone, alcohols, methacrylates, acrolein, polycyclic aromatic hydrocarbons, and pesticides.

   Sources: Paints, cleaning compounds, moth-balls, glues, photocopiers, "spirit" duplicators, signature machines, silicone caulking materials, insecticides, herbicides, combustion products, asphalt, gasoline vapors, tobacco smoke, dried out floor drains, cosmetics and other personal products.

   Acute health effects: Nausea; dizziness; eye, respiratory tract, and mucous membrane irritation; headache; fatigue.

9. Miscellaneous Inorganic Gases. Includes ammonia, hydrogen sulfide, sulfur dioxide.

   Sources: Microfilm equipment, window cleaners, acid drain cleaners, combustion products, tobacco smoke, blue-print equipment.

   Acute health effects: Eye, respiratory tract, mucous membrane irritation; aggravation of chronic respiratory diseases.

10. Asbestos.
    Sources: Insulation and other building materials such as floor tiles, dry wall compounds, reinforced plaster.

    Acute health effects: Asbestos is normally not a source of acute health effects. However, during renovation or maintenance operations, asbestos may be dislodged and become airborne. Evaluation of employee exposure to asbestos will normally be covered under the OSHA Asbestos standard.

11. Synthetic Fibers.
    Sources: Fibrous glass and mineral wool.

    Acute health effects: Irritation to the eyes, skin and lungs; dermatitis.

12. Tobacco Smoke.
    Sources: Cigars, cigarettes, pipe tobacco.

    Acute health effects: Tobacco smoke can irritate the respiratory system and, in allergic or asthmatic persons, often results in eye and nasal irritation, coughing, wheezing, sneezing, headache, and related sinus problems. People who wear contact lenses often complain of burning, itching, and tearing eyes when exposed to cigarette smoke.[6] Tobacco smoke is a major contributor to indoor air quality problems. Tobacco smoke contains several hundred toxic substances including carbon monoxide, nitrogen dioxide, hydrogen sulfide, formaldehyde, ammonia, benzene, benzo(a)pyrene, tars, and nicotine. Most indoor air particulates are due to tobacco smoke and are in the respirable range.

13. Microorganisms and Other Biological Contaminants (Microbials).
Includes viruses, fungi, mold, bacteria, nematodes, amoeba, pollen, dander, and mites.

Sources: Air handling system condensate, cooling towers, water damaged materials, high humidity indoor areas, damp organic material and porous wet surfaces, humidifiers, hot water systems, outdoor excavations, plants, animal excreta, animals and insects, food and food products.

Acute health effects: Allergic reactions such as hypersensitivity diseases (hypersensitivity pneumonitis, humidifier fever, allergic rhinitis, etc.) and infections such as legionellosis are seen. Symptoms include chills, fever, muscle ache, chest tightness, headache, cough, sore throat, diarrhea, and nausea.

III. INVESTIGATION GUIDELINES.

A. EMPLOYER AND EMPLOYEE INTERVIEWS.

1. Employer Interview.

a. What is the magnitude and distribution of employee complaints or illnesses? Are any employees obtaining medical care?

b. What are the design and operational parameters of the heating, ventilating, and air-conditioning (HVAC) system, such as source and amount of fresh air per occupant delivered to the breathing zone; adjustable or local HVAC controls; type of humidifier and how controlled; recent ventilation changes; and areas serviced by various units?

c. Does the frequency and type of maintenance performed on the HVAC systems, such as cleaning and oiling, meet the HVAC manufacturer's recommendations: filter change; prevention of bacterial buildup by use of biocides; repair and cleanup of water leaks; operating fresh air intake damper; and system balance checks?

d. Is smoking allowed in the office, in adjacent areas or in areas serviced by the same ventilation system? Are there designated smoking areas that have separate, nonrecirculating exhaust systems?

e. What type of copying machines, signature machines, spirit duplicators, blueprint machines and other office machines are used in the vicinity of complaints or in areas serviced by the same ventilation system?

f. Has there been any recent renovation or maintenance that can be a source of contaminants, such as painting, carpet installation, air conditioning repairs, use of acid drain cleaners, carpet cleaning, disinfecting of HVAC system, pesticide application?

g. Has there been any recent renovation or maintenance that can alter air flow patterns such as installation of partitions or relocation of air intakes or exhausts?

2. Employee Interviews.

a. What are the complaints and associated symptoms experienced; when do they occur (season, time, days, frequency); where do they occur; how long do symptoms last; do they clear up after leaving work (how soon); have the symptoms been triggered by any specific event or in any specific area; what is the source of symptoms; was any medical diagnosis or care rendered?

b. What are the workers' characteristics, such as smoker, allergies, pre-existing illnesses and disabilities; are they taking any medication; what are the occupational contributors?

B. WALKAROUND INSPECTION. NIOSH has determined that inadequate ventilation is the main problem in 52% of their IAQ investigations. Therefore, ventilation surveys should be initially conducted. During the walk-around inspection, the investigator could determine the building characteristics, discuss with knowledgeable personnel the proper operation of the HVAC systems, verify information obtained from the employer and employee interviews, perform ventilation-system testing, and, if appropriate, collect screening samples to identify potential causes of the problem.

Evaluation and testing of the HVAC system should follow the procedure established in the Ventilation Investigation chapter of the OTM. Investigators may need to discuss the operation of the ventilation system with building engineers and perform ventilation testing to determine proper fresh air intake. A simple traverse of the fresh-air intake duct may provide adequate information to determine the fresh-air flow. Measurements should be made under maximum and minimum air-flow conditions to determine the range of fresh-air intake.

The walk-around inspection should cover all the affected areas. Factors to be evaluated include inside and outside contamination sources; the HVAC system, e.g., location of air source, contamination, and proper operation; and occupational contributors, such as those listed in Appendix III:2-1.

1. Potential Problem Areas. The following is a compilation of specific concerns in past investigations but may not apply in every situation.

   a. Are there sources of indoor contaminants that could lead to employee complaints (e.g., copy machines, signature machines, blueprint copiers, paints, cleaning compounds and disinfectants, tobacco smoke, adhesives and glues, off-gassing of construction material and building fabric, contaminants generated by construction or renovation, positive- or negative-pressure work areas, improperly vented gas appliances, air fresheners, pesticides)?

   b. Are there sources of outdoor contaminants that lead to employee complaints (e.g., vehicle exhaust, roofing materials, cooling towers, dust, or other contaminants from construction activity, industrial plant, or building exhaust; gasoline vapors, pollen, biological contaminants, atmospheric pollutants)?

   c. Are heating, ventilating, and air-conditioning systems being operated and maintained properly with respect to location of air intakes and exhausts, pressure differentials between rooms that may account for influx of contaminants, design for supplied outdoor air, flow and distribution of air, position of dampers, local exhaust ventilation, air-cleaning equipment, HVAC operating times, regular operation checks, equipment cleaning and disinfecting, presence of water leaks or standing water, water-damaged building materials, and bacteriological contamination?

2. Sample Collection.

   a. During the walk-around inspection, professional judgment must be exercised to determine if samples should be collected to evaluate potential sources and potential contaminants including gases, vapors, and particulates.

   b. Initial sampling will normally consist of collecting environmental data using grab or screening samples with direct reading equipment such as detector tubes, particulate monitors, air velocity measuring instruments, and psychrometers. Screening samples for airborne contaminants should be collected for formaldehyde, carbon dioxide, carbon monoxide, and VOC's which are common potential sources of contamination.

   c. Samples may be collected to monitor the possible buildup of contaminants during the workday. Detector tube samples can be collected for carbon dioxide early in the day and again toward the end of the day; direct reading instruments can monitor continuously using a strip chart recorder to obtain a hard copy of contaminant variations during the day.

   d. To evaluate thoroughly, collect samples at fresh-air intakes, near return-air ducts, adjacent to both indoor and outdoor potential sources of contaminants, and in employee work areas both for complaint and noncomplaint areas. Sampling methods and equipment are covered in Section IV.

C. ENVIRONMENTAL EVALUATION. Based on initial sampling, further investigations may be performed using standard OSHA sampling procedures listed in the OSHA Analytical Method Manual.

1. Microbiological Evaluation.

a. NIOSH found that 5% of its investigations of indoor air quality involved some type of microbiological contamination.[6]

b. The ACGIH Bioaerosols Committee's guidelines for assessing the role of bioaerosols7 contains information on sampling, analysis, and recommendations for remedial actions. Air sampling should be initiated only after medical or clinical reports indicate the existence of workplace-related illnesses, such as hypersensitivity and allergic disorders, that are likely due to bioaerosols. The Office of Occupational Medicine should be consulted before initiating any sampling.

c. At present, specialized bioaerosol sampling equipment is available through the OSHA Health Response Team. Use of this equipment requires advance arrangements for preparing culture media for sampling, specialized handling techniques for the samples, and arrangements for analysis by laboratories familiar with the handling and processing of biological samples. The OSHA Health Response Team may be consulted for further information.

IV. SAMPLING INSTRUMENTATION AND METHODS.

A. LOW CONTAMINANT LEVELS.

1. Choose sampling procedures that can determine concentrations of toxic materials that are much lower than are normally found in industrial investigations. Few procedures have been validated for these lower level contaminants. Contact the Salt Lake Technical enter (SLTC) with any sampling questions.

2. Present OSHA sampling and analytical procedures were developed to meet precision and accuracy requirements for airborne contaminants in the range of OSHA Permissible Exposure Limits (PEL's) and American Conference of governmental Industrial Hygienists (ACGIH) Threshold Limit Values (TLV's). These procedures are used for sampling 8-hour Time-Weighted Averages (TWA's) and Short-Term Exposure Limits (STEL's) of 15 or 30 minutes.

3. In many IAQ investigations, extensive air monitoring may not be warranted because inadequate introduction and/or distributions of fresh air may be the main problem.

B. GENERAL SCREENING. Use screening techniques to determine the potential sources that may require more sensitive and accurate evaluation or may require action as described in Section E, depending upon professional judgment.

1. Collect screening samples using detector tubes or direct reading instruments. For increased sensitivity, higher flow rates or longer sampling times may be used. Low range detector tubes are available from manufacturers. Appendix III:2-2 contains a table of screening methods, concentration range, validated testing methods, and contaminant types.

2. Based on screening results, validated sampling procedures may be required to further quantify employee exposures. Much of the information on validated sampling and analytical methods is contained in the OSHA Chemical Information Manual or in the OSHA Analytical Methods Manual.

3. Much of the specialized equipment is available through the OSHA Health Response Team (HRT), Cincinnati Technical Center (CTC), or Regional Offices.

C. OPTIONAL SCREENING FOR COMMON INDOOR AIR CONTAMINANTS, BASED UPON PROFESSIONAL JUDGMENT.

1. Acetic Acid. Use detector tubes (0-10 ppm) to evaluate complaints of eye, nose, and throat irritation. Low levels of acetic acid have been found from off-gassing of silicone caulking compounds and in hospitals where x-ray developing equipment is improperly ventilated.

2. Asbestos. Screening is not a routine procedure. Any requested screening should be done in accordance with the proper OSHA Standards.

3. Carbon Dioxide. Use low level detector tubes (0-2000 ppm) or portable infrared spectrometers to screen for indoor carbon dioxide levels. Carbon dioxide measurement is a useful screening technique which is often helpful in determining whether adequate quantities of outside fresh air have been introduced and distributed into the building.

   NIOSH recommendations:

250-350 ppm	normal outdoor ambient concentrations
600 ppm	minimal air quality complaints
600-1,000 ppm	less clearly interpreted
1,000 ppm	indicates inadequate ventilation; complaints such as headaches, fatigue, and eye and throat irritation will be more widespread; 1,000 ppm should be used as the upper limit for indoor levels

   These levels are only guidelines. If carbon dioxide levels exceed 1,000 ppm it does not necessarily indicate that the building is hazardous and should be evacuated. Rather this level should be used as a guideline that helps maximize comfort for all occupants. [6]

4. Carbon Monoxide. Low-level detector tubes (2-200 ppm). Direct-reading CO monitors.

5. Formaldehyde. Use low-level (0.04-1 ppm) detector tubes to evaluate complaints of eye, nose, and throat irritation which may be due to off-gassing from insulation, building materials, carpets, drapes, or glues and adhesives.

6. Nitrogen Oxides and Ozone. Detector tubes. Also collect outdoor samples since ambient levels of ozone may reach levels that are 1-3 times the PEL of 0.1 ppm during air-temperature inversions. If a more accurate or continuous ozone evaluation is required, a chemiluminescent monitor that is specific for ozone and can measure in the range of 0.01 to 10 ppm is available from the HRT.

7. Radon. A rapid, easy-to-use screening method for measuring radon gas concentrations is available from the SLTC. It is used for deciding if additional measurements are required or remedial actions should be undertaken. Additional longer-term quantitative procedures are available from the HRT if required. The HRT or Regional Offices may be contacted if sampling is to be initiated and for interpretation of the results.

   EPA Recommendations[8] for the results of screening samples:

<4 picocuries per liter of air (pCi/L)	follow-up measurements probably not required
>4 pCi/L	follow-up measurements should be performed

8. Airborne Particulates. Use a particle counting instrument capable of measuring concentrations as low as 2,000 particles/cubic centimeter (cc) of air for comparing particulates in various areas. The investigator may be able to determine where additional ventilation or air filtration is necessary to eliminate or minimize employee complaints. For example, if employee complaints are more prevalent in an area where the particulate concentration is 40,000 particles/cc, and other areas are below 15,000 particles/cc, the investigator may recommend that a high efficiency filter be installed or, if the area has a separate ventilation system, that the ventilation rate be increased.

9. Airborne Microorganisms. The ACGIH5 recommends a pre-assessment of the extent of microbial contamination prior to initiation of air sampling. Airborne microbials sampling equipment is available from the HRT if sampling is necessary.

   Before biological sampling, several precautions must be taken including making arrangements for preparing culture media for sampling, specialized shipping procedures, and making arrangements for analysis by a laboratory familiar with the handling and processing of biological samples. Contact the Directorate of Technical Support for information about laboratories experienced in the analysis of microbial samples and with knowledge of the health effects.

   Legionella pneumophila is often present in hot water tanks, washing systems, and pools of stagnant water, but health effects are not observed until the contaminants become aerosolized within the building confinements.

   The identification of predominant taxa, or at least fungi, is recommended in addition to determining the number of colony-forming units/m3 of air (cfu/m3). During growing seasons, outdoor fungus-spore levels can range from 1,000 to 100,000 cfu/m3 of air.

   Contamination indicators:[9]

   - 1,000 viable colony-forming units in a cubic meter of air
   - 1,000,000 fungi per gram of dust or material
   - 100,000 bacteria or fungi per milliliter of stagnant water or slime

   Levels in excess of the above do not necessarily imply that the conditions are unsafe or hazardous. The type and concentrations of the airborne microorganisms will determine the hazard to employees.

D. MISCELLANEOUS AIRBORNE CONTAMINANTS.

1. Use a portable infrared spectrometer to evaluate a wide variety of potential air contaminants including acetic acid, ammonia, carbon dioxide, carbon monoxide, nitric oxide, nitrogen dioxide, sulfur dioxide, and a number of volatile organic compounds. It can be connected to a strip chart recorder to obtain a hard copy showing variations of concentration during the day.

2. Take care in interpreting the results since the instrument is not always specific for one compound. Note: Equipment not generally available in the field, such as the particulate analyzer, infrared spectrometer, and airborne biological sampler is available through the HRT along with a written description of the equipment, operating manuals, and methods of analysis.

V. RECOMMENDATIONS FOR THE EMPLOYER.

The following are general recommendations which, where relevant, should be standard procedure. If followed, they will help prevent or alleviate many indoor air-quality problems.

A. ENGINEERING RECOMMENDATIONS.

1. Ventilation.

   a. Includes the use of natural, dilution, local exhaust, or increased ventilation efficiency. The most effective engineering control for prevention of indoor air quality problems is assuring an adequate supply of fresh outdoor air through natural or mechanical ventilation.

   b. SHRAE in its 62-1989 standard recommends 20 cubic feet per minute (CFM) of outdoor air per occupant for offices. For smoking lounges, up to 60 CFM of outdoor air per occupant should be provided.

c. When possible, use local exhaust ventilation and enclosure to capture and remove contaminants generated by specific processes. Room air in which contaminants are generated should be discharged directly outdoors rather than recirculated.

2. Efficiency.

a. Ventilation efficiency can be improved by:

- Ensuring that outdoor air-supply dampers and room air-vents are open;
- Removing or modifying partitions or obstructions that block fresh-air flow;
- Rebalancing the system to prevent inflow or outflow of contaminated air due to pressure differentials between rooms;
- Preventing poor distribution of make-up air by proper placement of air inlets and exhausts; and
- Using room fans to improve mixing and dilution of pollutants.

b. Outside air intakes should not be located in close proximity to potential sources of contamination (automobile garages, cooling towers, building exhausts, roadways).

3. Air Treatment. Air treatment is the removal of air contaminants and/or the control of room temperature and humidity. Recommendations for air treatment include:

- The use of filtration, electronic cleaners, chemical treatment with activated charcoal or other sorbents;
- Humidity control in the range of 20%-60%; and
- Temperature control in the range of 68-76 F.

4. Source Controls. Source controls include substitution, removal, encapsulation, local exhaust ventilation, and use of physical barriers.

B. ADMINISTRATIVE AND WORK PRACTICE RECOMMENDATIONS. Recommendations include programs that change the behavioral patterns of occupants.

1. Preventive Maintenance (PM). Preventive maintenance plans for humidifiers, water spray, and other HVAC system components should include:

- Checking damper positions and functioning belts, baffles, ductwork, and system balance;
- Measuring airflow and performing necessary adjustment if necessary to meet ASHRAE recommendations;
- Replacing filters on air handling units at regular intervals;
- Cleaning air distribution ducts and dampers; and
- Replacing damaged insulation.

2. Microbial Contamination.

a. Eliminate or control all known and potential sources of microbial contaminants by prompt cleanup and repair of all areas where water collection and leakage has occurred including floors, roofs, HVAC cooling coils, drain pans, humidifiers containing reservoirs of stagnant water, air washers, fan coil units, and filters.

b. Remove and discard porous organic materials that are contaminated (e.g., damp insulation in ventilation system, moldy ceiling tiles, and mildewed carpets).

c. Clean and disinfect nonporous surfaces where microbial growth has occurred with detergents, chlorine-generating slimicides, or other biocides and insuring that these cleaners have been removed before air handling units are turned on.

d. Maintain indoor air relative humidity below 60% (50% where cold surfaces are in contact with room air).

e. Adjust intake of outdoor air to avoid contamination from nearby soil, vegetable debris, cooling towers, or sanitary stacks unless air is adequately conditioned.

f. Adjust combustion sources such as furnaces or water heaters to assure proper burning and exhaust to an area where re-entrainment will not occur.

g. Minimize exposure by limiting occupancy of contaminated airspace, limiting use of offending sources to specific areas or times, or evacuating contaminated areas until they can be ventilated adequately.

h. Isolate, if feasible, areas of renovation, painting, carpet laying, pesticide application, etc., from occupied areas that are not under construction. If possible, perform this work during evenings and weekends. If ventilation is turned off during weekends or other periods, ensure that system is on so that contaminant concentrations are sufficiently diluted prior to occupancy.

i. Supply adequate ventilation during and after completion of work to assist in diluting the contaminant levels.

j. Personnel affected with hypersensitivity should be thoroughly evaluated and the problem identified and corrected before returning them to the workplace. If, after the remedial action, the illness persists in the workplace, the affected personnel should be considered for permanent reassignment to another area.

k. Eliminate or reduce contamination of the air supply with cigarette smoke by banning smoking or restricting smoking to designated areas which have their air discharged directly to the outdoor rather than recirculated.

VI. REFERENCES.

1. National Institute for Occupational Safety and Health (NIOSH), Feb., 1989. Personal Correspondence to Long Loo, Occupational Safety and Health Administration.

2. American Society for Heating, Refrigerating and Air-Conditioning Engineers (ASHRAE). 1973. ASHRAE Standard 62-73: *Standards for Natural and Mechanical Ventilation.* New York: ASHRAE.

3. American Society for Heating, Refrigerating and Air-Conditioning Engineers (ASHRAE). 1975. ASHRAE Standard 90-75: *Energy Conservation in New Building Design.* New York: ASHRAE.

4. American Society of Heating, Refrigerating and Air-Conditioning engineers (ASHRAE). 1989. ASHRAE 62-1989: *Standards for Acceptable Indoor Air Quality.* Atlanta: ASHRAE.

5. ACGIH Committee on Bioaerosols, American Conference of Governmental Industrial Hygienists. 1987. *Guidelines for Assessment and Sampling of Saprophytic Bioaerosols in the Indoor Environment.* Applied Industrial Hygiene 2(5):RIO to R16.

6. National Institute for Occupational Safety and Health (NIOSH). 1987. *Guidance for Indoor Air Quality Investigations.* Cincinnati: NIOSH.

7. ACGIH Committee on Bioaerosols, American Conference of Governmental Industrial Hygienists. 1986. *Rationale for Monitoring Viable Microorganisms in the Office Environment.* Applied Industrial Hygiene 1:R19-R23.

8. Environmental Protection Agency (EPA). 1986. *A Citizen's Guide to Radon.*

9. Brief, R. S. and T. Bernath. 1988. *Indoor Pollution: Guidelines for Prevention and Control of Microbiological Respiratory Hazards Associated with Air Conditioning and Ventilation System.* Applied Industrial Hygiene. 3(1):5-10.

VII. BIBLIOGRAPHY.

American Academy of Industrial Hygiene (AAIH). 1984. "Items to Consider When Investigating Complaints Relative to the Office Environment." Personal Correspondence to Edward Baier, Occupational Safety and Health Administration. Lansing, MI: AAIH Peer Review Committee.

American Society for Heating, Refrigerating and Air-conditioning Engineers (ASHRAE). 1981. *Thermal Environmental Conditions for Human Occupancy*. ASHRAE Standard 55-1981.

Burge, H. A., and Hodgson, M. 1988. *Health Risks of Indoor Pollutants*. ASHRAE, p. 34-38.

Fisk, W. J. et al. 1987. *Indoor Air Quality Control Techniques*. New Jersey: Noyes Data Corp.

International Agency for Research on Cancer (IARC). 1988. Man-Made Mineral Fibres and Radon. *IARC Monographs. Vol. 43*. Lyon, France: IARC.

Lioy, P. J. 1983. *Air Sampling Instruments for Evaluation of Atmospheric Contaminants* (6th ed.). Cincinnati, Ohio: ACGIH.

Maryland State Department of Education. 1987. *Indoor Air Quality*. Baltimore: Maryland Public Schools.

McLellan, R. K. 1983. The Health Hazards of Office Work. *Toxic Substances Journal* 5(3):162-181.

Meyer, B. 1983. *Indoor Air Quality*. Boston: Addison-Wesley.

Samet, J. M., et al. 1987, 1988. *Health Effects and Sources of Indoor Air Pollution. Parts I and II. American Review of Respiratory Disease 136:1486-1508 and 137:221-242.*

Wadden, R. A., and Scheff, P. A. 1983. *Indoor Air Pollution*. New York: John Wiley & Sons.

Walsh, P. J., Dudney, C. S., and Copenhover, E. D. 1984. *Indoor Air Quality*. Florida: CRC Press, Inc.

APPENDIX III:2-1. INVESTIGATING OFFICE-RELATED COMPLAINTS.

Investigations of office related complaints using industrial hygiene techniques often fail to identify the source of these problems. The combined effects of multiple, low-level air contaminants have not been investigated thoroughly and may be a cause of the problem.

In a recent NIOSH document, Stress Management in Work Settings, occupational stress is discussed in terms of assessment methods, stress management, and programs and training necessary to reduce occupational stress. The synergistic effect of multiple stressors appears to indicate that building-related problems may be more than an air quality problem. The combined effect of these multiple stressors may interact with employees and could result in acute adverse emotional or physical reactions. In the short term, these reactions may lead to decreased productivity, absenteeism, and high turnover rates and if prolonged may lead to a variety of illnesses including hypertension, coronary heart disease, ulcers, alcoholism and mental illness.

These office-related health problems can be evaluated by a consultant through employee interviews, analysis of job demands, and training employees. The following potential problems may need to be addressed:

- Physical hazards including noise from nearby sources such as air conditioning systems and printers, inadequate lighting, stress from the operation of video display terminals (VDT's), vibration sources, extremes of heat, cold and humidity, drafts, and poor air circulation.
- Ergonomic problems such as carpal tunnel syndrome or inflammatory disorders of the tendons and joints of keyboard operators due to tasks requiring repetitive motions. Proper design of fixed work stations where employees are required to perform repetitive tasks includes proper lighting to prevent glare, maintaining temperature and humidity in a comfortable range with minimum temperature variations, maximum

flexibility in work station design including adjustable chair, keyboard, and screen height, and a work-rest regimen that allows breaks to reduce psychological distress.

- Reduction of job stress by: (a) adequate flow of information from management to employees; (b) explanation of any changes introduced into the workplace including new chemicals, ventilation, production modification, and work schedules; (c) maximizing employee participation in planning and implementing changes; (d) stress reduction techniques including exercise, biofeedback, and assertiveness training; and (e) training workers to understand chemicals they may be working with and their health effects, dose/response relationships, and results of environmental evaluation.

APPENDIX III:2-2. SAMPLING AND ANALYTICAL METHODS.

Contaminant	***Concentration range***	***Screening method***	***Validated method***
Bioaerosols	0-1,000 cfu/m^3	Viable biological sampler	
Carbon dioxide	0-2,000 ppm	DT, IR	Sampling bag, GC/TCD *OSHA ID172*
Carbon monoxide	2-50 ppm	DT, meter	Sampling bag, meter
Formaldehyde	0.04-1 ppm	DT	Coated XAD-2, GC/NPD *OSHA-52*
Nitric oxide	0-25 ppm	DT	TEA tube with oxidizer, DPP *OSHA ID190*
Nitrogen dioxide	0-5 ppm	DT	TEA-Molecular Sieve Tube, IC *OSHA ID 182*
Particulates	0-40,000 particles/cc	Light scattering meter	
Pesticides	(See *OSHA Chemical Information Manual*)		
Ozone	0-0.1 ppm	DT, Chemiluminescent meter	
Radon	4-200 pCi/L	Radon Cartridge, Electrect	
VOC's	(See *OSHA Chemical Information Manual*)		

Key:

DPP	Differential pulse polarographic
DT	Detector tubes
GC	Gas chromatograph
IC	Ion chromatograph
IR	Infrared spectrometer
	Calibrated, direct reading meter available through laboratory, HRT, or area meter office
NPD	Nitrogen phosphorus detector
TCD	Thermal conductivity detector
TEA	Triethanolamine

NOTE: Referenced OSHA procedures can be found in the OSHA Analytical Methods Manual or the OSHA Chemical Information Manual. See individual manufacturer's literature for information on interferences to the screening or sampling method.

SECTION III: CHAPTER 3

VENTILATION INVESTIGATION

Contents:

I. INTRODUCTION.

Industrial ventilation generally involves the use of supply and exhaust ventilation to control emissions, exposures, and chemical hazards in the workplace. Traditionally, nonindustrial ventilation systems commonly known as heating, ventilating, and air-conditioning (HVAC) systems were built to control temperature, humidity, and odors.

A. Ventilation may be deficient in:

- confined spaces;
- facilities failing to provide adequate maintenance of ventilation equipment;
- facilities operated to maximize energy conservation;
- windowless areas; and
- areas with high occupant densities.

Any ventilation deficiency must be verified by measurement.

B. There are five basic types of ventilation systems:

- dilution and removal by general exhaust;
- local exhaust (see Figure III:3-1);
- makeup air (or replacement);
- HVAC (primarily for comfort); and
- recirculation systems.

C. Ventilation systems generally involve a combination of these types of systems. For example, a large local exhaust system may also serve as a dilution system, and the HVAC system may serve as a makeup air system (see Appendix

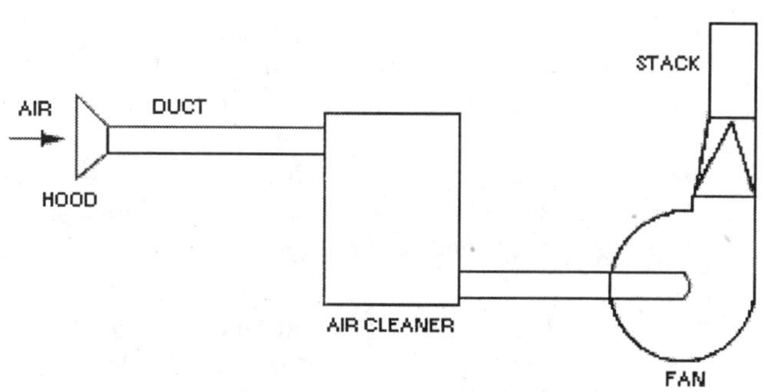

FIGURE III: 3-1.COMPONENTS OF A LOCAL EXHAUST SYSTEM.

III:3-1 for a primer and Appendix III:3-2 for an explanation of these terms).

II. HEALTH EFFECTS.

Inadequate or improper ventilation is the cause of about half of all indoor air quality (IAQ) problems in nonindustrial workplaces (see Section III, Chapter 2, Indoor Air Quality). This section of the manual addresses ventilation in commercial buildings and industrial facilities.

A. INDOOR AIR CONTAMINANTS include but are not limited to particulates, pollen, microbial agents, and organic toxins. These can be transported by the ventilation system or originate in the following parts of the ventilation system:

- wet filters;
- wet insulation;
- wet undercoil pans;
- cooling towers; and
- evaporative humidifiers.

People exposed to these agents may develop signs and symptoms related to "humidifier fever," "humidifier lung," or "air conditioner lung." In some cases, indoor air quality contaminants cause clinically identifiable conditions such as occupational asthma, reversible airway disease, and hypersensitivity pneumonitis.

B. VOLATILE ORGANIC AND REACTIVE CHEMICALS (for example, formaldehyde) often contribute to indoor air contamination. The facility's ventilation system may transport reactive chemicals from a source area to other parts of the building. Tobacco smoke contains a number of organic and reactive chemicals and is often carried this way. In some instances the contaminant source may be the outside air. Outside air for ventilation or makeup air for exhaust systems may bring contaminants into the workplace (e.g., vehicle exhaust, fugitive emissions from a neighboring smelter).

See Section III, Chapter 2, Indoor Air Quality, for a discussion of common indoor-air contaminants and their biological effects.

III. STANDARDS AND CODES.

A. CONSENSUS STANDARDS. Appendix III:3-3 is a compilation of OSHA and industry consensus standards. Foremost are those recommended by the Air Movement and Control Association (AMCA), the American Society of Heating, Refrigerating, and Air-Conditioning Engineers (ASHRAE), the American National Standards Institute (ANSI), the Sheet Metal and Air Conditioning Contractors National Association (SMACNA), the National Fire Protection Association (NFPA), and the American Conference of Governmental Industrial Hygienists (ACGIH). AMCA is a trade association that has developed standards and testing procedures for fans. ASHRAE is a society of heating and air conditioning engineers that has produced, through consensus, a number of standards related to indoor air quality, filter performance and testing, and HVAC systems. ANSI has produced several important standards on ventilation, including ventilation for paintspray booths, grinding exhaust hoods, and open-surface tank exhausts. Four ANSI standards were adopted by OSHA in 1971 and re codified in 29 CFR 1910.94; these standards continue to be important as guides to design. ANSI has recently published a new standard for laboratory ventilation (ANSI Z9.5). SMACNA is an association representing sheet metal contractors and suppliers. It sets standards for ducts and duct installation. NFPA has produced a number of recommendations (which become requirements when adopted by local fire agencies), e.g., NFPA 45 lists a number of ventilation requirements for laboratory fume hood use. The ACGIH has published widely used guidelines for industrial ventilation.

B. OSHA REGULATIONS. Ventilation criteria or standards are included in OSHA regulatory codes for job- or task-specific worker protection (see Appendix III:3-3). In addition, many OSHA health standards include ventilation requirements. The four standards in 29 CFR 1910.94 deal with local exhaust systems, and OSHA's construction standards (29 CFR 1926) contain ventilation standards for welding. OSHA's compliance policy regarding violation of ventilation standards is set forth in the Field Inspection Reference Manual.

IV. INVESTIGATION GUIDELINES.

A. INVESTIGATION PHASES. Workplace investigations of ventilation systems may be initiated by worker complaints of possible overexposures to air contaminants, possible risk of fire or explosion from flammable gas or vapor levels at or near the lower explosive limit (LEL), or indoor air quality complaints. The second phase of the investigation involves an examination of the ventilation system's physical and operating characteristics.

B. FAULTY VENTILATION CONDITIONS AND CAUSES. Common faulty ventilation conditions and their probable causes are listed in Table III:3-1. Specific points to consider during any investigation of a ventilation system include emission source, air behavior, and employee involvement. Points that should be included in a review of operational efficacy are shown in Table III:3-2. Appendix III:3-4 contains information on points to be checked in a troublesome exhaust system.

TABLE III:3-1. COMMON VENTILATION CONDITIONS AND CAUSES

Condition	***Possible cause(s)***
Worker complaints, improper use of system, nonuse of system, alteration of system by employees.	The hood interferes with work The hood provides poor control of contaminants.
Excessive employee exposures although flow volumes and capture velocities are at design levels.	Employee work practices need improvement. The ventilation system interferes with work or worker productivity and leads workers to bypass the system. Employee training is not adequate. Design of system is poor.
Constant plugging of duct.	Plugged ducts occur when transport velocity is inadequate or when vapor condenses in the duct, wets particles, and causes a build-up of materials. These problems are caused by poor design, open access doors close to the fan, fan problems, or other problems.
Reduced capture velocities or excessive fugitive emissions.	The cause of these conditions is usually reduced flow rate, unless the process itself has changed. Reduced flow rate occurs in the following situations: • plugged or dented ducts • slipping fan belts • open access doors • holes in ducts, elbows • closed blast gate to branch, or opened blast gates to other branches, or corroded and stuck blast gates • fan turning in reverse direction (This can occur when lead wires are reversed and cause the motor and fan to turn backwards. Centrifugal fans turning backwards may deliver up to only 50% of rated capacity.) • worn out fan blades • additional branches or hoods added to system since initial installation, or • clogged air cleaner.

TABLE III:3-2. PROBLEM CHARACTERIZATION

Emission source

- Where are all emission sources or potential emission sources located?
- Which emission sources actually contribute to exposure?
- What is the relative contribution of each source to exposure?
- Characterization of each contributor:
  - chemical composition
  - temperature
  - rate of emission
  - direction of emission
  - initial emission velocity
  - pattern of emission (continuous or intermittent)
  - time intervals of emission
  - mass of emitted material

Air behavior

- Air temperature
- Air movement (direction, velocity)
- Mixing potential
- Supply and return flow conditions, to include pressure differences between space and surrounding areas
- Sources of tempered and untempered make-up air
- Air changes per hour
- Influence of existing HVAC systems
- Effects of wind speed and direction
- Effects of weather and season

Employee

- Worker interaction with emission source
- Worker exposure levels
- Worker location
- Worker education, training, cooperation

C. BASIC TESTING EQUIPMENT might include:

- smoke tubes
- velometers, anemometers: - swinging vane anemometer- thermal or hot-wire nemometer
- pressure-sensing devices: - U-tube or electronic manometers- Pitot tube- thermal (thermal and swinging vane instruments measure static pressure indirectly) - aneroid ("bellows") gauges
- noise-monitoring equipment
- measuring tapes
- other: rags, flashlight, mirror, tachometer
- combustible gas meter or oxygen meter
- tubes for CO, CO2, formaldehyde, etc.

D. DOCUMENTATION. The characteristics of the ventilation system that must be documented during an investigation include equipment operability, physical measurements of the system, and use practices.

E. EQUIPMENT OPERABILITY. Before taking velocity or pressure measurements, note and record the operating status of the equipment. For example, are filters loaded or clean? Are variable-flow devices like dampers, variable-frequency drives, or inlet vanes in use? Are make-up units operating? Are system blueprints available?

F. MEASUREMENTS.

1. Duct diameters are measured to calculate duct areas. Inside duct diameter is the most important measurement, but an outside measurement is often sufficient for a sheet metal duct. To measure the duct, the tape should be thrown around the duct to obtain the duct circumference, and the number should be divided by (3.142) to obtain the diameter of the duct.

2. Hood and duct dimensions can be estimated from plans, drawings, and specifications. Measurements can be made with measuring tape. If a duct is constructed of 2½ or 4-foot sections, the sections can be counted (elbows and tees should be included in the length).

3. Hood-face velocities outside the hood or at the hood face can be estimated with velometers, smoke tubes, and swinging-vane anemometers, all of which are portable, reliable, and require no batteries.

   a. The minimum velocity that can be read by an anemometer is 50 feet per minute (fpm). The meter should always be read in the upright position, and only the tubing supplied with the equipment should be used.

   b. Anemometers often cannot be used if the duct contains dust or mist because air must actually pass through the instrument for it to work. The instrument requires periodic cleaning and calibration at least once per year. Hot-wire anemometers should not be used in airstreams containing aerosols.

   c. Hood-face velocity measurement involves the following steps:

      - mark off imaginary areas;
      - measure velocity at center of each area; and
      - average all measured velocities.

   d. Smoke is useful for measuring face velocity (see Figure III:3-2) because it is visible. Nothing convinces management and employees more quickly that the ventilation is not functioning properly than to show smoke drifting away from the hood, escaping the hood, or traveling into the worker's breathing zone. Smoke can be used to provide a rough estimate of face velocity:

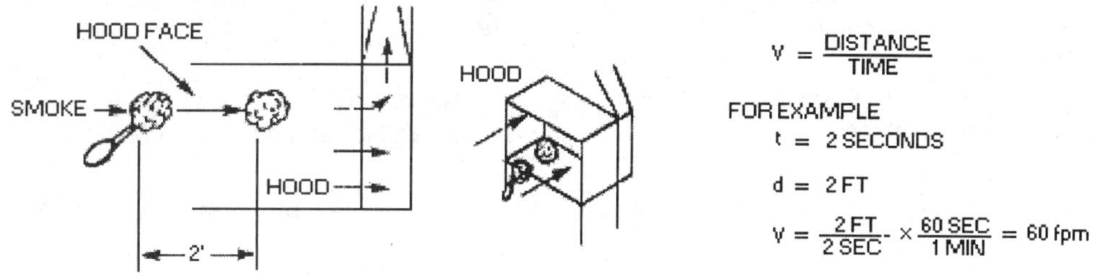

FIGURE III: 3-2. USE OF SMOKE TO DEMONSTRATE AIR FLOW.

Velocity = Distance/Time , or

$$V = \frac{D}{T}$$

Squeeze off a quick burst of smoke. Time the smoke plume's travel over a two-foot distance. Calculate the velocity in feet per minute. For example, if it takes two seconds for the smoke to travel two feet, the velocity is 60 fpm.

4. Hood static pressures (SPH) should be measured about 4-6 duct diameters downstream in a straight section of the hood take-off duct. The measurement can be made with a pitot tube or by a static pressure tap into the duct sheet metal (see Figure III:3-3).

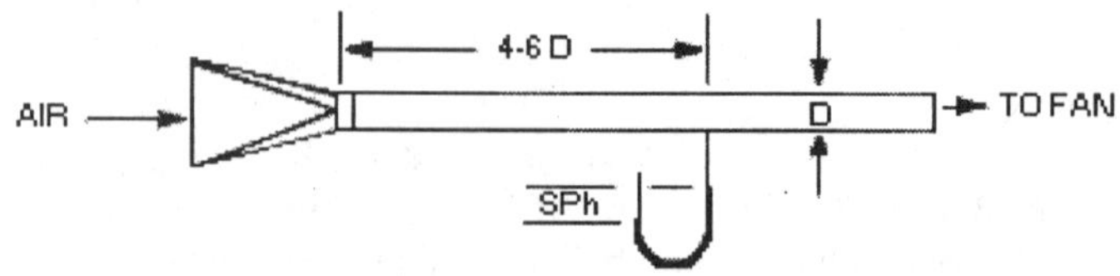

FIGURE III: 3-3. USE OF STATIC PRESSURE TAP INTO DUCT TO MEASURE HOOD STATIC PRESSURE

   a. Pressure gauges come in a number of varieties, the simplest being the U-tube manometer.

   b. Inclined manometers offer greater accuracy and greater sensitivity at low pressures than U-tube manometers. However, manometers rarely can be used for velocities less than 800 fpm (i.e. velocity pressures less than 0.05" w.g.). Aneroid-type manometers use a calibrated bellows to measure pressures. They are easy to read and portable but require regular calibration and maintenance.

5. Duct velocity measurements may be made directly (with velometers and anemometers) or indirectly (with manometers and pitot tubes) using duct velocity pressure.

   a. Air flow in industrial ventilation ducts is almost always turbulent, with a small, nonmoving boundary layer at the surface of the duct.

   b. Because velocity varies with distance from the edge of the duct, a single measurement may not be sufficient. However, if the measurement is taken in a straight length of round duct, 4-6 diameters downstream and 2-3 diameters upstream from obstructions or directional changes, then the average velocity can be estimated at 90% of the centerline velocity. (The average velocity pressure is about 81% of centerline velocity pressure.)

   c. A more accurate method is the traverse method, which involves taking six or ten measurements on each of two or three passes across the duct, 90° or 60° opposed. Measurements are made in the center of concentric circles of equal area.

   d. Density corrections (e.g., temperature) for instrument use should be made in accordance with the manufacturer's instrument instruction manual and calculation/correction formulas.

6. Air cleaner and fan condition measurements can be made with a pitot tube and manometer.

G. GOOD PRACTICES.

1. Hood placement must be close to the emission source to be effective. Maximum distance from the emission source should not exceed 1.5 duct diameters.

   a. The approximate relationship of capture velocity (Vc) to duct velocity (Vd) for a simple plain or narrow flanged hood is illustrated in Figure III:3-4. For example, if an emission source is one duct diameter in front of the hood and the duct velocity (Vd) = 3,000 feet per minute (fpm), then the expected capture velocity (Vc) is 300 fpm. At two duct diameters from the hood opening, capture velocity decreases by a factor of 10, to 30 fpm.

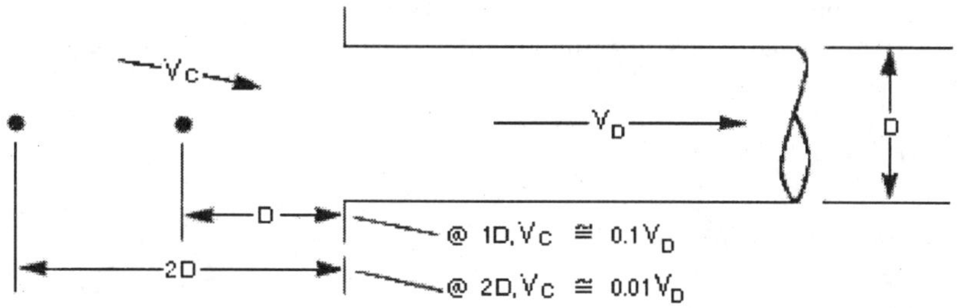

FIGURE III: 3-4. RELATIONSHIP OF CAPTURE VELOCITY (Vc) TO DUCT VELOCITY (Vd)

b. Figure III:3-5 shows a rule of thumb that can be used with simple capture hoods. If the duct diameter (D) is 6 inches, then the maximum distance of the emission source from the hood should not exceed 9 in. Similarly, the minimum capture velocity should not be less than 50 fpm.

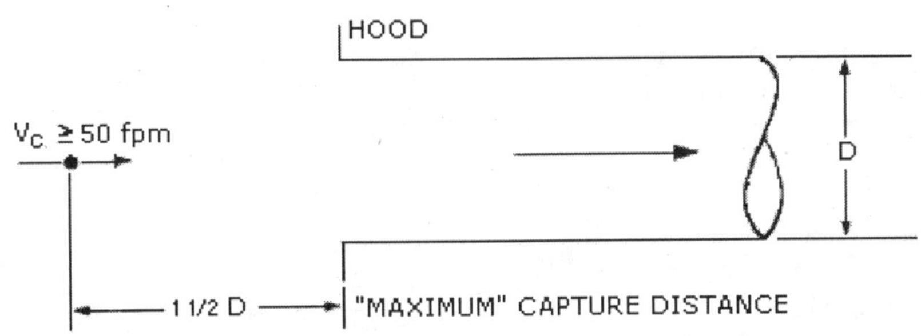

FIGURE III: 3-5. RULE OF THUMB FOR SIMPLE CAPTURE HOODS: MAXIMUM CAPTURE DISTANCE SHOULD NOT BE MORE THAN 1.5 TIMES THE DUCT DIAMETER.

c. Figure III:3-6 provides a guide for determining an effective flange width.

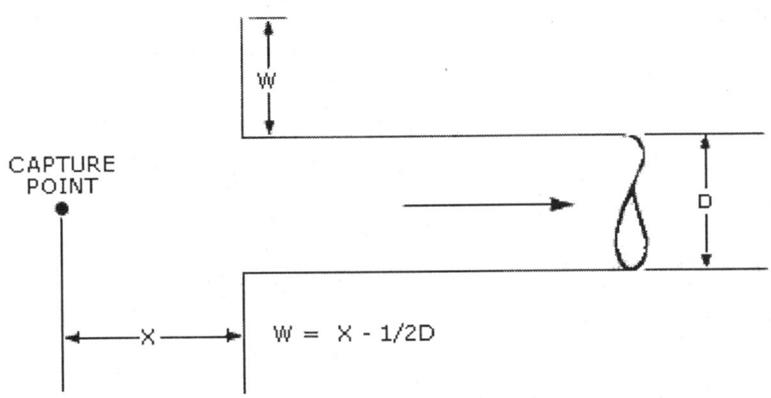

FIGURE III: 3-6. EFFECTIVE FLANGE WIDTH (W).

2. System effect loss, which occurs at the fan, can be avoided if the necessary ductwork is in place.

   a. Use of the six-and-three rule ensures better design by providing for a minimum loss at six diameters of straight duct at the fan inlet and a minimum loss at three diameters of straight duct at the fan outlet (Figure II:3-7).

   b. System effect loss is significant if any elbows are connected to the fan at inlet or outlet. For each 2½ diameters of straight duct between the fan inlet and any elbow, CFM loss will be 20%.

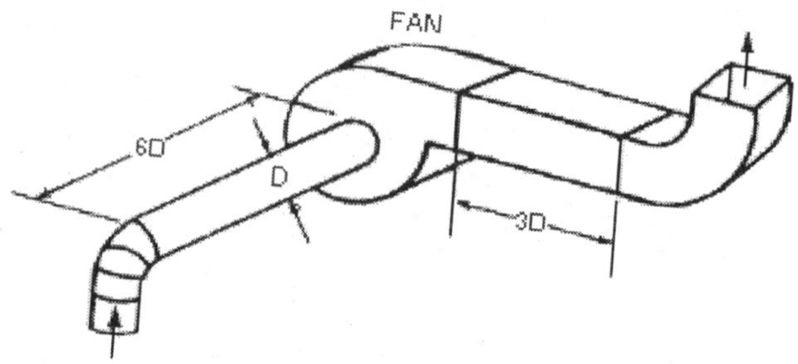

FIGURE III: 3-7. AN ILLUSTRATION OF THE SIX-AND-THREE RULE.

3. Stack height should be 10 ft higher than any roof line or air intake located within 50 ft of the stack (Figure III:3-8). For example, a stack placed 30 ft away from an air intake should be at least 10 ft higher than the center of the intake.

4. Ventilation system drawings and specifications usually follow standard forms and symbols, e.g., as described in the Uniform Construction Index (UCI).

   a. Plan sections include electrical, plumbing, structural, or mechanical drawings (UCI, Section 15). The drawings come in several views: plan (top), elevation (side and front), isometric, or section.

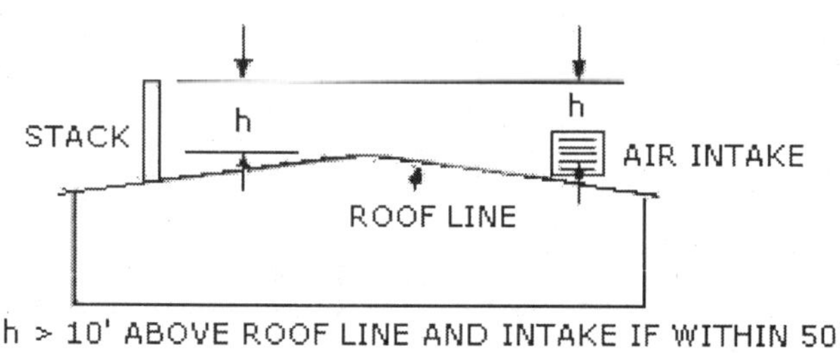

FIGURE III: 3-8. MINIMUM STACK HEIGHT IN FELATION TO IMMEDIATE ROOF LINE OR CENTER OF ANY AIR INTAKE ON THE SAME ROOF

   b. Elevations (side and front views) give the most detail. An isometric drawing is one that illustrates the system in three dimensions. A sectional drawing provides duct or component detail by showing a cross-section of the component.

   c. Drawings are usually drawn to scale. (Check dimensions and lengths with a ruler or a scale to be sure that this is the case. For example, 1/8 inch on the sheet may represent one foot on the ground.) Good practices to follow when reviewing plans and specifications are listed in Table III:3-3.

TABLE III: 3-3. GOOD PRACTICES FOR REVIEWING PLANS AND SPECIFICATIONS

- Investigate the background and objectives of the project.
- Understand the scope of the project. What is to be included and why?
- Look for conciseness and precision. Mark ambiguous phrases, "legalese," and repetition.
- Do the specifications spell out exactly what is wanted? What is expected?
- Do plans and specifications adhere to appropriate codes, standards, requirements,policies, and do they recommend good practice as established by the industry?
- Will the designer be able to design, or the contractor to build, the system from the plans and specifications?
- Will the project meet OSHA requirements if it is built as proposed?

V. PREVENTION AND CONTROL.

A well-designed system and a continuing preventive maintenance program are key elements in the prevention and control of ventilation system problems.

A. ELEMENTS OF A GOOD MAINTENANCE PROGRAM.

PUT IT ON PAPER

1. Establish a safe place to file drawings, specifications, fan curves, operating instructions, and other papers generated during design, construction, and testing.

2. Establish a program of periodic inspection.

   a. The types and frequencies of inspections depend on the operation of the system and other factors.

      - Daily: Visual inspection of hoods, ductwork, access and clean-out doors, blast gate positions, hood static pressure, pressure drop across air cleaner, and verbal contact with users. ("How is the system performing today?")
      - Weekly: Air cleaner capacity, fan housing, pulley belts.
      - Monthly: Air cleaner components.

   b. A quick way to check for settled material in a duct is to take a broomstick and tap the underside of all horizontal ducts. If the tapping produces a "clean" sheet metal sound, the duct is clear. If the tapping produces heavy, thudding sounds and no sheet metal vibration, liquids or settled dust may be in the duct.

3. Establish a preventive maintenance program. Certain elements of any ventilation system should be checked on a regular schedule and replaced if found to be defective.

4. Provide worker training. Workers need to be trained in the purpose and functions of the ventilation system. For example, they need to know how to work safely and how best to utilize the ventilation system. Exhaust hoods do little good if the welder does not know that the hood must be positioned close to the work.

5. Keep written records. Maintain written documentation not only of original installations but also of all modifications as well as problems and their resolution.

B. DEALING WITH MICRO-ORGANISMS. If you suspect microbial agents, check for stagnant water in the ventilation system. The presence of mold or slime is a possible sign of trouble. Table III:3-4 lists preventive measures for controlling microbial problems in ventilation systems.

TABLE II: 3-4. PREVENTIVE MEASURES FOR REDUCING MICROBIAL PROBLEMS IN BUILDINGS

- Prevent buildup of moisture in occupied spaces (relative humidity of 60% or less).
- Prevent moisture collection in HVAC components.
- Remove stagnant water and slime from mechanical equipment.
- Use steam for humidifying.
- Avoid use of water sprays in HVAC systems.
- Use filters with a 50-70% collection efficiency rating.
- Find and discard microbe-damaged furnishings and equipment.
- Provide regular preventive maintenance.

C. VOLATILE ORGANIC OR REACTIVE CHIICALS. If an organic or reactive chemical (e.g., formaldehyde) is believed to be the primary agent in an IAQ problem, potential controls to consider include additional dilution ventilation, removal or isolation of the offending material, and the transfer of sensitized employees.

D. TOBACCO SMOKE IN AIR. OSHA has published a proposed rule for IAQ (including tobacco smoke in the workplace), and this rulemaking is likely to be completed in the near future. Smoking policies should include provisions for dedicated smoking areas. Dedicated smoking areas should be configured so that migration of smoke into nonsmoking areas will not occur. Such areas should:

- have floor-to-ceiling walls of tight construction;
- be under negative pressure relative to adjacent areas; AND
- be exhausted outside the building and not recirculated.

For more information on investigation of complaints, CSHO's should consult the NIOSH Guidance for Indoor Air Quality Investigation and the EPA guide Building Air Quality (1991).

VI. BIBLIOGRAPHY.

American Conference of Governmental Industrial Hygienists (ACGIH). 1988. *Industrial Ventilation, a Manual of Recommended Practice.* 20th ed. Cincinnati, OH: American Conference of Governmental Industrial Hygienists.

Air Movement and Control Association (AMCA). 1988. *AMCA Publication One*. Arlington Heights, IL: Air Movement and Control Association.

American Society of Heating, Refrigerating, and Air-Conditioning Engineers (ASHRAE). Handbooks and Standards. Atlanta, GA: American Society of Heating, Refrigerating, and Air-Conditioning Engineers.

Sheet Metal and Air Conditioning Contractors National Association (SMACNA). *SMACNA Publications*. Arlington, VA: Sheet Metal and Air Conditioning Contractors National Association.

American National Standards Institute (ANSI) Standards:

Z9.1 - *Open Surface Tanks Operation*
Z9.2 - *Fundamentals Covering the Design and Operation of Local Exhaust Systems*
Z9.3 - *Design, Construction, and Ventilation of Spray Finishing Operations*
Z9.4 - *Ventilation and Safe Practice of Abrasive Blasting Operations*
Z9.5 - *Laboratory Ventilation*. Fairfax, VA: American Industrial Hygiene Association.

Burgess, W. A. et al. 1989. *Ventilation and Control of the Work Environment*. New York: Wiley Interscience.

Burton, D. J. 1989. *Industrial Ventilation Workbook*. Salt Lake City, UT: IVE, Inc.

Burton, D. J. 1990. *Indoor Air Quality Workbook*. Salt Lake City, UT: IVE, Inc.

Jorgensen, R. et al. 1983. *Fan Engineering*. 8th ed. Buffalo, NY: Buffalo Forge Co.

Homeon, W. C. L. 1963. *Plant and Process Ventilation*. New York: Industrial Press.

National Institute for Occupational Safety and Health (NIOSH). 1987. *Guidance for Indoor Air Quality Investigations*. Cincinnati: NIOSH.

OSHA Field Operations Manual. 1992. OSHA Instruction CPL 2.45B. Washington, D.C.: U.S. Government Printing Office.

U.S. Environmental Protection Agency (EPA). 1991. *Building Air Quality*.

APPENDIX III:3-1. VENTILATION PRIMER.

SELECTION. Before an appropriate ventilation system can be selected, the employer should study emission sources, worker behavior, and air movement in the area. In some cases the employer may wish to seek the services of an experienced professional ventilation engineer to assist in the data gathering. Table III:3-5 shows factors to consider when selecting a ventilation system. Combinations of controls are often employed for HVAC purposes.

TABLE III:3-5. SELECTION CRITERIA FOR GENERAL AND LOCAL EXHAUST SYSTEMS

General exhaust ventilation (dilution ventilation) is appropriate when:

- Emission sources contain materials of relatively low hazard. (The degree of hazard is related to toxicity, dose rate, and individual susceptibility);
- Emission sources are primarily vapors or gases, or small, respirable-size aerosols (those not likely to settle);
- Emissions occur uniformly;
- Emissions are widely dispersed;
- Moderate climatic conditions prevail;
- Heat is to be removed from the space by flushing it with outside air;
- Concentrations of vapors are to be reduced in an enclosure; and
- Portable or mobile emission sources are to be controlled.

Local exhaust ventilating is appropriate when:

- Emission sources contain materials of relatively high hazard;
- Emitted materials are primarily larger-diameter particulates (likely to settle);
- Emissions vary over time;
- Emission sources consist of point sources;
- Employees work in the immediate vicinity of the emission source;
- The plant is located in a severe climate; and
- Minimizing air turnover is necessary.

GENERAL EXHAUST (DILUTION) VENTILATION SYSTIS. General exhaust ventilation, also called dilution ventilation, is different from local exhaust ventilation because instead of capturing emissions at their source and removing them from the air, general exhaust ventilation allows the contaminant to be emitted into the workplace air and then dilutes the concentration of the contaminant to an acceptable level (e.g., to the PEL or below). Dilution systems are often used to control evaporated liquids.

To determine the correct volume flow rate for dilution (Qd), it is necessary to estimate the evaporation rate of the contaminant (qd) according to the following equation:

$$q_d = \frac{(387)(lbs)}{(MW)(\min)(d)}$$

where:	qd	=	evaporation rate in acfm
	387	=	volume in cubic feet formed by the evaporation of one lb-mole of a substance, e.g., a solvent
	MW	=	molecular weight of emitted material
	lbs	=	lbs of material evaporated
	min	=	time of evaporation
	d	=	density of correction factor

The appropriate dilution volume flow rate for toxics is:

$$Q_d = \frac{(q_d)(K_m)(10^6)}{C_a}$$

where:	Qd	=	volume flow rate of air, in acfm
	qd	=	evaporation rate, in acfm
	Km	=	mixing factor to account for poor or random mixing (Note: Km = 2 to 5; Km = 2 is optimum
	Ca	=	acceptable airborne concentration of the material (typically half of the PEL).

The number of air changes per hour is the number of times one volume of air is replaced in the space per hour. In practice, replacement depends on mixing efficiency. When using dilution ventilation:

- position exhausts as close to emission sources as possible;
- use auxiliary fans for mixing;
- make sure employees are upwind of the dilution zone; and
- add make-up air where it will be most effective.

LOCAL EXHAUST VENTILATION SYSTEMS. A typical local exhaust ventilation system is composed of five parts: fans, hoods, ducts, air cleaners, and stacks. Local exhaust ventilation is designed to capture an emitted contaminant at or near its source, before the contaminant has a chance to disperse into the workplace air.

FAN SELECTION. To choose the proper fan for a ventilation system, this information must be known:

- air volume to be moved;
- fan static pressure;
- type and concentration of contaminants in the air (because this affects the fan type and materials of construction); and
- the importance of noise as a limiting factor.

Once this information is available, the type of fan best suited for the system can be chosen. Many different fans are available, although they all fall into one of two classes: axial flow fans and centrifugal fans. For a detailed explanation of fans, see the ACGIH Industrial Ventilation Manual.

HOODS. The hood captures, contains, or receives contaminants generated at an emission source. The hood converts duct static pressure to velocity pressure and hood entry losses (e.g., slot and duct entry losses).

Hood entry loss (He) is calculated according to the following equation:

$$H_e = (K)(VP) = |SP_h| = VP$$

where:

K	=	loss factor
VP	=	velocity pressure in duct
{SP h}	=	absolute static pressure about 5 duct diameters down the duct from the hood

A hood's ability to convert static pressure to velocity pressure is given by the coefficient of entry (Ce), as follows:

$$C_e = \frac{Q_{ideal}}{Q_{actual}} = \sqrt{\frac{VP}{SP_h}} = \sqrt{\frac{1}{1+K}}$$

To minimize air-flow requirements, the operation should be enclosed as much as possible, either with a ventilated enclosure, side baffles, or curtains. This helps both to contain the material and to minimize the effect of room air currents.

When using a capture or receiving hood, the hood should be located as close to the contaminant source as possible. Reducing the amount of contaminants generated or released from the process reduces ventilation requirements.

The hood should be designed to achieve good air distribution into the hood openings so that all the air drawn into the hood helps to control contaminants. Avoid designs that require that the velocities through some openings be very high in order to develop the minimum acceptable velocity through other openings or parts of the hood.

The purpose of most ventilation systems is to prevent worker inhalation of contaminants. For this reason, the hood should be located so that contaminants are not drawn through the worker's breathing zone. This is especially important where workers lean over an operation such as an open-surface tank or welding bench.

Hoods must meet the design criteria in the ACGIH Industrial Ventilation Manual or applicable OSHA standards. Most hood design recommendations account for cross-drafts that interfere with hood operation. Strong cross-drafts can easily reduce a hood's effectiveness by 75%. Standard hood designs may not be adequate to contain highly toxic materials.

The hood should be designed to cause minimum interference with the performance of work. Positioning access doors inside an enclosure that must be opened and closed often means that in practice the doors will be left open, and locating capture hoods too close to the process for the worker's convenience often means that the hood will be disassembled and removed. Hoods should never increase the likelihood of mechanical injury by interfering with a worker's freedom to move around machinery.

Two common misconceptions about hoods that are a part of local exhaust systems are:

- Hoods draw air from a significant distance away from the hood opening, and therefore they can control contaminants released some distance away. It is easy to confuse a fan's ability to blow a jet of air with its ability to draw air into a hood. Hoods must be close to the source of contamination to be effective.
- Heavier-than-air vapors tend to settle to the workroom floor and therefore can be collected by a hood located there. A small amount of contaminant in the air (1,000 ppm means 1,000 parts of contaminant plus 999,000 parts of air) has a resulting density close to that of air, and random air currents will disperse the material throughout the room. DUCTS. Air flows turbulently through ducts at between 2,000-6,000 feet per minute (fpm). Ducts can be made of galvanized metal, fiberglass, plastic, and concrete. Friction losses vary according to ductwork type, length of duct, velocity of air, duct area, density of air, and duct diameter.

AIR CLEANERS. The design of the air cleaner depends on the degree of cleaning required. Regular maintenance of air cleaners increases their efficiency and minimizes worker exposure. Different types of air cleaners are made to remove particulates (e.g., precipitators, cyclones, etc.) and gases and vapors (e.g., scrubbers).

STACKS. Stacks disperse exhaust air into the ambient environment. The amount of reentrainment depends on exhaust volume, wind speed and direction, temperature, location of intakes and exhausts, etc. When installing stacks:

- Provide ample stack height (a minimum of 10 ft above adjacent rooflines or air intakes);
- Place stack downwind of air intakes;
- Provide a stack velocity of a minimum of 1.4 times the wind velocity;
- Place the stack as far from the intake as possible (50 ft is recommended);
- Place the stack at least 10 ft high on most roofs to avoid recirculation; and
- Avoid rain caps if the air intake is within 50 ft of the stack.

MAKE-UP AIR SYSTIS. Exhaust ventilation systems require the replacement of exhausted air. Replacement air is often called make-up air. Replacement air can be supplied naturally by atmospheric pressure through open doors, windows, wall louvers, and adjacent spaces (acceptable), as well as through cracks in walls and windows, beneath doors, and through roof vents (unacceptable). Make-up air can also be provided through dedicated replacement air systems. Generally, exhaust systems are interlocked with a dedicated make-up air system.

Other reasons for designing and providing dedicated make-up air systems are that they:

- Avoid high-velocity drafts through cracks in walls, under doors, and through windows;
- Avoid differential pressures on doors, exits, and windows; and
- Provide an opportunity to temper the replacement air.

If make-up air is not provided, a slight negative pressure will be created in the room and air flow through the exhaust system will be reduced.

HVAC (heating, ventilating, and air-conditioning) is a common term that can also include cooling, humidifying or dehumidifying, or otherwise conditioning air for comfort and health. HVAC also is used for odor control and the maintenance of acceptable concentrations of carbon dioxide.

Air-conditioning has come to include any process that modifies the air for a work or living space: heating or cooling, humidity control, and air cleaning. Historically, air-conditioning has been used in industry to improve or protect machinery, products, and processes. The conditioning of air for humans has become normal and expected. Although the initial costs of air conditioning are high, annual costs may account only for about 1% to 5% of total annual operating expenses. Improved human productivity, lower absenteeism, better health, and reduced housekeeping and maintenance almost always make air-conditioning cost effective.

Mechanical air-handling systems can range from simple to complex. All distribute air in a manner designed to meet ventilation, temperature, humidity, and air-quality requirements established by the user. Individual units may be installed in the space they serve, or central units can serve multiple areas.

HVAC engineers refer to the areas served by an air handling system as zones. The smaller the zone, the greater the likelihood that good control will be achieved; however, equipment and maintenance costs are directly related to the number of zones. Some systems are designed to provide individual control of rooms in a multiple-zone system.

Both the provision and distribution of make-up air are important to the proper functioning of the system. The correct amount of air should be supplied to the space. Supply registers should be positioned to avoid disruption of emission and exposure controls and to aid dilution efforts.

Considerations in designing an air-handling system include volume flow rate, temperature, humidity, and air quality. Equipment selected must be properly sized and may include:

- outdoor air plenums or ducts
- filters
- supply fans and supply air systems
- heating and cooling coils
- humidity control equipment
- supply ducts
- distribution ducts, boxes, plenums, and registers
- dampers
- return air plenums
- exhaust air provisions
- return fans
- controls and instrumentation

RECIRCULATION. Although not generally recommended, recirculation is an alternative to air exchanging. Where used, recirculation should incorporate air cleaners, a by-pass or auxiliary exhaust system, regular maintenance and inspection, and devices to monitor system performance. Key points to consider in the use of recirculation are shown in Table III:3-6.

TABLE III:3-6. RECIRCULATION CRITERIA

- Protection of employees must be the primary design consideration.
- The system should remove as much of the contaminant as can economically be separated from exhaust air.
- The system should not be designed simply to achieve PEL levels of exposure.
- The system should never allow recirculation to significantly increase existing exposures.
- Recirculation should not be used if a carcinogen is present.
- The system should have fail-safe features, e.g., warning devices on critical parts, back-up systems.
- Cleaning and filtering devices that ensure continuous and reliable collection of the contaminant should be used.
- The system should provide a by-pass or auxiliary exhaust system for use during system failure.
- The system should include feedback devices that monitor system performance, e.g., static pressure taps, particulate counters, amperage monitors.
- The system should be designed not to recirculate air during equipment malfunction.
- The employer should train employees in the use and operation of the system.

APPENDIX III:3-2. GLOSSARY.

acfm Actual cubic feet per minute of gas flowing at existing temperature and pressure. (See also scfm.)

ACH, AC/H (air changes per hour) The number of times air is replaced in an hour.

AIR DENSITY The weight of air in lbs per cubic foot. Dry standard air at T=68° F (20° C) and BP = 29.92 in Hg (760 mm Hg) has a density of 0.075 lb/cu ft.

ANIOMETER A device that measures the velocity of air. Common types include the swinging vane and the hot-wire anemometer.

AREA (A) The cross-sectional area through which air moves. Area may refer to the cross-sectional area of a duct, a window, a door, or any space through which air moves.

ATMOSPHERIC PRESSURE The pressure exerted in all directions by the atmosphere. At sea level, mean atmospheric pressure is 29.92 in Hg, 14.7 psi, 407 in w.g., or 760 mm Hg.

BRAKE HORSEPOWER (bhp) The actual horsepower required to move air through a ventilation system against a fixed total pressure plus the losses in the fan. bhp=ahp × 1/eff, where eff is fan mechanical efficiency.

BRANCH In a junction of two ducts, the branch is the duct with the lowest volume flow rate. The branch usually enters the main at an angle of less than 90.

CANOPY HOOD (Receiving Hood) A one- or two-sided overhead hood that receives rising hot air or gas.

CAPTURE VELOCITY The velocity of air induced by a hood to capture emitted contaminants external to the hood.

COEFFICIENT OF ENTRY (Ce) A measure of the efficiency of a hood's ability to convert static pressure to velocity pressure; the ratio of actual flow to ideal flow.

DENSITY CORRECTION FACTOR A factor applied to correct or convert dry air density of any temperature to velocity pressure; the ratio of actual flow to ideal flow.

DILUTION VENTILATION (General Exhaust Ventilation) A form of exposure control that involves providing enough air in the workplace to dilute the concentration of airborne contaminants to acceptable levels.

ENTRY LOSS See Hood Entry Loss or Branch Entry Loss.

EVASE (pronounced eh-va-say) A cone-shaped exhaust stack that recaptures static pressure from velocity pressure.

FAN A mechanical device that moves air and creates static pressure.

FAN LAWS Relationships that describe theoretical, mutual performance changes in pressure, flow rate, rpm of the fan, horsepower, density of air, fan size, and sound power.

FAN CURVE A curve relating pressure and volume flow rate of a given fan at a fixed fan speed (rpm).

FRICTION LOSS The static pressure loss in a system caused by friction between moving air and the duct wall, expressed as in w.g./100 ft, or fractions of VP per 100 ft of duct (mm w.g./m; Kpa/m).

GAUGE PRESSURE The difference between two absolute pressures, one of which is usually atmospheric pressure.

GENERAL EXHAUST See Dilution Ventilation.

HEAD Pressure, e.g. "The head is 1 in w.g."

HOOD A device that encloses, captures, or receives emitted contaminants.

HOOD ENTRY LOSS (He) The static pressure lost (in inches of water) when air enters a duct through a hood. The majority of the loss usually is associated with a vena contracta formed in the duct.

HOOD STATIC PRESSURE (SPh) The sum of the duct velocity pressure and the hood entry loss; hood static pressure is the static pressure required to accelerate air at rest outside the hood into the duct at velocity.

HVAC (HEATING, VENTILATION, AND AIR CONDITIONING) SYSTIS Ventilating systems designed primarily to control temperature, humidity, odors, and air quality.

INDOOR AIR QUALITY (IAQ), SICK-BUILDING SYNDROME, TIGHT-BUILDING SYNDROME The study, examination, and control of air quality related to temperature, humidity, and airborne contaminants.

in. w.g. (inches of water) A unit of pressure. One inch of water is equal to 0.0735 in. of mercury, or 0.036 psi. Atmospheric pressure at standard conditions is 407 in. w.g.

INDUSTRIAL VENTILATION (IV) The equipment or operation associated with the supply or exhaust of air by natural or mechanical means to control occupational hazards in the industrial setting.

LAMINAR FLOW (also Streamline Flow) Air flow in which air molecules travel parallel to all other molecules; laminar flow is characterized by the absence of turbulence.

LOCAL EXHAUST VENTILATION An industrial ventilation system that captures and removes emitted contaminants before dilution into the ambient air of the workplace.

LOSS Usually refers to the conversion of static pressure to heat in components of the ventilation system, e.g., "the hood entry loss."

MAKE-UP AIR See Replacement and Compensating Air.

MANOMETER A device that measures pressure difference; usually a U-shaped glass tube containing water or mercury.

MINIMUM TRANSPORT VELOCITY (MTV). The minimum velocity that will transport particles in a duct with little settling; MTV varies with air density, particulate loading, and other factors.

OUTDOOR AIR (OA) Outdoor air is the "fresh" air mixed with return air (RA) to dilute contaminants in the supply air.

PITOT TUBE A device used to measure total and static pressures in an airstream.

PLENUM A low-velocity chamber used to distribute static pressure throughout its interior.

PRESSURE DROP The loss of static pressure across a point; for example, "the pressure drop across an orifice is 2.0 in. w.g."

REPLACEMENT AIR (also, Compensating Air, Make-Up Air) Air supplied to a space to replace exhausted air.

RETURN AIR Air that is returned from the primary space to the fan for recirculation.

scfm Standard cubic feet per minute. A measure of air flow at standard conditions, i.e., dry air at 29.92 in. Hg (760 mm Hg) (gauge), 68° F (20° C).

SLOT VELOCITY The average velocity of air through a slot. Slot velocity is calculated by dividing the total volume flow rate by the slot area (usually, Vs = 2,000 fpm).

STACK A device on the end of a ventilation system that disperses exhaust contaminants for dilution by the atmosphere.

STANDARD AIR, STANDARD CONDITIONS Dry air at 68° F (20° C), 29.92 in Hg (760 mm Hg).

STATIC PRESSURE (SP) The pressure developed in a duct by a fan; the force in inches of water measured perpendicular to flow at the wall of the duct; the difference in pressure between atmospheric pressure and the absolute pressure inside a duct, cleaner, or other equipment; SP exerts influence in all directions.

SUCTION PRESSURE (See Static Pressure.) An archaic term that refers to static pressure on the upstream side of the fan.

TOTAL PRESSURE (TP) The pressure exerted in a duct, i.e., the sum of the static pressure and the velocity pressure; also called Impact Pressure, Dynamic Pressure.

TRANSPORT VELOCITY See Minimum Transport Velocity.

TURBULENT FLOW Air flow characterized by transverse velocity components as well as velocity in the primary direction of flow in a duct; mixing velocities.

VELOCITY (V) The time rate of movement of air; usually expressed as feet per minute.

VELOCITY PRESSURE (VP) The pressure attributed to the velocity of air.

VOLUME FLOW RATE (Q) Quantity of air flow in cfm, scfm, or acfm.

APPENDIX III:3-3. OSHA AND CONSENSUS STANDARDS.

I. OSHA STANDARDS.

A. HEALTH-RELATED VENTILATION STANDARDS. This list includes some, but not necessarily all, OSHA standards that address the control of employee exposure to recognized contaminants.)

General industry

29 CFR 1910.94(a)	Abrasive blasting
29 CFR 1910.94(b)	Grinding, polishing and buffing operations
29 CFR 1910.94(d)	Open surface tanks
29 CFR 1910.252(f)(2)(i)(a) and (b);(f)(2)(ii)	Ventilation for general welding and cutting--General
29 CFR 1910.252(f)(3)	Local exhaust hoods and booths
29 CFR 1910.252(f)(5)(ii)	Fluorine compounds--Maximum allowable concentration
29 CFR 1910.252(f)(12)	Cutting of stainless steels
29 CFR 1910.1003 to .1016	Carcinogens
29 CFR 1910.1025(e)(5)	Lead
29 CFR 1910.1027(f)(3)	Cadmium

Construction

29 CFR 1926.57(a)	Ventilation--General
29 CFR 1926.62(e)(3)	Lead
29 CFR 1926.63(f)(4)	Cadmium
29 CFR 1926.154(a)(1)	Temporary heating devices--Ventilation
29 CFR 1926.353(e)(1)	Ventilation and protection in welding, cutting and heating--General welding, cutting, and heating

Maritime

29 CFR 1915.32(a)(2)	Toxic cleaning solvents
29 CFR 1915.51(f)(1)	Ventilation and protection in welding, cutting and heating--General welding, cutting, and heating
29 CFR 1918.93(a)(1)(iii)	Ventilation and atmospheric conditions

B. HEALTH-RELATED VENTILATION STANDARDS OTHER THAN AIRFLOW. This list includes some, but not necessarily all, OSHA standards that do not contain airflow requirements but are located in the health-related ventilation standards.

General Industry

29 CFR 1910.94(a)(3)(i)(d)	Abrasive blasting--Blasting cleaning
29 CFR 1910.94(a)(5)	Abrasive blasting--Personal protective equipment
29 CFR 1910.94(a)(6)	Abrasive blasting--Air supply and air compressors
29 CFR 1910.94(a)(7)	Abrasive blasting--Operational procedures and general safety
29 CFR 1910.94(d)(9)	Open surface tanks--Personal protection
29 CFR 1910.94(d)(10)	Open surface tanks--Special precautions for cyanide
29 CFR 1910.94(d)(11)	Open surface tanks--Inspection, installation and maintenance
29 CFR 1910.94(d)(12)	Open surface tanks--Vapor degreasing tanks

C. FIRE AND EXPLOSION-RELATED VENTILATION STANDARDS. This list includes some, but not necessarily all, OSHA standards that are intended to prevent fire and explosions.

General Industry

29 CFR 1910.94(c)	Ventilation--Spray finishing operations
29 CFR 1910.103(b)(3)(ii)(b)	Hydrogen--Gaseous hydrogen systems--Separate buildings
29 CFR 1910.103(b)(3)(iii)(b)	Hydrogen--Gaseous hydrogen systems--Special rooms
29 CFR 1910.103(c)(3)(ii)(b)	Hydrogen--Liquid hydrogen systems--Separate buildings
29 CFR 1910.103(c)(3)(iii)(b)	Hydrogen--Liquid hydrogen systems--Special rooms
29 CFR 1910.104(b)(3)(xii)	Oxygen--Bulk oxygen systems--Ventilation
29 CFR 1910.104(b)(8)(vii)	Oxygen--Bulk oxygen systems--Venting
29 CFR 1910.106(d)(4)(iv)	Flammable and combustible liquids--Container and portable tank storage--Design and construction of inside storage room--Ventilation
29 CFR 1910.106(e)(3)(v)	Flammable and combustible liquids--Industrial plants--Unit physical operations--Ventilation
29 CFR 1910.106(f)(2)(iii)(a)	Flammable and combustible liquids--Bulk plants--Building--Ventilation
29 CFR 1910.106(h)(3)(iii)	Flammable and combustible liquids--Processing plants--Processing building--Ventilation
29 CFR 1910.107(b)(5)(i)	Spray finishing using flammable and combustible materials--Spray booths--Dry type overspray collectors
29 CFR 1910.107(d)(1) and (2)	Spray finishing using flammable and combustible materials--Ventilation--Conformance--General
29 CFR 1910.107(i)(9)	Spray finishing using flammable and combustible materials--Electrostatic hand spraying equipment--Ventilation
29 CFR 1910.108(b)(1) and (2)	Dip tanks containing flammable combustible liquids--Ventilation--Ventilation combined with drying
29 CFR 1910.307	Hazardous (classified) locations

D. EXCEPTIONS TO 25% OF THE LEL FOR FIRE AND EXPLOSION-RELATED STANDARDS. This list includes but is not limited to OSHA standards that allow concentrations of flammable materials no greater than 10% of the LEL.

Maritime

29 CFR 1915.12(a)(2)	Precautions before entering--Flammable atmospheres and residues
29 CFR 1915.13(a)(2)	Cleaning and other cold work (flammable vapors)
29 CFR 1915.35(b)(1), (2), (3)	Painting--Paints and tanks coatings dissolved in highly volatile, toxic and/or flammable solvents
29 CFR 1915.36(a)(2)	Flammable liquids ventilation

Construction

29 CFR 1926.803(i)(2)	Compressed Air--Ventilation and air quality--(Tunnels)

E. SPECIAL CONDITIONS STANDARDS. This list includes some but not necessarily all OSHA standards that involve confined space operations and/or high-hazard contaminants specifically designated in he standard.

General Industry

29 CFR 1910.252(c)(2)(i)(c)	Welding, cutting and brazing--Health protection and ventilating--Ventilation for general welding and cutting--General
29 CFR 1910.252(c)(4)	Welding, cutting and brazing--Health protection and ventilating--Ventilation in confined spaces
29 CFR 1910.252(c)(5)(i)	Welding, cutting and brazing--Fluorine compounds
29 CFR 1910.252(c)(6)(i)	Welding, cutting and brazing--Zinc--Confined spaces
29 CFR 1910.252(c)(7)(i)	Welding, cutting and brazing--Lead--Confined spaces
29 CFR 1910.252(c)(8)	Welding, cutting and brazing--Beryllium
29 CFR 1910.252(c)(9)	Welding, cutting and brazing--Cadmium
29 CFR 1910.252(c)(10)	Welding, cutting and brazing--Mercury

Construction

29 CFR 1926.154(a)(2)	Temporary heating devices--Ventilation
29 CFR 1926.353(b)(1)	Ventilation and protection in welding, cutting and heating--Welding, cutting and heating in confined spaces
29 CFR 1926.353(c)(1) and (2)	Ventilation and protection in welding, cutting and heating--Welding, cutting or heating of metals of toxic significance
29 CFR 1926.800(k)	Tunnels and shafts--Air quality and ventilation

Maritime

29 CFR 1915.12(b)(2)	Precautions before entering--Toxic atmospheres and residues
29 CFR 1915.12(c)(2)	Precautions before entering--Oxygen deficient atmospheres
29 CFR 1915.12(d)	Precautions before entering--Exceptions
29 CFR 1915.34(a)(4)	Mechanical paint removers--Power tools--(paint dust)
29 CFR 1915.51(c)(3)	Ventilation and protection in welding, cutting and heating--Welding, cutting and heating confined spaces
29 CFR 1915.51(d)(1) and (2)	Ventilation and protection in welding, cutting and heating--cutting or heating of metals of toxic significance.

II. CONSENSUS STANDARDS.

Standard	Source	Title
Air filters		
ASHRAE 52-76	ASHRAE	Methods of Testing Air-Cleaning Devices Used in General Ventilation for Removing Particulate Matter
Exhaust systems		
ANSI Z33.1-1982 NFPA 91-1983	NFPA	Installation of Blower and Exhaust Systems for Dust, Stock, Vapor Removal or Conveying (1983)
ANSI Z9.2-1979	AIHA	Fundamentals Governing the Design and Operation of Local Exhaust Systems
ANSI Z9.1-1977	AIHA ASHRAE	Practices for Ventilation and Operation of Open-Surface Tanks
ANSI Z9.3-1964	ANSI	Safety Code for Design, Construction, and Ventilation of Spray Finishing Operations (reaffirmed 1971)
ANSI Z9.4-1979 ANSI Z9.4A-1981	ANSI	Ventilation and Safe Practices of Abrasives Blasting Operations
ANSI Z9.5-1992	AIHA	Laboratory Ventilation
Fans		
AMCA 99-83 ANSI/UL 507-1976	AMCA UL	Standards Handbook Electric Fans (1977)
ASHRAE 51-75 AMCA 210-74	ASHRAE	Laboratory Methods of Testing Fans for Rating
ANSI/ASHRAE 87.7-1983	ASHRAE	Methods of Testing Dynamic Characteristics of Propeller Fans--Aerodynamically Excited Fan Vibrations and Critical Speeds
AMCA 210-74	AMCA	Laboratory Methods of Testing Fans for Rating Purposes
AMCA 99-2404-78	AMCA	Drive Arrangement for Centrifugal Fans
AMCA 99-2406-83	AMCA	Designation for Rotation and Discharge of Centrifugal Fans
AMCA 99-2407-66	AMCA	Motor Positions for Belt or Chain Drive Centrifugal Fans
AMCA 99-2410-82	AMCA	Drive Arrangement for Tubular Centrifugal Fans

Industrial duct		
SMACNA	SMACNA	Round Industrial Duct Construction
SMACNA	SMACNA	Rectangular Industrial Duct Construction
Venting		
NFPA 68	NFPA	Guide for Explosion Venting
NFPA 204M	NFPA	Guide for Smoke and Heat Venting
SMACNA	SMACNA	Guide for Steel Stack Design and Construction (1983)
Ventilation		
NFPA 96	NFPA	Vapor Removal from Cooking Equipment (1984)
NFPA-88A, 88B	NFPA	Parking Structures (1979); Repair Garages (1979)
ASHRAE 62-1989	ASHARAE	Ventilation for Acceptable Indoor Air Quality
ACGIH	ACGIH	Industrial Ventilation

III. SOURCES OF CONSENSUS STANDARDS.

Copies of the consensus standards are published and available directly from the organization issuing the standard. A minimal fee is often required.

Source	***Organization***
ACGIH	American Conference of Governmental Industrial Hygienists 6500 Glenway Ave., Bldg. D-5 Cincinnati, OH 45211
AIHA	American Industrial Hygiene Association 2700 Prosperity Ave., Suite 250 Fairfax, VA 22031-4319
AMCA	Air Movement and Control Association 30 W. University Dr. Arlington Heights, IL 60004
ANSI	American National Standards Institute 1430 Broadway New York, NY 10018
ASHRAE	American Society of Heating, Refrigeration and Air-Conditioning Engineers, Inc. 1791 Tullie Circle, N.E., Atlanta, GA 30329
NFPA	National Fire Protection Association Batterymarch Park Quincy, MA 02269
SMACNA	Sheet Metal and Air Conditioning Contractors' National Association 8224 Old Courthouse Rd. Vienna, VA 22180
UL	Underwriters Laboratories Inc. 333 Pfingsten Rd. Northbrook, IL 60062

APPENDIX III:3-4. TROUBLESHOOTING AN EXHAUST SYSTEM--SOME HELPFUL HINTS.

Most of the following checks can be made by visual observation and do not require extensive measurements.

If air flow is low in hoods, check:

- Fan rotation (reversed polarity will cause fan to run backwards; a backward-running centrifugal fan delivers only 30-50% of rated flow);
- Fan RPM;
- Slipping belt;
- Clogged or corroded fan wheel and casing;
- Clogged ductwork (high hood static pressure and low air flow may indicate restricted ducts; open clean-out doors and inspect inside ducts);
- Closed dampers in ductwork;
- Clogged collector or air cleaning devices;
- Weather cap too close to discharge stack (a 3/4 duct- diameter gap should exist between cap and stack; weather caps are not recommended);
- Poorly designed ductwork (short radius elbows); (branch entries enter main duct at sharp angles); (duckwork diameter too small for the air-flow needed; and
- Lack of make-up air (high negative pressures affect propeller fan system output; lack of supplied make-up air causes high airflow velocities at doors and windows).

If air flow is satisfactory in a hood but contaminant control is poor, check:

- Crossdrafts (from process air movements); (worker-cooling fans and air-supply systems); (open doors and windows);
- Capture velocity (work operation too far from hood opening);
- Hood enclosure: (door, baffles, or sides may be open or removed); and
- Hood type: (canopy hoods are inappropriate for toxic materials).

SECTION III: CHAPTER 4

HEAT STRESS

Contents:

I. INTRODUCTION.

Operations involving high air temperatures, radiant heat sources, high humidity, direct physical contact with hot objects, or strenuous physical activities have a high potential for inducing heat stress in employees engaged in such operations. Such places include: iron and steel foundries, nonferrous foundries, brick-firing and ceramic plants, glass products facilities, rubber products factories, electrical utilities (particularly boiler rooms), bakeries, confectioneries, commercial kitchens, laundries, food canneries, chemical plants, mining sites, smelters, and steam tunnels.

Outdoor operations conducted in hot weather, such as construction, refining, asbestos removal, and hazardous waste site activities, especially those that require workers to wear semipermeable or impermeable protective clothing, are also likely to cause heat stress among exposed workers.

A. CAUSAL FACTORS.

1. Age, weight, degree of physical fitness, degree of acclimatization, metabolism, use of alcohol or drugs, and a variety of medical conditions such as hypertension all affect a person's sensitivity to heat. However, even the type of clothing worn must be considered. Prior heat injury predisposes an individual to additional injury.

2. It is difficult to predict just who will be affected and when, because individual susceptibility varies. In addition, environmental factors include more than the ambient air temperature. Radiant heat, air movement, conduction, and relative humidity all affect an individual's response to heat.

B. DEFINITIONS.

1. The American Conference of Governmental Industrial Hygienists (1992) states that workers should not be permitted to work when their deep body temperature exceeds 38°C (100.4°F).

2. Heat is a measure of energy in terms of quantity.

3. A calorie is the amount of heat required to raise 1 gram of water 1°C (based on a standard temperature of 16.5 to 17.5°C).

4. Conduction is the transfer of heat between materials that contact each other. Heat passes from the warmer material to the cooler material. For example, a worker's skin can transfer heat to a contacting surface if that surface is cooler, and vice versa.

5. Convection is the transfer of heat in a moving fluid. Air flowing past the body can cool the body if the air temperature is cool. On the other hand, air that exceeds 35°C (95°F) can increase the heat load on the body.

6. Evaporative cooling takes place when sweat evaporates from the skin. High humidity reduces the rate of evaporation and thus reduces the effectiveness of the body's primary cooling mechanism.

7. Radiation is the transfer of heat energy through space. A worker whose body temperature is greater than the temperature of the surrounding surfaces radiates heat to these surfaces. Hot surfaces and infrared light sources radiate heat that can increase the body's heat load.

8. Globe temperature is the temperature inside a blackened, hollow, thin copper globe.

9. Metabolic heat is a by-product of the body's activity.

10. Natural wet bulb (NWB) temperature is measured by exposing a wet sensor, such as a wet cotton wick fitted over the bulb of a thermometer, to the effects of evaporation and convection. The term natural refers to the movement of air around the sensor.

11. Dry bulb (DB) temperature is measured by a thermal sensor, such as an ordinary mercury-in-glass thermometer, that is shielded from direct radiant energy sources.

II. HEAT DISORDERS AND HEALTH EFFECTS.

A. HEAT STROKE occurs when the body's system of temperature regulation fails and body temperature rises to critical levels. This condition is caused by a combination of highly variable factors, and its occurrence is difficult to predict. Heat stroke is a medical emergency. The primary signs and symptoms of heat stroke are confusion; irrational behavior; loss of consciousness; convulsions; a lack of sweating (usually); hot, dry skin; and an abnormally high body temperature, e.g., a rectal temperature of 41°C (105.8°F). If body temperature is too high, it causes death. The elevated metabolic temperatures caused by a combination of work load and environmental heat load, both of which contribute to heat stroke, are also highly variable and difficult to predict.

If a worker shows signs of possible heat stroke, professional medical treatment should be obtained immediately. The worker should be placed in a shady area and the outer clothing should be removed. The worker's skin should be wetted and air movement around the worker should be increased to improve evaporative cooling until professional methods of cooling are initiated and the seriousness of the condition can be assessed. Fluids should be replaced as soon as possible. The medical outcome of an episode of heat stroke depends on the victim's physical fitness and the timing and effectiveness of first aid treatment.

Regardless of the worker's protests, no employee suspected of being ill from heat stroke should be sent home or left unattended unless a physician has specifically approved such an order.

B. HEAT EXHAUSTION. The signs and symptoms of heat exhaustion are headache, nausea, vertigo, weakness, thirst, and giddiness. Fortunately, this condition responds readily to prompt treatment. Heat exhaustion should not be dismissed lightly, however, for several reasons. One is that the fainting associated with heat exhaustion can be dangerous because the victim may be operating machinery or controlling an operation that should not be left unattended; moreover, the victim may be injured when he or she faints. Also, the signs and symptoms seen in heat exhaustion are similar to those of heat stroke, a medical emergency.

Workers suffering from heat exhaustion should be removed from the hot environment and given fluid replacement. They should also be encouraged to get adequate rest.

C. HEAT CRAMPS are usually caused by performing hard physical labor in a hot environment. These cramps have been attributed to an electrolyte imbalance caused by sweating. It is important to understand that cramps can be caused by both too much and too little salt. Cramps appear to be caused by the lack of water replenishment. Because sweat is a hypotonic solution (±0.3% NaCl), excess salt can build up in the body if the water lost through sweating is not replaced. Thirst cannot be relied on as a guide to the need for water; instead, water must be taken every 15 to 20 minutes in hot environments.

Under extreme conditions, such as working for 6 to 8 hours in heavy protective gear, a loss of sodium may occur. Recent studies have shown that drinking commercially available carbohydrate-electrolyte replacement liquids is effective in minimizing physiological disturbances during recovery.

D. HEAT COLLAPSE ("Fainting"). In heat collapse, the brain does not receive enough oxygen because blood pools in the extremities. As a result, the exposed individual may lose consciousness. This reaction is similar to that of heat exhaustion and does not affect the body's heat balance. However, the onset of heat collapse is rapid and unpredictable. To prevent heat collapse, the worker should gradually become acclimatized to the hot environment.

E. HEAT RASHES are the most common problem in hot work environments. Prickly heat is manifested as red papules and usually appears in areas where the clothing is restrictive. As sweating increases, these papules give rise to a prickling sensation. Prickly heat occurs in skin that is persistently wetted by unevaporated sweat, and heat rash papules may become infected if they are not treated. In most cases, heat rashes will disappear when the affected individual returns to a cool environment.

F. HEAT FATIGUE. A factor that predisposes an individual to heat fatigue is lack of acclimatization. The use of a program of acclimatization and training for work in hot environments is advisable. The signs and symptoms of heat fatigue include impaired performance of skilled sensorimotor, mental, or vigilance jobs. There is no treatment for heat fatigue except to remove the heat stress before a more serious heat-related condition develops.

III. INVESTIGATION GUIDELINES.

These guidelines for evaluating employee heat stress approximate those found in the 1992-1993 ACGIH publication, Threshold Limit Values for Chemical Substances and Physical Agents and Biological Exposure Indices.

A. EMPLOYER AND EMPLOYEE INTERVIEWS.

1. The inspector will review the OSHA 200 Log and, if possible, the OSHA 101 forms for indications of prior heat stress problems.

2. Following are some questions for employer interviews: What type of action, if any, has the employer taken to prevent heat stress problems? What are the potential sources of heat? What employee complaints have been made?

3. Following are some questions for employee interviews: What heat stress problems have been experienced? What type of action has the employee taken to minimize heat stress? What is the employer's involvement, i.e., does employee training include information on heat stress? (Appendix III:4-1 lists factors to be evaluated when reviewing a heat stress situation, and Appendix III:4-2 contains a follow-up checklist.)

B. WALKAROUND INSPECTION. During the walk-around inspection, the investigator will: determine building and operation characteristics; determine whether engineering controls are functioning properly; verify information obtained from the employer and employee interviews; and perform temperature measurements and make other determinations to identify potential sources of heat stress. Investigators may wish to discuss any operations that have the potential to cause heat stress with engineers and other knowledgeable personnel. The walk-around inspection should cover all affected areas. Heat sources, such as furnaces, ovens, and boilers, and relative heat load per employee should be noted.

C. WORK-LOAD ASSESSMENT.

1. Under conditions of high temperature and heavy workload, the CSHO should determine the work-load category of each job (Table III:4-1 and Figure III:4-1). Work-load category is determined by averaging metabolic rates for the tasks and then ranking them:

- Light work: up to 200 kcal/hour
- Medium work: 200-350 kcal/hour
- Heavy work: 350-500 kcal/hour

2. Cool Rest Area: Where heat conditions in the rest area are different from those in the work area, the metabolic rate (M) should be calculated using a time-weighted average, as follows:

$$\text{Average}\,M = \frac{(M_1)(t_1)+(M_2)(t_2)+\ldots+(M_n)(t_n)}{t_1+t_2+\ldots+t_n}$$

EQUATION III: 4-1. AVERAGE METABOLIC RATE

where: M = metabolic rate
t = time in minutes

In some cases, a videotape is helpful in evaluating work practices and metabolic load.

- Light hand work: writing, hand knitting
- Heavy hand work: typewriting
- Heavy work with one arm: hammering in nails (shoemaker, upholsterer)
- Light work with two arms: filing metal, planing wood, raking the garden
- Moderate work with the body: cleaning a floor, beating a carpet
- Heavy work with the body: railroad track laying, digging, barking trees

Sample Calculation: Assembly line work using a heavy hand tool

Walking along	2.0 kcal/min
Intermediate value between heavy work with two arms and light work with the body	3.0 kcal/min
Add for basal metabolism	1.0 kcal/min
Total:	6.0 kcal/min

FIGURE III: 4-1. ACTIVITY EXAMPLES

Source: ACGIH 1992.

TABLE III:4-1. ASSESSMENT OF WORK

Body position and movement		*kcal/min**
Sitting		0.3
Standing		0.6
Walking		2.0-3.0
Walking uphill		add 0.8 for every meter (yard) rise
Type of work	***Average kcal/min***	***Range kcal/min***
Hand work		
Light	0.4	0.2-1.2
Heavy	0.9	
Work: One arm		
Light	1.0	0.7-2.5
Heavy	1.7	
Work: Both arms		
Light	1.5	1.0-3.5
Heavy 2.5		
Work: Whole body		
Light	3.5	2.5-15.0
Moderate	5.0	
Heavy	7.0	
Very heavy	9.0	

* For a "standard"worker of 70 kg body weight (154 lbs) and 1.8m^2 body surface (19.4 ft^2).

Source: ACGIH 1992.

IV. SAMPLING METHODS.

A. BODY TEMPERATURE MEASUREMENTS. Although instruments are available to estimate deep body temperature by measuring the temperature in the ear canal or on the skin, these instruments are not sufficiently reliable to use in compliance evaluations.

B. ENVIRONMENTAL MEASUREMENTS. Environmental heat measurements should be made at, or as close as possible to, the specific work area where the worker is exposed. When a worker is not continuously exposed in a single hot area but moves between two or more areas having different levels of environmental heat, or when the environmental heat varies substantially at a single hot area, environmental heat exposures should be measured for each area and for each level of environmental heat to which employees are exposed.

C. WET BULB GLOBE TEMPERATURE INDEX.

1. Wet Bulb Globe Temperature (WBGT) should be calculated using the appropriate formula in Appendix III:4-2. The WBGT for continuous all-day or several hour exposures should be averaged over a 60-minute period. Intermittent exposures should be averaged over a 120-minute period. These averages should be calculated using the following formula:

$$Average\ WBGT = \frac{(WBGT_1)(t_1) + (WBGT_2)(t_2) + \ldots + (WBGT_n)(t_n)}{t_1 + t_2 + \ldots + t_n}$$

EQUATION III: 4-2. AVERAGE WET BULB GLOBE TEMPERATURE (WBGT)

For indoor and outdoor conditions with no solar load, WBGT is calculated as:

$$WBGT = 0.7NWB + 0.3GT$$

For outdoors with a solar load, WBGT is calculated as

$$WBGT = 0.7NWB + 0.2GT + 0.1DB$$

where:

where:	WBGT	=	Wet Bulb Globe Temperature Index
	NWB	=	Nature Wet-Bulb Temperature
	DB	=	Dry-Bulb Temperature
	GT	=	Globe Temperature

2. The exposure limits in Table III:4-2 are valid for employees wearing light clothing. They must be adjusted for the insulation from clothing that impedes sweat evaporation and other body cooling mechanisms. Use Table III:4-3 to correct Table III:4-2 for various kinds of clothing.

3. Use of Table III:4-2 requires knowledge of the WBGT and approximate workload. Workload can be estimated using the data in Table III:4-1, and sample calculations are presented in Figure III:4-1.

D. MEASUREMENT. Portable heat stress meters or monitors are used to measure heat conditions. These instruments can calculate both the indoor and outdoor WBGT index according to established ACGIH Threshold Limit Value equations. With this information and information on the type of work being performed, heat stress meters can determine how long a person can safely work or remain in a particular hot environment. See Appendix III:4-2 for an alternate method of calculation.

E. OTHER THERMAL STRESS INDICES.

1. The Effective Temperature index (ET) combines the temperature, the humidity of the air, and air velocity. This index has been used extensively in the field of comfort ventilation and air-conditioning. ET remains a useful measurement technique in mines and other places where humidity is high and radiant heat is low.

2. The Heat-Stress Index (HSI) was developed by Belding and Hatch in 1965. Although the HSI considers all environmental factors and work rate, it is not completely satisfactory for determining an individual worker's heat stress and is also difficult to use.

TABLE III:4-2. PERMISSIBLE HEAT EXPOSURE THRESHOLD LIMIT VALUES

	------------- Work Load* ------------		
Work/rest regimen	***Light***	***Moderate***	***Heavy***
Continuous work	30.0°C (86°F)	26.7°C (80°F)	25.0°C (77°F)
75% Work, 25% rest, each hour	30.6°C (87°F)	28.0°C (82°F)	25.9°C (78°F)
50% Work, 50% rest, each hour	31.4°C (89°F)	29.4°C (85°F)	27.9°C (82°F)
25% Work, 75% rest, each hour	32.2°C (90°F)	31.1°C (88°F)	30.0°C (86°F)

*Values are in °C and °F, WBGT.

These TLV's are based on the assumption that nearly all acclimatized, fully clothed workers with adequate water and salt intake should be able to function effectively under the given working conditions without exceeding a deep body temperature of 38°C (100.4° F). They are also based on the assumption that the WBGT of the resting place is the same or very close to that of the workplace. Where the WBGT of the work area is different from that of the rest area, a time-weighted average should be used (consult the ACGIH *1992-1993 Threshold Limit Values for Chemical Substances and Physical Agents and Biological Exposure Indices* (1992).

These TLV's apply to physically fit and acclimatized individuals wearing light summer clothing. If heavier clothing that impedes sweat or has a higher insulation value is required, the permissible heat exposure TLV's in Table III:4-2 must be reduced by the corrections shown in Table III:4-3.

Source: ACGIH 1992.

TABLE III: 4-3. WBGT CORRECTION FACTORS IN DEGREES CELSIUS

Clothing type	***Clo* value***	***WBGT correction***
Summer lightweight working clothing	0.6	0
Cotton coveralls	1.0	-2
Winter work clothing	1.4	-4
Water barrier, permeable	1.2	-6

*Clo: Insulation value of clothing. One clo = 5.55 kcal/m^2/hr of heat exchange by radiation and convection for each degree °C difference in temperature between the skin and the adjusted dry bulb temperature.

Note: Deleted from the previous version are trade names and "fully encapsulating suit, gloves, boots and hood" including its clo value of 1.2 and WBGT correction of -10.

Source: ACGIH 1992.

V. CONTROL.

Ventilation, air cooling, fans, shielding, and insulation are the five major types of engineering controls used to reduce heat stress in hot work environments. Heat reduction can also be achieved by using power assists and tools that reduce the physical demands placed on a worker.

However, for this approach to be successful, the metabolic effort required for the worker to use or operate these devices must be less than the effort required without them. Another method is to reduce the effort necessary to operate power assists. The worker should be allowed to take frequent rest breaks in a cooler environment.

A. ACCLIMATIZATION.

1. The human body can adapt to heat exposure to some extent. This physiological adaptation is called acclimatization. After a period of acclimatization, the same activity will produce fewer cardiovascular demands. The worker will sweat more efficiently (causing better evaporative cooling), and thus will more easily be able to maintain normal body temperatures.

2. A properly designed and applied acclimatization program decreases the risk of heat-related illnesses. Such a program basically involves exposing employees to work in a hot environment for progressively longer periods. NIOSH (1986) says that, for workers who have had previous experience in jobs where heat levels are high enough to produce heat stress, the regimen should be 50% exposure on day one, 60% on day two, 80% on day three, and 100% on day four. For new workers who will be similarly exposed, the regimen should be 20% on day one, with a 20% increase in exposure each additional day.

B. FLUID REPLACEMENT. Cool (50°-60°F) water or any cool liquid (except alcoholic beverages) should be made available to workers to encourage them to drink small amounts frequently, e.g., one cup every 20 minutes. Ample supplies of liquids should be placed close to the work area. Although some commercial replacement drinks contain salt, this is not necessary for acclimatized individuals because most people add enough salt to their summer diets.

C. ENGINEERING CONTROLS.

1. General ventilation is used to dilute hot air with cooler air (generally cooler air that is brought in from the outside). This technique clearly works better in cooler climates than in hot ones. A permanently installed ventilation system usually handles large areas or entire buildings. Portable or local exhaust systems may be more effective or practical in smaller areas.

2. Air treatment/air cooling differs from ventilation because it reduces the temperature of the air by removing heat (and sometimes humidity) from the air.

3. Air conditioning is a method of air cooling, but it is expensive to install and operate. An alternative to air conditioning is the use of chillers to circulate cool water through heat exchangers over which air from the ventilation system is then passed; chillers are more efficient in cooler climates or in dry climates where evaporative cooling can be used.

4. Local air cooling can be effective in reducing air temperature in specific areas. Two methods have been used successfully in industrial settings. One type, cool rooms, can be used to enclose a specific workplace or to offer a recovery area near hot jobs. The second type is a portable blower with built-in air chiller. The main advantage of a blower, aside from portability, is minimal set-up time.

5. Another way to reduce heat stress is to increase the air flow or convection using fans, etc. in the work area (as long as the air temperature is less than the worker's skin temperature). Changes in air speed can help workers stay cooler by increasing both the convective heat exchange (the exchange between the skin surface and the surrounding air) and the rate of evaporation. Because this method does not actually cool the air, any increases in air speed must impact the worker directly to be effective.

If the dry bulb temperature is higher than 35°C (95°F), the hot air passing over the skin can actually make the worker hotter. When the temperature is more than 35°C and the air is dry, evaporative cooling may be improved by air movement, although this improvement will be offset by the convective heat. When the temperature exceeds 35°C and the relative humidity is 100%, air movement will make the worker hotter. Increases in air speed have no effect on the body temperature of workers wearing vapor-barrier clothing.

6. Heat conduction methods include insulating the hot surface that generates the heat and changing the surface itself.

7. Simple engineering controls, such as shields, can be used to reduce radiant heat, i.e. heat coming from hot surfaces within the worker's line of sight. Surfaces that exceed 35°C (95°F) are sources of infrared radiation that can add to the worker's heat load. Flat black surfaces absorb heat more than smooth, polished ones. Having cooler surfaces surrounding the worker assists in cooling because the worker's body radiates heat toward them.

   With some sources of radiation, such as heating pipes, it is possible to use both insulation and surface modifications to achieve a substantial reduction in radiant heat. Instead of reducing radiation from the source, shielding can be used to interrupt the path between the source and the worker. Polished surfaces make the best barriers, although special glass or metal mesh surfaces can be used if visibility is a problem.

   Shields should be located so that they do not interfere with air flow, unless they are also being used to reduce convective heating. The reflective surface of the shield should be kept clean to maintain its effectiveness.

D. ADMINISTRATIVE CONTROLS AND WORK PRACTICES.

1. Training is the key to good work practices. Unless all employees understand the reasons for using new, or changing old, work practices, the chances of such a program succeeding are greatly reduced.

2. NIOSH (1986) states that a good heat stress training program should include at least the following components:

   - Knowledge of the hazards of heat stress;
   - Recognition of predisposing factors, danger signs, and symptoms;
   - Awareness of first-aid procedures for, and the potential health effects of, heat stroke;
   - Employee responsibilities in avoiding heat stress;
   - Dangers of using drugs, including therapeutic ones, and alcohol in hot work environments;
   - Use of protective clothing and equipment; and
   - Purpose and coverage of environmental and medical surveillance programs and the advantages of worker participation in such programs.

3. Hot jobs should be scheduled for the cooler part of the day, and routine maintenance and repair work in hot areas should be scheduled for the cooler seasons of the year.

E. WORKER MONITORING PROGRAMS.

1. Every worker who works in extraordinary conditions that increase the risk of heat stress should be personally monitored. These conditions include wearing semipermeable or impermeable clothing when the temperature exceeds 21°C (69.8°F), working at extreme metabolic loads (greater than 500 kcal/hour), etc.

2. Personal monitoring can be done by checking the heart rate, recovery heart rate, oral temperature, or extent of body water loss.

3. To check the heart rate, count the radial pulse for 30 seconds at the beginning of the rest period. If the heart rate exceeds 110 beats per minute, shorten the next work period by one third and maintain the same rest period.

4. The recovery heart rate can be checked by comparing the pulse rate taken at 30 seconds (P1) with the pulse rate taken at 2.5 minutes (P3) after the rest break starts. The two pulse rates can be interpreted using Table III:4-4.

5. Oral temperature can be checked with a clinical thermometer after work but before the employee drinks water. If the oral temperature taken under the tongue exceeds 37.6°C, shorten the next work cycle by one third.

6. Body water loss can be measured by weighing the worker on a scale at the beginning and end of each work day. The worker's weight loss should not exceed 1.5% of total body weight in a work day. If a weight loss exceeding this amount is observed, fluid intake should increase.

F. OTHER ADMINISTRATIVE CONTROLS. The following administrative controls can be used to reduce heat stress:

- Reduce the physical demands of work, e.g., excessive lifting or digging with heavy objects;
- Provide recovery areas, e.g., air-conditioned enclosures and rooms;
- Use shifts, e.g., early morning, cool part of the day, or night work;
- Use intermittent rest periods with water breaks;
- Use relief workers;
- Use worker pacing; and
- Assign extra workers and limit worker occupancy, or the number of workers present, especially in confined or enclosed spaces.

TABLE III:4-4. HEART RATE RECOVERY CRITERIA

Heart rate recovery pattern	***P_3***	***Difference between P_1 and P_3***
Satisfactory recovery	<90	--
High recovery (Conditions may require further study)	90	10
No recovery (May indicate too much stress)	90	<10

VI. PERSONAL PROTECTIVE EQUIPMENT.

A. REFLECTIVE CLOTHING, which can vary from aprons and jackets to suits that completely enclose the worker from neck to feet, can stop the skin from absorbing radiant heat. However, since most reflective clothing does not allow air exchange through the garment, the reduction of radiant heat must more than offset the corresponding loss in evaporative cooling. For this reason, reflective clothing should be worn as loosely as possible. In situations where radiant heat is high, auxiliary cooling systems can be used under the reflective clothing.

B. AUXILIARY BODY COOLING.

1. Commercially available ice vests, though heavy, may accommodate as many as 72 ice packets, which are usually filled with water. Carbon dioxide (dry ice) can also be used as a coolant. The cooling offered by ice packets lasts only 2 to 4 hours at moderate to heavy heat loads, and frequent replacement is necessary. However, ice vests do not encumber the worker and thus permit maximum mobility. Cooling with ice is also relatively inexpensive.

2. Wetted clothing is another simple and inexpensive personal cooling technique. It is effective when reflective or other impermeable protective clothing is worn. The clothing may be wetted terry cloth coveralls or wetted two-piece, whole-body cotton suits. This approach to auxiliary cooling can be quite effective under conditions of high temperature and low humidity, where evaporation from the wetted garment is not restricted.

3. Water-cooled garments range from a hood, which cools only the head, to vests and "long johns," which offer partial or complete body cooling. Use of this equipment requires a battery-driven circulating pump, liquid-ice coolant, and a container.

Although this system has the advantage of allowing wearer mobility, the weight of the components limits the amount of ice that can be carried and thus reduces the effective use time. The heat transfer rate in liquid cooling systems may limit their use to low-activity jobs; even in such jobs, their service time is only about 20 minutes per pound of cooling ice. To keep outside heat from melting the ice, an outer insulating jacket should be an integral part of these systems.

4. Circulating air is the most highly effective, as well as the most complicated, personal cooling system. By directing compressed air around the body from a supplied air system, both evaporative and convective cooling are improved. The greatest advantage occurs when circulating air is used with impermeable garments or double cotton overalls.

One type, used when respiratory protection is also necessary, forces exhaust air from a supplied-air hood ("bubble hood") around the neck and down inside an impermeable suit. The air then escapes through openings in the suit. Air can also be supplied directly to the suit without using a hood in three ways:

- by a single inlet;
- by a distribution tree; or
- by a perforated vest.

In addition, a vortex tube can be used to reduce the temperature of circulating air. The cooled air from this tube can be introduced either under the clothing or into a bubble hood. The use of a vortex tube separates the air stream into a hot and cold stream; these tubes also can be used to supply heat in cold climates. Circulating air, however, is noisy and requires a constant source of compressed air supplied through an attached air hose.

One problem with this system is the limited mobility of workers whose suits are attached to an air hose. Another is that of getting air to the work area itself. These systems should therefore be used in work areas where workers are not required to move around much or to climb. Another concern with these systems is that they can lead to dehydration. The cool, dry air feels comfortable and the worker may not realize that it is important to drink liquids frequently.

C. RESPIRATOR USAGE. The weight of a self-contained breathing apparatus (SCBA) increases stress on a worker, and this stress contributes to overall heat stress. Chemical protective clothing such as totally encapsulating chemical protection suits will also add to the heat stress problem.

VII. BIBLIOGRAPHY.

American Conference of Governmental Industrial Hygienists (ACGIH). 1990. *Documentation of the Threshold Limit Values and Biological Exposure Indices*. 6th ed. Cincinnati: American Conference of Governmental Industrial Hygienists.

American Conference of Governmental Industrial Hygienists (ACGIH). 1992. 1992-*1993 Threshold Limit Values for Chemical Substances and Physical Agents and Biological Exposure Indices.* Cincinnati: American Conference of Governmental Industrial Hygienists.

American Industrial Hygiene Association (AIHA). 1975. *Heating and Cooling for Man in Industry*. 2nd ed. Akron, OH: American Industrial Hygiene Association.

Electric Power Research Institute (EPRI). 1987. *Heat-Stress Management Program for Nuclear Power Plants*. Palo Alto, CA: Electric Power Research Institute.

Eastman Kodak Company. 1983. *Ergonomic Design for People at Work.* Vol. II. Belmont, CA: Lifetime Learning Publications.

National Institute for Occupational Safety and Health. *Criteria for a Recommended Standard--Occupational Exposure to Hot Environments*. DHHS (NIOSH) Publication No. 86-113, April 1986.

National Institute for Occupational Safety and Health. *Occupational Safety and Health Guidance Manual for Hazardous Waste Site Activities*. DHHS (NIOSH) Publication No. 85-115, 1985.

National Institute for Occupational Safety and Health. *Standards for Occupational Exposures to Hot Environments: Proceedings of a Symposium*. DHHS (NIOSH) Publication No. 76-100, January 1976.

National Institute for Occupational Safety and Health. *Working in Hot Environments*. DHHS (NIOSH) Publication No. 86-112. Revised 1986.

National Safety Council. 1985. *Pocket Guide to Heat Stress*. Chicago, IL: National Safety Council.

Ramsey, J. D., Buford, C. L., Beshir, M.Y., and Jensen, R .C. *Effects of Workplace Thermal Conditions on Safe Work Behavior*. Journal of Safety Research 14:105-114, 1983.

Zenz, C. 1988. *Occupational Medicine: Principles and Practical Applications*. 2nd ed. St. Louis, MO: Mosby Year Book, Inc.

APPENDIX III:4-1. HEAT STRESS: GENERAL WORKPLACE REVIEW.

NOTE: Listed below are sample questions that the Compliance Officer may wish to consider when investigating heat stress in the workplace.

WORKPLACE DESCRIPTION.

A. Type of business

B. Heat-producing equipment or processes used

C. Previous history (if any) of heat-related problems

D. At "hot" spots:

- Is the heat steady or intermittent?
- Number of employees exposed?
- For how many hours per day?
- Is potable water available?
- Are supervisors trained to detect/evaluate heat stress symptoms?

ARE EXPOSURES TYPICAL FOR A WORKPLACE IN THIS INDUSTRY?

A. Weather at Time of Review

- Temperature
- Humidity
- Air velocity

B. Is Day Typical of Recent Weather Conditions? (Get information from the Weather Bureau)

C. Heat-Reducing Engineering Controls

- Ventilation in place?
- Ventilation operating?
- Air conditioning in place?
- Air conditioning operating?

- Fans in place?
- Fans operating?
- Shields or insulation between sources and employees?
- Are reflective faces of shields clean?

WORK PRACTICES TO DETECT, EVALUATE, AND PREVENT OR REDUCE HEAT STRESS.

A. Training program?

- Content?
- Where given?
- For whom?

B. Liquid replacement program?

C. Acclimatization program?

D. Work/rest schedule?

E. Scheduling of work (during cooler parts of shift, cleaning and maintenance during shut-downs, etc.)

F. Cool rest areas (including shelter at outdoor work sites)?

G. Heat monitoring program?

H. Personal Protective Equipment

- Reflective clothing in use?
- Ice and/or water-cooled garments in use?
- Wetted undergarments (used with reflective or impermeable clothing) in use?
- Circulating air systems in use?

I. First Aid Program

- Trained personnel?
- Provision for rapid cool-down?
- Procedures for getting medical attention?
- Transportation to medical facilities readily available for heat stroke victims?

J. Medical Screening and Surveillance Program

- Content?
- Who manages program?

K. Additional Comments

(Use additional pages as needed.)

APPENDIX III: 4-2. HEAT STRESS-RELATED ILLNESS OR ACCIDENT FOLLOW-UP.

A. Describe events leading up to the episode.

B. Evaluation/comments by other workers at the scene.

- Work at time of episode (heavy, medium, light)?
- How long was affected employee working at site prior to episode?
- Medical history of affected worker, if known.

C. Appropriate engineering controls in place?

D. Appropriate engineering controls in operation?

E. Appropriate work practices used by affected employee(s)?

F. Appropriate personal protective equipment available?

G. Appropriate personal protective equipment in use?

H. Medical screening for heat stress and continued surveillance for signs of heat stress given other employees?

I. Additional comments regarding specific episode(s): (Use additional pages as needed.)

APPENDIX III: 4-3. MEASUREMENT OF WET BULB GLOBE TEMPERATURE.

Measurement is often required of those environmental factors that most nearly correlate with deep body temperature and other physiological responses to heat. At the present time, the Wet Bulb Globe Temperature Index (WBGT) is the most used technique to measure these environmental factors. WBGT values are calculated by the following equations:

EQUATION III: 4-4. INDOOR OR OUTDOOR WET BULB GOBE TEMPERATURE INDEXES (WBGI)

Indoor or outdoors with no solar load

$$WBGT = 0.7NWB + 0.3GT$$

Outdoors with solar load

$$WBGT = 0.7NWB + 0.2GT + 0.1DB$$

where:

where:	WBGT	=	Wet Bulb Globe Temperature Index
	NWB	=	Natural Wet-Bulb Temperature
	DB	=	Dry-Bulb (air) Temperature
	GT	=	Globe Thermometer Temperature

The determination of WBGT requires the use of a black globe thermometer, a natural (static) wet-bulb thermometer, and a dry-bulb thermometer. The measurement of environmental factors shall be performed as follows:

1. The range of the dry and the natural wet-bulb thermometers should be -5°C to +50°C, with an accuracy of ±0.5°C. The dry bulb thermometer must be shielded from the sun and the other radiant surfaces of the environment without restricting the airflow around the bulb. The wick of the natural wet bulb thermometer should be kept wet with distilled water for at least one-half hour before the temperature reading is made. It is

not enough to immerse the other end of the wick into a reservoir of distilled water and wait until the whole wick becomes wet by capillarity. The wick must be wetted by direct application of water from a syringe one-half hour before each reading. The wick must cover the bulb of the thermometer and an equal length of additional wick must cover the stem above the bulb. The wick should always be clean, and new wicks should be washed before using.

2. A globe thermometer, consisting of a 15 cm (6-inch) in diameter hollow copper sphere painted on the outside with a matte black finish, or equivalent, must be used. The bulb or sensor of a thermometer (range -5°C to +100°C with an accuracy of ±0.5°C) must be fixed in the center of the sphere. The globe thermometer should be exposed at least 25 minutes before it is read.

3. A stand should be used to suspend the three thermometers so that they do not restrict free air flow around the bulbs and the wet-bulb and globe thermometer are not shaded.

4. It is permissible to use any other type of temperature sensor that gives a reading similar to that of a mercury thermometer under the same conditions.

5. The thermometers must be placed so that the readings are representative of the employee's work or rest areas, as appropriate.

Once the WBGT has been estimated, employers can estimate workers' metabolic heat load (see Tables III:4-1 and III:4-2) and use the ACGIH method to determine the appropriate work/rest regimen, clothing, and equipment to use to control the heat exposures of workers in their facilities.

SECTION III: CHAPTER 5

NOISE MEASUREMENT

Contents:

I. DEFINITIONS.

A. THRESHOLD AND CRITERION LEVELS.

1. The threshold level is the A-weighted sound level at which a personal noise dosimeter begins to integrate noise into a measured exposure. For example, if the threshold level on a sound level meter is set at 80 decibels (dB), it will capture and integrate into the computation of dose all noise in the employee's hearing zone that exceeds 80 dB. Sound levels below this threshold would not be included in the computation of noise dose.

2. The criterion level is the continuous equivalent A-weighted sound level that constitutes 100% of an allowable noise exposure. In other words, the criterion level is the permissible exposure limit (PEL). For OSHA purposes, this is 90 decibels, averaged over an 8-hour period on the A scale of a standard sound level meter set on slow response. Noise measurements taken with an instrument set on the A weighting scale are expressed as dBA.

3. Paragraphs 29 CFR 1910.95(a) and (b) of the OSHA Occupational Noise Exposure Standard date back to the 1969 Walsh-Healey Act. Because this early standard predated noise dosimetry and OSHA had no instructions for taking noise measurements, the first dosimeters that were developed used 90 dBA both as the threshold and criterion levels.

4. Paragraph 1910.95(c) of the 1983 Hearing Conservation Amendment to the Occupational Noise Exposure Standard requires employers to administer a continuing, effective hearing conservation program for all employees whose noise exposures equal or exceed an 8-hour TWA (time-weighted average) of 85 dBA or, equivalently, a noise dose that is equal to 50 percent of the PEL. The standard requires that all continuous, intermittent, and impulsive sound levels from 80 dB to 130 dB be included in the measurement of dose. In other words, the threshold level for noise measurement purposes is 80 dB.

5. Differences between sound-measuring instruments must be taken into account when measuring employee noise exposure. For example, because a dosimeter with an 80 dBA threshold integrates all noise above 80 dB into the dose, such a dosimeter will report a higher noise dose than the dose reported by a dosimeter with a 90 dBA threshold if both instruments are used side-by-side to evaluate the same noise exposure (see Table III:5-1).

6. Dosimeters can be used to calculate both the continuous equivalent A-weighted sound level (LA) and the 8-hour TWA for the time period sampled, using the following formulas:

$$L_A = 16.61 \log_{10} \frac{D}{12.5t} + 90$$

EQUATION III: 5-1a LA SOUND CALCULATION

$$TWA = 16.61 \log_{10} \frac{D}{100} + 90$$

EQUATION III: 5-1b TWA SOUND CALCULATION

where:			
	LA	=	the continuous equivalent A-weighted sound level in decibels for the time period sampled
	D	=	the dosimeter readout in percent noise dose
	t	=	the sampling time in hours
	TWA	=	the 8-hour time-weighted average in decibels

Equation III:5-1b is used for enforcement purposes, and Equation III:5-1a can be used to assist in evaluating hearing protectors and engineering controls.

B. EXCHANGE RATE.

1. The exchange rate is the increase or decrease in decibels corresponding to twice (or half) the noise dose. This means that the sound level at 90 dB is twice that at 85 dB (assuming that duration is held constant). The OSHA exchange rate is 5 dB (see Table D-2 of the construction noise standard, 29 CFR 1926.52, and Tables G-16 and G-16a of the general industry noise standard, 29 CFR 1910.95).

2. Only instruments using a 5 dB exchange rate may be used for OSHA compliance measurements. CSHO's should be aware that noise dosimeters used by the Department of Defense use a 4 dB exchange rate, while instruments used by the Environmental Protection Agency and most foreign governments use a 3 dB exchange rate. Additionally, the ACGIH Physical Agents Threshold Limit Values (TLV) Committee recently revised its noise TLV to also use the 3 dB exchange rate.

3. The hypothetical exposure situations shown in Table III:5-1 illustrate the relationship between criterion level, threshold, and exchange rate and show the importance of using a dosimeter with an 80 dBA threshold to characterize an employee's noise exposure. For example, an instrument with a 90 dBA threshold will not capture any noise below that level, and will thus give a readout of 0% even if the employee being measured is actually being exposed to 89 dBA for 8 hours (i.e., to 87% of the allowable noise dose over any 8-hour period).

TABLE III: 5-1. DOSIMETER READOUT, IN PERCENT OF MEASURED DOSE

Exposure conditions	Dosimeter with threshold set at 90 dBA	Dosimeter with threshold set at 80 dBA
90 dBA for 8 hours	100.0%	100.0%
89 dBA for 8 hours	0.0%	87.0%
85 dBA for 8 hours	0.0%	50.0%
80 dBA for 8 hours	0.0%	25.0%
79 dBA for 8 hours	0.0%	0.0%
90 dBA for 4 hours plus 80 dBA for 4 hours	50.0%	62.5%
90 dBA for 7 hours plus 89 dBA for 1 hour	87.5%	98.4%
100 dBA for 2 hours plus 89 dBA for 6 hours	100.0%	165.3%

* Assumes 5 dB exchange rate, 90 dBA PEL, ideal threshold activation, and continuous sound levels.

II. EFFECTS.

A. AUDITORY EFFECTS.

1. Chronic noise-induced hearing loss is a permanent sensorineural condition that cannot be treated medically. It is initially characterized by a declining sensitivity to high-frequency sounds, usually at frequencies above 2,000 Hz.

2. Exposure of a person with normal hearing to workplace noise at levels equal to or exceeding the PEL may cause a shift in the worker's hearing threshold. Such a shift is called a standard (or significant) threshold shift and is defined as a change in hearing thresholds of an average 10 dB or more at 2,000, 3,000, and 4,000 Hz in either ear. Workers experiencing significant threshold shifts are required by 29 CFR 1910.95(g)(8) to be fitted with hearing protectors and to be trained in their use.

B. EXTRA-AUDITORY EFFECTS. In addition to effects on hearing, noise:

- Interferes with speech;
- Causes a stress reaction;
- Interferes with sleep;
- Lowers morale;
- Reduces efficiency;
- Causes annoyance;
- Interferes with concentration; and
- Causes fatigue.

III. INSTRUMENT PERFORMANCE.

A. EFFECTS OF THE ENVIRONMENT. Temperature, humidity, atmospheric pressure, wind, and dust can all affect the performance of noise-measuring instruments and their readings. Magnetic fields can also affect the performance of instruments. Each of these factors is discussed below.

1. Temperature. Sound-measuring equipment should perform within design specifications over an ambient temperature range of -20°F to 140°F (-29°C to 60°C). If the temperature at the measurement site is outside this

range, refer to the manufacturer's specifications to determine if the sound level meter or dosimeter is capable of performing properly.

Sound-measuring instruments should not be stored in automobiles during hot or cold weather because this may cause warm-up drift, moisture condensation, and weakened batteries, all of which can affect instrument performance.

2. Humidity. OSHA noise instruments will perform accurately as long as moisture does not condense or deposit on the microphone diaphragm. If excessive moisture or rain is a problem in a given exposure situation, the Assistant Regional Administrator (ARA) for Technical Support should be consulted.

3. Atmospheric Pressure. Both atmospheric pressure and temperature affect the output of sound level calibrators; atmospheric pressure is the more important of these two factors. When checking an acoustical calibrator, always apply the corrections for atmospheric pressure that are specified in the manufacturer's instruction manual.

   In general, if the altitude of the measurement site is less than 10,000 feet above sea level, no pressure correction is needed. If the measurement site is at an altitude higher than 10,000 feet, or if the site is being maintained at greater-than-ambient pressure (e.g., in underwater tunnel construction), use the following equation to correct the instrument reading:

$$C = 10\log\left[\left(\frac{460+t}{528}\right)^{0.5}\frac{30}{B}\right]$$

EQUATION III: 5-2. AIR PRESSURE CORRECTION

where:	C	=	correction, in decibels, to be added to or subtracted from the measured sound level
	t	=	temperature in degrees Fahrenheit
	B	=	barometric pressure in inches of mercury

NOTE: For high altitude locations, C will be positive; in hyperbaric conditions, C will be negative.

4. Wind or Dust. Wind or dust blowing across the microphone of the dosimeter or sound level meter produces turbulence, which may cause a positive error in the measurement. A wind screen should be used for all outdoor measurements and whenever there is significant air movement or dust inside a building (e.g. when cooling fans are in use or wind is gusting through open windows).

5. Magnetic Fields. Certain equipment and operations, such as heat sealers, induction furnaces, generators, transformers, electromagnets, arc welding, and radio transmitters generate electromagnetic fields that can induce current in the electronic circuitry of sound level meters and noise dosimeters and cause erratic readings. If sound level meters or dosimeters must be used near such devices or operations, the extent of the field's interference should be determined by consulting the manufacturer's instructions.

B. EFFECTS OF SOUND. For sound level meters and noise dosimeters equipped with omnidirectional microphones, the effects of microphone placement and orientation are negligible in a typically reverberant environment. If the measurement site is nonreverberant and/or the noise source is highly directional, the manufacturer's literature should be consulted to determine proper microphone placement and orientation.

   For determining compliance with the impulse noise provision of 29 CFR 1910.95(b)(1) or 29 CFR 1926.52(e), the unweighted peak mode setting of the GenRad 1982 or 1983, Quest 155, or equivalent impulse precision sound level meter should be used.

IV. NOISE MEASUREMENTS.

A. INSTRUMENTS. Several sound measuring instruments are available to CSHO's. These include noise dosimeters, sound level meters, and octave-band analyzers. The uses and limitations of each kind of instrument are discussed below.

1. Noise Dosimeter. The noise dosimeters used by OSHA meet the American National Standards Institute (ANSI) Standard S1.25-1978, "Specifications for Personal Noise Dosimeters," which set performance and accuracy tolerances. For OSHA use, the dosimeter must have a 5 dB exchange rate, use a 90 dBA criterion level, be set at slow response, and use either an 80 dBA or 90 dBA threshold gate, whichever is appropriate for the evaluation.

2. Sound Level Meter.

a. All sound level meters used by OSHA meet ANSI Standard S1.4-1971 (R1976) or S1.4-1983, "Specifications for Sound Level Meters," which set performance and accuracy tolerances. Sound level meters may be used for five purposes:

- To spot-check noise dosimeter performance;
- To determine an employee's noise dose whenever a noise dosimeter is unavailable or inappropriate;
- To identify and evaluate individual noise sources for abatement purposes;
- To aid in the determination of the feasibility of engineering controls for individual noise sources for abatement purposes; and
- To evaluate hearing protectors.

b. For practical purposes, this procedure should be followed for all sound level measurements:

- The microphone should be in the monitored employee's hearing zone. OSHA defines the hearing zone as a sphere with a two-foot diameter surrounding the head.
- Considerations of practicality and safety will dictate the actual microphone placement at each survey location.
- When noise levels at an employee's two ears are different, the higher level must be used for compliance determinations.

Sound level readings in a nonreverberant environment should be taken in accordance with the manufacturer's instructions.

3. Octave-Band Noise Analyzers.

a. The Type 1 sound level meters (such as the GenRad 1982 and 1983 and the Quest 155) used by OSHA have built-in octave band analysis capability. These devices can be used to determine the feasibility of controls for individual noise sources for abatement purposes and to evaluate hearing protectors.

b. Octave-band analyzers segment noise into its component parts. Octave-band filter sets provide filters with the following geometric mean frequencies: 31.5; 63; 125; 250; 500; 1,000; 2,000; 4,000; 8,000; and 16,000 Hz.

c. The special signature of a given noise can be obtained by taking sound level meter readings at each of these settings (assuming that the noise is fairly constant over time). The results indicate those octave-bands that contain the majority of the total radiated sound power.

d. Octave-band noise analyzers can assist CSHO's in determining the adequacy of various types of frequency-dependent noise controls. They also can be used to determine the efficacy of hearing protectors because they can measure the amount of attenuation offered by the protectors in the octave-bands responsible for most of the sound energy in a given situation.

e. (Section VI-B of this chapter discusses the evaluation of hearing protection using overall A-weighted exposures and the Noise Reduction Rating.) The Assistant Regional Administrator for Technical Support can provide assistance in the use of octave-band analyzers.

For compliance measurements, the A-weighted network and slow response setting shall be used exclusively. All continuous, intermittent, and impulse noise is included in the computation of noise dose for compliance with all OSHA noise standards.

NOTE: In general, for enforcement purposes a 90 dBA threshold dosimeter is necessary to establish noncompliance with Table G-16 or D-2. However, in unusual situations, results obtained with an 80-dBA threshold dosimeter may be discussed with the ARA for Technical Support to determine whether they sufficiently demonstrate noncompliance with Table G-16 or D-2. (See Table III:5-3 for exposures exceeding 230%.)

B. ACCURACY. For compliance purposes, readings with an ANSI Type 2 sound level meter and readings with a noise dosimeter are considered to have an accuracy of ±2 dBA. Readings with D-2 and ANSI Type 1 sound level meter are considered to have an accuracy of ±1 dBA. For unusual measurement situations, refer to the manufacturer's instructions and appropriate ANSI standards for guidance in interpreting instrument accuracy.

C. CALIBRATION. Calibrate all noise-measuring instruments according to the manufacturer's instructions before and after each day of use and whenever the temperature or relative humidity changes significantly.

D. SAMPLING STRATEGY. The results of the walk-around survey are used to assist the CSHO in planning the workshift noise sampling strategy.

1. Walk-around Survey. When screening for noise exposures, sound level meter measurements and estimates of exposure duration are sufficient. The resulting spot readings can be used to determine the need for a more complete evaluation. An integrating sound level meter may be useful for this purpose.

2. Workshift Sampling.

a. If the results of the walk-around survey indicate that OSHA's noise standard may be exceeded, the noise exposures of representative employees from each job classification potentially overexposed should be sampled.

b. The CSHO should use a dosimeter with a threshold of 90 dBA as well as one with a threshold of 80 dBA to measure the noise exposure of employees identified as potentially overexposed during the walk-around.

c. The CSHO shall use the 80 dBA threshold dosimeter to measure the noise dose of those employees identified during the walk-around as having noise exposures that are in compliance with Table G-16 of 29 CFR 1910.95 but whose exposure may exceed the levels specified in Table G-16a. The CSHO shall use the dosimeter with a threshold of 90 dBA to measure the noise dose of those employees identified during the walk-around as having potential noise exposures that exceed the sound levels in Table G-16 of 1910.95 or Table D-2 of 1926.52.

NOTE: 29 CFR 1926 does not contain the monitoring requirements of the hearing conservation amendment (29 CFR 1910.95, paragraph (c)). In other words, 29 CFR 1926 addresses only allowable noise doses.

d. As a minimum strategy, the CSHO shall conduct sampling for the time necessary to establish that an exposure above the limits permitted by Tables G-16, G-16a, or D-2 exists (in general industry or construction workplaces, respectively), taking instrument accuracy into account.

TABLE III:5-2. COMPARISON TABLE OF DURATION PER DAY IN HOURS TO ALLOWABLE SOUND LEVEL IN dBA (SLOW-RESPONSE SPL)

Duration per day, hours	*Sound level, dBA, slow-response*
8.0	90
6.0	92
4.0	95
2.0	100
1.0	105
0.5	110
0.25	115

Source: 29 CFR 1910.95, Table G-16.

E. SAMPLING PROTOCOL.

1. CSHO's evaluating employee exposures to workplace noise should:

    a. Inform the monitored employee that the dosimeter will not interfere with his/her normal duties, and emphasize that the employee should continue to work in a routine manner.

    b. Explain to each employee being sampled the purpose of the dosimeter, and emphasize that the dosimeter is not a speech recording device.

    c. Instruct the employee being sampled not to remove the dosimeter unless absolutely necessary and not to cover the microphone with a coat or outer garment. Inform the employee when and where the dosimeter will be removed. When the dosimeter is positioned (generally in the shirt pocket or at the waist), clip the microphone to the employee's shirt collar at the shoulder, close to the employee's ear. Care should be taken to ensure that the microphone is in the vertical position. Clips should be placed in accordance with the manufacturer's instructions.

    d. Position and secure any excess microphone cable to avoid snagging or inconveniencing the employee.

    e. Check the dosimeter periodically to ensure that the microphone is oriented properly.

    f. Obtain and note sound level meter readings during different phases of the work performed by the employee during the shift. Take enough readings to identify work cycles. For statistical reasons, more readings should be taken when noise levels fluctuate widely.

    g. Record the information required on the OSHA-92.

2. Some dosimeters indicate when a 115 dBA sound level has been exceeded. This indication is not to be used for compliance determination.

V. ULTRASONICS.

Ultrasound is high frequency sound that is inaudible to the human ear. The frequency audibility limit of the human ear is approximately 20 kHz. However, this limit is not fixed, and some individuals may have higher or lower (usually lower) limits. The frequency limit normally declines with age.

Most of the audible noise associated with ultrasonic sources (e.g., ultrasonic welders or ultrasonic cleaners) is actually comprised of subharmonics of the machine's fundamental ultrasonic frequencies. For example, many ultrasonic welders have a fundamental operating frequency of 20 kHz, a sound that is at the human frequency audibility limit. However, a good deal of noise is also present at 10 kHz, the first subharmonic frequency of the 20 kHz operating frequency, and is therefore audible to most persons.

TABLE III: 5-3. CONVERSION FROM PERCENT NOISE EXPOSURE OR DOSE TO 8-HOUR TIME-WEIGHTED AVERAGE SOUND LEVEL

Dose or percent	*TWA (dBA)*
.50	85.0
.55	85.7
.60	86.3
.65	86.9
.70	87.4
.75	87.9
.80	88.4
.85	88.8
.90	89.2
.95	89.6
100	90.0
105	90.4
110	90.7
115	91.1
120	91.3
125	91.6
Assumes 5 dB exchange rate and 90 dBA PEL.	

Source: 29 CFR 1910.95.

A. APPLICABILITY OF 29 CFR 1910.95. 29 CFR 1910.95, OSHA's Occupational Noise Exposure Standard, addresses airborne ultrasonic noise because of OSHA's criterion parameters. At 20 kHz, an A-weighted level is 10 dB below an unweighted sound pressure level. At 50 kHz, an A-weighted level is 25 dB below the corresponding unweighted level. In other words, if an A-weighted instrument is used to measure a 50 kHz, 110 dB tone, the instrument will indicate 85 dBA. [However, note that the ANSI S1.4-1976 or S1.4-1983 standard, which sets the performance and accuracy tolerances that all OSHA sound-level meters meet, requires a Type II microphone accuracy of only ±5 to 4 at frequencies above 10,000 Hz.]

B. HEALTH EFFECTS.

1. Research indicates that ultrasonic noise has little effect on general health unless there is direct body contact with a radiating ultrasonic source. Reported cases of headache and nausea associated with airborne ultrasonic exposures appear to have been caused by high levels of audible noise from source subharmonics. The American Conference of Governmental Industrial Hygienists (ACGIH) has established permissible ultrasound exposure levels (Table III:5-4). The latest edition of the ACGIH publication, 1998 Threshold Limit Values (TLV's) and Biological Exposure Indices (BEI's), adopted TLV's that had been proposed in the 1997 publication as a "nature of intended changes." Therefore, Table III:5-4 represents the newly adopted TLV's for ultrasound. These recommended limits (set at the middle frequencies of the one-third octave bands from 10 kHz to 50 kHz) are designed to prevent possible hearing loss caused by the subharmonics of the set frequencies rather than the ultrasonic sound itself.

2. To measure any source suspected of producing sound at levels exceeding the ACGIH recommended limits requires the use of a precision sound level meter equipped with a suitable microphone of adequate frequency

response and a portable third-octave filter set. Consult with the Assistant Regional Administrator for Technical Support for guidance.

TABLE III:5-4. TLV's FOR ULTRASOUND

Notice of Intended Change - Ultrasound

These TLV's represent conditions under which it is believed that nearly all workers may be repeatedly exposed without adverse effect on their ability to hear and understand normal speech. Previous TLV's for the frequencies 10 kilohertz (kHz) to 20 kHz, set to prevent subjective effects, are referenced in a cautionary note below (1). The 8-hour TWA values are an extension of the TLV for Noise, which is an 8-hour TWA of 85 dBA for sound below 10 kHz. The ceiling values may be verified by using an integrating sound level meter with slow detection and 1/3 octave bands. All instrumentation should have adequate frequency response and should meet the specifications of ANSI S1.4-1983 and IEC 804.

	One-third Octave-Band Level		
	Measured in Air in dB re: 20 µ Pa; Head in Air		***Measured in Water in dB re: 1 µ Pa; Head in Water***
Mid-Frequency of Third-Octave Band (kHz)	***Ceiling Values***	***8-Hour TWA***	***Ceiling Values***
10	105[1]	88	167
12.5	105[1]	89	167
16	105[1]	92	167
20	105[1]	94	167
25	110[2]	--	172
31.5	115[2]	--	177
40	115[2]	--	177
50	115[2]	--	177
63	115[2]	--	177
80	115[2]	--	177
100	115[2]	--	177

1. Subjective annoyance and discomfort may occur in some individuals at levels between 75 and 105 dB for the frequencies from 10 kHz to 20 kHz, especially if they are tonal in nature. Hearing protection or engineerin controls may be needed to prevent subjective effects. Tonal sounds in frequencies below 10 kHz might also need to be reduced to 80 dB.

2. These values assume that human coupling with water or other substrate exists. These thresholds may be raised by 30 dB when there is no possibility that the ultrasound can couple with the body by touching water o some other medium. [When the ultrasound source directly contacts the body, the values in the table do not apply. The vibration level at the mastoid bone must be used.] Acceleration Values 15 dB above the reference of 1g RMS should be avoided by reduction of exposure or isolation of the body from the coupling source. (g = acceleration due to the force of gravity, 9.80665 meters/second: RMS = root-mean-square.)

Source: ACGIH, 1997, p.81.

C. CONTROLS.

1. High frequency noise is very directional and is relatively easily reflected or blocked by any type of barrier. The wavelength of a 16 kHz tone, for example, is about ¾ inch, so a barrier of 1 to 2 inches is generally sufficient to

reflect noise of approximately the same frequency away from a nearby worker. Such barriers are inherent in some of the machine surfaces themselves.

2. High frequency audible noise is also easily absorbed by any of the so-called acoustical materials (e.g., glass fiber or foam). Moreover, higher frequency sound is absorbed readily by ambient air, which means that any high frequency noise problem is almost certain to be a very local problem in any workplace layout.

VI. GENERAL NOISE INSPECTION DATA.

A. INFORMATION TO BE COLLECTED. In addition to the information required for a health inspection, collect the following information where pertinent and where necessary to document a violation of 29 CFR 1910.95, 1926.52, or 1926.101, as follows.

1. Employee Data.

- Distance from the employee to the primary noise source(s);
- Need for the employee to be present or in close proximity to the noise field;
- Employee exposure time pattern;
- Existence of any known employee auditory problems (e.g., ear infections, ringing in the ears, or trouble with hearing after the work shift);
- Employee's opinion of the practicality of potential noise controls (considering machine operation), where relevant;
- Hearing protection provided and any problems with its use or acceptance. (Octave band levels and/or dBC and dBA data may aid in the evaluation of hearing protectors.); and
- Time since the last audiometric examination and frequency of such examinations.

2. Machine and/or Process Data.

- Type of machine, and a brief description of it and/or the process, including identifying numbers, sketches, and photographs whenever possible;
- Condition of the machine (e.g., age and maintenance status);
- Machine operation (e.g., speed, cycle times, parts/minute, and materials used);
- Apparent existing noise and/or vibration controls;
- Source(s) and characteristics of the noise (e.g., fan noise--discrete and broad band components, continuous or noncontinuous). (Octave band analyzers, real time analyzers, and narrow band analyzers may be useful in determining sources of noise.); and
- Practical engineering and/or administrative controls and estimated costs of such controls.

3. Building Data.

- Size and shape of the room;
- Layout of equipment, work stations and break areas;
- Surface materials (e.g., ceiling/steel; walls/cinderblock; floor/concrete);
- Existing acoustical treatment;
- Potential acoustical treatment;
- Noise from other sources (spill-over noise); and
- Presence of barriers or enclosures.

4. Employer Data.

- What has been done to control the noise (e.g., have consultants been used, is plant noise monitored, and are controls implemented)?
- What is planned for the future?
- Are administrative controls utilized? How are they enforced?

5. Hearing Loss Data.

   Document the following when hearing loss is used to support a citation:

   - The amount of the threshold shift and date it was recorded;
   - Employee's exposure level;
   - Frequency and duration of employee's exposure;
   - Length of employment;
   - Explanation of any follow-up measures taken; and
   - Duration of audiometric testing program.

B. EVALUATION OF HEARING PROTECTION.

The actual effectiveness of any individual hearing protector cannot be determined under workplace conditions; however, OSHA's noise standards (29 CFR 1910.95(j)(2) and 1926.52(b)) require that personal hearing protection be worn to attenuate the occupational noise exposure of employees to within the limits shown in Tables G-16, G-16a, and D-2, respectively. Hearing protectors are evaluated under rigorous laboratory conditions specified by the American National Standards Institute in ANSI Z24.22-1957 (R1971) and ANSI S3.19-1974. However, OSHA's experience and the published scientific literature indicate that laboratory-obtained real ear attenuation for hearing protectors can seldom be achieved in the workplace.

1. 29 CFR 1910.95(c)-(p). Appendix B of 29 CFR 1910.95 provides information on how to determine the adequacy of hearing protector attenuation using the noise reduction rating (NRR) of a given hearing protector.

2. Field Attenuation of Hearing Protection. To estimate the attenuation afforded to a noise-exposed employee in an actual work environment by muffs, plugs, or a combination of both, proceed as follows.

   a. For muffs or plugs:

      - Determine the laboratory-based noise attenuation afforded by the given hearing protector. This is based on laboratory testing using human subjects and is provided by the manufacturer. It is referred to as the Noise Reduction Rating or NRR and should be on the packaging.
      - Subtract 7dB from the NRR to correct for using
      - A-weighted measurements.
      - To adjust for workplace conditions, applying a safety factor of 50% is recommended. This is because the field use of ear protectors does not afford the same degree of protection achieved in the laboratory using well-trained subjects under ideal test conditions.

   b. For dual protection (i.e. muffs and plugs):

      - Determine the laboratory-based noise attenuation (NRR) for the higher rated hearing protector, subtract 7dB, and (it is recommended that you) apply a safety factor of 50%.
      - Then add 5 dB to the field-adjusted NRR to account for the use of the second hearing protector.

Figure III:5-1. Case Scenarios, gives examples of hearing protector attenuation calculations.

VII. EVALUATION OF THE HEARING CONSERVATION PROGRAM.

When employees are exposed to sound levels that exceed the limits shown in Tables D-2 of 29 CFR 1926.52 or G-16a of 29 CFR 1910.95, as appropriate, a continuing effective hearing conservation program is required. The program must include all employees whose work brings them, either routinely or intermittently, into areas in which the sound levels exceed those permitted by the appropriate table.

A. AGRICULTURAL WORKSITES. Since paragraph 29 CFR 1928.21(a) does not reference the general industry noise standard (29 CFR 1910.95), hearing conservation programs are not mandatory in agricultural operations. However, if the CSHO inspects such operations and determines that they are likely to cause employees to be exposed to noise in excess of an 8-hour, time-weighted average of 85 dBA, the employer should be advised that it is good practice to reduce the noise level or provide ear protection and to train employees in the proper use and fit of ear protection and in the hazards of noise exposure. Wherever it is practical, periodic audiometric testing should be encouraged to ensure the effectiveness of hearing protection.

The Hearing Conservation Amendment (HCA), does not require employers to apply a 50% safety factor to the Noise Reduction Rating (NRR) when evaluating hearing protector adequacy but it is recommended. Therefore, to determine the protected TWA for a given muff or plug, obtain the Noise Reduction Rating (NRR), and subtract 7 dB from the NRR for spectral uncertainty pursuant to Appendix B of 1910.95. **Note:** When the TWA exposure is less than 100dB, hearing protectors may be used in lieu of engineering controls if certain conditions are met as referenced in CPL 2-2.35A, Appendix A.

The following show how the safety factor is used:

Example 1: TWA_8 = 100 dBA and Muff NRR = 19 dB
Approximate Field Attenuation is: (19-7) × 50% = 6 dB
100 dBA TWA_8 - 6 dB = 94 dBA

Conclusion: The protected TWA can be assumed to be 94 dBA. Feasible engineering controls must be implemented. Since the HCA does not require applying a 50% correction factor, the protected TWA is 88 dBA. Therefore, better hearing protection (with greater attenuation) is not required by the standard.

Example 2: Given TWA_8 = 98 dBA and Plug NRR = 29 dB
Approximate Field Attenuation is: (29-7) × 50% = 11 dB
98 dBA TWA_8 - 11 dB = 87 dBA

Conclusion: The protected TWA can be assumed to be 87 dBA. Unless the employee exhibits an STS, the employer meets one of the necessary conditions for being excused from the requirement to implement engineering controls. Better hearing protection (with greater attenuation) is not required since the protected TWA, without the safety factor, is 76 dBA. If an employee exhibits an STS, better hearing protection should be recommended.

Example 3: Given TWA_8 = 110 dBA, Plug NRR = 29 and Muff NRR = 25 dB
Employer requires dual protection.

- Calculate field-adjusted NRR for the higher-rated protector: (29-7) × 50% = 11 dB
- Add 5 dB to this field-adjusted NRR: 11+5 = 16 dB
- Calculate the protected TWA_8: 110 dBA TWA_8 - 16 dB = 94 dBA

Conclusion: The protected TWA_8 can be assumed to be 94 dBA. If feasible, engineering controls must be implemented. The protected TWA_8, without using the 50% safety factor, is 83 dBA. Therefore, better hearing protection is not required by the HCA. Hearing protectors with greater attenuation should be recommended.

FIGURE III: 5-1. CASE SCENARIOS

B. MARITIME WORKSITES. Shipyard and longshoring operations come under the requirements of the general industry noise standard; therefore, employers in such operations must meet the elements of the general industry Hearing Conservation Amendment (29 CFR 1910.95(c) through (p)).

C. CONSTRUCTION WORKSITES. Construction employees are not covered by the Hearing Conservation Amendment. However, certain aspects of hearing conservation are covered by the construction noise standards (29 CFR 1926.52 and 1926.101). In evaluating hearing conservation programs in construction workplaces, CSHO's should consider the information in the following paragraphs.

1. Monitoring. In order for the employer to know who must wear hearing protection and to be able to evaluate the adequacy of the hearing protection used, the noise created by different machines and operations should be known. This information can be obtained by a variety of means, such as sound surveys, full-shift dosimetry, utilization of the OSHA-funded consultation program, use of insurance carriers, or reliance on information from equipment manufacturers and the technical literature. The sophistication of the monitoring performed will be limited by the size of the employer's business and the practicalities of the job site. Although lack of monitoring does not constitute a violation, it is indicative of a less than fully effective hearing conservation program.

2. Training. Training and motivation are crucial to the effectiveness of hearing protection and the hearing conservation program in general. Due to the transient work force in construction employment, training is much more difficult to provide than is the case in general industry or maritime operations. However, such training is required, as outlined in 29 CFR 1926.21(b)(2).

3. Hearing Protection. 29 CFR 1926.52(b) requires hearing protection to be provided and used to reduce the sound levels reaching the cochlea to within the limits specified in Table D-2. In addition, paragraph 1926.101(b) requires all insert hearing protectors to be fitted individually to each overexposed employee by a competent person. A competent person is defined as a person trained in ear protection fitting and one who is able to recognize the difference between a good fit and a poor fit.

4. Audiometric Testing. When employees are overexposed to noise, periodic audiometric testing is required. To the extent practicable, the only way to ensure that a hearing conservation program is a continuing, effective one is to perform periodic audiograms.

D. GENERAL INDUSTRY WORKSITES.

1. The Hearing Conservation Amendment (29 CFR 1910.95(c)-(p)) establishes the minimum requirements for a continuing, effective hearing conservation program. This program is mandated for all employees exposed at or above a noise dose of 50%.

2. The standard threshold shift (STS) is an indicator of a change in hearing threshold. As defined in 29 CFR 1910.95(g)(10), a standard threshold shift is "a change in hearing threshold relative to the baseline audiogram of an average of 10 dB or more at 2,000, 3,000, and 4,000 Hz in either ear." Figure III:5-2 shows how to compute standard threshold shifts.

VIII. CONTROL.

Noise control should minimize sources of noise; preclude the propagation, amplification, and reverberation of noise; and protect workers from excessive noise. Engineering controls include antivibration machine mountings, acoustical enclosures, and so forth. Administrative practices may require shift rotation or exposure limitation. Personal protective equipment such as ear muffs or ear plugs or both can be used.

IX. NEW DEVELOPMENTS IN HEARING PROTECTION LABELING.

The paragraphs below are intended to merely inform readers of developments in hearing protection since the OTM's original publication.

Example 1

Frequency (Hz)	*Baseline audiogram threshold (dB)*	*Annual audiogram threshold (dB)*	*Change*
500	5	5	0
1,000	5	5	0
2,000	0	10	+10
3,000	5	20	+15
4,000	10	35	+25
5,000	10	15	+5
6,000			

Considering the values for 2,000, 3,000, and 4,000 Hz, there are changes in hearing threshold of 10, 15, and 25 dB, respectively. Thus:

STS = (10 + 15 + 25)/3 = 50/3 = 16.7 dB

Conclusion: The STS is +16.7 dB; hearing has deteriorated, the employee must be notified in writing within 21 days, and, depending on professional discretion, the employer may elect to revise the baseline audiogram.

Example 2

Frequency (Hz)	*Baseline audiogram threshold (dB)*	*Annual audiogram threshold (dB)*	*Change*
500	5	5	0
1,000	5	0	-5
2,000	0	-10	-10
3,000	5	-5	-10
4,000	10	-5	-15
5,000	10	5	-5
6,000			

Again, considering the values for 2,000, 3,000, and 4,000 Hz, the hearing threshold is lowered by 10, 10, and 15 dB, respectively. Thus:

STS = (-10 -10 -15)/3 = -35/3 = -11.6 dB

Conclusion: The STS is -11.6 dB; hearing has improved, the employee should be notified, and, depending on professional discretion, the baseline audiogram may be revised.

FIGURE III: 5-2. COMPUTING THE STANDARD THRESHOLD SHIFT

A. BACKGROUND. When OSHA promulgated its Hearing Conservation Amendment in 1983, it incorporated the EPA labeling requirements for hearing protectors (40 CFR 211), which required manufacturers to identify the noise reduction capability of all hearing protectors on the hearing protector package. This measure is referred to as the noise reduction rating or NRR. It is a laboratory-derived numerical estimate of the attenuation that is provided by the protector. The standard also prescribes various methods to estimate the adequacy of hearing protection attenuation using the NRR. By December 1983, however, it became evident that the amount of protection users were receiving in the workplace with the prescribed hearing protectors did not correlate with the attenuation indicated by the NRR. OSHA acknowledged that in most cases, this number overstated the protection afforded to workers and issued a change in compliance policy in its compliance instruction CPL 2-2.35A, Appendix A, requiring the application of a safety factor of 50 percent to the NRR, above and beyond the 7 dB subtraction called for when using A-weighted measurements for certain circumstances (see section VI.B, Figure III:5-1).

Prompted by the concerns that hearing protection users may not be adequately protected from noise exposure, the National Hearing Conservation Association (NHCA) created a task force on hearing protector effectiveness. The task force was made up of professional organizations, government, industry, and accredited standards working groups. Their task was to develop new guidelines for the proper selection, use and labeling of hearing protectors. In 1997, while the task force was working on their recommended changes, ANSI published a new test method (subject-fit) for measuring the real ear attenuation of hearing protectors (ANSI S12.6-1997). This method may provide more representative estimates of the real world performance of hearing protectors. It is an attempt to better approximate the protection attained in real workplaces because the untrained subjects in this test method (the only instruction they had was the instruction that comes with the package) closely replicate real world users.

Some manufacturers of hearing protectors may be testing their products according to the subject-fit method of ANSI S12.6-1977 standard. You may contact the manufacturer for information they may have on products tested to this standard.

In the future, hearing protector manufacturers who voluntarily choose to test their product according to the subject-fit method may publish the protector's attenuation data.

B. TASK FORCE RECOMMENDED CHANGES AND OTHER RELEVANT PUBLICATIONS.

1. NHCA task-force recommendations on changes in hearing protector labeling and noise reduction rating methodologies based on ANSI S12.6-1997.

    a. For further information on the suggested labeling procedures contact the NHCA. Phone number: (303) 224-9022 Website: http://www.hearingconservation.org Email address: nhca@gwami.com

    b. Publication of NHCA task-force recommendation: Royster LH: Recommendations for the labeling of hearing protectors Sound and Vibration, 29(7), 16-19, 1995.

2. Subject-fit versus real-world data: Berger, E.H., Franks, J.R., Behar, A., Casali, J.G., Dixon-Ernst, C., Kieper, R.W., Merry, C.J., Mozo, B.T., Nixon, C.W., Ohlin, D., Royster, J.D., and Royster, L.H. (1998). "Development of a new standard laboratory protocol for estimating the field attenuation of hearing protection devices. Part III. The validity of using subject-fit data, "J. Acoust. Soc. Am. 103, 665-672.

3. Recommendations of the National Institute for Occupational Safety and Health in reference to the subject-fit method and the NRR(SF): CRITERIA FOR A RECOMMENDED STANDARD Occupational Noise Exposure Revised Criteria 1998 DHHS (NIOSH) Pub. No 98-126.

    For further information on this NIOSH Criteria Document, contact NIOSH.
    Phone number: 1 800 35 NIOSH Website: http://www.cdc.gov/niosh/homepage.html

X. BIBLIOGRAPHY.

Alberti, P.W. 1982. *Personal Hearing Protection in Industry*. New York: Raven Press.

Beranek, L.L. 1971. *Noise and Vibration Control*. New York: McGraw-Hill.

Berger, E.H., W.D. Ward, J.C. Morrill, and L.H. Royster. 1986. *Noise and Hearing Conservation Manual*. 4th ed. Akron, OH: American Industrial Hygiene Association.

Code of Federal Regulations. 1992. Title 29, Parts 1900 to 1990. Washington, D.C.: U.S. Government Printing Office.

Harris, C.M. 1957. *Handbook for Noise Control.* 2nd ed. New York: McGraw-Hill.

Harris, C.M. 1988. *Shock and Vibration Handbook.* 3rd ed. New York: McGraw-Hill.

Kryter, K.D. 1970. *The Effects of Noise on Man.* New York: Academic Press.

Kryter, K.D. 1985. *The Effects of Noise on Man. 2nd ed.* New York: Academic Press.

Olishifski, P.E. and E.R. Harford. 1975. *Industrial Noise and Hearing Conservation.* Chicago: National Safety Council.

OSHA *Field Operations Manual.* 1992. OSHA Instruction CPL 2.45B. Washington, D.C.: U.S. Government Printing Office.

Sataloff, J. and P. Michael. 1973. *Hearing Conservation.* Springfield, IL: Charles C Thomas.

Sataloff, R.T. and J.T. Sataloff. 1987. *Occupational Hearing Loss.* New York: Marcel Dekker.

APPENDIX III:5-1. EVALUATING NOISE EXPOSURE OF EMPLOYEES WEARING SOUND-GENERATING HEADSETS.

EMPLOYEES AT RISK.

Noise overexposure in the workplace can occur where employees wear a communications headset as part of their employment. Clerical personnel, aircraft pilots and other cockpit personnel, air traffic controllers, reservation clerks, receptionists, and telephone operators are just a few examples of the more than three million workers who can be exposed to high noise levels via communications headsets. For a person wearing a sound-generating headset, the sound/noise exists predominantly between the ear drum and the headset. Because of the amplification properties of the human ear, the sound that exists inside the ear while wearing a headset is quite different from ambient levels.

Probe microphones and similar devices allow sound levels to be measured inside the ear. Most people, however, find that inserting a probe microphone into their ear canal is uncomfortable and object to wearing a probe for an 8-hour workday. In addition, there is a possibility of damaging the ear drum, which means that the person inserting the probe requires careful training. For these reasons, probe microphones should not be used for compliance purposes.

METHODOLOGY.

A method of monitoring employee exposure without invading the ear canal has been developed. This sampling method evaluates the noise dose that an employee receives during the actual work day while wearing an insert type headset, a monaural or binaural muff, or a monaural or binaural foam headset. The technique measures the electrical signal input to the worker's headset using a conventional Bruel & KjaerTM noise dosimeter and a set of filters (audio equalizer). The filters take into account the electrical-to-acoustical conversion efficiency of the headset and the differences between the sound levels occurring in the ear canal and the free field. The method is safe, easy to operate, and convenient for field use. Current analysis indicates that the accuracy of the system is ±1.5 dB.

Prior to sampling, the actual headset or an identical model of the headset that is worn by the employee must be sent to the Salt Lake Technical Center to calibrate the equipment. The headset is placed on a head and ear simulator, which simulates the acoustic response of a median human ear. White noise is input to the headset and the transfer function for the headset is determined at all frequencies over the range of human hearing.

A digitally programmable one-third octave filter set (audio equalizer) is preprogrammed with the transfer functions obtained in the earlier laboratory headset evaluation. Employee monitoring can then be conducted in the workplace utilizing the OSHA sampling procedures outlined in Section II, Chapter 4. The input signal to the employee's headset is split, with one branch going into an audio equalizer attached to a Bruel & KjaerTM dosimeter, and the second branch going into the employee's headset. After the sampling period (which is normally an 8-hour workday), the employee's noise dose (in percent exposure) is recorded from the dosimeter. Any citations for employee overexposure shall be issued in accordance with procedures outlined in the OSHA Field Inspection Reference Manual.

ACOUSTIC LIMITING DEVICES.

Laboratory evaluations have determined that headsets can be categorized in two basic groups:

Those without any form of electronic limiting device. Those with some form of limiting device built into the headset.

Most modern telecommunication headsets use sophisticated limiting circuits, while headsets for special applications (e.g., Walkman™ head-sets) do not. Those headsets that contain acoustic limiting devices that are functioning as they were designed to do have been shown in both laboratory and field tests to provide sufficient protection to maintain employee noise exposures below OSHA permissible noise levels. However, headsets without limiting devices have, in some work environments, caused employee noise exposures to exceed the levels permitted by OSHA.

For more information, contact the OSHA Health Response Team at the Salt Lake Technical Center.

SECTION III: CHAPTER 6

LASER HAZARDS

Contents:

I. INTRODUCTION.

The term LASER is an acronym for Light Amplification by Stimulated Emission of Radiation. Light can be produced by atomic processes which generate laser light. A laser consists of an optical cavity, a pumping system, and an appropriate lasing medium (Figure III:6-1).

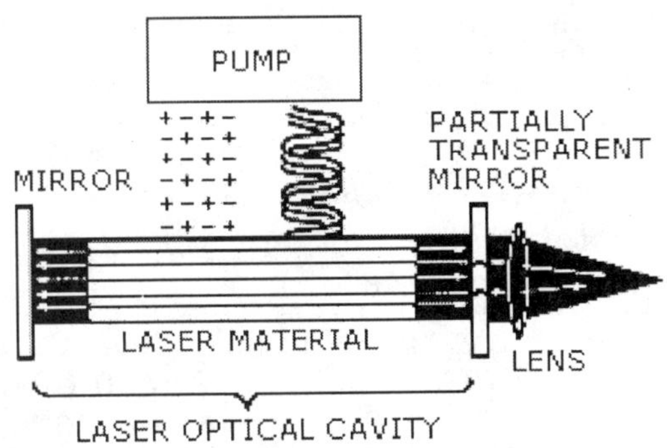

FIGURE III: 6-1 COMPONENTS OF A LASER

A. The optical cavity contains the media to be excited with mirrors to redirect the produced photons back along the same general path.

B. The pumping system uses photons from another source as a xenon gas flash tube (optical pumping) to transfer energy to the media, electrical discharge within the pure gas or gas mixture media (collision pumping), or relies upon the binding energy released in chemical reactions to raise the media to the metastable or lasing state.

C. The laser medium can be a solid (state), gas, dye (in liquid), or semiconductor. Lasers are commonly designated by the type of lasing material employed.

1. Solid state lasers have lasing material distributed in a solid matrix, e.g., the ruby or neodymium-YAG (yttrium aluminum garnet) lasers. The neodymium-YAG laser emits infrared light at 1.064 micrometers.

2. Gas lasers (helium and helium-neon, HeNe, are the most common gas lasers) have a primary output of a visible red light. CO2 lasers emit energy in the far-infrared, 10.6 micrometers, and are used for cutting hard materials.

3. Excimer lasers (the name is derived from the terms excited and dimers) use reactive gases such as chlorine and fluorine mixed with inert gases such as argon, krypton, or xenon. When electrically stimulated, a pseudomolecule or dimer is produced and when lased, produces light in the ultraviolet range.

4. Dye lasers use complex organic dyes like rhodamine 6G in liquid solution or suspension as lasing media. They are tunable over a broad range of wavelengths.

5. Semiconductor lasers, sometimes called diode lasers, are not solid-state lasers. These electronic devices are generally very small and use low power. They may be built into larger arrays, e.g., the writing source in some laser printers or compact disk players.

D. The wavelength output from a laser depends upon the medium being excited. Table III:6-1 lists most of the laser types and their wavelength output defined by the medium being excited. Laser use today is not restricted to the laboratory or specialized industries. Table III:6-2 lists some of the major uses of lasers.

TABLE III:6-1. WAVELENGTHS OF MOST COMMON LASERS

Laser type	*Wavelength (μmeters)*	*Laser type*	*Wavelength (μmeters)*
Argon fluoride (Excimer-UV)	0.193	Helium neon (yellow)	0.594
Krypton chloride (Excimer-UV)	0.222	Helium neon (orange)	0.610
Krypton fluoride (Excimer-UV)	0.248	Gold vapor (red)	0.627
Xenon chloride (Excimer-UV)	0.308	Helium neon (red)	0.633
Xenon fluoride (Excimer-UV)	0.351	Krypton (red)	0.647
Helium cadmium (UV)	0.325	Rohodamine 6G dye (tunable)	0.570-0.650
Nitrogen (UV)	0.337	Ruby ($CrAlO_3$) (red)	0.694
Helium cadmium (violet)	0.441	Gallium arsenide (diode-NIR)	0.840
Krypton (blue)	0.476	Nd:YAG (NIR)	1.064
Argon (blue)	0.488	Helium neon (NIR)	1.15
Copper vapor (green)	0.510	Erbium (NIR)	1.504
Argon (green)	0.514	Helium neon (NIR)	3.39
Krypton (green)	0.528	Hydrogen fluoride (NIR)	2.70
Frequency doubled Nd YAG (green)	0.532	Carbon dioxide (FIR)	9.6
Helium neon (green)	0.543	Carbon dioxide (FIR)	10.6
Krypton (yellow)	0.568		
Copper vapor (yellow)	0.570		

Key: UV = ultraviolet (0.200-0.400 μm)
VIS = visible (0.400-0.700 μm)
NIR = near infrared (0.700-1.400 μm)

II. NONBEAM LASER HAZARDS.

In some laser operations, particularly in the research laboratory, general safety and health guidelines should be considered.

A. INDUSTRIAL HYGIENE. Potential hazards associated with compressed gases, cryogenic materials, toxic and carcinogenic materials and noise should be considered. Adequate ventilation shall be installed to reduce noxious or potentially hazardous fumes and vapors, produced by laser welding, cutting and other target interactions, to levels below the appropriate threshold limit values, e.g., American Conference of Governmental industrial Hygienists (ACGIH) threshold limit values (TLV's) or Occupational Safety and Health Administration's (OSHA) permissible exposure limits (PEL's).

TABLE III:6-2. MAJOR CATEGORIES OF LASER USE

Alignment	Drilling	Plasma diagnostics
Annealing	Entertainment	Spectroscopy
Balancing	Heat treating	Velocimetry
Biomedical	Holography	Lidar
Cellular research	Information handling	Special photography
Dental	Copying	Scanning microscopy
Diagnostics	Displays	Military
Dermatology	Plate making	Distance ranging
Ophthalmology	Printing	Rifle simulation
Surgery	Reading	Weaponry
Communications	Scanning	Nondestructive training
Construction	Typesetting	Scanning
Alignment	Videodisk	Sealing
Ranging	Marking	Scribing
Surveying	Laboratory instruments	Soldering
Cutting	Interferometry	Welding
Displays	Metrology	

B. EXPLOSION HAZARDS. High-pressure arc lamps and filament lamps or laser welding equipment shall be enclosed in housings which can withstand the maximum pressures resulting from lamp explosion or disintegration. The laser target and elements of the optical train which may shatter during laser operation shall also be enclosed.

C. NONBEAM OPTICAL RADIATION HAZARDS. This relates to optical beam hazards other than laser beam hazards. Ultraviolet radiation emitted from laser discharge tubes, pumping lamps and laser welding plasmas shall be suitably shielded to reduce exposure to levels below the ANSI Z 136.1 (extended source), OSHA PEL's, and/or ACGIH TLV's.

D. COLLATERAL RADIATION. Radiation, other than laser radiation, associated with the operation of a laser or laser system, e.g., radio frequency (RF) energy associated with some plasma tubes, x-ray emission associated with the high voltage power supplies used with excimer lasers, shall be maintained below the applicable protection guides. The appropriate protection guide for RF and microwave energy is that given in the American National Standard "Safety levels with respect to human exposure to radio frequency electromagnetic fields, 300 kHz to 100 GHz," ANSI C95.1; the appropriate protection guides for exposure to X-ray emission is found in the Department of Labor Occupational Safety and Health Standards, 29 CFR Part 1910.1096 and the applicable State Codes. Lasers and laser systems which, by design, would be expected to generate appreciable levels of collateral radiation, should be monitored.

E. ELECTRICAL HAZARDS. The intended application of the laser equipment determines the method of electrical installation and connection to the power supply circuit (for example, conduit versus flexible cord). All equipment shall be installed in accordance with the National Electrical Code and the Occupational Safety and Health Act. [Additional specific recommendations can be found in Section 7.4 of ANSI Z 136.1 (1993)].

F. FLAMMABILITY OF LASER BEAM ENCLOSURES. Enclosure of Class IV laser beams and terminations of some focused Class IIIB lasers, can result in potential fire hazards if the enclosure materials are exposed to irradiances exceeding 10 W/cm2. Plastic materials are not precluded as an enclosure material, but their use and potential for flammability and toxic fume release following direct exposure should be considered. Flame-resistant materials and commercially available products specifically designed for laser enclosures should also be considered.

III. BIOLOGICAL EFFECTS OF THE LASER BEAM.

A. EYE INJURY. Because of the high degree of beam collimation, a laser serves as an almost ideal point source of intense light. A laser beam of sufficient power can theoretically produce retinal intensities at magnitudes that are greater than conventional light sources, and even larger than those produced when directly viewing the sun. Permanent blindness can be the result.

B. THERMAL INJURY. The most common cause of laser-induced tissue damage is thermal in nature, where the tissue proteins are denatured due to the temperature rise following absorption of laser energy.

1. The thermal damage process (burns) is generally associated with lasers operating at exposure times greater than 10 microseconds and in the wavelength region from the near ultraviolet to the far infrared (0.315 μm-103 μm). Tissue damage may also be caused by thermally induced acoustic waves following exposures to sub-microsecond laser exposures.

2. With regard to repetitively pulsed or scanning lasers, the major mechanism involved in laser-induced biological damage is a thermal process wherein the effects of the pulses are additive. The principal thermal effects of laser exposure depend upon the following factors:

   - The absorption and scattering coefficients of the tissues at the laser wavelength.
   - See Table III:6-1 for a summary of more common laser types and wavelengths.
   - Irradiance or radiant exposure of the laser beam.
   - Duration of the exposure and pulse repetition characteristics, where applicable.
   - Extent of the local vascular flow.
   - Size of the area irradiated.

C. OTHER.

1. Other damage mechanisms have also been demonstrated for other specific wavelength ranges and/or exposure times. For example, photochemical reactions are the principal cause of threshold level tissue damage following exposures to either actinic ultraviolet radiation (0.200 μm-0.315 μm) for any exposure time or "blue light" visible radiation (0.400 μm-0.550 μm) when exposures are greater than 10 seconds.

2. To the skin, UV-A (0.315 μm-0.400 μm) can cause hyperpigmentation and erythema.

3. Exposure in the UV-B range is most injurious to skin. In addition to thermal injury caused by ultraviolet energy, there is the possibility of radiation carcinogenesis from UV-B (0.280 mm - 0.315 mm) either directly on DNA or from effects on potential carcinogenic intracellular viruses.

4. Exposure in the shorter UV-C (0.200 μm-0.280 μm) and the longer UV-A ranges seems less harmful to human skin. The shorter wavelengths are absorbed in the outer dead layers of the epidermis (stratum corneum) and the longer wavelengths have an initial pigment-darkening effect followed by erythema if there is exposure to excessive levels. These biological effects are summarized in Table III:6-3.

5. The hazards associated with skin exposure are of less importance than eye hazards; however, with the expanding use of higher-power laser systems, particularly ultraviolet lasers, the unprotected skin of personnel may be exposed to extremely hazardous levels of the beam power if used in an unenclosed system design.

   NOTE: The primary purpose of an exiting laser beam, e.g. cutting or welding of hard materials, must not be forgotten! Some laser beams designed for material alteration may be effective some distance from their intended impact point.

TABLE III:6-3. SUMMARY OF BASIC BIOLOGICAL EFFECTS OF LIGHT

Photobiological spectral domain	*Eye effects*	*Skin effects*
Ultraviolet C (0.200-0.280 μm)	Photokeratitis	Erythema (sunburn) Skin cancer
Ultraviolet B (0.280-315 μm)	Photokeratitis	Accelerated skin aging Increased pigmentation
Ultraviolet A (0.315-0.400 μm)	Photochemical UV cataract	Pigment darkening Skin burn
Visible (0.400-0.780 μm)	Photochemical and thermal retinal injury	Photosensitive reactions Skin burn
Infrared A (0.780-1.400 μm)	Cataract, retinal burns	Skin burn
Infrared B (1.400-3.00 μm)	Corneal burn Aqueous flare IR cataract	Skin burn
Infrared C (3.00-1000 μm)	Corneal burn only	Skin burn

IV. LASER HAZARD CLASSIFICATIONS.

A. INTRODUCTION.

1. The intent of laser hazard classification is to provide warning to users by identifying the hazards associated with the corresponding levels of accessible laser radiation through the use of labels and instruction. It also serves as a basis for defining control measures and medical surveillance.

2. Lasers and laser systems received from manufacturers are required by federal law, 21 CFR Part 1000, to be classified and appropriately labeled by the manufacturer. It should be stressed, however, that the classification may change whenever the laser or laser system is modified to accomplish a given task.

3. It should also be stressed that an agency such as the Food and Drug Administration's Center for Devices and Radiological Health (FDA/CDRH) does not "approve" laser systems for medical use. The manufacturer of the laser system first classifies the laser and then certifies that it meets all performance requirements of the Federal Laser Product Performance Standard (FLPPS). The forms submitted by the manufacturer to FDA/CDRH are reviewed for technical accuracy, omissions, and errors. If none are found, the manufacturer is notified only that the submission appears to be complete. Therefore, all lasers and laser systems that are manufactured by a company, or purchased by a company and relabeled and placed into commerce, or incorporated into a system and placed into commerce, shall be classified.

B. LASER HAZARD CLASSES.

1. Virtually all of the U.S. domestic as well as all international standards divide lasers into four major hazard categories called the laser hazard classifications. The classes are based upon a scheme of graded risk. They are based upon the ability of a beam to cause biological damage to the eye or skin. In the FLPPS, the classes are established relative to the Accessible Emission Limits (AEL) provided in tables in the standard. In the ANSI Z 136.1 standard, the AEL is defined as the product of the Maximum Permissible Exposure (MPE) level and the

area of the limiting aperture. For visible and near infrared lasers, the limiting aperture is based upon the "worst-case" pupil opening and is a 7 mm circular opening.

2. Lasers and laser systems are assigned one of four broad Classes (I to IV) depending on the potential for causing biological damage. The biological basis of the hazard classes are summarized in Table III:6-4.

    a. Class I: cannot emit laser radiation at known hazard levels (typically continuous wave: cw 0.4 w at visible wavelengths). Users of Class I laser products are generally exempt from radiation hazard controls during operation and maintenance (but not necessarily during service).

       Since lasers are not classified on beam access during service, most Class I industrial lasers will consist of a higher class (high power) laser enclosed in a properly interlocked and labeled protective enclosure. In some cases, the enclosure may be a room (walk-in protective housing) which requires a means to prevent operation when operators are inside the room.

    b. Class I.A.: a special designation that is based upon a 1000-second exposure and applies only to lasers that are "not intended for viewing" such as a supermarket laser scanner. The upper power limit of Class I.A. is 4.0 mW. The emission from a Class I.A. laser is defined such that the emission does not exceed the Class I limit for an emission duration of 1000 seconds.

    c. Class II: low-power visible lasers that emit above Class I levels but at a radiant power not above 1 mW. The concept is that the human aversion reaction to bright light will protect a person. Only limited controls are specified.

    d. Class IIIA: intermediate power lasers (cw: 1-5 mW). Only hazardous for intrabeam viewing. Some limited controls are usually recommended.

       NOTE: There are different logotype labeling requirements for Class IIIA lasers with a beam irradiance that does not exceed 2.5 mW/cm2 (Caution logotype) and those where the beam irradiance does exceed 2.5 mW/cm2 (Danger logotype).

    e. Class IIIB: moderate power lasers (cw: 5-500 mW, pulsed: 10 J/cm2 or the diffuse reflection limit, whichever is lower). In general Class IIIB lasers will not be a fire hazard, nor are they generally capable of producing a hazardous diffuse reflection. Specific controls are recommended.

    f. Class IV: High power lasers (cw: 500 mW, pulsed: 10 J/cm2 or the diffuse reflection limit) are hazardous to view under any condition (directly or diffusely scattered) and are a potential fire hazard and a skin hazard. Significant controls are required of Class IV laser facilities.

C. HOW TO DETERMINE THE CLASS OF LASERS DURING INSPECTION.

1. The classification of a laser or laser product is, in some instances, a rather detailed process. It can involve determination of the AEL, measurement of the laser emission, measurement/determination of the emission pulse characteristics (if applicable), evaluation of various performance requirements (protective housing, interlocks, etc.) as specified by the FLPPS and/or ANSI standards.

2. It should be stressed that classification is a required specification provided by the laser manufacturer and the label that specifies the class is found in only one location on the laser product. The class of the laser will be specified only on the lower left-hand corner (position three) of the warning logotype label.

   The logotype is the rectangular label that has the laser "sunburst" symbol and the warning statement of CAUTION (Class II and some Class IIIA) or DANGER (some Class IIIA, all Class IIIB and Class IV). This label will also have the type of laser designated (HeNe, Argon, CO2, etc.) and the power or energy output specified (1 mW CW/MAX, 100 mJ pulsed, etc.).

3. Class I lasers have no required labeling indicating the Class I status. Although the FLPPS requires no classification labeling of Class I lasers it does require detailed compliance with numerous other performance requirements (i.e., protective housing, identification and compliance labeling, interlocking, etc.)

TABLE III:6-4. LASER CLASSIFICATIONS--SUMMARY OF HAZARDS

	Applies to --- wavelength ranges ---				------------ Hazards ------------		
Class	***UV***	***VIS***	***NIR***	***IR***	***Direct ocular***	***Diffuse ocular***	***Fire***
I	X	X	X	X	No	No	No
IA	--	X*	--	--	Only after 1000 sec	No	No
II	--	X	--	--	Only after 0.25 sec	No	No
IIIA	X	X**	X	X	Yes	No	No
IIIB	X	X	X	X	Yes	Only when laser output is near Class IIIB limit of 0.5 Watt	No
IV	X	X	X	X	Yes	Yes	Yes

Key: X = Indicates class applies in wavelength range.
* = Class **IA** applicable to lasers "not intended for viewing" *ONLY*.
** = CDRH Standard assigns Class **IIIA** to visible wavelengths *ONLY*. ANSI Z 136.1 assigns Class IIIA to all wavelength ranges.

D. ANSI Z 136.2 OPTICAL FIBER SERVICE GROUP DESIGNATIONS.

1. Optical Fiber Communication Systems (OFCS) and the associated optical test sets use semiconductor lasers or LED transmitters that emit energy at wavelengths typically in the range from 0.650 to 1.20 mm into the light-guide fiber-optic cables.

2. All OFCS are designed to operate with the beam totally enclosed within the fiber-optic and associated equipment and, therefore, are always considered as Class I in normal operation.

3. The only risk for exposure would occur during installation and service when light-guide cables are disconnected or during an infrequent accidental cable break.

4. Under the requirements of the ANSI Z 136.2 (1988) Standard "For the Safe Use of Optical Fiber Communication Systems Utilizing Laser Diode and LED Sources," Optical Fiber Communication Systems (OFCS) are assigned into one of four service group (SG) designations: SG1, SG2, SG3a, SG3b, depending on the potential for an accessible beam to cause biological damage.

5. The service group designations relate to the potential for ocular hazards to occur only during accessible beam conditions. This would normally occur only during periods of service to a OFCS. Such designations apply only during periods of service in one of the following four service groups:

   a. Service Group 1: An OFCS that is SG1 has a total output power that is less than the Accessible Emission Limit (AEL) for Class I and there is no risk of exceeding the Maximum Permissible Irradiance (MPI) when viewing the end of a fiber with a microscope, an eye-loupe or with the unaided eye.

b. Service Group 2: An OFCS is SG2 only if wavelengths between 0.400 and 0.700 mm are emitted and is potentially hazardous if viewed for more than 0.25 second. (Note: At present there are virtually no OFCS's that operate in this wavelength range.)

c. Service Group 3A: A SG 3A OFCS is not hazardous when viewed with the unaided eye and is hazardous only when viewed with a microscope or an eye-loupe.

d. Service Group 3B: OFCS that meet none of the above criteria are designated as SG 3B.

NOTE: OFCS's where the total power is at or above 0.5W do not meet the criteria for optical fiber service group designation. In this case, the OFCS's are treated as a standard laser system.

V. INVESTIGATIONAL GUIDELINES.

A. REQUIREMENTS OF LASER STANDARDS. In the United States, several organizations concern themselves with laser safety. These organizations include the American National Standards Institute (ANSI); the Center for Devices and Radiological Health (CDRH) of the Food and Drug Administration (FDA); the Department of Labor's Occupational Safety and Health Administration (OSHA); and the Council of Radiation Control Program Directors (CRCPD). Several state governments and the CRCPD have developed a model state standard for laser safety.

1. OSHA Regulatory Practice. At the present time, OSHA does not have a comprehensive laser standard, though 29 CFR 1926.54 is applicable to the construction industry. A standard for personal protective equipment (Subpart I) may apply in some cases.

   The construction standard 29 CFR 1926.102(b)(2), for eye and face protection, states that "employees whose occupation or assignment requires exposure to laser beams shall be furnished suitable laser safety goggles which will protect for the specific wavelength of the laser and be of optical density (O.D.) adequate for the energy involved."

   OSHA citations are issued by invoking the general duty clause or, in some cases, Subpart I. In such cases, the employers are required to revise their reportedly unsafe work place using the recommendations and requirements of such industry consensus standards as the ANSI Z 136.1 Standard. See also Table III:6-8.

2. Specific and Model State Laser Regulations. A few states currently have laser regulations. Requirements are generally concerned with the registration of lasers and the licensing of operators and institutions. Physician-used and other medical lasers are generally exempt from state requirements.

   The complexity of state laser regulations may change in the future pending adoption of the "Suggested State Regulation for Lasers" promulgated through the Conference of Radiation Control Program Directors. This model state standard has been adopted in part, for example, by Arizona and Florida. Several other states have enacted some form of regulation. Table III:6-5 summarizes state regulations.

3. FDA Center for Devices and Radiological Health Performance Requirements.

   a. The CDRH of the Department of Health and Human Services was chartered by Congress to standardize the manufacture of lasers in interstate commerce after August 2, 1976. CDRH also has the responsibility for enforcing compliance with the medical devices legislation. All manufacturers of surgical lasers must obtain premarket approval of their devices through the CDRH.

   b. FDA sanctions the exploratory use of lasers for specific procedures through a process known as an Investigational Device Exemption (IDE). Approval of an IDE permits the limited use of a laser expressly for the purpose of conducting an investigation of the laser's safety and effectiveness. Once an IDE has been prepared and approved by the CDRH, the manufacturer may then actively market the laser for that specific medical or surgical procedure.

c. The FDA/CDRH Federal Laser Product Performance Standard (FLPPS) regulates the manufacturer of commercial laser products, not the user. The standard does not contain specific design specifications, but is a conceptual, performance standard which the designer of laser products must consider. The intent is to insure laser product safety.

TABLE III:6-5. SUMMARY OF CURRENT STATE LASER REGULATIONS

State	***Department***	***Regulation***
Alaska	Environmental Conservation	Title 18, Article 7
Arizona*	Radiation Regulatory Agency	
Arkansas	Division of Radiation Control & Emergency Management	Act 460
Florida*	Department of Health & Rehabilitative Services	Non-Ionizing Chapter: 10D-89
Georgia	Department of Public Health	Chapter: 270-5-27
Illinois	Department of Nuclear Safety	Chapter: 111 ½
Massachusetts	Department of Public Health	105 CMR 21
Montana	Health & Environmental Services	92-003
New York	Department of Labor	Code Rule 50
Pennsylvania	Environmental Resources	Chapter: 203, Title 25
Texas	Department of Health	Radiation Control Act Parts 50, 60, 70
Washington	Labor & Industry	Chapter 296-62-WAC

* Using CRCPD "Model State" laser standard as basis.

d. FLPSS is applicable to lasers or laser systems sold by a company within or imported into the U.S. In some cases it can also apply when a laser or laser system is transferred within a company for internal use within the U.S. The compliance procedure requires implementation of the procedures and requirements as set forth in the U.S. Federal Laser Product Performance Standard: 21 CFR Part 1000 [parts 1040.10 and 1040.11].

e. Under the requirements of the FLPPS, the manufacturer is first required to classify the laser as either a Class-I, Class-II, Class-I.A., Class-IIIA, Class-IIIB, or Class-IV laser product and then to certify (by means of a label on the product) as well as submit a report demonstrating that all requirements (performance features) of the compliance standard are met. Specific performance features include:

- protective housing;
- protective housing warning labels and logotype labels;
- product identification label and certification statement;
- safety interlocks;
- emission indicator;
- remote interlock connector;
- key control;
- beam attenuator;
- specification of control locations;
- viewing optic limitations;
- scanning beam safeguards; and
- manual reset of beam cutoff.

f. FDA/CDRH performance requirements are tabulated in Appendix III:6-1. An outline to assist in evaluating FLPPS laser system performance requirements is included in Appendix III:6-2.

4. The American National Standard Institute (ANSI). An American National Standard implies a consensus of those substantially concerned with its scope and provisions. These standards are intended as a guide for

manufacturers, consumers, and the general public. However, there is no inherent requirement for any person or company to adhere to an ANSI standard. Compliance is voluntary unless specifically required by an organization. For example, the U.S. Department of Energy requires adherence to the ANSI Z 136.1 by their staff as well as by all contractor organizations. Appendix III:6-3 summarizes ANSI Standards applicable to laser safety.

B. LASER EXPOSURE LIMITS. At present either the FDA criteria for medical lasers or the following ANSI standards can be useful in evaluating laser safety.

1. FDA Long-Term Exposure Limits. The FDA/CDRH Federal Laser Product Performance Standard (FLPPS) assumes a linearly additive biological effect for exposures to visible light between 10 and 104 seconds (2.8 hours). The standard accepts that a cumulative radiant energy exposure of 3.85 millijoules (mJ) will not cause a biological effect. Hence a 10-second total accumulated exposure corresponds to an average power entering a 7-mm aperture of 385 microwatts (mW). For an exposure of 104 seconds, the average power would be 0.385 mW. In the FLPPS, the power level of 0.385 mW is referred to as the Class I Accessible Emission Limit (AEL) for a visible CW laser.

2. ANSI Z 136.1, Long-Term Exposure Limits.

   a. The ANSI Z 136.1 (1993) standard is a "user" standard and therefore provides maximum permissible exposure (MPE) limits. These were derived by normalizing the power (or pulse energy) data derived from biological research studies relative to a defined limiting aperture. For example, in the visible and near-infrared spectra, the limiting aperture is based upon the diameter of a fully dilated pupil of the human eye, 7 mm. The area of a 7-mm pupil is 0.385 cm2. Hence, the irradiance limit for long-term ocular exposure is computed by dividing the AEL value of 0.385 mW by the area of the limiting aperture of 0.385 cm2. This yields the worst-case MPE value of 1.0 mW/cm2 for long-term exposure in the wavelength range of 0.400 to 0.550 mm.

   b. The ANSI Z 136 and FDA/CDRH allowable-exposure limits for CW lasers (ClassI limits) are essentially identical for wavelengths between 0.400 and 0.550 mm. The ANSI limits are, however, more relaxed for wavelengths between 0.550 and 1.40 mm. ANSI recognizes a decreased biological hazard in the red and infrared regions that is not recognized by the CDRH.

   c. The ANSI Z 136 MPE level for a very long term exposure by a helium-neon laser is, in fact, seventeen times greater than the CDRH standard. In the 1976 revision, ANSI Z 136 introduced the correction factor CB which has a value of 17.5 at the 0.633-μm HeNe laser wavelength, and, thus, permitted a radiant exposure of 185 mJ/cm2 accumulated exposure for times from T1 = 453 seconds to 104 seconds, and about 18 w/cm2 (7 w in a 7-mm limiting aperture) for continuous operation of exposure durations exceeding 104 seconds.

3. ANSI Z 136.1, Repetitively Pulsed Exposures.

   a. The ANSI Z 136 standard requires a decrease in the maximum permissible exposure (MPE) for scanned or repetitive-pulse radiation as compared to continuous-wave radiation for pulse repetition frequencies (PRF) in the general range of 1000-15000 Hz. Because of pulse additivity, scanned or repetitively pulsed radiation with repetition rates less than 15 KHz have lower retinal damage threshold levels than CW radiation of comparable power.

   b. The ANSI Z 136 Standard includes a reduction factor of the threshold for each of the single pulses based on biological data that are not yet well explained by any theory. The FDA/CDRH standard does not recognize this repetitive-pulse correction factor. However, some experts envision the possibility of a repetitively pulsed laser which is Class I by the FDA/CDRH standard could be rated Class II or even Class IIIB by the ANSI Z 136 standard.

   c. The ANSI standard requires that multiple-pulse (scanning) lasers operating from 1 to 15,000 Hz have a correction to the single pulse MPE. The correction factor is determined by taking the fourth root of the total number of pulses (N) in a pulse train. Then, the correction factor is calculated such that the MPE radiant exposure or integrated radiance of an individual pulse within the train is reduced by a factor $N^{-1/4}$.

4. ANSI Z 136.1, Maximum Permissible Exposure Limits.

   a. A summary of Maximum Permissible Exposure (MPE) limits for direct ocular exposures for some of the more common lasers is presented in Table III:6-6. For further information on MPE values, refer to the ANSI Z 136.1 "Safe Use of Lasers" Standard.

   b. The information in Table III:6-6 provides the MPE value for different lasers operating for different overall exposure times. The times chosen were:

      - 0.25 second: The human aversion time for bright-light stimuli (the blink reflex). Thus, this becomes the "first line of defense" for unexpected exposure to some lasers and is the basis of the Class II concept.
      - 10 seconds: The time period chosen by the ANSI Z 136.1 committees represents the optimum "worst-case" time period for ocular exposures to infrared (principally near-infrared) laser sources. It was argued that natural eye motions dominate for periods longer than 10 seconds.
      - 600 seconds: The time period chosen by the ANSI Z 136.1 committees represents a typical worst-case period for viewing visible diffuse reflections during tasks such as alignment.
      - 30,000 seconds: The time period that represents a full 1-day (8-hour) occupational exposure. This results from computing the number of seconds in 8 hours; e.g.: 8 hours × 60 minutes/hour × 60 seconds/minute = 28,800 seconds. Rounded off, it becomes 30,000 seconds.

   c. The "safety" exposure limits (MPE's) in Table III:6-6 are expressed in irradiance terms (W/cm2) that would be measured at the cornea. Note that they vary by wavelength and exposure time.

TABLE III:6-6. SUMMARY: MAXIMUM PERMISSIBLE EXPOSURE LIMITS*

	Wavelength	***MPE level (W/cm²)***			
Laser type	***(μm)***	***0.25 sec***	***10 sec***	***600 sec***	***30,000 sec***
CO2 (CW)	10.6	---	100.0×10^{-3}	---	100.0×10^{-3}
Nd: YAG (CW)	1.33	---	5.1×10^{-3}	---	1.6×10^{-3}
Nd: YAG (CW)	1.064	---	5.1×10^{-3}	---	1.6×10^{-3}
Nd: YAG (Q-switched)	1.064	---	17.0×10^{-6}	---	2.3×10^{-6}
GaAs (Diode/CW)	0.840	---	1.9×10^{-3}	---	610.0×10^{-6}
HeNe (CW)	0.633	2.5×10^{-3}	---	293.0×10^{-6}	17.6×10^{-6}
Krypton (CW)	0.647	2.5×10^{-3}	---	364.0×10^{-6}	28.5×10^{-6}
	0.568	31.0×10^{-6}	---	2.5×10^{-3}	18.6×10^{-6}
	0.530	16.7×10^{-6}	---	2.5×10^{-3}	1.0×10^{-6}
Argon (CW)	0.514	2.5×10^{-3}	---	16.7×10^{-6}	1.0×10^{-6}
XeFl (Excimer/CW)	0.351	---	---	---	33.3×10^{-6}
XeCl (Excimer/CW)	0.308	---	---	---	1.3×10^{-6}

* Source: ANSI Z 136.1 (1993)

C. LASER HAZARD COMPUTATIONS.

1. NHZ Definition, Use, and Values.

a. The Nominal Hazard Zone (NHZ) describes the space within which the level of direct, reflected, or scattered radiation during normal operation exceeds the MPE. The NHZ associated with open-beam Class IIIB and Class IV laser installations can be useful in assessing area hazards and implementing controls.

b. It is often necessary in some applications where open beams are required (e.g., industrial processing, laser robotics, surgical uses) to define the area where the possibility exists for potentially hazardous exposure. This is done by determining the NHZ. Consequently, persons outside the NHZ boundary would be exposed below the MPE level and are considered to be in a non-hazardous location.

c. The NHZ boundary may be defined, for example, by direct beams (intrabeam) and diffusely scattered laser beams, as well as beams transmitted from fiber optics and/or through lens arrays. The NHZ perimeter is the envelope of MPE exposure levels from any specific laser installation geometry.

d. The purpose of an NHZ evaluation is to define that space where control measures are required. This is an important factor since, as the scope of laser uses has expanded, controlling lasers by total enclosure in a protective housing or interlocked room is limiting and, in many instances, an expensive overreaction to the real hazards. The following factors are required in NHZ computations:

- laser power or energy output;
- beam diameter;
- beam divergence;
- pulse repetition frequency (prf) (if applicable);
- wavelength;
- beam optics and beam path; and
- maximum anticipated exposure duration.

e. Note that the ANSI Z 136 MPE value is required in all NHZ calculations. Examples of NHZ calculations can be found in the appendix of ANSI Z 136.1 (1993). In addition, computer software is also available to assist in the computations for NHZ, optical densities of protective eye wear, and other aspects of laser hazard analysis.

2. NHZ Example Summary. The intrabeam (direct) hazard for a Nd:YAG laser extends from 792 meters to 1410 meters, depending upon whether a 10-second or 8-hour criterion is used, as summarized in Table III:6-7. Similarly, with a lens on the laser, the hazard for a Nd:YAG laser exists over a range from 6.3 meters to 11.3 meters. The diffuse reflection zone for this laser type is, however, markedly smaller, 0.8 meter to 1.4 meters. Nonetheless, the analysis suggests that operating personnel and support staff close to the laser still need eye protection even for diffuse reflections.

Other calculations are also presented in Table III:6-7 for a 500-Watt CO2 and a 5-Watt argon laser. Note that the NHZ's do not vary for the CO2 laser (because the MPE values are nearly identical for 10-second and 8-hour criteria). Also note that the diffuse reflection NHZ's are very small except for the 8-hour criterion for the argon laser. In most cases, 0.25 second can be used with visible lasers unless intentional staring is required or intended.

TABLE III:6-7. NHZ DISTANCE VALUES FOR VARIOUS LASERS

Laser type	*Exposure criteria*	----- *Hazard range (meters)* ----- *Diffuse*	*Lens-on-laser*	*Direct*
Nd:YAG				
100 Watt	8 hours	1.4	11.3	1410
1.064 μm	10 seconds	0.8	6.3	792
CO_2				
500 Watt	8 hours	0.4	5.3	309
10.6 μm	10 seconds	0.4	5.3	390
Argon				
5.0 Watt	8 hours	12.6	1.7×10^3	25.2×10^3
0.488 μm	0.25 seconds	0.25	33.3	240

Laser criteria used for NHZ distance calculations:

Laser parameter	*Nd-YAG*	*CO_2*	*Argon*
Wavelength (μm)	1.064	10.6	0.488
Beam power (Watts)	100.0	500.0	5.0
Beam divergence (mrad)	2.0	2.0	1.0
Beam size at aperture (mm)	2.0	20.0	2.0
Beam size at lens (mm)	6.3	30.0	3.0
Lens focal length (mm)	25.4	200.0	200.0
MPE for 8 hours (w/cm^2)	1.6×10^3	1.0×10^5	1.0
MPE for 10 seconds (w/cm^2) 5.1×10^3	1.0×10^5	---	
MPE for 0.25 second (w/cm^2)	---	---	2.5×10^3

Source: ANSI A 136.1 (1993)

D. INTRABEAM OPTICAL DENSITY DETERMINATION.

1. Based upon these typical exposure conditions, the optical density required for suitable filtration can be determined. Optical density (OD) is a logarithmic function defined by:

$$OD = \log_{10} \frac{H_0}{MPE}$$

EQUATION III: 6-1. OPTICAL DENSITY

where: Ho = Anticipated worst-case exposure (J/cm² or W/cm²
MPE = Maximum permissible exposure level expressed in the same units as Ho

2. Based upon the worst case exposure conditions outlined above, one can determine the optical density recommended to provide adequate eye protection for this laser. For example, the minimum optical density at the 0.514 µm argon laser wavelength for a 600-second direct intrabeam exposure to the 5-watt maximum laser output can be determined as follows:

where: f = 5 watts
MPE = 16.7 W/cm² (using 600-second criterion)
d = 7mm (worst-case pupil size)

Computing the worst-case exposure Ho:

Ho = [Power/Area] = f/A =4f/ pd²
Ho = [(4)(5.0)/ p(0.7)²]
Ho = 12.99 W/cm²

Substitution gives:

OD = log10 [(12.99)/16.7 x 10 to the -6)]
OD = 5.9

3. The most conservative approach would be to choose an 8-hour (occupational) exposure. In this case, the optical density at 0.514 µm is increased to OD = 7.1 for a 5.0-watt intrabeam exposure because the 8-hour (30,000 sec.) MPE is reduced to 1.0 × 10-6 W/cm2. The OD values for various lasers, computed for various appropriate exposure times, are presented in Table III:6-8. It should be stressed these values are for intrabeam viewing (worst case) only. Viewing Class IV diffuse reflections (such as during alignment tasks) require, in general, less OD. These should be determined for each situation and would be dependent upon the laser parameters and viewing distance.

VI. CONTROL MEASURES AND SAFETY PROGRAMS.

The specific control measures specified in the ANSI Z 136.1 standard are summarized in Table III:6-9. The details of these controls are outlined the following sections.

A. CONTROL MEASURES: OVERVIEW.

1. There are four basic categories of controls useful in laser environments. These are engineering controls, personal protective equipment, administrative and procedural controls, and special controls. The controls to be reviewed here are based upon the recommendations of the ANSI Z 136.1 standard.

2. Important in all controls is the distinction between the functions of operation, maintenance, and service. First, laser systems are classified on the basis of level of the laser radiation accessible during operation. Maintenance is defined as those tasks specified in the user instructions for assuring the performance of the product and may include items such as routine cleaning or replenishment of expendables. Service functions are usually performed with far less frequency than maintenance functions (e.g., replacing the laser resonator mirrors or repair of faulty components) and often require access to the laser beam by those performing the service functions. The safety procedures required for such beam access during service functions should be clearly delineated in the laser product's service manual.

B. LASER SAFETY OFFICER (LSO).

1. The LSO has the authority to monitor and enforce the control of laser hazards and effect the knowledgeable evaluation and control of laser hazards. The LSO administers the overall laser safety program where the duties include, but are not limited to, items such as confirming the classification of lasers, doing the NHZ evaluation, assuring that the proper control measures are in place and approving substitute controls, approving standard operating procedures (SOP's), recommending and/or approving eye wear and other protective equipment, specifying appropriate signs and labels, approving overall facility controls, providing the proper laser safety training as needed, conducting medical surveillance, and designating the laser and incidental personnel categories.

TABLE III:6-8. OPTICAL DENSITIES FOR PROTECTIVE EYEWEAR FOR VARIOUS LASER TYPES

Laser type and power	Wavelength (mm)	---- Optical density for exposure durations ---- 0.25 sec.	10 sec.	600 sec.	30,000 sec.
XeCl 50 Watts	0.308[a]	--	6.2	8.0	9.7
XeFl 50 Watts	0.351[a]	--	4.8	6.6	8.3
Argon 1.0 Watts	0.514	3.0	3.4	5.2	6.4
Krypton 1.0 Watt	0.530	3.0	3.4	5.2	6.4
Krypton 1.0 Watt	0.568	3.0	3.4	4.9	6.1
HeNe 0.005 Watt	0.633	0.7	1.1	1.7	2.9
Krypton 1 Watt	0.647	3.0	3.4	3.9	5.0
GaAs 50 mW	0.840[c]	--	1.8	2.3	3.7
Nd: YAG 100 Watt	1.064[a]	--	4.7	5.2	5.2
Nd: YAG (Q-switch)[b]	1.064[a]	--	4.5	5.0	5.4
Nd: YAG[c] 50 Watts	1.33[a]	--	4.4	4.9	4.9
CO_2 1000 Watts	10.6[a]	--	6.2	8.0	9.7

a. Repetitively pulsed at 11 Hertz, 12-nanosecond pulses, 20 mJ/pulse.
b. OD for UV and FIR beams computed using a 1-mm limiting aperture, which presents a "worst-case" scenario. All visible and NIR computations assume a 7-mm limiting aperture.
c. Nd:YAG operating at a less-common 1.33 μm wavelength.
Note: All OD values determined using MPE criteria of ANSI Z 136.1 (1993).

2. The LSO should receive detailed training including laser fundamentals, laser bioeffects, exposure limits, classifications, NHZ computations, control measures (including area controls, eye wear, barriers, etc.), and medical surveillance requirements.

3. In many industrial situations, the LSO functions will be a part-time activity, depending on the number of lasers and general laser activity. The individual is often in the corporate industrial hygiene department or may be a laser engineer with safety responsibility. Some corporations implement an internal laser policy and establish safety practices based upon the ANSI Z 136.1 standard as well as their own corporate safety requirements.

C. CLASS I, CLASS II, CLASS I.A., AND CLASS IIIA LASERS. Accident data on laser usage have shown that Class I, Class II, Class I.A., and Class IIIA lasers are normally not considered hazardous from a radiation standpoint unless illogically used.

> Direct exposure on the eye by a beam of laser light should *always* be avoided with any laser, *no matter how low the power.*

D. BEAM PATH CONTROLS. There are some uses of Class IIIB and Class IV lasers here the entire beam path may be totally enclosed, other uses where the beam path is confined by design to significantly limit access and yet other uses where the beam path is totally open. In each case, the controls required will vary as follows:

TABLE III:6-9. ENGINEERING CONTROL MEASURES FOR THE FOUR LASER CLASSES [ANSI Z 136.1 (1993)]

Control measures	Class					
	I	**IA**	**II**	**IIIA**	**IIIB**	**IV**
Protective housing	X	X	X	X	X	X
Without protective housing	-- LSO shall establish alternate controls --					
Interlocks on protective housing	a	a	a	X	X	X
Service access panel	b	b	b	b	b	X
Key switch master	_	_	_	_	▪	X
Viewing portals	_	_	◊	◊	◊	◊
Collecting optics	_	_	◊	◊	◊	◊
Totally open beam path	_	_	_	_	X	X
Limited open beam path	_	_	_	_	X	X
Remote interlock connector	_	_	_	_	▪	X
Beam stop or attenuator	_	_	_	▪	▪	X
Activation warning system	_	_	_	_	▪	X
Emission delay	_	_	_	_	_	▪
Class IIIB laser controlled area	_	_	_	_	X	_
Class IV laser controlled area	_	_	_	_	_	X
Laser outdoor controls	_	_	_	_	X	X
Temporary laser controlled area	b	b	b	b	_	_
Remote firing & monitoring	_	_	_	_	_	▪
Labels	_	X	X	X	X	X
Area posting	_	_	▪	▪	X	X
Administrative & procedural controls	_	X	X	X	X	X
Standard operating procedures	_	_	_	_	▪	X
Output emission limitations	_	_	_	--LSO determines--		
Education and training	_	_	_	X	X	X
Authorized personnel	_	_	_	_	X	X
Alignment procedures	_	_	X	X	X	X
Eye protection	_	_	_	_	▪	X
Spectator control	_	_	_	_	▪	X
Service personnel	b	b	b	b	X	X
Laser demonstration	_	_	X	X	X	X
Laser fiber optics	_	_	X	X	X	X

Key: X = Shall.
a. = Shall if embedded Class IIIA, Class IIIB, Class IV.
b. = Shall if embedded Class IIIB or Class IV.
_ = No requirement.
▪ = Should.
◊ = Shall if MPE is exceeded.

1. Enclosed (Total) Beam Path.

   a. Perhaps the most common form of a Class I laser system is a high-power laser that has been totally enclosed (embedded) inside a protective enclosure equipped with appropriate interlocks and/or labels on all removable panels or access doors. Beam access is prevented, therefore, during operation and maintenance.

   b. Such a completely enclosed system, if properly labeled and properly safeguarded with protective housing interlocks (and all other applicable engineering controls), will fulfill all requirements for a Class I laser and may be operated in the enclosed manner with no additional controls for the operator.

   c. It should be noted that during periods of service or maintenance, controls appropriate to the class of the embedded laser may be required (perhaps on a temporary basis) when the beam enclosures are removed and beam access is possible. Beam access during maintenance or service procedures will not alter the Class I status of the laser during operation.

2. Limited Open Beam Path.

   a. It is becoming an accepted work practice, particularly with industrial materials-processing lasers, to build an enclosure that completely surrounds the laser-focusing optics and the immediate area of the workstation. Often a computer-controlled positioning table is located within this enclosure. The design often allows a gap of less than one quarter of an inch between the bottom of the enclosure and the top of the material to be laser processed. Such a design enables the part to be laser cut or welded to move while the laser delivery optics remain stationary.

   b. Such a system might not meet the stringent "human access" requirements of the FLPPS for a Class I laser, but the real laser hazards are well confined. Such a design provides what can be called a limited open beam path. In this situation, the ANSI Z 136.1 standard recommends that the LSO shall conduct a laser hazard analysis and establish the extent of the NHZ.

   c. In many system designs, (such as described above), the NHZ will be extremely limited, and procedural controls (rather than elaborate engineering controls) will be sufficient to ensure safe use. In many cases, the laser units may be reclassified by the LSO as Class I under the specifications of the ANSI Z 136 standard.

   d. Such an installation will require a detailed standard operating procedure (SOP). Training is also needed for the system operator commensurate with the class of the embedded laser.

   e. Protective equipment (eye protection, temporary barriers, clothing and/or gloves, respirators, etc.) would be recommended, for example, only if the hazard analysis indicated a need or if the SOP required periods of beam access such as during setup or infrequent maintenance activities. Temporary protective measures for service can be handled in a manner similar to the service of any embedded Class IV laser.

3. Totally Unenclosed Beam Path. There are several specific application areas where high power (Class IIIB and Class IV) lasers are used in an unenclosed beam condition. This would include, for example, open industrial processing systems (often incorporating robotic delivery), laser research laboratory installations, surgical installations, etc. Such laser uses will require that the LSO conduct a hazard analysis and NHZ assessment. Controls are chosen to reflect the magnitude of hazards associated with the accessible beam.

E. LASER-CONTROLLED AREA. When the entire beam path from a Class IIIB or Class IV laser is not sufficiently enclosed and/or baffled to ensure that radiation exposures will not exceed the MPE, a "laser-controlled area" is required. During periods of service, a controlled area may be established on a temporary basis. The controlled area will encompass the NHZ. Those controls required for both Class IIIB and Class IV installations are as follows:

1. Posting with Appropriate Laser Warning Signs.

   a. Class IIIA (beam irradiance 2.5 mW/cm2), Class IIIB and Class IV lasers: Require the ANSI DANGER sign format: white back-ground, red laser symbol with black outline and black lettering (see Appendix

III:6-4). Note that under ANSI Z 136.1 criteria, area posting is required only for Class IIIB and Class IV lasers.

b. Class II or Class IIIA areas (if area warning is deemed unnecessary by the LSO): All signs (and labels) associated with these lasers (when beam irradiance for Class IIIA does not exceed 2.5 mW/cm2) use the ANSI CAUTION format: yellow background, black symbol and letters.

c. During times of service and other times when a temporary laser-controlled area is established, an ANSI NOTICE sign format is required: white background, red laser symbol with blue field and black lettering. This sign is posted only during the time when service is in progress. Examples of area warning signs and logotype designs are given in Appendix III:6-4.

2. Operated by Qualified and Authorized Personnel. Training of the individuals in aspects of laser safety is required for Class IIIB and Class IV laser installations.

3. Transmission from Indoor Controlled Area. The beams shall not, under any circumstances, be transmitted from an indoor laser-controlled area unless for specific purposes (such as testing). In such cases, the operator and the LSO must assure that the beam path is limited to controlled air space.

F. CLASS IV LASER CONTROLS--GENERAL REQUIREMENTS. Those items recommended for Class IIIB but required for Class IV lasers are as follows:

- Supervision directly by an individual knowledgeable in laser safety.
- Entry of any noninvolved personnel requires approval.
- A beam stop of an appropriate material must be used to terminate all potentially hazardous beams.
- Use diffusely reflecting materials near the beam, where appropriate.
- Appropriate laser protective eye wear must be provided all personnel within the laser controlled area.
- The beam path of the laser must be located and secured above or below eye level for any standing or seated position in the facility.
- All windows, doorways, open portals, etc., of an enclosed facility should be covered or restricted to reduce any escaping laser beams below appropriate ocular MPE level.
- Require storage or disabling of lasers when not in use.

G. ENTRYWAY CONTROL MEASURES (CLASS IV). In addition, there are specific controls required at the entryway to a Class IV laser controlled area. These can be summarized as follows:

- All personnel entering a Class IV area shall be adequately trained and provided proper laser protective eye wear.
- All personnel shall follow all applicable administrative and procedural controls.
- All class IV area and entryway controls shall allow rapid entrance and exit under all conditions.
- The controlled area shall have a clearly marked "Panic Button" (nonlockable disconnect switch) that allows rapid deactivation of the laser.

Class IV areas also require some form of area and entryway controls. In the past, doorway interlocking was customary for Class IV installations. The ANSI Z 136 Standard now provides four options that allow the LSO to provide an entryway control suited for the installation. The options include:

1. Nondefeatable Entryway Controls. A nondefeatable control, such as a magnetic switch built into the entryway door which cuts the beam off when the door is opened, is one option. In this case, training is required only for those persons who regularly work in the laser area.

2. Defeatable Entryway Controls.

a. Defeatable controls may be used at an entryway, for example, during long-term testing in a laser area. In this case the controls may be temporarily made inactive if it is clearly evident that there is no hazard at the point of entry. Training is required for all personnel who may frequently require entry into the area.

b. Such defeatable controls shall be designed to allow both rapid egress by the laser personnel at all times and admittance to the laser controlled area in an emergency condition. A readily accessible "panic button" or control/disconnect switch shall be available for deactivating the laser under such emergency conditions.

c. Under conditions where the entire beam path is not completely enclosed, access to the laser-controlled area shall be limited only to persons wearing proper laser protective eye wear when the laser is capable of emitting a beam. In this case, all other optical paths (for example, windows) from the facility shall be covered or restricted in such a way as to reduce the transmitted intensity of the laser radiation to levels at or below the MPE for direct irradiation of the eye.

3. Procedural Entryway Controls. A blocking barrier, screen, or curtain that can block or filter the laser beam at the entryway may be used inside the controlled area to prevent the laser light from exiting the area at levels above the applicable MPE level. In this case, a warning light or sound is required outside the entryway that operates when the laser is energized and operating. All personnel who work in the facility shall be appropriately trained.

4. Entryway Warning Systems. In order to safely operate a Class IV laser or laser system, a laser warning system shall be installed as described:

- A laser activation warning light assembly shall be installed outside the entrance to each laser room facility containing a Class IV laser or laser system. ▪In lieu of a blinking entryway warning, the entryway light assembly may alternatively be interfaced to the laser in such a manner that a light will indicate when the laser is not operational (high voltage off) and by an additional light when the laser is powered up (high voltage applied) but not operating and by an additional (flashing) light when the laser is operating.

A laser warning sign shall be posted both inside and outside the laser-controlled area.

H. TEMPORARY LASER-CONTROLLED AREA. Should overriding interlocks become necessary during periods of special training, service, or maintenance, and access to Class IIIB or Class IV lasers is required, a temporary laser-controlled area shall be devised following specific procedures approved by the LSO. These procedures shall outline all safety requirements necessary during such operation.

Such temporary laser-controlled areas, which by nature will not have the built-in protective features as defined for a laser-controlled area, shall nevertheless provide all of the safety requirements for all personnel, both within and without the temporary laser-controlled area during periods of operation when the interlocks are defeated.

I. ADMINISTRATIVE AND PROCEDURAL CONTROLS.

1. Standard Operating Procedures. One of the more important of the administrative and procedural controls is the written Standard Operating Procedure (SOP). The ANSI Z 136.1 standard requires an SOP for a Class IV laser and recommends SOP's for Class IIIB lasers.

   The key to developing an effective SOP is the involvement of those individuals who operate, maintain and service the equipment under guidance of the LSO. Most laser equipment comes with instructions for safe operation by the manufacturers; however, sometimes the instructions are not well suited to a specific application due to special use conditions.

2. Alignment Procedures. Many laser eye accidents occur during alignment. The procedures require extreme caution. A written SOP is recommended for all recurring alignment tasks.

3. Limitations on Spectators. Persons unnecessary to the laser operation should be kept away. For those who do enter a laser area with unenclosed Class IIIB or Class IV beam paths, appropriate eye protection and instruction is required.

4. Protective Equipment. Protective equipment for laser safety generally means eye protection in the form of goggles or spectacles, clothing, and barriers and other devices designed for laser protection.

   a. Laser Protective Eyewear and Clothing.

      - Eye-protection devices designed to protect against radiation from a specific laser system shall be used when engineering controls are inadequate to eliminate the possibility of potentially hazardous eye exposure (i.e., whenever levels of accessible emission exceed the appropriate MPE levels.) This generally applies only to Class IIIB and Class IV lasers. All laser eye wear shall be clearly labeled with OD values and wavelengths for which protection is afforded. Skin protection can best be achieved through engineering controls. If the potential exists for damaging skin exposure, particularly for ultraviolet lasers (0.200-0.400 m), then skin covers and or sun-screen creams are recommended. For the hands, gloves will provide some protection against laser radiation. Tightly woven fabrics and opaque gloves provide the best protection. A laboratory jacket or coat can provide protection for the arms. For Class IV lasers, flame-resistant materials may be best.
      - In general, other controls should serve as primary protection rather than depending on employees to use protective eye wear. Many accidents have occurred when eye wear was available but not worn. This may be because laser protective eye wear is often dark, uncomfortable to wear, and limits vision.

   b. Laser Barriers and Protective Curtains.

      - Area control can be effected in some cases using special barriers specifically designed to withstand either direct or diffusely scattered beams. The barrier will be described with a barrier threshold limit (BTL): the beam will penetrate the barrier only after some specified exposure time, typically 60 seconds. The barrier is located at a distance from the laser source so that the BTL is not exceeded in the worst-case exposure scenario.
      - Currently available laser barriers exhibit BTL's ranging from 10 to 350 W/cm2 for different laser wavelengths and power levels. An analysis conducted in a manner similar to the NHZ evaluations described previously can establish the recommended barrier type and installation distances for a given laser. It is essential that the barrier also not support combustion or be itself consumed by flames during or following a laser exposure.

J. ENGINEERING CONTROLS. Engineering controls are normally designed and built into the laser equipment to provide for safety. In most instances, these will be included on the equipment (provided by the laser manufacturer) as part of the "performance requirements" mandated by the FLPPS. Specifics on some of the more important engineering controls recommended in the ANSI Z 136.1 standard are detailed as follows:

1. Protective Housing. A laser shall have an enclosure around it that limits access to the laser beam or radiation at or below the applicable MPE level. A protective housing is required for all classes of lasers except, of course, at the beam aperture. In some cases, the walls of a properly enclosed room area can be considered as the protective housing for an open beam laser. Such a "walk-in" enclosure can also be a FDA/CDRH Class I provided that controls preclude operation with personnel within the room (viz.: pressure sensitive floor-mat switches, IR sensors, door interlocks, etc.)

2. Master Switch Control. All Class IV lasers and laser systems require a master switch control. The switch can be operated by a key or computer code. When disabled (key or code removed), the laser cannot be operated. Only authorized system operators are to be permitted access to the key or code. Inclusion of the master switch control on Class IIIB lasers and laser systems is also recommended but not required.

3. Optical Viewing System Safety. Interlocks, filters, or attenuators are to be incorporated in conjunction with beam shutters when optical viewing systems such as telescopes, microscopes, viewing ports, or screens are used

to view the beam or beam-reflection area. For example, an electrical interlock could prevent laser system operation when a beam shutter is removed from the optical system viewing path. Such optical filter interlocks are required for all except Class I lasers.

4. Beam Stop or Attenuator. Class IV lasers require a permanently attached beam stop or attenuator which can reduce the output emission to a level at or below the appropriate MPE level when the laser system is on "standby." Such a beam stop or attenuator is also recommended for Class IIIA and Class IIIB lasers.

5. Laser Activation Warning System. An audible tone or bell and/or visual warning (such as a flashing light) is recommended as an area control for Class IIIB laser operation. Such a warning system is mandatory for Class IV lasers. Such warning devices are to be activated upon system start-up and are to be uniquely identified with the laser operation. Verbal "countdown" commands are an acceptable audible warning and should be a part of the SOP.

6. Service Access Panels. The ANSI Z 136.1 standard requires that any portion of the protective housing that permits direct access to an embedded Class IIIB or Class IV laser (intended for removal only by service personnel) must have either an interlock or require a tool in the removal process. If an interlock is used and is defeatable, a warning label indicating this fact is required on the housing near the interlock. The design shall not allow replacement of a removed panel with the interlock in the defeated condition.

   The FDA/CDRH Federal Laser Product Performance Standard requires warning labels on removable protective housing panels under all conditions.

7. Protective Housing Interlock Requirements.

   a. Interlocks, which cause beam termination or reduction of the beam to MPE levels, must be provided on all panels intended to be opened during operation and maintenance of all Class IIIA, Class IIIB, and Class IV lasers. The interlocks are typically electrically connected to a beam shutter. The removal or displacement of the panel closes the shutter and eliminates the possibility of hazardous exposures.

   b. Under the requirements of the ANSI Z 136 Standard, for embedded Class IIIB and Class IV lasers only, the interlocks are to be "fail-safe." This usually means that dual, redundant, electrical series-connected interlocks are associated with each removable panel.

   c. Adjustments or procedures during service on the laser shall not cause the safety interlocks to become inoperative or the laser radiation outside a Class I laser protective housing to exceed the MPE limits, unless a temporary laser-controlled area is established. The interlocking requirements under the FLPPS are detailed and summarized in Appendix III:6-2.

8. Remote Interlock Connector. All Class IV lasers or laser systems must have a remote interlock connector to allow electrical connections to an emergency master disconnect ("panic button") interlock or to room, door or fixture interlocks. When open circuited, the interlock shall cause the accessible laser radiation to be maintained below the appropriate MPE level. The remote interlock connector is also recommended for Class IIIB lasers.

K. LASER USE WITHOUT PROTECTIVE HOUSING (ALL CLASSES). In some circumstances, such as during the manufacture of lasers and during research and development, operation of an unenclosed laser or laser system may become necessary. In such cases, the LSO shall determine the hazard and ensure that controls are instituted appropriate to the class of maximum accessible emission to ensure safe operation. Such controls may include but are not limited to:

- access restriction;
- eye protection;
- area controls;
- barriers, shrouds, beam stops, etc.;
- administrative and procedural controls; and
- education and training.

L. OPTICAL FIBER (LIGHT WAVE) COMMUNICATION SYSTEMS (OFCS).

1. Under normal operation such systems are completely enclosed (Class I) with the optical fiber and optical connectors forming the enclosure. During installation or servicing, or when an accidental break in the cable occurs, the system can no longer be considered enclosed. If engineering controls limit the accessible emission to levels below the applicable MPE (irradiance), no controls are necessary. If the accessible emission is above the MPE, the following requirements shall apply:

   a. Only authorized trained personnel shall be permitted to perform service on light wave transmission systems if access to laser emission is required.

   b. Only authorized trained personnel shall be permitted to use the laser test equipment (Optical Loss Test Set, Optical Time Domain Reflectometer, etc.) during installation and/or service.

   c. All unauthorized personnel shall be excluded from the immediate area of access to laser radiation during service and installation when there is a possibility that the system may become energized. The immediate area shall be considered a temporary laser-controlled area.

   d. Staring into the end of any broken, severed, or unterminated optical fiber or cable shall be avoided.

   e. The end of any broken, severed, or unterminated optical fiber shall not be viewed with unfiltered optical instruments (microscopes, telescopes, etc.) An exception to this is the use of indirect image converters such as an infrared image converter or closed-circuit television system for verification that a fiber is not energized.

   f. During a splicing operation (either installation or service), if it is required that the ends of the fiber be examined with an eye-loupe for a satisfactory cut, only an eye-loupe containing an appropriate filter shall be used. If a fusion splicer is used, the appropriate operating safety procedures shall be rigidly adhered to.

VII. BIBLIOGRAPHY.

American National Standards Institute, *American National Standard for the Safe Use of Lasers*: ANSI Z 136.1 (1993), Laser Institute of America, New York, NY (1993).

American National Standards Institute, *American National Standard for the Safe Use of Optical Fiber Communication Systems Utilizing Laser Diode and LED Sources*: ANSI Z 136.2 (1988), Laser Institute of America, Orlando, Florida, 1988.

American National Standards Institute, *American National Standard for the Safe Use Lasers in the Health Care Environment:* ANSI Z 136.3 (1988), Laser Institute of America, Orlando, Florida, 1988.

Food and Drug Administration: *Regulations for the Administration and Enforcement of the Radiation Control for Health and Safety Act of 1968 (Lasers),* Center for Devices and Radiological Health, Food and Drug Administration (DHHS), 21 CFR Chapter 1, 1968.

Food and Drug Administration: *Performance Standard for Laser Products*, Center for Devices and Radiological Health, Food and Drug Administration (DHHS), 50 CFR 161: 33682-33702, 1985.

Food and Drug Administration: *Compliance Guide for Laser Products*, Center for Devices and Radiological Health, Food and Drug Administration (DHHS), FDA 86-8260.

Doyle, Daryl and Kokasa, John, Laser Processing of Kevlar: *Hazardous Chemical By-products*, Proceedings of ICALEO, Laser Institute of America, Toledo, Ohio, 1986.

Rockwell, R. James, Jr., Ed., *Laser Safety Training Manual* - Sixth Edition, Rockwell Associates, Inc., Cincinnati, Ohio, 1989.

Sliney, David H. and Wolbarsh, Myron L., *Safety With Lasers and Other Optical Sources.* New York, Plenum, 1980.

Rockwell, R. James, Jr., "Ensuring safety in laser robotics," *Lasers and Applications*, 3(11): 65-69, November 1984.

Rockwell, R. James, Jr., "Fundamentals of Industrial Laser Safety." In *Industrial Laser Annual Handbook,* M. Levitt and D. Belforte, Eds. Penn Well Books, Tulsa, Okla., pp. 131-148, 1986.

Laser Institute of America, "Safety information on electrical hazards," *Laser News*, Laser Inst. of America, 6(5): 8-14, September 1984.

Doyle, D.J. and Kokosa, J.M., "Hazardous By-Products of Plastics Processing with Carbon Dioxide Lasers." In: *Laser Welding, Machining and Materials Processing*: C. Albright, Ed. Proceedings of ICALEO, IFS LTD., Bedford, U.K., pp. 201-203, 1985.

Rockwell, R. James, Jr. and Moss, C.E., "Optical radiation hazards of laser welding processes, part I: Nd:YAG laser," *The Journal of The American Industrial Hygiene Association*, 44(8): 572-579, August 1983.

Rockwell, R. James, Jr. and Moss, C.E., "Optical radiation hazards of laser welding processes, part II: carbon dioxide laser," *The Journal of The American Industrial Hygiene Association*, 50(8): 419-427, August 1989.

Rockwell, R. James, Jr., "Laser accidents: are they all reported and what can be learned from them?" *Journal of Laser Applications*, Laser Institute of America, Toledo, Ohio, pp. 53-57, October 1989.

Underwriters Laboratories Inc., UL 1740: *Standard for Industrial Robots and Robotic Equipment.* Underwriters Laboratories Inc. Northbrook, Ill., 1995.

APPENDIX III:6-1. FDA/CDRH REQUIREMENTS FOR LASER PRODUCTS

Requirements	Class[1]					
	I	**IA**	**II**	**IIIA**	**IIIB**	**IV**
Performance (all laser products)						
Protective housing	R[2]	R[2]	R[2]	R[2]	R[2]	R[2]
Safety interlock	R[3,4]	R[3,4]	R[3,4]	R[3,4]	R[3,4]	R[3,4]
Location of controls	_	R	R	_	R	R
Viewing optics	R	R	R	R	R	R
Scanning safeguard	R	R	R	R	R	R
Performance (laser systems)						
Remote control connector	_	_	_	_	R	R
Key control	_	_	_	_	R	R
Emission indicator	_	_	R	R	R[10]	R[10]
Beam attenuator	_	_	R	R	R	R
Reset	_	_	_	_	_	R[13]
Performance (specific purpose products)						
Medical	S	S	S	S[8]	S[8]	S[8]
Surveying, leveling, alignment	S	S	S	S	NP	NP
Demonstration	S	S	S	S	S[11]	S[11]
Labeling (all laser products)						
Certification and identification	R	R	R	R	R	R
Protective housing	D[5]	R[5]	R[5]	R[5]	R[5]	R[5]
Aperture	_	_	R	R	R	R
Class warning	_	R[6]	R[7]	R[9]	R[12]	R[12]
Information (all lasers)						
User information	R	R	R	R	R	R
Product literature	_	R	R	R	R	R
Service information	R	R	R	R	R	R

Key:
- R = required
- _ = not applicable
- S = same as other products of class
- NP = not permitted
- D = depends on level of interior radiation

Notes:

1. Based on highest level accessible during operation.
2. Required wherever and whenever human access to laser radiation above Class I limits is not needed for product to perform its function.
3. Required for protective housings opened during operation or maintenance, if human access thus gained is not always necessary when housing is opened.
4. Interlock requirements vary according to Class of internal radiation.
5. Wording depends on level and wavelength of laser radiation within protective housing.
6. Warning statement label.
7. **CAUTION** logotype.
8. Requires means to measure level of radiation intended to irradiate the body.

9. **CAUTION** if 2.5 mWcm-2 or less, **DANGER** if greater than 2.5 mWcm-2.
10. Delay required between indication and emission.
11. Variance required for Class IIIB or IV demonstration laser products and light shows.
12. **DANGER** logotype.
13. Required after August 20, 1986.

APPENDIX III:6-2. FDA/CDRH FEDERAL PRODUCT PERFORMANCE STANDARD EVALUATION OUTLINE

LASER EMISSIONS DESCRIPTION:

Type:

CLASSIFICATION DESIGNATION:

CLASS I	[ ]	CLASS IIIA	[ ]
CLASS IIA	[ ]	CLASS IIIB	[ ]
CLASS II	[ ]	CLASS IV	[ ]

PERFORMANCE REQUIREMENTS:	PRESENT: [YES]	[NO]
Protective Housing—All Classes—1040.10(f)(1)	[ ]	[ ]
Safety Interlocks—All Classes—1040.10(f)(2)		
Note: requirements are dependent on class of internal radiation		
Nondefeatable Interlocks	[ ]	[ ]
Defeatable Interlocks	[ ]	[ ]
Remote Interlock Connector—Classes IIIB & IV—1040.10(f)(3)	[ ]	[ ]
Key Control—Classes IIIB & IV—1040.10(f)(4)	[ ]	[ ]
Emission Indicator—Classes II, III & IV—1040.10(f)(5)		
Class II & IIIA: no delay—1040.10(f)(5)(i)	[ ]	[ ]
Class IIIB & IV: with delay—1040.10(f)(5)(ii)	[ ]	[ ]
Indicators on laser and controls, if separated by more than 2 meters. **Note:** Class IIA is exempt.	[ ]	[ ]
Beam Attenuator—Classes II, III & V—1040.10(f)(6)	[ ]	[ ]
Location of Controls—Classes IIA, II, III & IV—1040.10(f)(7)	[ ]	[ ]
Viewing Optics—All Classes—1040.10(f)(8)	[ ]	[ ]
Scanning Safeguard—All Classes—1040.10(f)(9)	[ ]	[ ]
Manual Reset Mechanism—Class IV—1040.10(f)(10)	[ ]	[ ]

LABEL REQUIREMENTS:	PRESENT: [YES]	[NO]
Certification Label—All Classes—1010.2	[ ]	[ ]
Identification Label—All Classes—1010.3	[ ]	[ ]
Class Designation and Warning Label— All Classes except Class I—1040.10(g)(1,2,& 3)	[ ]	[ ]
Radiation Output Information (Position 2 on label)— Classes II, III & IV— 1040.10(g)(4) **Note:** Class IIA is exempt.	[ ]	[ ]
Aperture Label—Classes II, III & IV—1040.10(g)(5) **Note:** Class IIA is exempt.	[ ]	[ ]
Noninterlocked Protective Housing Labels (i.e. service panels)— All Classes—1040.10(g)(6)	[ ]	[ ]
Defeatably Interlocked Protective Housing Labels—1040.10(g)(7)	[ ]	[ ]

Warning for Invisible Radiation (wherever applicable - all labels)—1040.10(g)(8)	[ ]	[ ]
Positioning of Labels—1040.10(g)(9)	[ ]	[ ]
Label Specifications—1040.10(g)(10)	[ ]	[ ]

<u>INFORMATION REQUIREMENTS</u>:	**PRESENT:**	
	[YES]	**[NO]**
User Information—1040.10(h)(1)	[ ]	[ ]
Operator & Maintenance Instructions—(h)(1)(i)	[ ]	[ ]
Statement of Parameters—(h)(1)(ii)	[ ]	[ ]
Label Reproductions—(h)(1)(iii)	[ ]	[ ]
Listing of Controls, Adjustments, and Procedures, including Warnings—(h)(1)(iv)	[ ]	[ ]
Service Information—1040.10(h)(2)(ii)	[ ]	[ ]

APPENDIX III:6-3. THE AMERICAN NATIONAL STANDARDS INSTITUTE (ANSI).

An American National Standard implies a consensus of those substantially concerned with its scope and provisions. These standards are intended as a guide to aid the manufacturer, the consumer and the general public. There is, however, no inherent requirement for anyone or any company to adhere to an ANSI standard. Compliance is voluntary unless specifically required by some alliance. For example, the Federal Department of Energy requires adherence to the ANSI Z 136.1 by their internal staff as well as all contractor organizations. At present, the following ANSI standards can be useful in laser safety matters:

ANSI Z 136.1 (1993): "FOR THE SAFE USE OF LASERS"

This is the master or "parent" standard in the Z 136 series. Originally published in 1973, this standard has gone through revisions in 1976, 1980, 1986, and 1993. Revisions are currently being considered by the committee for the next publication which should occur in 1991.

The Z 136.1 (1993) standard includes the basis of laser hazard assessment including the Nominal Hazard Zone (NHZ) concept and measurements, establishes Maximum Permissible Exposure (MPE) limits based on bioeffects of the eye and skin, introduces a general classification scheme, specifies the recommended control measures, outlines suggested medical surveillance practice, specifies training requirements and recommends practices for other (nonbeam) concerns.

ANSI Z 136.2 (1988): "FOR THE SAFE USE OF OPTICAL FIBER COMMUNICATION SYSTEMS UTILIZING LASER DIODE AND LED SOURCES"

This ANSI Standard provides guidance for the safe use of optical fiber communications systems (OFCS) utilizing laser diodes and/or LED sources by defining control measures for each of four service group (SG) classifications. This concept eliminates measurements and/or calculations. The standard is limited to OFCS that emit at power levels at or below 0.5 W.

The Z 136.2 standard addresses the same general topics as the base Z 136.1 standard. The significant variation between the two standards lies in the definition of a Maximum Permissible Irradiance (MPI) value based upon a 5 mm limiting aperture in the visible and near-infrared spectral regions. This is in contrast to the larger 7mm limiting aperture used in this spectral region in the ANSI Z 136.1 Standard.

ANSI Z 136.3 (1988): "FOR THE SAFE USE OF LASERS IN HEALTH CARE FACILITIES"

The ANSI Z 136.3 Standard provides guidance for the safe use of lasers for diagnostic and therapeutic uses in health care facilities. The publication includes an extensive appendix that details suggested practice in fourteen medical subspecialty areas. (Although the appendix is not technically a part of the standard.)

Major emphasis is included on the associated hazard topics of airborne contaminants ("plume") and electrical and fire hazards. The information in the base of the standard has been kept less mathematical and refers the reader frequently to the ANSI Z 136.1 Standard if such detail is needed. Summary charts provide information on MPE and NHZ values. Control measures are identical, in most cases, to the master standards, with items special to medical laser systems (foot pedals, output calibration, QC and safety audits, etc.) given additional coverage.

ANSI/NFPA 70-1990: THE NATIONAL ELECTRICAL CODE - 1990 HANDBOOK

The fifth edition of the National Electrical Code Handbook is based on the new edition of The National Fire Protection Association's NFPA-70, The National Electrical Code. This document, authored by the NFPA's Electrical Code Committee, is unquestionably the most authoritative and comprehensive document on electrical safety and addresses the topic from both a fire and electrocution point of view. The Handbook includes the complete text of the Code provisions along with explanations. It provides a basis for safety procedures for all electrical hazards associated with lasers.

ANSI Z 87.1 (1989): "PRACTICE FOR OCCUPATIONAL AND EDUCATIONAL EYE AND FACE PROTECTION"

This ANSI standard is included in this listing mainly because a high percentage of industrial facilities require that the eye protection used in their plants meet the ANSI Z 87 requirements. In the past, this basically referred to the "drop ball test" in which a 1-inch diameter ball weighing 2.4 ounces and dropped 50 inches must not fracture the protective lens.

The new Z 87.1 (1989) requires a high-velocity impact test wherein small (1/4 in) diameter steel balls weighing 0.04 ounces are projected at high velocity at the protective filter. In addition, there is a high-mass impact test in which a 1-inch diameter, pointed, heat-treated steel projectile (30-degree cone) that weighs 17.6 ounces is dropped at specified distances. These new requirements place far more stringent performance requirements on safety eye wear.

APPENDIX III:6-4. WARNING SIGNS.

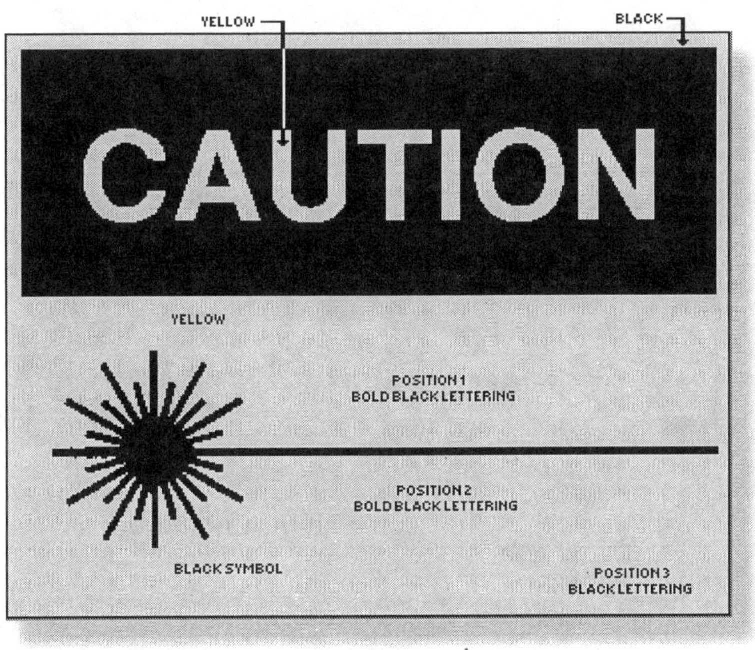

FIGURE III: 6-2. CAUTION

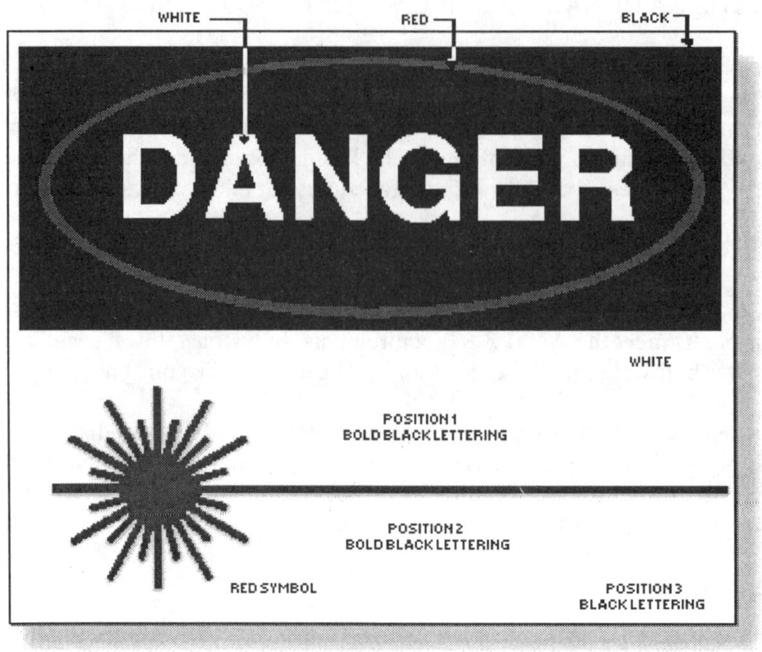

FIGURE III: 6-3. DANGER

FIGURE III: 6-4. LASER REPAIR NOTICE

APPENDIX III:6-5. GLOSSARY OF LASER TERMS.

Absorb To transform radiant energy into a different form, with a resultant rise in temperature.

Absorption Transformation of radiant energy to a different form of energy by the interaction of matter, depending on temperature and wavelength.

Accessible Emission Level The magnitude of accessible laser (or collateral) radiation of a specific wavelength or emission duration at a particular point as measured by appropriate methods and devices. Also means radiation to which human access is possible in accordance with the definitions of the laser's hazard classification.

Accessible Emission Limit (AEL) The maximum accessible emission level permitted within a particular class. In ANSI Z 136.1, AEL is determined as the product of accessible emission Maximum Permissible Exposure limit (MPE) and the area of the limiting aperture (7 mm for visible and near-infrared lasers).

Aperture An opening through which radiation can pass.

Argon A gas used as a laser medium. It emits blue-green light primarily at 448 and 515 nm.

Attenuation The decrease in energy (or power) as a beam passes through an absorbing or scattering medium.

Aversion Response Movement of the eyelid or the head to avoid an exposure to a noxious stimulant, bright light. It can occur within 0.25 seconds, and it includes the blink reflex time.

Beam A collection of rays that may be parallel, convergent, or divergent.

Beam Diameter The distance between diametrically opposed points in the cross section of a circular beam where the intensity is reduced by a factor of e-1 (0.368) of the peak level (for safety standards). The value is normally chosen at e-2 (0.135) of the peak level for manufacturing specifications.

Beam Divergence Angle of beam spread measured in radians or milliradians (1 milliradian = 3.4 minutes of arc or approximately 1 mil). For small angles where the cord is approximately equal to the arc, the beam divergence can be closely approximated by the ratio of the cord length (beam diameter) divided by the distance (range) from the laser aperture.

Blink Reflex See aversion response.

Brightness The visual sensation of the luminous intensity of a light source. The brightness of a laser beam is most closely associated with the radio-metric concept of radiance.

Carbon Dioxide Molecule used as a laser medium. Emits far energy at 10,600 nm (10.6 μm).

Closed Installation Any location where lasers are used which will be closed to unprotected personnel during laser operation.

CO2 Laser A widely used laser in which the primary lasing medium is carbon dioxide gas. The output wavelength is 10.6 μm (10600 nm) in the far infrared spectrum. It can be operated in either CW or pulsed.

Coherence A term describing light as waves which are in phase in both time and space. Monochromaticity and low divergence are two properties of coherent light.

Collimated Light Light rays that are parallel. Collimated light is emitted by many lasers. Diverging light may be collimated by a lens or other device.

Collimation Ability of the laser beam to not spread significantly (low divergence) with distance.

Continuous Mode The duration of laser exposure is controlled by the user (by foot or hand switch).

Continuous Wave (CW) Constant, steady-state delivery of laser power.

Controlled Area Any locale where the activity of those within are subject to control and supervision for the purpose of laser radiation hazard protection.

Diffuse Reflection Takes place when different parts of a beam incident on a surface are reflected over a wide range of angles in accordance with Lambert's Law. The intensity will fall off as the inverse of the square of the distance away from the surface and also obey a Cosine Law of reflection.

Divergence The increase in the diameter of the laser beam with distance from the exit aperture. The value gives the full angle at the point where the laser radiant exposure or irradiance is e-1 or e-2 of the maximum value, depending upon which criteria is used.

Embedded Laser A laser with an assigned class number higher than the inherent capability of the laser system in which it is incorporated, where the system's lower classification is appropriate to the engineering features limiting accessible emission.

Emission Act of giving off radiant energy by an atom or molecule.

Enclosed Laser Device Any laser or laser system located within an enclosure which does not permit hazardous optical radiation emission from the enclosure. The laser inside is termed an "embedded laser."

Energy (Q) The capacity for doing work. Energy is commonly used to express the output from pulsed lasers and it is generally measured in Joules (J). The product of power (watts) and duration (seconds). One watt second = one Joule.

Excimer "Excited Dimer" A gas mixture used as the active medium in a family of lasers emitting ultraviolet light.

Fail-safe Interlock An interlock where the failure of a single mechanical or electrical component of the interlock will cause the system to go into, or remain in, a safe mode.

Gas Discharge Laser A laser containing a gaseous lasing medium in a glass tube in which a constant flow of gas replenishes the molecules depleted by the electricity or chemicals used for excitation.

Gas Laser A type of laser in which the laser action takes place in a gas medium.

Helium-Neon (HeNe) Laser A laser in which the active medium is a mixture of helium and neon. Its wavelength is usually in the visible range. Used widely for alignment, recording, printing, and measuring.

Infrared Radiation (IR) Invisible electromagnetic radiation with wavelengths which lie within the range of 0.70 to 1000 μm. These wavelengths are often broken up into regions: IR-A (0.7-1.4 μm), IR-B (1.4-3.0 μm) and IR-C (3.0-1000 μm).

Intrabeam Viewing The viewing condition whereby the eye is exposed to all or part of a direct laser beam or a specular reflection.

Irradiance (E) Radiant flux (radiant power) per unit area incident upon a given surface. Units: Watts per square centimeter. (Sometimes referred to as power density, although not exactly correct).

Laser An acronym for light amplification by stimulated emission of radiation. A laser is a cavity with mirrors at the ends, filled with material such as crystal, glass, liquid, gas or dye. It produces an intense beam of light with the unique properties of coherency, collimation, and monochromaticity.

Laser Accessories The hardware and options available for lasers, such as secondary gases, Brewster windows, Q-switches and electronic shutters.

Laser Controlled Area See Controlled Area.

Laser Device Either a laser or a laser system.

Laser Medium (Active Medium) Material used to emit the laser light and for which the laser is named.

Laser Rod A solid-state, rod-shaped lasing medium in which ion excitation is caused by a source of intense light, such as a flash lamp. Various materials are used for the rod, the earliest of which was synthetic ruby crystal.

Laser Safety Officer (LSO) One who has authority to monitor and enforce measures to control laser hazards and effect the knowledgeable evaluation and control of laser hazards.

Laser System An assembly of electrical, mechanical and optical components which includes a laser. Under the Federal Standard, a laser in combination with its power supply (energy source).

Lens A curved piece of optically transparent material which, depending on its shape, is used to either converge or diverge light.

Light The range of electromagnetic radiation frequencies detected by the eye, or the wavelength range from about 400 to 760 nm. The term is sometimes used loosely to include radiation beyond visible limits.

Limiting Aperture The maximum circular area over which radiance and radiant exposure can be averaged when determining safety hazards.

Maintenance Performance of those adjustments or procedures specified in user information provided by the manufacturer with the laser or laser system, which are to be performed by the user to ensure the intended performance of the product. It does not include operation or service as defined in this glossary.

Maximum Permissible Exposure (MPE) The level of laser radiation to which a person may be exposed without hazardous effect or adverse biological changes in the eye or skin.

Nd:Glass Laser A solid-state laser of neodymium:glass offering high power in short pulses. A Nd-doped glass rod used as a laser medium to produce 1064 nm light.

Nd:YAG Laser Neodymium:Yttrium Aluminum Garnet. A synthetic crystal used as a laser medium to produce 1064 nm light.

Neodymium (Nd) The rare earth element that is the active element in Nd:YAG laser and Nd:Glass lasers.

Nominal Hazard Zone (NHZ) The nominal hazard zone describes the space within which the level of the direct, reflected or scattered radiation during normal operation exceeds the applicable MPE. Exposure levels beyond the boundary of the NHZ are below the appropriate MPE level.

Optical Cavity (Resonator) Space between the laser mirrors where lasing action occurs.

Optical Density A logarithmic expression for the attenuation produced by an attenuating medium, such as an eye protection filter.

Optical Fiber A filament of quartz or other optical material capable of transmitting light along its length by multiple internal reflection and emitting it at the end.

Optical Pumping The excitation of the lasing medium by the application of light rather than electrical discharge.

Optical Radiation Ultraviolet, visible, and infrared radiation (0.35-1.4 μm) that falls in the region of transmittance of the human eye.

Output Power The energy per second measured in watts emitted from the laser in the form of coherent light.

Power The rate of energy delivery expressed in watts (Joules per second). Thus: 1 Watt = 1 Joule × 1 Sec.

Protective Housing A protective housing is a device designed to prevent access to radiant power or energy.

Pulse A discontinuous burst of laser, light or energy, as opposed to a continuous beam. A true pulse achieves higher peak powers than that attainable in a CW output.

Pulse Duration The "on" time of a pulsed laser, it may be measured in terms of milliseconds, microseconds, or nanoseconds as defined by half-peak-power points on the leading and trailing edges of the pulse.

Pulsed Laser Laser which delivers energy in the form of a single or train of pulses.

Pump To excite the lasing medium. See Optical Pumping or Pumping.

Pumped Medium Energized laser medium.

Pumping Addition of energy (thermal, electrical, or optical) into the atomic population of the laser medium, necessary to produce a state of population inversion.

Radiant Energy (Q) Energy in the form of electromagnetic waves usually expressed in units of Joules (watt-seconds).

Radiant Exposure (H) The total energy per unit area incident upon a given surface. It is used to express exposure to pulsed laser radiation in units of J/cm2.

Reflection The return of radiant energy (incident light) by a surface, with no change in wavelength.

Refraction The change of direction of propagation of any wave, such as an electromagnetic wave, when it passes from one medium to another in which the wave velocity is different. The bending of incident rays as they pass from one medium to another (e.g., air to glass).

Resonator The mirrors (or reflectors) making up the laser cavity including the laser rod or tube. The mirrors reflect light back and forth to build up amplification.

Ruby The first laser type; a crystal of sapphire (aluminum oxide) containing trace amounts of chromium oxide.

Scanning Laser A laser having a time-varying direction, origin or pattern of propagation with respect to a stationary frame of reference.

Secured Enclosure An enclosure to which casual access is impeded by an appropriate means (e.g., door secured by lock, magnetically or electrically operated latch, or by screws).

Semiconductor Laser A type of laser which produces its output from semiconductor materials such as GaAs.

Service Performance of adjustments, repair or procedures on a non-routine basis, required to return the equipment to its intended state.

Solid Angle The ratio of the area on the surface of a sphere to the square of the radius of that sphere. It is expressed in steradians (sr).

Source The term source means either laser or laser-illuminated reflecting surface, i.e., source of light.

Tunable Laser A laser system that can be "tuned" to emit laser light over a continuous range of wavelengths or frequencies.

Tunable Dye Laser A laser whose active medium is a liquid dye, pumped by another laser or flash lamps, to produce various colors of light. The color of light may be tuned by adjusting optical tuning elements and/or changing the dye used.

Ultraviolet (UV) Radiation Electromagnetic radiation with wavelengths between soft X-rays and visible violet light, often broken down into UV-A (315-400 nm), UV-B (280-315 nm), and UV-C (100-280 nm).

Visible Radiation (light) Electromagnetic radiation which can be detected by the human eye. It is commonly used to describe wavelengths in the range between 400 nm and 700-780 nm.

Wavelength The length of the light wave, usually measured from crest to crest, which determines its color. Common units of measurement are the micrometer (micron), the nanometer, and (earlier) the Angstrom unit.

YAG Yttrium Aluminum Garnet, a widely used solid-state crystal composed of yttrium and aluminum oxides and a small amount of the rare earth neodymium.

SECTION III: CHAPTER 7

LEGIONNAIRES' DISEASE

Contents:

I. INTRODUCTION.

This chapter provides information to assist industrial hygienists in the assessment of work sites for potential Legionnaires' disease. It provides information on disease recognition, investigation procedures to identify probable water sources, and control strategies. The primary focus of this document is on the control and prevention of contaminated water sources, not on case identification, an area of expertise primarily exercised by local health departments frequently in conjunction with the Centers for Disease Control and Prevention (CDC) in Atlanta. Appendices include details on conducting an employee awareness program, water sampling protocols and guidelines for acceptable levels of the organism in water, procedures for identifying new cases of the disease, and water treatment and control strategies for facilities where an outbreak has occurred.

II. DISEASE RECOGNITION.

A. CAUSATIVE AGENT.

1. Legionella pneumophila was first identified in 1977 by the CDC as the cause of an outbreak of pneumonia that caused 34 deaths at a 1976 American Legion Convention in Philadelphia. L. pneumophila had undoubtedly caused previous pneumonia outbreaks, but the organism's slow growth and special growth requirements prevented earlier discovery.

2. The diseases produced by Legionella are called legionellosis. More than 34 species of Legionella have been identified, and more than 20 linked with human diseases. L. pneumophila causes the pneumonia known as Legionnaires' disease and the flu-like Pontiac fever. L. pneumophila has also been implicated in wound infections, pericarditis, and endocarditis without the presence of pneumonia. Because the majority of legionellosis is caused by L. pneumophila, this chapter will deal exclusively with that organism. Cases where other species of Legionella are involved in disease require actions similar to those to control Legionnaires' disease.

3. The L. pneumophila bacteria are gram-negative rods that exist in a number of distinguishable serogroups. Each serogroup contains further subtypes that have different surface structures on the cell membrane and can be distinguished by special tests. Evidence indicates that some Legionella serogroups are more virulent than others. L. pneumophila serogroup 1 is the most frequently identified form of the bacterium isolated from patients with

Legionnaires' disease, although other serogroups and subtypes of the bacterium are frequently isolated from water sources. Serogroups 4 and 6 are the next most frequently linked with disease.

B. SYMPTOMS.

1. Legionnaires' disease has an incubation period of two to ten days. Severity ranges from a mild cough and low fever to rapidly progressive pneumonia and coma. Early symptoms include malaise, muscle aches, and slight headache. Later symptoms include high fever (up to 105°F), a dry cough, and shortness of breath. Gastrointestinal symptoms including vomiting, diarrhea, nausea, and abdominal pain are common. The disease is treated with erythromycin or a combination of erythromycin and rifampin.

2. Pontiac fever is a non-pneumonia, flu-like disease associated with, and likely caused by, the Legionella bacterium. This disease has an "attack rate" of 90% or higher among those exposed, and a short incubation period, 1-3 days. Complete recovery usually occurs in 2-5 days without medical intervention. The factors that cause the same organism to produce two illnesses with major differences in attack rate and severity are not known.

C. INCIDENCE. In the U.S., Legionnaire's disease is considered to be fairly common and serious, and the Legionella organism is one of the top three causes of sporadic, community-acquired pneumonia. Because it is difficult to distinguish this disease from other forms of pneumonia, many cases go unreported. Approximately 1,000 cases are reported annually to the CDC, but it is estimated that over 25,000 cases of the illness occur each year and cause more than 4,000 deaths.

D. RISK FACTORS. Legionnaires' disease is frequently characterized as an "opportunistic" disease that most frequently attacks individuals who have an underlying illness or weakened immune system. The most susceptible include persons who are elderly, smokers, and immunosuppressed. Individuals with chronic obstructive pulmonary disease (COPD), organ transplant patients, and persons taking corticosteroid therapy are also at elevated risk. The attack rate for the average population is approximately 5% or less. The fatality rate is similar to that of other forms of pneumonia, approximately 15%.

E. DIAGNOSIS. CDC guidelines define two types of cases of Legionelloses, probable and confirmed. A probable case of Legionnaire's disease is a person who experienced an illness clinically compatible with Legionnaire's and has a single antibody titer of 256 or higher (discussed below), and can be associated with a population of individuals who have experienced confirmed cases of the disease (outbreak). A confirmed case of Legionella requires a physician's diagnosis of pneumonia based on a chest x-ray and positive laboratory test results. A laboratory test is necessary for confirmation because the symptoms and x-ray evidence of Legionnaires' disease resemble those of other types of pneumonia. Various methods are used to confirm the presence of the disease.

1. Culture of the Organism. The definitive laboratory method of confirming the presence of the disease is by culturing viable cells of Legionella from sputum, bronchial washing, or autopsy on special media. Further identification of the cultured cells will identify the species and serogroup. Special tests may determine subtype of certain isolates. The sensitivity of this test to detect the disease is reported to be about 70%.

2. Urine Antigen Test. The detection of antigen from L. pneumophila in the urine is considered a reliable measure of the disease. These antigenic materials may include L. pneumophila cells or portions of cells in the urine during and after the disease. The presence of antigen in the urine is a strong indicator of the disease, and a patient may have a positive response for several months following the disease. The sensitivity of this test is limited because the only commercially available urinary antigen test detects only serogroup 1 forms of L. pneumophila. The CDC recommends only the radioimmunoassay (RIA) test because the latex antigen (LA) test has a high false-positive rate. Fortunately, 80-90% of the clinically diagnosed cases are caused by serogroup 1. The absence of a positive urinary test is not proof that a patient did not have Legionnaires' disease, but merely indicates the absence of antigen in the urine at the time of the test.

3. Direct Fluorescent Antibody (DFA) Staining. Direct fluorescent antibody staining of lung aspirates can detect L. pneumophila. However, this test is frequently negative during the initial stages of the disease because few organisms are present in the aspirate or sputum. This test also requires an antigen-specific reagent. There are a

multitude of serogroups and subtypes of L. pneumophila, and a test will be negative if the exact antigen-specific reagent is not included.

4. Serology (Antibody Titers).

   a. An increase in the antibody level in the serum of infected persons occurs several weeks after the onset of the disease. A fourfold increase in the antibody titer coupled with a physician's diagnosis of pneumonia is considered a reliable indicator of disease. This is measured by comparing the antibody level four to eight weeks after onset (convalescent titer) to an initial (acute) titer at the beginning of the disease. Pontiac fever also produces an elevated antibody titer, but the flu-like symptoms of this disease do not match those of Legionnaires' disease.

   b. Frequently only a convalescent titer has been measured from individuals who had symptoms of the disease. For situations in which these cases are associated with an outbreak of Legionnaires' disease, a single titer of 256 to 1 or higher is generally used as a presumptive indication of disease (probable case). Antibody strength is determined by the number of dilutions of serum which elicit a positive antibody response. The reciprocal value of the number of dilutions is the antibody titer. For example, an antibody titer of 256 means a positive antibody test of the patient's serum following serial dilutions of 1:2, then 1:4, then 1:16, etc., until the 1:256 dilution point is reached.

   c. The indirect fluorescent antibody (IFA) test is the accepted diagnostic tool for demonstrating L. pneumophilia exposure. Another widely used test of antibody response is the enzyme-linked immunosorbent assay method (ELISA). CDC believes that direct comparison of results between IFA and ELISA is not reliable because there are insufficient data to compare the two. The ELISA method has gained wide medical acceptance as a useful means of demonstrating exposure to Legionella.

F. TRANSMISSION.

1. The likelihood of contracting Legionnaires' disease is related to the level of contamination in the water source, the susceptibility of the person exposed, and the intensity of exposure to the contaminated water. Disease transmission usually occurs via inhalation of an aerosol of water contaminated with the organism. Aspiration of contaminated water into the lungs may also cause the disease. In the Philadelphia Legionnaires' disease outbreak, the hotel's cooling tower was identified as the likely source of the disease, although domestic water sources were not evaluated.

2. The disease has been associated with domestic hot-water systems in a number of outbreaks. In many instances it has been difficult to identify a likely source for aerosolization of the suspected water source. Although transmission of the disease other than through direct inhalation of aerosols may occur, the mechanisms are not clearly understood. The organism requires water, and the disease cannot occur in the absence of a contaminated water source. There is no evidence that the disease can be transmitted from one person to another.

III. SOURCE IDENTIFICATION.

A. CONDITIONS THAT PROMOTE GROWTH.

1. L. pneumophila bacteria are widely distributed in water systems. They tend to grow in biofilms or slime on the surfaces of lakes, rivers and streams, and they are not eradicated by the chlorination used to purify domestic water systems. Low and even nondetectable levels of the organism can colonize a water source and grow to high concentrations under the right conditions.

   Conditions that promote growth of the organism include heat, sediment, scale, and supporting (commensal) microflora in water. Common water organisms including algae, amoebae, and other bacteria appear to amplify Legionella growth by providing nutrients or harboring the organism. Because of its ability to remain viable in domestic water systems, it is capable of rapid multiplication under the proper conditions.

2. Water conditions that tend to promote the growth of Legionella include:

- stagnation;
- temperatures between 20° and 50°C (68° - 122°F) (The optimal growth range is 35° - 46°C [95° - 115°F]);
- pH between 5.0 and 8.5;
- sediment that tends to promote growth of commensal microflora; and
- micro-organisms including algae, flavobacteria, and Pseudomonas, which supply essential nutrients for growth of Legionella or harbor the organism (amoebae, protozoa).

B. COMMON SOURCES OF CONTAMINATED WATER.

1. Water sources that frequently provide optimal conditions for growth of the organisms include:

    - cooling towers, evaporative condensers, and fluid coolers that use evaporation to reject heat. These include many industrial processes that use water to remove excess heat;
    - domestic hot-water systems with water heaters that operate below 60°C (140°F) and deliver water to taps below 50°C (122°F);
    - humidifiers and decorative fountains that create a water spray and use water at temperatures favorable to growth;
    - spas and whirlpools;
    - dental water lines, which are frequently maintained at temperature above 20°C (68°F) and sometimes as warm as 37°C (98.6°F) for patient comfort; and
    - other sources including stagnant water in fire sprinkler systems and warm water for eye washes and safety showers.

2. Water stored below 20°C (68°F) is generally not a source for amplified L. pneumophila levels. However, high levels of bacteria have been measured in the water supplying ice machines. The source of amplification in this case was thought to be heat from the condenser coil of the ice maker to the cold water supply. However, no cases of Legionnaires' disease have been linked to consumption of ice made from contaminated water.

C. MONITORING.

1. Air. An air sample applied to special culture plates by an Andersen-type sampler sometimes demonstrates the presence of the organism in the air. However, negative results are frequent because of the difficulty in maintaining viability of the organism on the culture plates. Air sampling for Legionella is strongly not recommended as a means of measuring potential exposure because of the high likelihood of false negatives.

2. Water. Analysis of water samples from a source suspected of being contaminated with L. pneumophila is a valuable means of identifying potential sources of the disease. A qualified microbiological laboratory experienced in Legionella detection can determine the number of organisms present in colony forming units (CFU) per volume of water and can identify the different serogroups of Legionella pneumophila in the sample. Appendix III:7-2 provides details on the collection, storage, and shipping of water samples.

D. MICROBIOLOGICAL ANALYSIS OF WATER SAMPLES.

1. Cultured Samples. Water samples are cultured on special buffered charcoal yeast extract (BCYE) culture media. Selective isolation processes to eliminate other microbial overgrowth can determine the number of CFU of L. pneumophila per milliliter of water. This process of growth and isolation is time-consuming, and results typically require 7-14 days from the time of submission.

    Cultured samples can also be analyzed to identify specific serogroups. Matching the same serogroup and subtype of organism in the patient as found in a water source is considered strong evidence of an associated link.

2. Direct Fluorescence Antibody (DFA). The number of organisms in a water sample can also be determined via direct fluorescence antibody (DFA) conjugate tests that stain the organism with a fluorescent dye. This test is unable to distinguish between live and dead bacteria and may also have some cross-reactivity with other bacteria. Sample results can be available in one or two days, and this method can be useful in screening water

samples. Use caution, however, in interpreting the results since the potential exists for both false positive and negative results.

3. DNA Amplification. A relatively new method for rapid, specific detection of the organism in water employs a polymerase chain reaction (PCR) process to amplify and then detect portions of DNA unique to L. pneumophila. This method can produce results in 1 day, and preliminary evidence indicates that its sensitivity and specificity are comparable to those of cell culture, which can take 10-14 days to obtain results. Further testing may lead to acceptance of this technique as the method of choice for monitoring water sources for contamination.

E. INTERPRETATION OF SAMPLE RESULTS.

1. The probability of infection with L. pneumophila is a function both of the intensity of the exposure dose and the level of host susceptibility. Because total eradication of Legionella may not be possible, an acceptable control strategy is to minimize the number of organisms present in a water source. Ample evidence indicates that Legionella levels are readily controllable. A survey of over 1,000 cooling towers indicates that approximately 60% contained nondetectable levels of L. pneumophila when measured by DFA analysis for the number of organisms per milliliter of water (detection limit is 10 bacteria per milliliter of water). In another survey of 663 cooling towers, 57% contained Legionella that were not detected when measured by culture (detection limit less than 1 CFU/mL). Other studies of domestic hot-water sources indicate that although the organism is common, especially in large hot-water systems, practical control measures can limit the potential for amplification.

2. A private consulting firm and microbiological laboratory, PathCon Inc. of Norcross, Georgia, has introduced suggested guidelines for control of the organism based on the number of CFU of L. pneumophila per milliliter of water (Appendix III:7-3). These guidelines vary depending on the water source, a recognition by the authors that dose is related both to the potential for exposure and to concentration. For example, recommended exposure limits for contaminated water from a humidifier, which would involve direct exposure to an aerosol are lower than for a cooling tower where the opportunity for exposure is normally less. Work operations such as maintenance on cooling towers may involve direct exposure to cooling tower mist, and precautions to minimize exposure are always necessary. The authors recognize that these guidelines are based on limited data, but they represent the best available information and must suffice until the dose effect of L. pneumophila is better understood.

IV. INVESTIGATION PROTOCOL.

A. COMMUNITY HEALTH CONCERNS. It is important to remember that an outbreak of Legionnaires' disease among workers may have its origin in the community and may not be related to the work environment. A Legionnaires' outbreak is both an occupational and a public-health concern, and the investigation may include local public health departments and the Centers for Disease Control (CDC). To minimize employee risk and maximize the effectiveness of effort, close coordination among OSHA, other public agencies, and the employer is imperative.

B. TYPES OF INVESTIGATIONS.

1. The course of action chosen during an investigation of a facility should be based on the degree of certainty that the site is the source of a reported illness. For this reason, two investigation protocols are based on differing levels of suspected risk for exposure to Legionella. It is important to remember that these procedures are provided only to assist in the investigation of potential Legionnaires' cases. Individual circumstances may require changes in the investigation. All cases require sound professional judgment in deciding the appropriate course of action.

2. A level-one investigation may be initiated when there is a probable basis for suspecting that workplace water sources are contaminated with Legionella, or when there is information that one case of Legionnaires' disease may exist. A level-two investigation should be conducted when more then one possible case of Legionnaires' disease has been reported at a facility.

3. If two or more cases of the disease can be attributed to a work site, assume that a Legionnaires' disease outbreak has occurred. If evidence indicates that the outbreak is still in progress (that is, at least one of the cases has

occurred in the last 30 days), prompt actions should be undertaken to provide maximum protection to employees and eliminate the hazard. Appendix III:7-5 includes examples of actions required to control water sources where an outbreak has occurred.

4. Both investigations follow the same general pattern and include a preliminary opening conference, a walk-through of the facility to conduct a physical assessment of the water systems, a more detailed examination of the systems including a review of maintenance records, assessment of findings, and a closing conference to present control actions based on the findings.

C. LEVEL-ONE INVESTIGATION. Use the following procedure when Legionnaires' disease may be related to the work environment.

1. Step 1. Obtain an overview of all water systems at the facility.

a. A facilities engineer or experienced member of the building maintenance staff should be available to explain system operation and assist in the walkthrough investigation. This person should have a working knowledge of the system's design and current operation.

b. The overview of the water systems should include plumbing systems, heating-ventilating-air-conditioning (HVAC) systems, and other water reservoirs. A review of the plumbing system should include both hot and cold domestic water systems, water heaters, distribution pipes, water coolers, water treatment equipment, connections to process water systems protected (or unprotected) by backflow preventers, and storage tanks.

c. The HVAC system review should include cooling towers, evaporative condensers, fluid coolers, humidifiers, direct evaporative air-cooling equipment, indirect evaporative air-cooling equipment, air washers for filtration, etc. Note the location of the fresh-air intakes of the building's air-handling units relative to water sources such as the cooling towers.

d. Investigate other potential sources of employee exposure including decorative fountains, plant misters, whirlpools, spas, tepid-water eye-washes and safety showers, humidifiers, and water for cooling industrial processes.

e. Review maintenance records on water systems including water heaters and cooling towers. The records should include temperature checks of domestic water, visual and physical checks of cooling towers, and reports of cooling-tower water-quality assessment and chemical treatment.

f. Identify the locations of portions of the system in which water is allowed to stagnate such as storage tanks or unused plumbing pipe sections ("dead legs"), or infrequently used faucets. Check for cross-connections between domestic and process water systems, and note the condition and type of back-flow prevention devices.

g. Investigate recent major maintenance or changes in the system's operation. Determined if there were recent or frequent losses of water pressure from the incoming water supply due to line breakage or street repairs. The failure of a back-flow prevention device under loss of pressure can contaminate the system.

2. Step 2. Conduct a walk-through investigation of the facility.

a. Equipment you will need includes a thermometer for measuring water temperatures, a flashlight, and a film or video camera to record observations. Measure and record the temperature of water drawn from each storage-type water heater in the facility. This temperature may be significantly below the water heater's gauge temperature because of heat stratification. Note the presence of rust and scale in this water.

b. Record the maximum temperature of water at faucets connected to each water heater on the system. Record temperatures at locations near, intermediate, and distant from the heaters. It may be necessary to run the water for several minutes before it reaches a temperature maximum.

c. Examine the water temperature and the potential for stagnation of cold-water storage tanks used for reserve capacity or to maintain hydrostatic pressure. These should be protected from temperature extremes and covered to prevent contamination. Record the temperature of the domestic cold-water lines at various locations within the facility. Note both the initial temperature and the final equilibrium temperature on the cold-water line, and record the time required to reach equilibrium, because this can be an indicator of the amount of stagnation in the system.

d. Evaluate cooling towers, evaporative condensers, and fluid coolers for biofilm growth, scale buildup, and turbidity. Record the location of the tower relative to fresh-air intakes, kitchen exhausts, leaves, plant material, or other sources of organic material that might contribute to the growth of the organism.

e. Record the general condition of the cooling tower. Determine the presence and condition of drift eliminators, which are designed to limit the vapor release from the units, along with the basin temperature of the water in the cooling tower if it is currently being operated. If the cooling tower is operating and is suspected of being contaminated, wear appropriate respiratory protection in the form of a half-face piece respirator equipped with a HEPA or similar type of filter capable of effectively collecting one-micron particles during the examination of the system.

f. Note the location and evaluate the condition of the sumps for the cooling tower(s), evaporative condenser(s), and fluid cooler(s). These sumps are sometimes located indoors to protect them from freezing. Record the locations of any cross-connections between the cooling tower water system and any domestic water system. These may supply a back-up source of cool water to refrigeration condenser units or serve to supply auxiliary cooling units. The lack of a regular maintenance schedule or water-treatment program for a cooling tower or evaporative condenser system strongly suggests a potential for Legionella contamination.

3. Step 3. Assess the results of the walkthrough investigation to determine the course of action. If no potential problems are identified, the operating temperatures measured at water heaters are 60°C (140°F) or above, and the delivery temperature at distant faucets is 50°C (122°F) or higher, no further action will be necessary. However, if the system is poorly maintained and operating temperatures are below recommended minimums, then recommendations for corrective action should be made.

4. Step 4. Recommend control actions.

a. Details of suggested control actions are discussed in Section E. These actions may include disinfection of the domestic water system via heat treatment, chlorination, or other means, and cleaning and disinfecting the cooling tower system according to the Wisconsin Division of Health protocol for "Control of Legionella in Cooling Towers" or a similar process for cleaning heat rejection systems that follows sound practices to minimize potential for Legionella growth.

b. Additional actions may include eliminating dead legs in the plumbing system, insulating plumbing lines and installing heat tracing to maintain proper temperatures in the system, eliminating rubber gaskets, and removing or frequently cleaning fixtures such as aerators and shower heads.

c. Corrective actions limited to raising the water heater temperature without evaluating the system for points of stagnation, heat loss and gain, cross-contamination, and other factors that contribute to growth are generally not sufficient.

d. For a level-one investigation it is not recommended that water samples be collected to confirm the presence of Legionella in the system. The absence of proper operating conditions alone is sufficient for assuming that the water system can pose an unnecessary risk to the employees. Take water samples after the completion of the control actions to confirm that the corrective measures were successful. The employer may also want to obtain samples before starting corrective actions to assess the extent of the problem.

e. The employer should take necessary corrective actions even if the results of presampling are negative. Water sampling can produce false negatives, a contaminated portion of the system may have been missed, and the absence of Legionella organisms at the time of sampling does not insure that the system will remain negative.

f. If, after control actions, the Legionella levels in a water source exceed the guidelines in Appendix III:7-3, re-examine the water system to determine if potential contamination points within the system were overlooked and reassess control procedures to determine if they were performed properly. Repeat the procedures as needed until contamination levels meet the guidelines.

D. LEVEL-TWO INVESTIGATION. A level-two investigation is similar to a level-one investigation with several additional steps. Supplemental actions include:(1) medical surveillance of all employees currently on sick leave to identify any new cases, (2) employee awareness training on the disease to minimize employee concerns and aid in early recognition of new cases, (3) assessment of past sick-leave absences for undetected cases of the disease, and (4) collection of water samples during the walkthrough assessment.

1. Step 1. Assess water systems as described for a level-one investigation.

2. Step 2. Conduct a second walkthrough survey of the facility and collect water samples. Estimate the size of the building and the number of water services during the initial walkthrough and prearrange supply and shipping of the required number of sterile sample containers with the appropriate laboratory. (See Appendix III:7-2 for water sampling procedures.)

3. Step 3. Initiate an employee awareness program and monitor current sick leave for new cases. It is important to ensure that employees understand the early symptoms of the disease and seek medical assistance promptly. It is imperative not to alarm the workers. It is equally important to stress the importance of the need to know the health status of all employees on sick leave. (See Appendix III:7-1, Employee Awareness Program.)

4. Step 4. Review worker absences to detect other cases. This requires identification of all employees who took three or more consecutive days of sick leave from approximately six weeks before the case of Legionnaires' disease was identified up to the present. Request those employees who may have had pneumonia during this period to undergo additional voluntary tests for evidence of Legionnaires' disease. (See Appendix III:7-4, Case Identification.)

5. Step 5. Assess results of worker absence survey and analysis of water systems. If evidence indicates more than one case of Legionnaires' disease at the workplace, then the site should be treated as having an outbreak. Take immediate control of all water sources to eliminate potential for exposure, and take measures to eliminate the hazard. (See Appendix III:7-5.)

   No action is necessary if the results of the investigation are negative, that is, (1) all water and HVAC systems are well maintained and in good operating condition; (2) all water sample results are negative or acceptably low (Appendix III:7-3); and (3) no new cases of the disease have been identified at the work site. Under these circumstances, assume that the site is not the origin of the identified case.

6. Step 6. For recommended control actions, see the level-one investigation.

7. Ongoing Outbreak.

   a. If the evidence indicates that two or more cases of Legionnaires' disease have occurred at a site, and at least one of the cases was within the last 30 days, assume that an outbreak is in progress and requires a high-priority investigation and prompt action. Conduct a level-two investigation as outlined above, and take the following precautions to protect building occupants.

   b. Immediately initiate control measures to prevent additional exposures to all water systems that have a reasonable potential for worker exposure including hot and cold domestic water, cooling towers, humidifiers, and any other potential sources of water exposure. Collect appropriate water samples to determine Legionella levels before shutting down the water systems (Appendix III:7-2). These sample results will be invaluable in establishing the cause of the outbreak. A member of the building maintenance or engineering staff who has a working knowledge of the system's design and current operation can explain how the water system operates and the proper procedure for a controlled shutdown.

These control actions need not require facility shutdown. Temporary provisions can allow work to continue: bottled water can be supplied for drinking, shutting off water heaters can eliminate hot-water access, and temporary cooling towers can allow work to continue.

V. CONTROLS.

A. GENERAL DISCUSSION. This section contains background information on water system operations and proper controls to prevent Legionella amplification. This discussion encompasses a variety of water systems, some of which have not been implicated with outbreaks of Legionnaires' disease. Nevertheless, it is important to remember that any water system can be a source of disease if the water in it is subjected to conditions that promote growth of the organism. Remember, however, that the primary sources of exposure to contaminated water are heat rejection systems (cooling towers, fluid coolers, etc.) and domestic hot-water systems.

B. COOLING TOWERS, EVAPORATIVE CONDENSERS, AND FLUID COOLERS. The function of cooling towers, evaporative condensers, and fluid coolers is to reject heat from system fluids through evaporation. Cooling towers remove heat from condenser water via direct-contact evaporation in a wet airstream. This cooled water circulates through the condenser side of a mechanical refrigeration unit to absorb heat. Evaporative condensers operate similarly, except that the refrigerant condenser coils are directly inside the wet air stream and water passing over the coils directly cools the refrigerant. Fluid coolers are employed to reject heat from industrial processes, computer-room air conditioners, etc. Like evaporative condensers, fluid coolers have heat-exchanger coils directly in the wet air stream.

Because all of these systems use a fan to move air through a recirculated water system, a considerable amount of water vapor is introduced into the surroundings despite the presence of drift eliminators designed to limit vapor release. In addition, this water may be in the ideal temperature range for Legionella growth, 20° - 50°C, 68° - 122°F.

1. Inspection and Maintenance. Visual inspection and periodic maintenance of the system are the best ways to control growth of Legionella and related organisms. Good maintenance is necessary both to control Legionella growth and for effective operation. The system should be properly monitored and maintained to prevent buildup of scale and sediment and bio-fouling, all of which support Legionella growth and reduce operating efficiency.

2. Biocide. Unfortunately, measurements of water quality such as total bacterial counts, total dissolved solids, and pH have not proven to be good indicators of Legionella levels in cooling towers. Periodic use of biocides is needed to ensure control of Legionella growth.

   a. Little information exists on the demonstrated effectiveness of many commercial biocides for preventing Legionella growth in actual operations. Recent Australian studies indicate that Fentichlor [2,2'-thiobis (4-chlorophenol)] used weekly for 4 hours at 200 ppm, or bromo-chloro-dimethyl-hydantoin (BCD) in a slow-release cartridge at an initial concentration of 300 ppm are effective in controlling the growth of Legionella. There are no U.S. suppliers of Fentichlor, although the chemical is licensed by the EPA for water treatment in cooling towers. Towerbrom 60MTM, a chlorotriazine and sodium bromide salt mixture, has been reported to be effective when alternated with BCD for control of Legionella in U.S. studies of Legionella contamination of cooling towers. The Australian study also indicates that quaternary ammonium compounds, widely used for control of bio-fouling in cooling towers, are not effective in controlling Legionella.

   b. Traditional oxidizing agents such as chlorine and bromine have been proven effective in controlling Legionella in cooling towers. Continuous chlorination at low free residual levels can be effective in controlling Legionella growth. It is important, however, that the proper oxidant level be established and maintained because free residual chlorine above 1 ppm may be corrosive to metals in the system and may damage wood used in cooling towers; free residual levels below 1 ppm may not adequately control Legionella growth. Chlorine also combines with organic substances in water to form toxic by-products that are of environmental concern. Frequent monitoring and control of pH is essential for maintaining adequate levels of free residual chlorine. Above a pH of 8.0, chlorine effectiveness is greatly reduced. Proper control of pH will maintain the effectiveness of chlorination and minimize corrosion.

c. Bromine is an effective oxidizing biocide. It is frequently added as a bromide salt and generated by reaction with chlorine. Bromine's effectiveness is less dependent than chlorine on the pH of the water; it is less corrosive; and it also produces less toxic environmental by-products.

d. The effectiveness of any water-treatment regimen depends on the use of clean water. High concentrations of organic matter and dissolved solids in the water will reduce the effectiveness of any biocidal agent. Each sump should be equipped with a "bleed," and make-up water should be supplied to reduce the concentration of dissolved solids.

3. Design.

a. One of the most effective means of controlling the growth of Legionella is to maintain sump water at a low temperature. Sump-water temperatures depend on tower design, heat load, flow rate, and ambient dry-bulb and wet-bulb temperatures. Under ideal conditions, sump-water temperatures in evaporative devices approach the ambient wet-bulb temperature, and that may be low enough to limit Legionella amplification. System design should recognize the value of operating with low sump-water temperatures.

b. High-efficiency drift eliminators are essential for all cooling towers. Older systems can usually be retrofitted with high-efficiency models. A well-designed and well-fitted drift eliminator can greatly reduce water loss and potential for exposure. Other important design features include easy access or easily disassembled components to allow cleaning of internal components including the packing (fill). Enclosure of the system will prevent unnecessary drift of water vapor, and other design features to minimize the spray generated by these systems are also desirable.

4. Frequency of Cleaning. Cooling towers should be cleaned and disinfected at least twice a year. Normally this maintenance will be performed before initial start-up at the beginning of the cooling season and after shut-down in the fall. Systems with heavy bio-fouling or high levels of Legionella may require additional cleaning. Any system that has been out of service for an extended period should be cleaned and disinfected. New systems require cleaning and disinfecting because construction material residue can contribute to Legionella growth.

5. Wisconsin Protocol. Acceptable cleaning procedures include those described in the Wisconsin Protocol. This procedure calls for an initial shock treatment with 50 ppm free residual (total) chlorine, addition of detergent to disperse bio-fouling, maintenance of 10 ppm chlorine for 24 hours, and a repeat of the cycle until there is no visual evidence of biofilms. To prevent exposure during cleaning and maintenance, wear proper personal protective equipment: a Tyvek-type suit with a hood, protective gloves, and a properly fitted respirator with a high-efficiency particulate (HEPA) filter or a filter effective at removing one-micron particles.

6. Recordkeeping. A description of the operating system (which includes all components cooled by the system) and details of the make-up water to the system should be available. Written procedures for proper operation and maintenance of the system should indicate the use of scale and corrosion inhibitors, antifoaming agents, and biocides or chlorine use and should be readily available. Log books should list dates of inspections and cleanings, water-quality test results, and maintenance.

C. DOMESTIC HOT-WATER SYSTEMS.

1. Background. Domestic hot-water systems are frequently linked to Legionnaires' outbreaks. The term "domestic" applies to all nonprocess water used for lavatories, showers, drinking fountains, etc., in commercial, residential, and industrial settings. Disease transmission from domestic hot water may be by inhalation or aspiration of Legionella-contaminated aerosolized water. Water heaters that are maintained below 60°C (140°F) and contain scale and sediment tend to harbor the bacteria and provide essential nutrients for commensal micro-organisms that foster growth of L. pneumophila. Large water heaters like those used in hospitals or industrial settings frequently contain cool zones near the base where cold water enters and scale and sediment accumulate. The temperature and sediment in these zones can provide ideal conditions for amplification of the organism. Dead legs (i.e., sections of piping or plumbing that have been altered or capped such that water cannot flow through)

and nonrecirculated plumbing lines that allow hot water to stagnate also provide areas for growth of the organism.

2. Design. Water systems designed to recirculate water and minimize dead legs will reduce stagnation. If potential for scalding exists, appropriate, fail-safe scald-protection equipment should be employed. For example, pressure-independent, thermostatic mixing valves at delivery points can reduce delivery temperatures. Point-of-use water heaters can eliminate stagnation of hot water in infrequently used lines. Proper insulation of hot-water lines and heat tracing of specific lines can help maintain distribution and delivery temperatures.

3. Maintenance.

    a. To minimize the growth of Legionella in the system, domestic hot water should be stored at a minimum of 60°C (140°F) and delivered at a minimum of 50°C (122°F) to all outlets. The hot-water tank should be drained periodically to remove scale and sediment and cleaned with chlorine solution if possible. The tank must be thoroughly rinsed to remove excess chlorine before reuse.

    b. Eliminate dead legs when possible, or install heat tracing to maintain 50°C (122°F) in the lines. Rubber or silicone gaskets provide nutrients for the bacteria, and removing them will help control growth of the organism. Frequent flushing of these lines should also reduce growth.

    c. Domestic hot-water recirculation pumps should run continuously. They should be excluded from energy conservation measures.

4. Control.

    a. Raising the water-heater temperature can control or eliminate Legionella growth. Pasteurize the hot water system by raising the water-heater temperature to a minimum of 70°C (158°F) for 24 hours and then flushing each outlet for 20 minutes. It is important to flush all taps with the hot water because stagnant areas can "re-seed" the system. Exercise caution to avoid serious burns from the high water temperatures used in Pasteurization.

    b. Periodic chlorination of the system at the tank to produce 10 ppm free residual chlorine and flushing of all taps until a distinct odor of chlorine is evident is another means of control. In-line chlorinators can be installed in the hot water line; however, chlorine is quite corrosive and will shorten the service life of metal plumbing. Control of the pH is extremely important to ensure that there is adequate residual chlorine in the system.

    c. Alternative means to control Legionella growth include the use of metal ions such as copper or silver (which have a biocidal effect) in solution. Ozonization injects ozone into the water. Ultraviolet (UV) radiation also kills microorganisms. Commercial, in-line UV systems are effective and can be installed on incoming water lines or on recirculating systems, but stagnant zones may diminish the effectiveness of this treatment. Scale buildup on the UV lamp surface can rapidly reduce light intensity and requires frequent maintenance to ensure effective operation.

D. DOMESTIC COLD-WATER SYSTEMS.

1. Domestic cold water systems are not a major problem for Legionella growth. Maintaining cold-water lines below 20°C will limit the potential for amplification of the bacteria. It is surprising, however, that elevated levels of Legionella have been measured in ice machines in hospitals. Cold-water lines near heat sources in the units are believed to have caused the amplification.

2. Dental water lines have recently been recognized as common sources of water contaminated with high concentrations of microorganisms including Legionella. However, to date an increased risk of disease among dental staff or patients has not been demonstrated. Dental water line operating conditions are especially appropriate for Legionella proliferation because the water is stagnant a majority of the time, the narrow plastic tubing encourages biofilm formation, and the water temperature is usually 20°C (68°F) or higher--some systems

maintain water at 37°C (98.6°F). Filtration of water at the point of use with replaceable, in-line, FDA-cleared, 0.22-microns pore sizes filters is recommended for minimizing risk to patients and staff in a dental facility.

3. Water tanks that allow water to remain uncirculated for long periods can also promote growth of bacteria. Such tanks should be eliminated or designed to reduce storage time to a day or less. They should also be covered to prevent contamination and protected from temperature extremes.

4. Cross-contaminations of the domestic cold-water system with other systems should always be suspected. All connections to process water should be protected by a plumbing code-approved device (e.g., back-flow preventer, air gap, etc.). If significant contamination of the domestic cold water system occurs, the source of contamination must be determined. Inspect the system for "dead legs" and areas where water may stagnate. Elimination of these sections or frequent flushing of taps to drain the stagnant areas may be necessary to limit growth of the organism. Insulate cold-water lines that are close to hot-water lines to reduce the temperature in the line.

5. If the cold-water lines have significant contamination, hyperchlorination can eradicate Legionella. Free chlorine levels of 20 to 50 ppm are allowed to remain for one hour at 50 ppm, or two hours at 20 ppm. Faucets are then allowed to run until the odor of chlorine is present, and the water is allowed to remain for approximately two hours.

E. HVAC SYSTEMS.

1. HVAC systems are not normally amplification sites for Legionella. The organism cannot survive without water, and a properly operated, well-maintained HVAC system is unlikely to be a source of problems. However, the HVAC system can disseminate contaminated water aerosols.

2. Water-aerosol sources are classified as either external or internal.

   a. External sources may emit contaminated aerosolized water that is drawn into a system's fresh-air intake. Mist discharged from cooling towers, evaporative condensers, and fluid coolers can be ingested by the HVAC fresh air intake. When evaluating this path, you should consider:

      - prevailing wind direction and velocity;
      - building effects (e.g., low-pressure zones on leeward sides of buildings and on roof);
      - architectural screen walls; and
      - distance from tower to intake.

   b. Fresh-air intake areaways, typically concrete plenums located at grade level, supply fresh air to air handlers in the basement or lower levels of buildings and can collect organic material (e.g. leaves, dirt, etc.) and water from rain or irrigation.

   c. Do not ignore direct paths such as through an open window. The transmission path through the HVAC system is tortuous, and the bacteria may die from desiccation in the airstream or impact on internal surfaces like filters, duct lining, etc. When evaluating external sources, examine the potential for direct transmission.

   d. Internal sources may provide contaminated aerosolized water that is then disseminated by the air-distribution system. Contaminated water can leak from pipes into HVAC ducts, where it can be aerosolized and distributed by the system. Potential sources of contaminated water include domestic water systems, fire-sprinklers, refrigeration condensers, etc.

   e. HVAC system humidifiers can be hazards. Four types are common:

      - Heated-pan humidifiers use a heat source to evaporate water from a pan open to the air stream. Intermittent use of the device coupled with a warm pan of water may support Legionella growth. Contaminant-free water is essential.
      - Direct steam-type humidifiers inject boiler-generated steam directly into the air stream. These systems normally operate above 70°C (158°F), and Legionella cannot survive at that temperature.

- Atomizing humidifiers use mechanical devices or pneumatic air to create a water mist that evaporates into the air stream. A contaminant-free water source is essential.

f. Direct evaporative air coolers may be used as humidifiers. These devices mix water and air in direct contact to create a cool, wet air stream by evaporation. These devices include sumps, which may stagnate when not in use.

g. When draining properly, the water that passes through the condensate pans of cooling coils in an air handler is normally not a source of growth for the organism because of the low temperature of water condensate.

h. Indirect evaporative air cooling in systems designed for dryer climates. One common design circulates cool water from a cooling tower sump through a water coil in the supply air stream. If the coil develops a leak, then pumped cooling tower water will be injected directly into the supply air stream with potentially deleterious effects if the sump water is contaminated with Legionella. Indirect evaporative air cooling is also found in air-to-air heat exchangers. One side of the heat exchanger is an evaporative-cooled wet air stream, and the other side supplies air for the conditioned space. If the heat exchanger leaks, the wet air stream can mix with supply air and cause problems if the wet air stream is contaminated with Legionella.

i. Many air-handling systems designed for dryer climates employ direct evaporative air cooling. Wet evaporative coolers, slinger air coolers and rotary air coolers common in commercial applications. These devices mix water and air in direct contact to create a cool, wet air stream by evaporation. If these systems are using 100% outside air in a dry climate, the water sump temperature may be low and will not represent a significant risk. However, improperly operated and maintained systems that use warm, stagnant sump water can present problems.

j. Other equipment may also be potential sources of Legionella:

- Residential humidifiers are small, free-standing, portable units that use an internal fan and wet media to disseminate a wet air stream. The sumps of these devices are frequently contaminated with Legionella. Daily cleaning is necessary to maintain acceptable water quality, but these units seldom receive appropriate maintenance, and their use in the commercial or industrial workplace is strongly discouraged.
- Computer room air conditioners typically include humidifiers and frequently are not well maintained. They may contain a sump filled with contaminated water.

3. The following are issues to consider when designing HVAC systems to minimize risk from Legionella contamination. Most apply to all types of microbial contamination.

a. Minimize use of water reservoirs, sumps, and pans. Chemically untreated, stagnant, warm-water sources provide an ideal environment for Legionella growth.

b. Provide a way to drain water sumps when not in use, e.g., an electric solenoid valve on the sump drain. If an HVAC sump is used during the hours when a building is occupied, drain the sump during unoccupied hours.

c. Provide a "bleed" for water sumps so that dissolved solids do not form sediments in the sump.

d. Slope and drain sumps from the bottom so that all the water can drain out and allow the pan to dry.

e. Locate HVAC fresh-air intakes so that they do not draw the mist from a cooling tower, evaporative condenser, or fluid cooler into the system. The American Conference of Governmental Industrial Hygienists publishes "Guidelines For The Assessment Of Bioaerosols In The Indoor Environment," which lists recommended minimum distances between cooling towers and fresh-air intakes.

f. Design indirect evaporative cooling systems with the knowledge that the failure of the heat exchanger will allow wet systems to mix with the air-distribution systems.

g. Use steam or atomizing humidifiers instead of units that use recirculated water. Do not use raw steam from the central heating boiler because it contains corrosion inhibitors and anti-scaling chemicals. Atomizing humidifiers must have contaminant-free water.

4. Operate all HVAC equipment as originally designed, and maintain it so that it can perform as designed. Test all HVAC equipment periodically to insure that it is performing as designed. Inactive sumps must be properly drained and bled to prevent accumulation of sediments. Maintenance failures can produce contaminated, stagnant water that can become an ideal environment for Legionella growth if heated (e.g., by sunlight).

VI. BIBLIOGRAPHY.

American Water Works Association, *A Procedure for Disinfecting Water Mains* AWWA C601 1981; Denver CO.

Best, M., A. Goetz, and V.L. Yu. "Heat eradication measures for control of nosocomial Legionnaires' disease." *American Journal of Infection Control*, 12, (1), 1984, pp. 26-30.

Broadbent, C.R. "Legionella in Cooling Towers: Practical Research, Design, Treatment and Control Guidelines." Delivered at 1992 Inter. Symp. on Legionella, Amer. Society for Micro., Jan. 26-29, 1992, Orlando Fl.

Chartered Institution of Building Services Engineers, *Minimizing The Risk of Legionnaires' Disease*, Delta House, 222 Balham High Rd., London 1987.

England, A.C. et al. "Sporadic legionellosis in the U.S.: the first 1000 cases." *Ann. Inter. Med.* 94, 1981, p. 164.

Gilpin, R.W., A.M. Kaplan, and E.F. Goldstein "Quantitation of Legionella pneumophila in one thousand commercial and industrial cooling towers." Proceedings 48th Inter. Water Conference, Oct. 24-26, 1988, Pittsburgh, pp.13-19.

Health and Safety Executive (UK), "The Control of legionellosis including Legionnaires' disease." Health and Safety Series Booklet, HS (G)70, Library and Information Services, Broad Lane, Sheffield S37HQ, Tel: (0742) 752539.

Health Department Victoria; Melbourne Australia, "Guidelines for the Control of Legionnaires' Disease" in *Environmental Health Standards*, 1989.

Morris, G.K. and B.G. Shelton. *Legionella in Environmental Samples: Hazard Analysis and Suggested Remedial Actions*. March 1991, Pathogen Control Assoc., Norcross, Georgia.

Muder, R.R., V.L. Yu, and A.H. Woo. "Mode of transmission of Legionella pneumophila." *Arch Intern Med*, 146, (1986), pp. 1607-1612.

Muraca, P., J.E. Stout, and V.L. Yu. "Comparative assessment of chlorine, heat, ozone, and UV light for killing Legionella pneumophila within a model plumbing system." *Applied and Environmental Microbiology*, 53, (2), 1987, p. 447-453.

Muraca, P.W., V.L. Yu, and R.N. Goetz. "Disinfection of water distribution systems for Legionella: a review of application procedures and methodologies." *Infect. Control Hosp. Epidemiol.*, 11, (2), 1990, pp. 79-88.

Muraca, P.W., J.E. Stout, V.L. Yu, and Y.C. Ying. "Legionnaires' disease in the work environment: implications for environmental health." *Am. Ind. Hyg. Assoc.*, 49, (11), 1988, pp. 584-590.

Nalco Chemical Company. *Cooling Water Chlorination*, Technifax, TF-132, Nalco Chemical Co., Naperville, Illinois, 1986.

Nguyen, M.H., J.E. Stout, and V.L. Yu. "Legionellosis." *Lower Respiratory Tract Infections*, 5, (3), September 1991, pp. 561-584.

Stout, J.E., V.L. Yu, and M.S. Muraca. "Isolation of Legionella pneumophila from the cold water of hospital ice machines: implications for origin and transmission of the organism." *Infection Control*, 7786, (4), 1985, pp. 141-146.

Stout, J.E., V.L. Yu, M.S. Muraca, J. Joly, N. Troup, and L.S. Tompkins. "Potable water as a cause of sporadic cases of community-acquired Legionnaires' disease." *New England Journal of Medicine*, 326, January 16, 1992, pp. 151-155.

Wisconsin Division of Health. Control of Legionella in Cooling Towers: *Summary Guidelines*, June 1987, Wisconsin Department of Health and Social Sciences.

Williams, J.F. et. al. "Microbial contamination of dental unit waterlines: prevalence, intensity and microbiological characteristics." J. of *American Dental Association*, 124, October 1993, pp. 59-65.

APPENDIX III:7-1. EMPLOYEE AWARENESS PROGRAM.

The purpose of an employee awareness program is to inform the employees of the potential outbreak, and to educate them about the disease. This educational program should be part of a level-two investigation or for a Legionnaires' disease outbreak. This program is of critical importance to aid in early recognition of the disease. It is also important to help alleviate employee concerns about the disease. This program should supplement the case identification program to discover previously undetected cases of the illness at the work site.

The employer should implement the following elements of this program immediately upon recognition of more than one probable or confirmed case of disease in the work place:

- An initial employee training session which provides basic information about the disease and actions being taken to investigate the problem;
- An ongoing general information service to provide updates and answer questions that may arise among employees; and
- Medical and psychological counseling services when an outbreak has occurred.

Below is a sample letter and supplemental information on the disease that the employer can use for informing employees of a potential or actual outbreak.

SAMPLE LETTER FROM EMPLOYER TO EMPLOYEES

DATE:

MEMO TO: ALL EMPLOYEES

FROM: MANAGEMENT OFFICIAL

SUBJECT: Legionnaires' Disease

On __________________, we were notified that one of the employees of our company had contracted legionellosis, commonly referred to as Legionnaires' disease. The employee is assigned to __________________ on ___________ shift.

We want to share with you some general information concerning the disease. In addition, we want to tell you what we are currently doing here at ______________________ to ensure all necessary steps are taken to address health concerns.

Legionellosis, or Legionnaire's disease, is a type of pneumonia caused by *Legionella* bacteria. Legionnaires' disease is not contagious, and you cannot catch it from another person. The bacteria are common and grow in water. People often receive low-level exposure in the environment without getting sick. Legionellosis usually occurs only when someone who is already susceptible receives concentrated exposure to the bacteria. Persons who are heavy smokers, elderly, or whose ability to resist infection is reduced are more likely to contract Legionnaires' disease than healthy nonsmokers. According to the Centers for Disease Control in Atlanta, there are between 10,000 and 50,000 cases of Legionnaire's disease every year in the U.S. We are cooperating fully with local health officials who are investigating this matter. Most cases of legionellosis are isolated and are not associated with an outbreak. To date, _____ case(s) of the disease has/have occurred among employees in this facility.

To identify any other cases, we will review sick-leave records for the period ____________ to _____________. Employees who took more than three consecutive days of sick leave will be identified, and we will attempt to determine if any in that group experienced pneumonia-like symptoms (fever, shortness of breath, cough). Those who used three or more consecutive days of sick leave during this period can expect to be contacted by a representative of our company for an interview. If you experienced a pneumonia-like illness in the past two months but used fewer then three consecutive days of sick leave, contact _________________ to arrange an interview.

To assure that you are being protected during the interim, we are also instituting a medical surveillance program to identify any new or old cases. Part of this surveillance will be to ask you a few questions about your illness when you call in sick to your supervisor. In addition, we are offering counseling and employee information services. If you would like to take advantage of these services or want more information, contact your manager. For the present, please pay attention to the following important points:

WHAT YOU SHOULD DO NOW:

1. If you are not sick, there is no need for you to see a doctor.
2. If you are now sick with a cough and fever:
   A. See your private doctor or contact ___________ to arrange to see a _________________ physician.

   B. Tell the physician that you work in a building that may be involved in a Legionnaires' disease outbreak.

   C. If you see a physician, notify _________________ so that your illness can be tracked.

If you have any concerns or questions concerning this issue, please ask your manager. Your health and safety are of great concern to us, and we will be grateful for your cooperation in this matter. As further information develops we will keep you informed.

SAMPLE INFORMATION TO BE OBTAINED BY INTERVIEW WITH EMPLOYEES CALLING IN ON SICK LEAVE

Interviewer:_____________________________Date:_________________

SUPERVISOR SURVEY FORM

We are screening employee illnesses as a result of our Legionnaire's disease incident. You are not obligated to participate in the survey, but your participation will help you and your fellow workers.

We recommend that you see a physician if you currently have pneumonia-like symptoms such as severe chills, high fever, a cough, and difficult breathing.

Are you currently experiencing these symptoms?

Yes______ No______ Prefer not to answer_______

- If the answer to the question is "No," do not complete the rest of this form. Thank you for your cooperation.
- If the answer is "Yes," please read the statement below and complete the bottom half of this form (Employee name, etc).
- If you answer is "Prefer not to answer," please complete ONLY the bottom half of this form (Employee name, etc).

STATEMENT: You will be contacted by ______________ to obtain additional information necessary to complete our survey.

Thank you!

Employee's Name ____
Work Telephone Number ________________________________
Home Telephone Number ________________________________
Shift: Day ___ Swing ___ Graveyard ___ Rotating ___
Branch ________________________/Organization Code ________________
Employee's Supervisor ____
Telephone Number ____
Date _______________________________________

PLEASE FORWARD TO __________________ BY 10:00 am EACH DAY

LEGIONNAIRES' DISEASE: QUESTIONS AND ANSWERS

BACKGROUND.

Legionnaires' disease is a common name for one of the several illnesses caused by Legionella bacteria. Legionnaires' disease is an infection of the lungs that is a form of pneumonia. A person can develop Legionnaires' disease by inhaling water mist contaminated with Legionella.

Legionella bacteria are widely present at low levels in the environment: in lakes, streams, and ponds. At low levels the chance of getting Legionnaires' disease from a water source is very slight. The problem arises when high concentrations of the organism grow in water sources. Water heaters, cooling towers, and warm, stagnant water can provide ideal conditions for the growth of the organism.

Scientists have learned much about the disease and about the Legionella bacteria since it was first discovered in 1976. The following questions and answers will help you learn more of what is currently known about Legionnaires' disease.

Q.What are the symptoms of Legionnaires' disease?

A.Early symptoms of the illness are much like the flu. After a short time (in some cases a day or two), more severe pneumonia-like symptoms may appear. Not all individuals with Legionnaires' disease experience the same symptoms. Some may have only flu-like symptoms, but to others the disease can be fatal.

Early flu-like symptoms:

- slight fever

- headache
- aching joints and muscles
- lack of energy, tired feeling
- loss of appetite

Common pneumonia-like symptoms:

- high fever (102° to 105°F, or 39° to 41°C)
- cough (dry at first, later producing phlegm)
- difficulty in breathing or shortness of breath
- chills
- chest pains

Q.How common is Legionnaires' disease?

A.It is estimated that in the United States there are between 10,000 and 50,000 cases each year.

Q.How does a person get Legionnaires' disease?

A.A person must be exposed to water contaminated with Legionella bacteria. This exposure may happen by inhaling or drinking water contaminated with the Legionella bacteria. For example, inhaling contaminated water mist from a cooling tower, a humidifier, or even a shower or sink can cause the disease.

Q.How soon after being exposed will a person develop symptoms of the disease?

A.If infection occurs, disease symptoms usually appear within 2 to 10 days.

Q.Are some people at a higher risk of developing Legionnaires' disease?

A.Yes, some people have lower resistance to disease and are more likely to develop Legionnaires' disease. Some of the factors that can increase the risk of getting the disease include:

- organ transplants (kidney, heart, etc.);
- age (older persons are more likely to get disease);
- heavy smoking;
- weakened immune system (cancer patients, HIV-infected individuals);
- underlying medical problem (respiratory disease, diabetes, cancer, renal dialysis, etc.);
- certain drug therapies (corticosteroids); and
- heavy consumption of alcoholic beverages.

Q.Is Legionnaires' disease spread from person to person?

A.No. Legionnaires' disease is not contagious and cannot be transmitted from one person to another.

Q.What causes Legionnaires disease?

A.Legionnaires' disease is caused by inhaling water contaminated with rod-shaped bacteria called Legionella pneumophila. There are over 30 different species of Legionella, many of which can cause disease. Legionella pneumophila is the most common species that causes disease.

Q.Does everyone who inhales Legionella into the lungs develop Legionnaires' disease?

A.No. Most people have resistance to the disease. It is thought that fewer than 5 out of 100 persons exposed to water contaminated with Legionella will develop Legionnaires' disease.

Q.Is Legionnaires' disease easy to diagnose?

A.No. The pneumonia caused by Legionella is not easy to distinguish from other forms of pneumonia. A number of diagnostic tests allow a physician to identify the disease. These tests can be performed on a sample of sputum, blood, or urine.

Q.How is Legionnaires' disease treated?

A.Erythromycin is currently the antibiotic of choice. Early treatment reduces the severity and improves chances for recovery. In many instances this antibiotic may be prescribed without the physician's knowledge that the disease is Legionnaires' because erythromycin is effective in treating a number of types of pneumonia.

Q.How did Legionnaires' disease get its name?

A.Legionnaires' disease got its name from the first outbreak in which the organism was identified as the cause. This outbreak occurred in 1976, in a Philadelphia hotel where the Pennsylvania American Legion was having a convention. Over 200 Legionnaires and visitors at this convention developed pneumonia, and some died. From lung tissue, a newly discovered bacterium was found to be the cause of the pneumonia and was named Legionella pneumophila.

Q.Is Legionnaire's disease a new disease?

A.No, Legionnaires' disease is not new, but it has only recently been identified. Unsolved pneumonia outbreaks that occurred before 1976 are now known to have been Legionnaires' disease. Scientists are still studying this disease to learn more about it.

Q.Are Legionella bacteria widespread in the environment?

A.Yes, studies have shown that these bacteria can be found in both natural and man-made water sources. Natural water sources including streams, rivers, freshwater ponds and lakes, and mud can contain the organism in low levels.

Q.Could I get the disease from natural water sources?

A.It is unlikely. In the natural environment the very low levels of this organism in water sources probably cannot cause disease.

Q.What water conditions are best for growth of the organism?

A.Warm, stagnant water provides ideal conditions for growth. At temperatures between 68° and 122°F the organism can multiply. Temperatures of 90°-105°F are ideal for growth. Rust (iron), scale, and other micro-organisms can also promote the growth of Legionella.

Q.What common types of water are of greatest concern?

A.Water mist from cooling towers or evaporative condensers, evaporative coolers (swamp coolers), humidifiers, misters, showers, faucets, and whirlpool baths can be contaminated with the organism and if inhaled or swallowed can cause the disease.

Q.Can Legionnaires' disease be prevented?

A.Yes. Avoiding water conditions that allow the organism to grow to high levels is the best means of prevention. Specific preventive steps include:

- Regularly maintain and clean cooling towers and evaporative condensers to prevent growth of Legionella. This should include twice-yearly cleaning and periodic use of chlorine or other effective biocide. ▪Maintain domestic water heaters at 140°F (60°C). The temperature of the water should be 122°F or higher at the faucet.

- Avoid conditions that allow water to stagnate. Large water-storage tanks exposed to sunlight can produce warm conditions favorable to high levels of Legionella.
- Frequent flushing of unused water lines will help alleviate stagnation.

Q.Do you recommend that I operate my home water heater at 140°F?

A.Probably not if you have small children or infirm elderly persons who could be at serious risk of being scalded by the hot water. However, if you have persons living with you who are at high risk of contracting the disease, then operating the water heater at a minimum temperature of 140F is probably a good idea.

Q.What can be done if a water system is already contaminated or is suspected of being contaminated?

A.Special cleaning procedures can eliminate Legionella from water sources. In many cases these procedures involve the use of chlorine-producing chemicals or high water temperatures. Professional assistance should be sought before attempting to clean a water system.

Q.Can my home water heater also be a source of Legionella contamination?

A.Yes, but evidence indicates that smaller water systems such as those used in homes are not as likely to be infected with Legionella as larger systems in work places and public buildings.

Q.Can Legionella bacteria cause other diseases?

A.Yes. In addition to Legionnaires' disease, the same bacteria also cause a flu-like disease called Pontiac fever.

Q.How does Pontiac fever differ from Legionnaires' disease?

A.Unlike Legionnaires disease, which can be a serious and deadly form of pneumonia, Pontiac fever produces flu-like symptoms that may include fever, headache, tiredness, loss of appetite, muscle and joint pain, chills, nausea, and a dry cough. Full recovery occurs in 2 to 5 days without antibiotics. No deaths have been reported from Pontiac fever.

Q.Are there other differences between Legionnaires' disease and Pontiac fever?

A.Yes. Unlike Legionnaires' disease, which occurs in only a small percentage of persons who are exposed, Pontiac fever will occur in approximately 90% of those exposed. In addition, the time between exposure to the organism and appearance of the disease (called the incubation period) is generally shorter for Pontiac fever than for Legionnaires' disease. Symptoms of Pontiac fever can appear within one to three days after exposure.

APPENDIX III:7-2. PHYSICAL SURVEY AND WATER SAMPLING PROTOCOL.

Arrange with the appropriate laboratory for supply and shipment of sterile sampling containers, and for analysis of water samples. During the initial walk-through, estimate the size of the building and the number of water services at the facility to determine the number of samples and the size of the purchase order. When investigating the water services within a building, it will be helpful to obtain or prepare a simple schematic diagram of the water services. Note the following features:

- The location of the incoming supply and/or private source.
- The location of storage tanks, water treatment systems, and pumps.
- The location of water heaters and boilers.
- The type of fittings used in the system (e.g., taps, showers, valves) and the material from which the pipework is made.
- The location of all cooling towers, evaporative condensers, and fluid coolers at the facility. The location and type of all systems served by the cooling tower water including sump tanks, condensers, and indirect evaporative cooling coils in air handling units.
- The location of any evaporative cooling systems or humidifiers.
- he location of ornamental fountains, whirlpools, eye washes, safety showers, or other water sources within or near the facility.

Trace the route of the service from the point of entry of the water supply. Note the condition of pipes, jointing methods used, insulation, sources of heat, and the kind of insulation in water storage tanks. Also note carefully any disconnected fittings, "dead legs," and cross-connections with other services. Once you have identified these features, take water samples from:

- The incoming water supply.
- Each storage tank and water heater.
- A representative number of faucets for each of the hot and cold water systems in the facility.
- All cooling towers, evaporative condensers, humidifiers, spas, showers, etc.
- The water entering or leaving any other type of fitting or piece of equipment under particular suspicion.

It is important not to overlook any potential water sources in the building. Water from ice machines, hand spray bottles, decorative fountains, and for plastic injection molding equipment have been implicated in past outbreaks or have been found to be significantly contaminated. The ability to maintain an open mind is essential in conducting an investigation because of the variety of potential sources of contamination at a facility.

WATER SAMPLING PROCEDURE.

Wear appropriate respiratory protection in the form of a half-face piece respirator equipped with a HEPA filter or a similar type of filter media capable of effectively collecting particles in the one micron size range during the examination of water systems if a significant potential exists for exposure to high concentrations of contaminated aerosols.

Collect samples in polypropylene (nalgene) containers (250 mL-1 L) that have been autoclaved at 121°C for 15 minutes. The microbiological laboratory that will analyze the samples should be able to provide the bottles. A local hospital or state health department should be able to autoclave the bottles. It is important not to flush the system to be sampled before collecting samples. Collect at least a 250 mL sample. Measure the temperature of the sampled water. It is preferable to accomplished this by measuring the water stream flowing from the water source and not by placing the thermometer in the sample container. To avoid cross-contamination of the samples, sanitize the thermometer with isopropyl alcohol before measuring the temperature of each sample. When measuring temperature from faucets, showers, water fountains, etc., record the initial water temperature, and then allow the fixture to discharge until the temperature stabilizes. Record the initial and final temperatures, and the time needed to stabilize.

Domestic Water Heaters.

- Take a sample of water from the bottom drain.
- Collect a sample of water from the outlet pipe if the plumbing provides for access.

Faucets and Showers.

- Collect a "before-flush" (initial flow) sample of water.
- Collect an "after-flush" sample of water when the maximum temperature has been reached.

The initial (before-flush) sample is intended to indicate the level of contamination at the sample point or fitting, and the final sample should reveal the quality of the water being supplied to the fitting. Collect sterile-swab samples from faucets or shower heads by removing the fitting and vigorously swabbing the interior. Swab samples may be positive for Legionella even when water samples from the source are negative. Sterile test tubes containing sterilized swabs are available for convenient sampling and shipping.

Cooling Towers.

- Take a sample from the incoming supply to the tower.
- Take samples from any storage tanks or reservoirs in the system (i.e., chilled-water return tanks or header tanks).
- Take a sample from the basin of the cooling tower at a location distant from the incoming make-up water, and from the water returning from the circulation system at the point of entry to the tower.
- Take a sample of any standing water in the condensate trays or from the cooling coils.

Humidifiers, Swamp Coolers, and Spas.

- Take a sample from the water reservoirs. Sample the incoming water supply if it is accessible.
- For cooling towers, humidifiers, swamp coolers, and building water services, collect samples of sludge, slimes, or sediments, particularly where accumulations occur.
- Take swabs of shower heads, pipes, and faucets and rehydrate from water taken from the sampling site. Swab areas of scale buildup (i.e., remove shower heads, faucet screens, and aerators). Use prepackaged sterile swabs and small glass or polypropylene bottles (autoclaved) for this purpose.

SAMPLE TRANSPORTATION.

Prepare samples for shipment carefully, as follows:

- Wrap vinyl tape clockwise around the neck of each bottle to hold its screw cap firmly in place and seal the interface between the cap and the bottle.
- Wrap absorbent paper around bottles, and place the bottles in a sealable (zip-lock) plastic bag.
- Place the sealed plastic bag in an insulated container (Styrofoam chest or box).

Samples should not be refrigerated or shipped at reduced temperature. They should be protected from temperature extremes such as sunlight or other external heat or cold sources. Ship to laboratory using overnight mail. If shipping on a Friday, make arrangements for weekend receipt. The samples should be stored at room temperature (20° ± 5°C) and processed within 2 days.

APPENDIX III:7-3. WATER SAMPLING GUIDELINES.

[Adapted from George K. Morris, PhD, and Brian G. Shelton, Pathcon Technical Bulletin 1.3, Legionella in Environmental Samples: Hazard Analysis and Suggested Remedial Actions, June 1991, Pathogen Control Associates, 270 Scientific Dr., Suite 3, Norcross, GA 30092]

Use the following guidelines to assess the effectiveness of water system maintenance. These guidelines are based on limited data and are subject to change. They are intended to apply only to water systems being used by healthy individuals and are not necessarily protective for persons who are immunocompromised.

The levels requiring action vary for the source of exposure based on the assumption that some routes or exposure result in a greater dose to the lung. For this reason, humidifiers and similar devices such as misters and evaporative condensers which produce an aerosol mist that can be directly inhaled should be controlled to lower levels. Remember that these numbers are only guidelines, and the goal is zero detectable Legionella in a water source. Levels of Legionella equal to or greater than the values in the table constitute a need for action, as described below.

Action 1:Prompt cleaning and/or biocide treatment of the system.

Action 2:Immediate cleaning and/or biocide treatment. Take prompt steps to prevent employee exposure.

Table III:7-1. Colony forming units (CFU) of Legionella per milliliter

TABLE III: 7-1. COLONY FORMING UNITS (CFU) OF LEGIONELLA PER MILLILITER

Action	*Cooling tower*	*Domestic water*	*Humidifier*
1	100	10	1
2	1,000	100	10

APPENDIX III:7-4. LEGIONNAIRES' DISEASE CASE IDENTIFICATION.

The purpose of this phase of an investigation will be to identify cases of Legionnaires' disease among the workers. The investigation will include identification of all employees who took three or more consecutive days of sick leave days from six weeks before the Legionnaires' case was identified to the present. Following a screening process, all employees who have been identified as having had pneumonia, or potentially having had pneumonia, during this period will be requested to undergo voluntary medical testing to detect evidence of Legionnaires' disease. A physician's diagnosis of pneumonia or pneumonia-like symptoms that include a fever (101°F) and cough indicate a need for further evaluation. A sample program is described below.

Examine sick leave records to identify all employees who used three or more consecutive days of sick leave from 6 weeks before the earliest known case to the present. These employees will be interviewed. If it appears that an employee experienced a pneumonia-like illness, the attached surveillance questionnaire will be completed. Employees who feel that they might have had symptoms of Legionnaires' disease but did not use three or more consecutive days of sick leave should also be interviewed.

Employees who have experienced a pneumonia-like illness and have seen a physician should be requested to sign a medical release form to allow the company and/or OSHA to obtain additional information from the physician. The physicians of all employees who have seen a physician and have signed a medical release will be interviewed using the physician interview survey form (attached).

Employees participating in surveys such as the one described above must be informed of their Privacy Act rights as well as their right to protect their own medical information. Physician-patient confidentiality must not be violated. Necessary medical information may be communicated only with the patient's written permission. When seeking employees' permission, clearly inform them that the purpose of obtaining a proper diagnosis and sharing this information with the Agency is to protect them and their fellow workers against the potential threat of legionellosis. All medical records will be handled in accordance with 29CFR 1913.10. It may be necessary for the CSHO to obtain medical releases from the employees interviewed so that amplifying information can be obtained from a company health unit or the employee's physician.

Arrangements similar to that described above should be sought for permanent contract employees controlled by separate contractor organizations in the building, e.g., janitors, cafeteria workers, security personnel.

Based on an interview with the employee's physician, potential cases should be considered for a clinical test to detect additional cases. Most probably this will be a serological test to determine the antibody level of the individual. A single antibody titer of 1/256 or greater based on a physician's diagnosis of pneumonia should be interpreted as a probable case of Legionnaires' disease. In the event that an antibody titer level for Legionella was obtained at the time of illness, or if serum collected from the patient at the early phase of the illness (acute phase) is available, then an antibody titer level should be determined from this sample to determine the convalescent to acute titer ratio. A fourfold increase in this titer will be sufficient to confirm a case of Legionnaires' disease.

Other diagnostic tests may also be appropriate. If the potential case occurred recently, then a urine antigen test may detect Legionella pneumophila serogroup-1 antigen. A positive urine antigen test for a diagnosed pneumonia case is also accepted as evidence of a confirmed case. However, this test is available only for Legionella pneumophila serogroup-1 infections. Culture currently symptomatic individuals for Legionella. A positive culture indicates confirmation.

If this process detects one or more additional cases of disease, then the facility should be considered to have experienced an outbreak. The immediacy of the action will depend on whether the outbreak is ongoing or occurred 30 days or more in the past. Take prompt action to control exposure at the site if there is evidence that the outbreak is still occurring. Whatever the circumstances, initiate control procedures and continue medical surveillance of the workforce to detect any new cases of disease and identify the water source responsible for the outbreak.

HEALTH SURVEILLANCE QUESTIONNAIRE - LEGIONELLOSIS

Records show that you took sick leave for three consecutive days or more. We would like to ask a few questions.

1. Name: (last)_______________________, (first)____________________

   Age:_______ Sex: _______ Work Location: _______________________

   Home Phone:_____________ Work Phone:____________________________

2. Dates of absence(s):____

3. Stated reason for absence:______________________________________

Ask about the following symptoms:

4. Fever: Yes ____ No____ If yes, highest temperature _____.
5. Cough: Yes____ No ____
6. Headache: Yes_____ No_____
7. Diarrhea: Yes_____ No_____
8. Shortness of breath: Yes ____ No ____
9. Chest pain: Yes ____ No ____
10. Did you see a physician about these symptoms? Yes ___ No ___

    Was a chest x-ray taken? Yes_____ No_____

    Were you diagnosed as having pneumonia? Yes ___ No ___

    Were you tested for legionellosis? Yes_____ No_____

Physician's name:________________________ Phone:______________

Physician's Address:____

11. Were you admitted to a hospital? Yes ____ No ____

If yes, which hospital?____

Admission Date: __________________ Date released: __________

12. Interviewer:__________________________________ Date:______________

PHYSICIAN SURVEY QUESTIONNAIRE - LEGIONELLOSIS

We are calling to inform you that ________________________ is a patient of yours and an employee at ____________. He/she has signed a medical release giving us permission to contact you to obtain information about her/his recent illness. This questionnaire will be used to determine if your patient's recent illness could be classified as a pneumonia that may have been caused by exposure to *Legionella* at the workplace.

1. Name of Physician: ____
   Address:____
   Phone:_______________________________

2. Date of visit(s): (1st)________ (2nd)________ (3rd)________

3. What was the patient's complaint?:__________________________
   ____

	yes	no	unknown
Cough?	yes	no	unknown
Short of breath?	yes	no	unknown
History of fever?			

4. Physical Findings: ____
   ____
   Abnormal chest or lung findings: ____________________________
   ____

Rales?	yes	no	not examined
Dyspnea?	yes	no	not examined
Cyanosis?	yes	no	not examined
Temperature ______			
Other: ____			

5. Chest x-ray done? yes no

   Findings: ____

6. Sputum culture? yes no

Results: ______________________________

Laboratory: ______________________________

Sputum cultured for *Legionella?* yes no
Laboratory: ______________________________

7. Diagnostic testing? yes no

Type of test: Urine Antigen Test, Direct Fluorescent Antibody Serology Tests:
Indirect Fluorescent Antibody (IFA) _______
ELISA _________
Laboratory: ______________________________

8. Diagnosis or impression: ______________________________

EPIDEMIOLOGICAL QUESTIONNAIRE

Background

Employee's Name: ______________________ Age:_____ Gender:_____
(last, first)

Home: ______________________________
(city, zip)

Race/Ethnicity: white, black, native American, Hispanic, Asian, Other (circle one)

Are you currently taking any oral steroid medications?: Y/N

On what date did you first become ill?: ____ /____ /____

How many days were you ill?: ________

Was anyone else in your family ill?: Y/N

If Yes, who? ______________________________

What symptoms did they have? ________________________________

Since ____________, have any of your friends been diagnosed with pneumonia?: Yes/No

If Yes, who? ____

Work Exposure

During the 10 days prior to your illness:

Job Description: ____

Primary work area: ____

List all areas in _______ building where you spend any time:

Area	Hours per week
_____________________	_____________________
_____________________	_____________________
_____________________	_____________________
_____________________	_____________________
_____________________	_____________________

Did you shower at work?: Yes/No

If Yes, where and how may times per week?: __________________

List all places you eat lunch:_______________________________________

List all places where you take a break: ______________________________

List all restrooms you use: ____

Do you smoke in the restrooms (or spend "extra" time, i.e., if a lounge is present): Yes/No

If Yes, Where:_______________________________

Did you attend any training courses outside of the building?: Yes/No

If Yes, where were they held? _______________________________

Do you have a second job?: Yes/No

If Yes, what job and where:

____

Any other places that you have not mentioned where you spend time while on the job?:

____

Community Exposure *(During the 10 days prior to your illness)*

Did you use any health clubs?: Yes/No

If Yes, which ones?: _______________________________________

How many times?____

Did you use any hot tubs (whirlpool spas)?: Yes/No

If Yes, list which hot tubs and when used:

____

Did you attend any churches?: Yes/No

If Yes, where____

How many times?____________

Have you had any dental work performed?: Yes/No

If Yes, where____

How many times?____________

Which grocery stores did you go to?: _____________________________

How often?__________________

Did you go to the movies?: Yes/No

If Yes, which one? _________________________________

How often?____________

Did you go to any shopping malls?: Yes/No

If Yes, which one(s)?__________________________

Did you go to any other public places which you feel might be significant (i.e. public meetings, schools etc.)?: Yes/No

If Yes, where? ____

APPENDIX III:7-5. WATER TREATMENT PROTOCOLS FOR FACILITIES THAT HAVE EXPERIENCED A LEGIONNAIRES' OUTBREAK.

BACKGROUND.

This section describes actions required to abate the threat of further infection in a building in which an outbreak of Legionnaires' disease has occurred. For purposes of this document, an "outbreak of Legionnaires' disease" may be said to exist when medically confirmed cases of Legionnaires' disease are epidemiologically associated with a building or some portion of a building. This usually means that two or more confirmed cases of Legionnaires' disease have been identified within a six-week period at the site.

Under most circumstances evacuation of the building is not recommended. It will be necessary, following confirmation of an outbreak, to isolate individuals who are at high risk of contracting the disease from all potential sources of infection. Individuals at high risk include the immunosuppressed, such as persons who have had organ transplants, individuals receiving chemotherapy including corticosteroids, and other individuals in poor health. In addition, a medical monitoring program must be instituted to track all workers currently on sick leave.

Following these initial actions, the building must be inspected to identify all potential Legionella sources including the HVAC cooling systems (cooling towers, evaporative condensers), domestic water systems, humidifiers, and any sources of water that are maintained above 20°C (68°F) and has a potential for being aerosolized.

Before flushing or disinfecting the water in these suspected sources, take water samples for analysis to determine the predominant serotypes and subtypes of L. pneumophila in the water source and to determine the number of colony forming units (CFU) per unit of water. This information will be helpful in identifying the source of the disease if the subtype of L. pneumophila has been identified in the afflicted worker population. Because of the 10-day to two-week delay in obtaining sample results, corrective action should begin immediately.

Because sampling for Legionella can be inconclusive, sampling results alone should not determine the appropriate course of action in a building where an outbreak has occurred. ALL POTENTIAL SOURCES OF CONTAMINATION WILL BE ASSUMED TO BE CONTAMINATED AND TREATED ACCORDINGLY IN THE EVENT THAT AN OUTBREAK HAS OCCURRED. Water sampling and testing must be in accordance with currently accepted, state-of-the-art procedures.

Treatment of potential sources of contamination following sampling is described below. After the treatment collect and analyze water samples for CFU of L. pneumophila to determine the effectiveness of the treatment. Upon re-use of a water system following treatment, periodic maintenance and regular water sampling are essential to ensure that the maintenance continues to be effective. Included are proper maintenance procedures for controlling the organism in a facility's water sources.

COOLING TOWERS AND EVAPORATIVE CONDENSERS.

An HVAC condenser water system absorbs heat from the AC refrigeration units and rejects it to the atmosphere through evaporation in cooling towers. Evaporative condensers operate similarly to cooling towers except that refrigerant coils are inside the water path, and water passes over the coils to cool the refrigerant gas directly. Because both cooling towers and evaporative condensers use a fan system to move air through a recirculated water system, they introduce a considerable amount of water vapor into the surroundings even with drift eliminators designed to limit vapor release. In addition, this water is typically in the 20°-50°C (68°-122°F) range, ideal for L. pneumophila growth.

Water Sampling Protocol. Before starting decontamination, collect an adequate number of water samples in sterile containers. These samples should be cultured to determine the degree of contamination and the subtype of L. pneumophila before treatment. Collect at least three water samples (200 milliliters to 1 liter volume). Include water from the incoming make-up water supply, water from the basin of the unit most distant from the make-up water source, and recirculated water from the HVAC system at its point of return to the unit.

Clean-up Procedure.

1.Clean and disinfect the entire cooling system including attached chillers and/or storage tanks (sumps) following the "Wisconsin Protocol" Emergency Protocol, as follows:

- "Shock" treat cooling tower water at 50 ppm free residual chlorine.
- Add dispersant.
- Maintain 10 ppm chlorine for 24 hours.
- Drain system.
- Refill and repeat steps a through d.
- Inspect system for visual evidence of biofilm. If found, repeat steps a through d.
- Perform mechanical cleaning (cooling tower design may require modified procedures).
- Refill system, bring chlorine to 10 ppm, and circulate for one hour.
- Flush system.
- Refill with clean water in accordance with an effective water treatment program. The unit is now ready to be returned to service.

2. Identify and eliminate all water leaks into the cooling water system.

3. After completing step 1, sample the cooling water for analysis of CFU of L. pneumophila. The unit may be put into service provided the medical monitoring program has been implemented. If sample culture results indicate detectable levels of L. pneumophila, repeat chlorination and resample the water.

4. Once the nondetectable level for L. pneumophila has been achieved, institute maintenance as outlined in the Wisconsin Protocol to ensure continued safe and proper operation, as follows:

    - Inspect equipment monthly.
    - Drain and clean quarterly.
    - Treat circulating water for control of microorganisms, scale, and corrosion. This should include systematic use of biocides and rust inhibitors, preferably supplied by continuous feed, and monthly microbiologic analysis to ensure control of bacteria.
    - Document operation and maintenance in a log or maintenance records book.

5. Test cooling-system water at the following intervals to verify that there is no significant growth of Legionella, as follows:

    - Test weekly for the first month after return to operation.
    - Test every two weeks for the next two months.
    - Test monthly for the next three months.

6. The standard for Legionella concentration throughout the six months of monitoring is fewer than 10 CFU per milliliter (based on PathCon guidelines). If no water samples exceed this level, monitoring may be suspended. The maintenance program must continue indefinitely.

    If any sample contains 10 or more CFU Legionella per milliliter, take immediate steps to reduce levels to acceptable limits. These steps may include increased frequency of application or concentration of biocides, pH adjustment, additional "shock" treatments, or any other action that reduces Legionella levels. Take new water samples and begin the testing schedule again. Make the results of all water monitoring tests available to building occupants.

DOMESTIC WATER SYSTEMS.

Domestic water systems are designed to provide heated water for washing, cleaning, consumption, etc. A large building may have multiple independent systems. These systems usually include a boiler or heater, a recirculating piping system, and pipes terminating in taps and fixtures. Operating temperatures vary depending on system design, energy conservation programs, and intended use of the water. It is recommended that water heaters be kept at a minimum of 60°C (140°F) and all water be delivered at each outlet at a minimum of 50°C (122°F).

It is essential to identify all parts of the domestic water systems where water may stagnate (e.g., "dead legs" or laterals that have been capped off, storage tanks that have "dead zones" or are not frequently used). For treatment to be effective, the stagnant zones must be removed from the system. Rubber and plastic gaskets in the plumbing system may also serve as a Legionella growth medium. Eliminate or minimize use of these materials and substitute materials not conducive to Legionella growth. It is also important to identify and test the integrity of all backflow preventers to assure protection of domestic water from cross-contamination with process water through a building code-approved method.

WATER SAMPLING PROTOCOL.

Collect water samples before beginning treatment to determine potential contamination. Draw 200 milliliters to 1 liter of water from the draw-off valve of all water heaters into a sterile container. Check the temperature of the water in these units to determine if it is significantly lower than the set temperature. Sample a representative number of domestic hot-water faucets or outlets. It is important not to flush the faucet before taking a sample because the end section of the water system may be a source of contamination. Collect a 200 milliliter to 1 liter "preflush sample" of the first hot water drawn from the outlet. Allow the water to run and measure the temperature, and then collect a second, "postflush" sample when the water temperature is constant. Submit the water samples to a laboratory qualified to measure CFU of Legionella per milliliter of water.

Use the clean-up procedure below to treat all hot-water systems that have either been tested and found to contain detectable levels of Legionella or have been assumed to be contaminated.

1. Disinfect the system using any effective chemical, thermal, or other treatment method. For example:

    - Pasteurize the hot water system by heating the water to at least 70°C (158°F) and maintain this temperature for a minimum of 24 hours. While maintaining the temperature at 70°C (158°F), continuously flush each faucet on the system with hot water for 20 minutes.
    - Use an accepted chemical disinfectant such as chlorine or an acceptable biocide treatment to clean the system. Thoroughly flush the system after treatment to remove all traces of the corrosive and possibly toxic chemicals.
    - Follow any other technique that has demonstrated effectiveness and safety.

2. Maintain domestic water heaters at 60°C (140°F) and water delivered at the faucet at a minimum of 122°F (50°C). Where these temperatures cannot be maintained, control Legionella growth with a safe and effective alternative method.

3. After treatment, resample the hot water from each storage tank. If Legionella are detected, re-treat and resample the water system. If no measurable levels are found in this system and all other potential sources have also been addressed, go to the next step.

4. Test the domestic hot- or warm-water system for Legionella on the following schedule to assure that recontamination has not occurred:

    - Weekly for the first month after resumption of operation.
    - Every two weeks for the next two months.
    - Monthly for the next three months.

5. Use the Pathcon criteria for Legionella in domestic water systems during the monitoring period. If 10 or more CFU per milliliter of water are present, re-treat the system according to steps 1-3 above. Resume weekly testing (step 4) after retreatment. If levels remain below 1 CFU per milliliter, no further monitoring is necessary. If the levels are between 1 and 9 CFU per milliliter, continue monthly sampling of the water indefinitely and continue efforts to determine the source of contamination. Make test results available to building residents.

TEPID WATER SYSTEMS. Warm-water systems or tepid water systems dilute domestic hot water from a water heater with cold water upstream from the outlet source are not recommended. Warm water left in these lines is at ideal temperatures for amplification of L. pneumophila. Localized mixing at the source to temper very hot water is more acceptable. Another alternative is "instantaneous" point of delivery heating of water using individual steam heating systems at each outlet.

DOMESTIC COLD-WATER SYSTEMS. Domestic cold-water systems are designed to provide water for drinking, washing, cleaning, toilet flushing, etc. These systems have not been a major source of concern for Legionnaires' disease because L. pneumophila will not amplify at low temperatures. cold-water storage and delivery should be at less than 20°C (68°F) to minimize potential for growth. cold-water lines near hot-water lines should be insulated. Try to eliminate stagnant places in the system as dead legs or storage tanks that are not routinely used.

Detectable levels of L. pneumophila in the system may indicate contamination of the source water supply and should represent the maximum allowable level in the system. If sampling of the system indicates a level of contamination significantly greater than that of the incoming domestic water supply system, treat the system and identify the source of contamination or amplification. By definition, these systems have no provision for heating water, and therefore disinfection cannot be by heat treatment. Follow the clean-up procedure below if cold-water systems are shown to contain measurable Legionella or are assumed to be contaminated.

Clean-Up Procedure.

1. Clean and disinfect all cold water systems including storage tanks, drinking fountains, water lines, and water outlets, as follows:

    - Use an accepted chemical disinfectant such as chlorine or other acceptable biocide.
    - Use any other technology that has been shown to be safe and effective.

2. Ensure that cold-water systems are maintained so that conditions do not promote growth of Legionella. Maintain temperatures at 20°C (68°F) and keep residual chlorine in the range of 1-2 ppm. In practice this level of chlorination may be objectionable and may also be excessively corrosive to metal pipes and containers.

3. Take samples according to sampling guidelines. If analysis shows no detectable Legionella and all other potential sources have been addressed, go to the next step.

4. Flush all cold-water outlets and fountains for four minutes, twelve hours before re-entry.

5. When steps a through d have been successfully completed, return the building to normal operation but test the domestic cold-water system for Legionella according to the following schedule:

    - Weekly for the first month after resumption of operation.
    - Every two weeks for the next two months.
    - Monthly for the next three months.

6. The same criteria used for hot water systems described above will also be used for the cold-water system during the monitoring period. Ten or more CFU per milliliter of water require retreatment of the system according to steps 1-3 above. Following retreatment, resume weekly testing and repeat the schedule outlined in step 5. If Legionella levels remain below 1 CFU per milliliter, additional monitoring is not necessary. If levels are between 1-9 CFU per milliliter, continue monthly sampling of the water source indefinitely and try to identify the source of contamination. Make monitoring results available to building occupants.

HVAC AIR DISTRIBUTION SYSTEMS.

Under normal conditions HVAC systems are not likely to be sources of L. pneumophila unless water contaminated with the bacteria enters the system. Under normal conditions, condensate pans on coiling coils should not serve as a water source in which amplification of the bacteria can occur because the temperature of the water is below 20°C (68°F). Improperly drained condenser pans may produce tepid conditions that can encourage microbial and fungal growth. Proper maintenance will lessen problems related to other diseases such as humidifier fever and asthmatic responses, and will minimize the possibility of a Legionnaires' outbreak.

Most probably, for a Legionnaires' disease outbreak to be linked directly with the HVAC system, Legionella-contaminated water must continuously enter the system, be aerosolized, and be delivered to building occupants. Examine the systems to rule out this possibility, as follows:

- Inspect the entire air distribution system (including return and exhaust systems) for visual evidence of water accumulation.
- Eliminate all water leaks and remove any standing water found in the system. Replace or eliminate any water-damaged insulation in the system.
- Operate the HVAC system using 100% outside air for 8 hours before returning the building to normal operation.

Sampling of air in the ducts to prove that the duct system is free of Legionella is not required and would be pointless. No reliable way to detect Legionella in the air is available, and Legionella can live only in water. If the ducts are dry, they cannot serve as a source of Legionella.

Following return of the building to normal operation, keep outside-air supply rates as high as possible for one month. At a minimum, the outdoor air requirements of ASHRAE Ventilation Standard 62-1989 must be met.

HUMIDIFIERS AND MISTERS.

Many HVAC systems supply humidified air to building occupants to maintain comfort. Improperly maintained humidifiers can be both amplifiers and disseminators of a variety of bioaerosols; however, generally the cool temperatures in HVAC systems are not conducive to growth of L. pneumophila. cold-water humidifiers in HVAC systems must be connected to a domestic water source and provided with a drain line to remove the water. Stand-alone, console-type humidifiers that recirculate water for humidification should not be used because the water in these systems becomes contaminated with micro-organisms rapidly. These stand-alone units have been linked to an outbreak of Legionnaires' disease in a hospital. Ideally, HVAC humidifiers should use steam injection systems that eliminate potential microbe problems.

Cold-water humidifiers require rigorous maintenance to ensure that the water source does not contribute to potential problems. Since humidifiers discharge into HVAC air distribution systems, inspect for standing water and treat according to the HVAC Air Distribution System protocol above. Where water in humidifiers has been sampled and shown to contain measurable Legionella, or where such water has been assumed to be contaminated with Legionella, use the following protocol:

1. Disinfect water in piping or reservoirs feeding the humidifier with chlorine or other effective biocides.

2. Sample the humidifier water to assure "kill" of Legionella. Samples must have no detectable CFU of Legionella per milliliter of water. If one or more are detected, repeat treatment and sampling.

3. Ensure that an adequate maintenance program is in effect to reduce the growth of Legionella. Water storage temperatures should be above or below the 20° - 50°C (68° - 122°F) range, and the system must be kept clean.

4. Before using the humidifier, flush the piping and/or reservoir thoroughly to remove biocides.

5. When steps 1 through 4 have been successfully completed, return the humidifier to operation and test the unit's water system to detect recontamination with Legionella according to the schedule below:

    - Weekly for the first month.
    - Every two weeks for the next two months.
    - Monthly for the next three months.

The criterion for Legionella in humidifier water systems during monitoring is fewer than 1 CFU per milliliter. If no samples exceed the criterion, suspend monitoring and continue the maintenance program indefinitely.

If any sample shows one or more CFU of Legionella per milliliter, retreat and retest the system according to the schedule above (step 5). Make monitoring results available to building occupants.

SECTION IV: CHAPTER 1

OIL WELL DERRICK STABILITY: GUYWIRE ANCHOR SYSTEMS

Contents:

I. INTRODUCTION.

Work-over rigs are mast type devices that vary significantly from crane or other boom (mast) type equipment. Work-over rigs experience constant and varying dynamic loading conditions. They are subjected to various compression forces, along with jarring and wind loading. Other forces induced by pipe, tubing, etc. being stacked in the derrick and workers aloft on the derrick platform, as well as an ever-changing number of lateral and vertical forces are also present. Because of a work-over rig's dynamic environment, the health and safety of the operation is dependent upon the stability of the rig and its guy anchor system.

A. CAUSAL FACTORS.

1. There is no specific OSHA standard that addresses the stability of derricks in the oilwell drilling and servicing industry (Figure IV:1-1). But because of the fatality record there is a need for a guideline detailing the type of temporary stability systems according to the type of soil and its holding capacity, methods of installing guywire anchors, integrity of the system, and acceptable parameters in lieu of actual pull testing should be established.

2. Investigation into each fatal incident has determined that the cause of the upset was component failure rather than total system failure. This clearly illustrates the fact that the integrity of the system is no sounder than its weakest component.

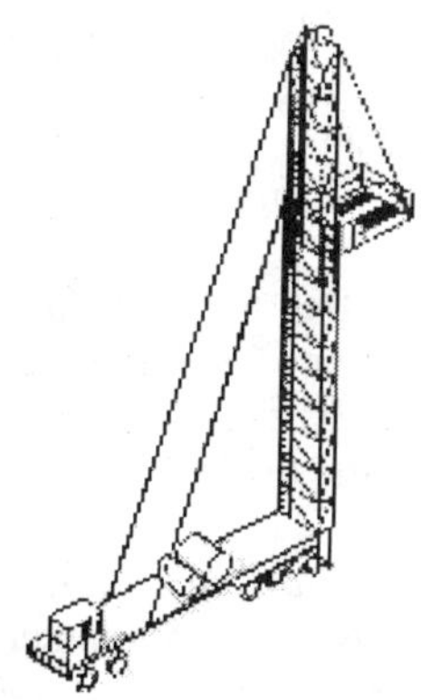

FIGURE IV: 1-1. OILWELL SERVICING DERRICK.

B. INDUSTRY RECOMMENDATIONS.

1. The American Petroleum Institute (API) in its Specification 4E "Specification for Drilling and Well Servicing Structures" sets forth a "Recommended Guying Pattern General Conditions." The Association of Oilwell Servicing Contractors (AOSC), in its publication Recommended Safe Procedures and Guidelines for Oil and Gas Well Servicing, recommends the same guying patterns as are set forth in API Specification 4E.

2. Though not present in the AOSC publication, the API Specification 4E provides a "Recommended Guyline Anchor Spacing and Load Chart." This is discussed in detail in the Guidelines on the Stability of Well Servicing Derricks. There has been considerable progress within the industry to design procedures to assure the integrity of the stability system without the necessity of conducting individual pull tests on each of the anchors.

C. APPLICATION. This chapter is intended to form the basis of a minimum safety guideline, for the use of Temporary Guywire Anchor Systems on derricks, in the oil well drilling and servicing industry. Recommended procedures, practices, equipment, and requirements have been developed based on availability, capability, adaptability, dependability, and reliability of the various types of systems.

II. TYPES OF GUYWIRE ANCHORS.

A. MANUFACTURED ANCHORS.

1. There are four basic types of manufactured anchors: the screw or helix anchor, the expanding plate anchor, the flat plate anchor, and the pivoting anchor. Holding capacity of these anchors varies; detailed information on holding capacity, comparison charts with illustrations, and characteristics specific to each design may be found in Section 2 of the support manual.

2. When installed in conformance with manufacturer specifications and evidence thereof is provided, this would satisfy the requirement for individual pull testing.

3. Screw(helix)-type anchors have a direct correlation between anchor capacity and the torque required to install the anchor. Following the manufacturer's specific recommendations as to torquing, with proof thereof, is a valid method of determining anchor holding capacity. Torquing according to manufacturer's specifications is an acceptable nonpull-test method of determining anchor capacity.

B. SHOP-MADE (IN-HOUSE FABRICATED) ANCHORS. These anchors should be designed by a registered engineer and conform to accepted engineering practices. Written procedures shall be established for installation. These manufactured anchors should be proof tested for structural integrity and holding capacity. Records shall be maintained of test protocols and holding capacity based on soil type. Individual pull testing will not be required if anchors are installed in accordance with written procedures. Proof thereof will be required of installation protocols and proof-tested holding capacities.

III. STABILITY CONSIDERATIONS.

A. FOUNDATION.

1. The area should be graded, leveled, and maintained so that oil, water, drilling fluid, and other fluids will drain away from the working area.

2. Safe Bearing Capacity shall be determined from the use of an appropriate table, soil core test, penetrometer test, flat-plate test, or other suitable soil test. When surface conditions are used to determine bearing capacity, care must be exercised to insure that the soil is homogeneous to a depth of at least twice the width of supplemental footing used to support the concentrated load.

3. Supplemental footing shall be provided to distribute the concentrated loads from the mast and rig support points. The manufacturer's load distribution diagram will indicate these locations. In the absence of a manufacturer's diagram, the supplemental footing shall be designed to carry the maximum anticipated hook load, the gross weight of the mast, the mast mount, the traveling equipment, and the vertical component of guywire tension under operational loading conditions. These footings must also support the mast and mast weight during mast erection.

4. Wellhead cellars present special foundation considerations. In addition to the obvious ones such as collecting water and fluids that can seep into the ground, cellars also require unique mast support considerations. These should be analyzed by a qualified person to insure that an adequate mast foundation is provided.

5. Small settlements (soil subsidence) at the beginning of rig-up is considered normal. External guywires should never be used for plumbing the mast. Rig foundations, guywire anchors and guywire tension should be checked at each tower (shift) change.

B. GUYWIRES.

1. All guywires, as indicated by the manufacturer's diagram, should be in position and properly tensioned prior to commencing any work. In the absence of manufacturer recommendations, or where mast manufacturer's recommendations cannot be implemented, the diagram in Figure IV:1-2 may be used.

2. Other guying patterns may be used; however, they must be based on sound engineering principles as determined by a qualified person. These recommendations should be posted on the mast in a weatherproof container and should state the loading conditions for which they were prepared. Guywires should be 6×19 or 6×37 class, regular lay, made of improved plow steel (IPS) or better with independent wire-rope core (IWRC) and not previously used for any other application. Double saddle clips should be used, and wire rope should be installed in accordance with the manufacturer's recommendations. In the absence of manufacturer recommendations, API RP 9B shall be followed.

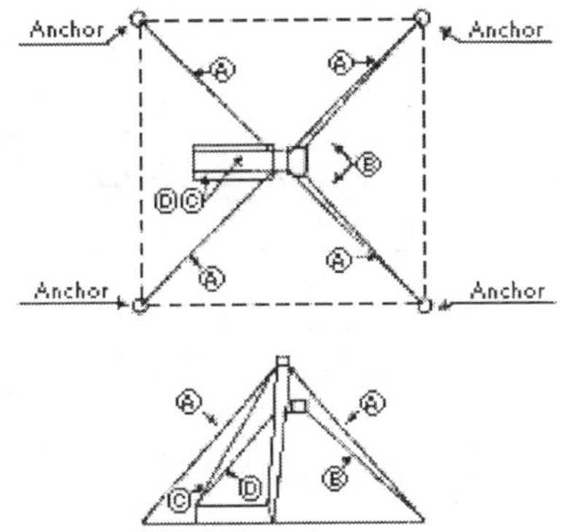

FIGURE IV: 1-2. ANCHOR LOCATION DIAGRAM.

C. GUYWIRE ANCHORS.

1. The mast manufacturer's recommendations shall be followed.

   In the absence of manufacturer recommendations the location diagram, Figure IV:1-3, may be used.

2. Each zone requires an anchor of different holding capacity. If anchors are located in more than one zone, then all anchors should be of the capacity required for the greater capacity zone. For example, if one anchor is located in "ZONE C" and the remaining anchors are located in "ZONE D," all anchors shall meet the holding capacity specified in the chart for "ZONE C." See Figure IV:1-4.

IV. OBSERVATIONS, DIRECTIONS, AND CONCLUSIONS.

A. VISUAL OBSERVATIONS.

1. There are characteristic visual observations that can serve as indicators of rig stability. They include, but are not limited to, the following:

   - The foundation supports the rig, substructure, and all applied loads while in an operational mode, without excessive movement, basically in a level and plumb configuration.
   - No large movement is observable between the mast support structure and the rotary/setback support structure when the slips are set and the load is removed from the mast, or vice versa.
   - The empty travel block hangs plumb with the centerline of the wellbore and the mast support structure remains level.
   - The mast support structure and/or substructure does not lean to one side more than the other when the load is applied. The guywire on one side becomes noticeably taut while the guywire on the opposite side becomes slack.
   - The guywire anchor(s) show(s) no visible signs of movement during the loading and unloading of the system while in operational mode.

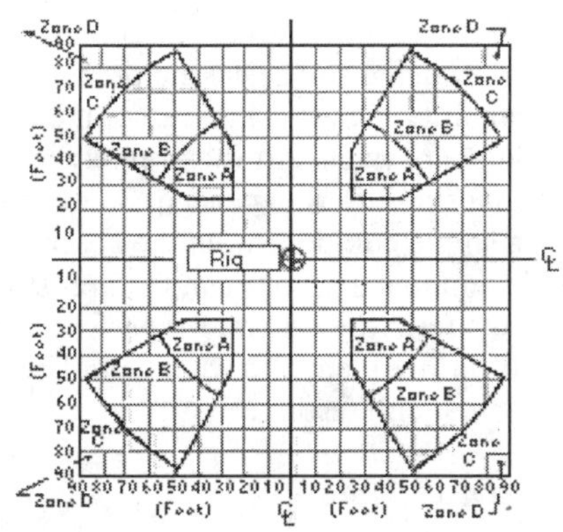

FIGURE IV: 1-3 RECOMMENDED ANCHOR LOCATIONS.

2. The chart presented in Figure IV:1-5 may be used as a guide to the pretensioning of guywires. This method is commonly referred to as the Catenary Method (guywire sag method).

ANCHOR CAPACITY (TONS)

Zone	Doubles Mast	Singles Mast	Post Mast
A	15.6	7.0	7.0
B	11.5	5.0	5.0
C	9.0	5.0	5.0
D	7.4	5.0	5.0

Anchor Capacities shown assume the following:
1. Adequate foundation support for mast and carrier.
2. Adequate crown-to-carrier internal load guys.
3. Maximum wind load -- 70 mph.
4. Maximum hook load, as described elsewhere in this chapter.
5. Full rod and tubing setback (N/A for Pole unit).

FIGURE IV: 1-4. ANCHOR CAPACITY REQUIREMENTS FOR EACH ZONE

B. SUPPORT MANUAL. The support manual, entitled Guideline on the Stability of Well Servicing Derricks, is divided into work sections and intended to supplement this chapter. It provides a detailed analysis of existing guides and standards along with state-of-the-art developments.

Section 3 provides the direction and guidance necessary to evaluate and select the proper system to assure rig stability. Section 4 discusses the installation of guywire anchor systems. It is extremely important to point out that stability is dependent on the entire system, and not In a single component. In the absence of support documentation or manufacturer specifications, Section 6 sets forth the criteria for performing effective pull testing. It further identifies what would be acceptable in lieu of actual pull testing.

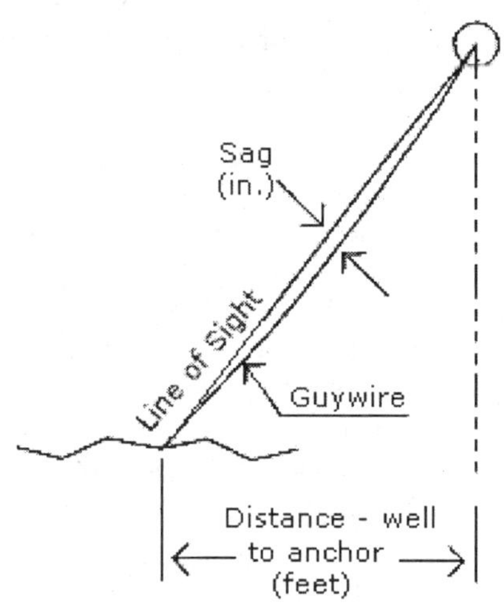

FIGURE IV: 1-5. CATENARY METHOD

C. CONCLUSION.

1. No set of observations or recommendations should be so restrictive or subjective as to preclude the use of innovative approaches to derrick stability systems.

   Properly designed substructures and base beams have been used effectively and safely as anchorages for guywires.

2. Engineering calculations based on sound engineering principles may also be used as evidence of an acceptable alternative to pull testing. Dead weight of equipment, fabricated components (i.e. padeyes) and other appurtenances are all considerations in determining rig stability.

3. The derrick manufacturer's specifications and recommendations should be the preferred and primary means of determining derrick stability. Guywire anchors, newly installed according to the manufacturer's specifications, may be used without the requirement for actual pull testing. (This would qualify as meeting the criteria as an acceptable alternative to pull testing.) If, however, there is a change in conditions, e.g. frozen ground to thawed

ground, or if use of the anchor has been interrupted, the anchor shall be pull tested, with documentation thereof, prior to being placed back in service.

	Guywire Sag (inches)					
	Pole Mast		Single Mast		Double Mast	
Distance Well to anchor (ft)	Tubing Board Guy	Crown-Ground Guy	Tubing Board Guy	Crown-Ground Guy	Tubing Board Guy	Crown-Ground Guy
40	-	4	4	4	6	5
60	-	6	8	6	12	8
80		10	15	10	17	11
100	-	14	22	14	26	15
120	-	18	32	18	32	21
Pre tension (Pounds)	500	1000	500	1000	500	1000

V. BIBLIOGRAPHY.

American Petroleum Institute (API). 1988. *Specification 4E: Specification for Drilling and Well Servicing Structures*. API: Washington, D.C.

Association of Oilwell Servicing Contractors (AOSC). 1988. *Recommended Safe Procedures and Guidelines for Oil and Gas Well Servicing*. AOSC: Dallas.

International Association of Drilling Contractors (IADC). 1990. *Accident Prevention Manual.* IADC: Houston.

International Association of Drilling Contractors. 1979. *Drilling Manual*. IADC: Houston.

Scardino, A.J. 1990. *Guidelines on the Stability of Well Servicing Derricks*. Sigma Associates Ltd.: Pass Christian, MS.

SECTION IV CHAPTER 2

PETROLEUM REFINING PROCESSES

Contents:

I. INTRODUCTION.

A. The petroleum industry began with the successful drilling of the first commercial oil well in 1859, and the opening of the first refinery two years later to process the crude into kerosene. The evolution of petroleum refining from simple distillation to today's sophisticated processes has created a need for health and safety management procedures and safe work practices. To those unfamiliar with the industry, petroleum refineries may appear to be complex and confusing places. Refining is the processing of one complex mixture of hydrocarbons into a number of other complex mixtures of hydrocarbons. The safe and orderly processing of crude oil into flammable gases and liquids at high temperatures and pressures using vessels, equipment, and piping subjected to stress and corrosion requires considerable knowledge, control, and expertise.

B. Safety and health professionals, working with process, chemical, instrumentation, and metallurgical engineers, assure that potential physical, mechanical, chemical, and health hazards are recognized and provisions are made for safe operating practices and appropriate protective measures. These measures may include hard hats, safety glasses and goggles, safety shoes, hearing protection, respiratory protection, and protective clothing such as fire resistant clothing where required. In addition, procedures should be established to assure compliance with applicable regulations and standards such as hazard communications, confined space entry, and process safety management.

C. This chapter of the technical manual covers the history of refinery processing, characteristics of crude oil, hydrocarbon types and chemistry, and major refinery products and by-products. It presents information on technology as normally practiced in present operations. It describes the more common refinery processes and includes relevant safety and health information. Additional information covers refinery utilities and miscellaneous supporting activities related to hydrocarbon processing. Field personnel will learn what to expect in various facilities regarding typical materials and process methods, equipment, potential hazards, and exposures.

D. The information presented refers to fire prevention, industrial hygiene, and safe work practices, and is not intended to provide comprehensive guidelines for protective measures and/or compliance with regulatory requirements. As some of the terminology is industry-specific, a glossary is provided as an appendix. This chapter does not cover petrochemical processing.

II. OVERVIEW OF THE PETROLEUM INDUSTRY.

A. BASIC REFINERY PROCESS: DESCRIPTION AND HISTORY. Petroleum refining has evolved continuously in response to changing consumer demand for better and different products. The original requirement was to produce kerosene as a cheaper and better source of light than whale oil. The development of the internal combustion engine led to the production of gasoline and diesel fuels. The evolution of the airplane created a need first for high-octane aviation gasoline and then for jet fuel, a sophisticated form of the original product, kerosene. Present-day refineries produce a variety of products including many required as feedstock for the petrochemical industry.

1. Distillation Processes. The first refinery, opened in 1861, produced kerosene by simple atmospheric distillation. Its by-products included tar and naphtha. It was soon discovered that high-quality lubricating oils could be produced by distilling petroleum under vacuum. However, for the next 30 years kerosene was the product consumers wanted. Two significant events changed this situation: (1) invention of the electric light decreased the demand for kerosene, and (2) invention of the internal combustion engine created a demand for diesel fuel and gasoline (naphtha).

2. Thermal Cracking Processes. With the advent of mass production and World War I, the number of gasoline-powered vehicles increased dramatically and the demand for gasoline grew accordingly. However, distillation processes produced only a certain amount of gasoline from crude oil. In 1913, the thermal cracking process was developed, which subjected heavy fuels to both pressure and intense heat, physically breaking the large molecules into smaller ones to produce additional gasoline and distillate fuels. Visbreaking, another form of thermal cracking, was developed in the late 1930's to produce more desirable and valuable products.

3. Catalytic Processes. Higher-compression gasoline engines required higher-octane gasoline with better antiknock characteristics. The introduction of catalytic cracking and polymerization processes in the mid- to late 1930's met the demand by providing improved gasoline yields and higher octane numbers.

   Alkylation, another catalytic process developed in the early 1940's, produced more high-octane aviation gasoline and petrochemical feedstock for explosives and synthetic rubber. Subsequently, catalytic isomerization was developed to convert hydrocarbons to produce increased quantities of alkylation feedstock. Improved catalysts and process methods such as hydrocracking and reforming were developed throughout the 1960's to increase gasoline yields and improve antiknock characteristics. These catalytic processes also produced hydrocarbon molecules with a double bond (alkenes) and formed the basis of the modern petrochemical industry.

4. Treatment Processes. Throughout the history of refining, various treatment methods have been used to remove nonhydrocarbons, impurities, and other constituents that adversely affect the properties of finished products or reduce the efficiency of the conversion processes. Treating can involve chemical reaction and/or physical separation. Typical examples of treating are chemical sweetening, acid treating, clay contacting, caustic washing, hydrotreating, drying, solvent extraction, and solvent dewaxing. Sweetening compounds and acids desulfurize crude oil before processing and treat products during and after processing.

   Following the Second World War, various reforming processes improved gasoline quality and yield and produced higher-quality products. Some of these involved the use of catalysts and/or hydrogen to change molecules and remove sulfur. A number of the more commonly used treating and reforming processes are described in this chapter of the manual.

B. BASICS OF CRUDE OIL.

1. Crude oils are complex mixtures containing many different hydrocarbon compounds that vary in appearance and composition from one oil field to another. Crude oils range in consistency from water to tar-like solids, and in color from clear to black. An "average" crude oil contains about 84% carbon, 14% hydrogen, 1%-3% sulfur, and less than 1% each of nitrogen, oxygen, metals, and salts. Crude oils are generally classified as paraffinic, naphthenic, or aromatic, based on the predominant proportion of similar hydrocarbon molecules. Mixed-base crudes have varying amounts of each type of hydrocarbon. Refinery crude base stocks usually consist of mixtures of two or more different crude oils.

2. Relatively simple crude oil assays are used to classify crude oils as paraffinic, naphthenic, aromatic, or mixed. One assay method (United States Bureau of Mines) is based on distillation, and another method (UOP "K" factor) is based on gravity and boiling points. More comprehensive crude assays determine the value of the crude (i.e., its yield and quality of useful products) and processing parameters. Crude oils are usually grouped according to yield structure.

3. Crude oils are also defined in terms of API (American Petroleum Institute) gravity. The higher the API gravity, the lighter the crude. For example, light crude oils have high API gravities and low specific gravities. Crude oils with low carbon, high hydrogen, and high API gravity are usually rich in paraffins and tend to yield greater

TABLE IV: 2-1. HISTORY OF REFINING

Year	Process name	Purpose	By-products, etc.
1862	Atmospheric distillation	Produce kerosene	Naphtha, tar, etc.
1870	Vacuum distillation	Lubricants (original) Cracking feedstocks (1930's)	Asphalt, residual coker feedstocks
1913	Thermal cracking	Increase gasoline	Residual, bunker fuel
1916	Sweetening	Reduce sulfur & odor	Sulfur
1930	Thermal reforming	Improve octane number	Residual
1932	Hydrogenation	Remove sulfur	Sulfur
1932	Coking	Produce gasoline basestocks	Coke
1933	Solvent extraction	Improve lubricant viscosity index	Aromatics
1935	Solvent dewaxing	Improve pour point	Waxes
1935	Cat. polymerization	Improve gasoline yield & octane number	Petrochemical feedstocks
1937	Catalytic cracking	Higher octane gasoline	Petrochemical feedstocks
1939	Visbreaking	Reduce viscosity	Increased distillate,tar
1940	Alkylation	Increase gasoline octane & yield	High-octane aviation gasoline
1940	Isomerization	Produce alkylation feedstock	Naphtha
1942	Fluid catalytic cracking	Increase gasoline yield & octane	Petrochemical feedstocks
1950	Deasphalting	Increase cracking feedstock	Asphalt
1952	Catalytic reforming	Convert low-quality naphtha	Aromatics
1954	Hydrodesulfurization	Remove sulfur	Sulfur
1956	Inhibitor sweetening	Remove mercaptan	Disulfides
1957	Catalytic isomerization	Convert to molecules with high octane number	Alkylation feedstocks
1960	Hydrocracking	Improve quality and reduce sulfur	Alkylation feedstocks
1974	Catalytic dewaxing	Improve pour point	Wax
1975	Residual hydrocracking	Increase gasoline yield from residual	Heavy residuals

proportions of gasoline and light petroleum products; those with high carbon, low hydrogen, and low API gravities are usually rich in aromatics.

4. Crude oils that contain appreciable quantities of hydrogen sulfide or other reactive sulfur compounds are called "sour." Those with less sulfur are called "sweet." Some exceptions to this rule are West Texas crudes, which are always considered "sour" regardless of their H2S content, and Arabian high-sulfur crudes, which are not considered "sour" because their sulfur compounds are not highly reactive.

TABLE IV: 2-2. TYPICAL APPROXIMATE CHARACTERISTICS AND PROPERTIES AND GASOLINE POTENTIAL OF VARIOUS CRUDES (Representative average numbers)

Crude source	Paraffins (% vol)	Aromatics (% vol)	Naphthenes (% vol)	Sulfur (% wt)	API gravity (approx.)	Napht. yield (% vol)	Octane no (typical)
Nigerian -Light	37	9	54	0.2	36	28	60
Saudi -Light	63	19	18	2	34	22	40
Saudi -Heavy	60	15	25	2.1	28	23	35
Venezuela -Heavy	35	12	53	2.3	30	2	60
Venezuela -Light	52	14	34	1.5	24	18	50
USA -Midcont. Sweet	-	-	-	0.4	40	-	-
USA -W. Texas Sour	46	22	32	1.9	32	33	55
North Sea -Brent	50	16	34	0.4	37	31	50

C. BASICS OF HYDROCARBON CHEMISTRY. Crude oil is a mixture of hydrocarbon molecules, which are organic compounds of carbon and hydrogen atoms that may include from one to 60 carbon atoms. The properties of hydrocarbons depend on the number and arrangement of the carbon and hydrogen atoms in the molecules. The simplest hydrocarbon molecule is one carbon atom linked with four hydrogen atoms: methane. All other variations of petroleum hydrocarbons evolve from this molecule.

Hydrocarbons containing up to four carbon atoms are usually gases, those with 5 to 19 carbon atoms are usually liquids, and those with 20 or more are solids. The refining process uses chemicals, catalysts, heat, and pressure to separate and combine the basic types of hydrocarbon molecules naturally found in crude oil into groups of similar molecules. The refining process also rearranges their structures and bonding patterns into different hydrocarbon molecules and compounds. Therefore it is the type of hydrocarbon (paraffinic, naphthenic, or aromatic) rather than its specific chemical compounds that is significant in the refining process.

1. Three Principal Groups or Series of Hydrocarbon Compounds that Occur Naturally in Crude Oil.

   a. Paraffins. The paraffinic series of hydrocarbon compounds found in crude oil have the general formula C_nH_{2n+2} and can be either straight chains (normal) or branched chains (isomers) of carbon atoms. The lighter, straight-chain paraffin molecules are found in gases and paraffin waxes. Examples of straight-chain molecules are methane, ethane, propane, and butane (gases containing from one to four carbon atoms), and pentane and hexane (liquids with five to six carbon atoms). The branched-chain (isomer) paraffins are usually found in heavier fractions of crude oil and have higher octane numbers than normal paraffins. These compounds are saturated hydrocarbons, with all carbon bonds satisfied, that is, the hydrocarbon chain carries the full complement of hydrogen atoms.

   b. Aromatics are unsaturated ring-type (cyclic) compounds which react readily because they have carbon atoms that are deficient in hydrogen. All aromatics have at least one benzene ring (a single-ring compound characterized by three double bonds alternating with three single bonds between six carbon atoms) as part of their molecular structure. Naphthalenes are fused double-ring aromatic compounds. The most complex aromatics, polynuclears (three or more fused aromatic rings), are found in heavier fractions of crude oil.

   c. Naphthenes are saturated hydrocarbon groupings with the general formula C_nH_{2n}, arranged in the form of closed rings (cyclic) and found in all fractions of crude oil except the very lightest. Single-ring naphthenes (monocycloparaffins) with five and six carbon atoms predominate, with two-ring naphthenes (dicycloparaffins) found in the heavier ends of naphtha.

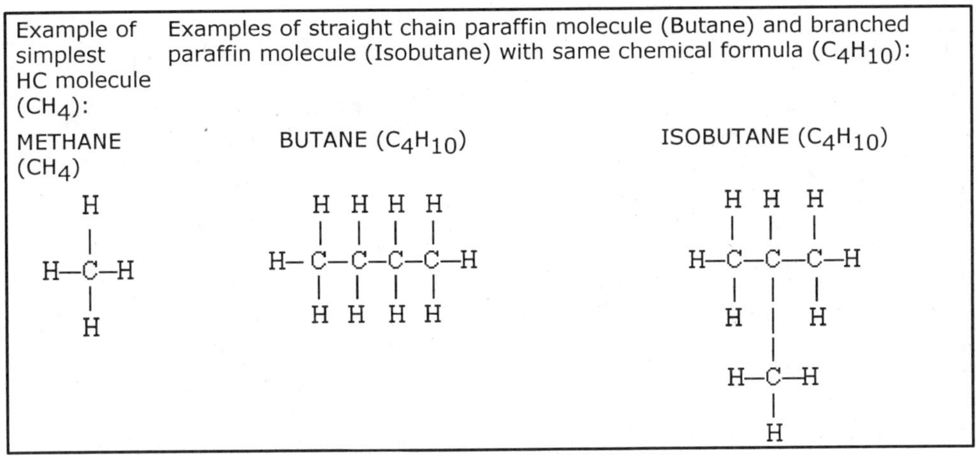

FIGURE IV: 2-1. TYPICAL PARAFFINS.

2. Other Hydrocarbons.

   a. Alkenes are mono-olefins with the general formula CnH2n and contain only one carbon-carbon double bond in the chain. The simplest alkene is ethylene, with two carbon atoms joined by a double bond and four hydrogen atoms. Olefins are usually formed by thermal and catalytic cracking and rarely occur naturally in unprocessed crude oil.

   b. Dienes and Alkynes. Dienes, also known as diolefins, have two carbon-carbon double bonds. The alkynes, another class of unsaturated hydrocarbons, have a carbon-carbon triple bond within the molecule. Both these series of hydrocarbons have the general formula CnH2n-2. Diolefins such as 1,2-butadiene and 1,3-butadiene, and alkynes such as acetylene,occur in C5 and lighter fractions from cracking. The olefins, diolefins, and alkynes are said to be unsaturated because they contain less than the amount of hydrogen necessary to saturate all the valences of the carbon atoms. These compounds are more reactive than paraffins or naphthenes and readily combine with other elements such as hydrogen, chlorine, and bromine.

3. Nonhydrocarbons.

   a. Sulfur Compounds. Sulfur may be present in crude oil as hydrogen sulfide (H2), as compounds (e.g. mercaptans, sulfides, disulfides, thiophenes, etc.) or as elemental sulfur. Each crude oil has different amounts and types of sulfur compounds, but as a rule the proportion, stability, and complexity of the compounds are greater in heavier crude-oil fractions. Hydrogen sulfide is a primary contributor to corrosion in refinery processing units. Other corrosive substances are elemental sulfur and mercaptans. Moreover, the corrosive sulfur compounds have an obnoxious odor.

   Pyrophoric iron sulfide results from the corrosive action of sulfur compounds on the iron and steel used in refinery process equipment, piping, and tanks. The combustion of petroleum products containing sulfur compounds produces undesirables such as sulfuric acid and sulfur dioxide. Catalytic hydrotreating processes such as hydrodesulfurization remove sulfur compounds from refinery product streams. Sweetening processes either remove the obnoxious sulfur compounds or convert them to odorless disulfides, as in the case of mercaptans.

   b. Oxygen Compounds. Oxygen compounds such as phenols, ketones, and carboxylic acids occur in crude oils in varying amounts.

   c. Nitrogen Compounds. Nitrogen is found in lighter fractions of crude oil as basic compounds, and more often in heavier fractions of crude oil as nonbasic compounds that may also include trace metals such as

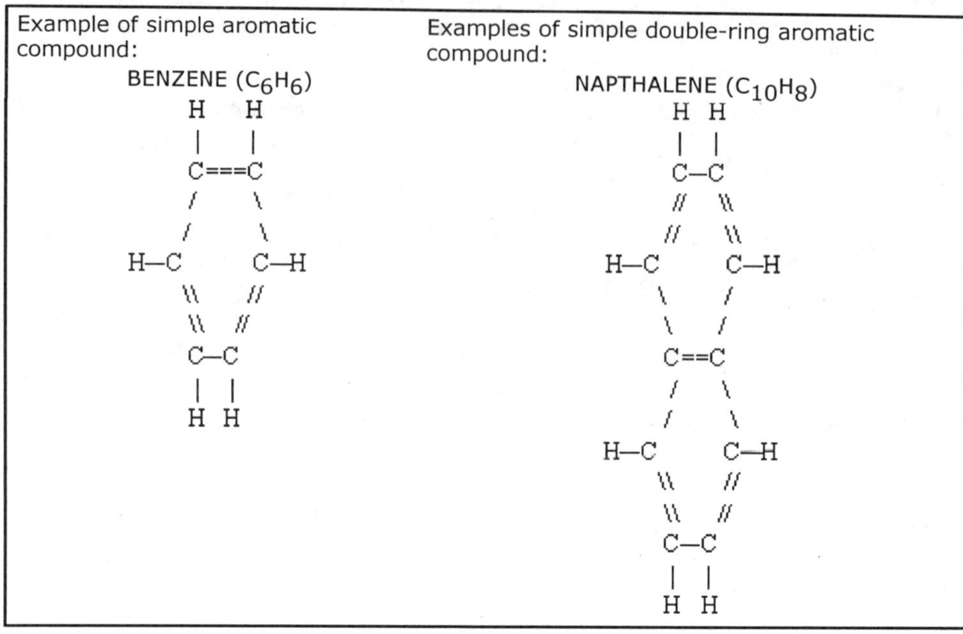

FIGURE IV: 2-2. TYPICAL AROMATICS

copper, vanadium, and/or nickel. Nitrogen oxides can form in process furnaces. The decomposition of nitrogen compounds in catalytic cracking and hydrocracking processes forms ammonia and cyanides that can cause corrosion.

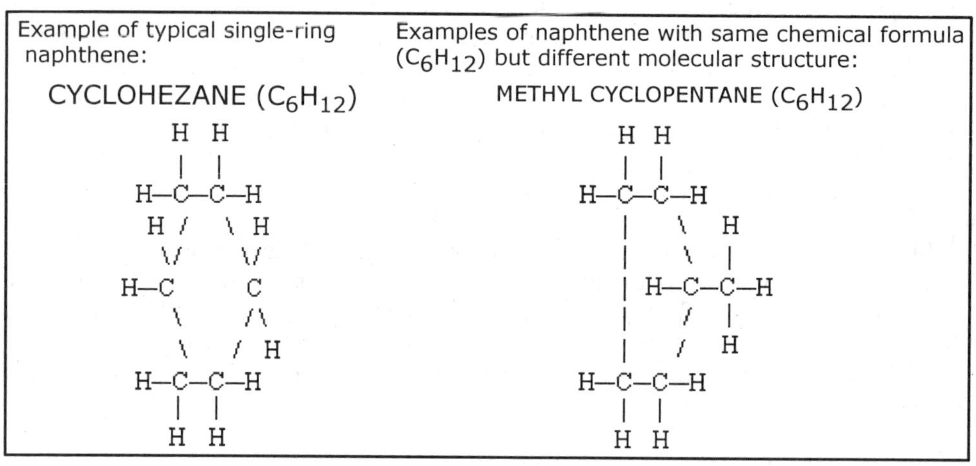

FIGURE IV: 2-3. TYPICAL NAPHTHENES.

d. Trace Metals. Metals, including nickel, iron, and vanadium are often found in crude oils in small quantities and are removed during the refining process. Burning heavy fuel oils in refinery furnaces and boilers can leave deposits of vanadium oxide and nickel oxide in furnace boxes, ducts, and tubes. It is also desirable to remove trace amounts of arsenic, vanadium, and nickel prior to processing as they can poison certain catalysts.

e. Salts. Crude oils often contain inorganic salts such as sodium chloride, magnesium chloride, and calcium chloride in suspension or dissolved in entrained water (brine). These salts must be removed or neutralized before processing to prevent catalyst poisoning, equipment corrosion, and fouling. Salt corrosion is caused by the hydrolysis of some metal chlorides to hydrogen chloride (HCl) and the subsequent formation of hydrochloric acid when crude is heated. Hydrogen chloride may also combine with ammonia to form ammonium chloride (NH4Cl), which causes fouling and corrosion.

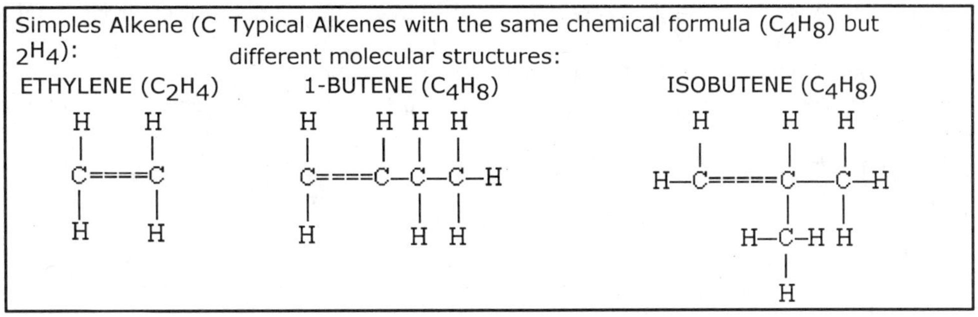

FIGURE IV: 2-4. TYPICAL ALKENES.

f. Carbon Dioxide. Carbon dioxide may result from the decomposition of bicarbonates present in or added to crude, or from steam used in the distillation process.

g. Naphthenic Acids. Some crude oils contain naphthenic (organic) acids, which may become corrosive at temperatures above 450° F when the acid value of the crude is above a certain level.

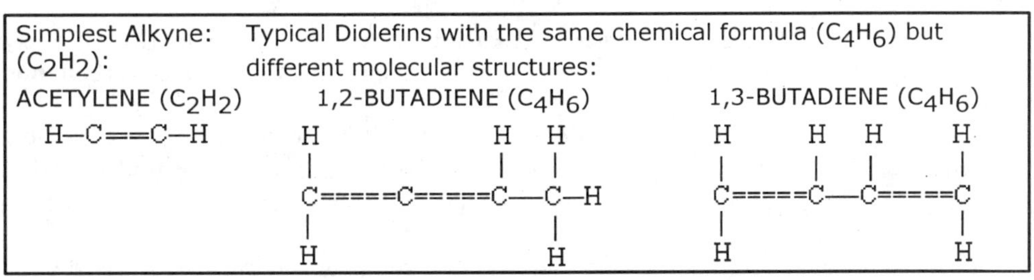

FIGURE IV: 2-5. TYPICAL DIOLEFINS AND LAKYNES.

D. MAJOR REFINERY PRODUCTS.

1. Gasoline. The most important refinery product is motor gasoline, a blend of hydrocarbons with boiling ranges from ambient temperatures to about 400 °F. The important qualities for gasoline are octane number (antiknock), volatility (starting and vapor lock), and vapor pressure (environmental control). Additives are often used to enhance performance and provide protection against oxidation and rust formation.

2. Kerosene. Kerosene is a refined middle-distillate petroleum product that finds considerable use as a jet fuel and around the world in cooking and space heating. When used as a jet fuel, some of the critical qualities are freeze point, flash point, and smoke point. Commercial jet fuel has a boiling range of about 375°-525° F, and military jet fuel 130°-550° F. Kerosene, with less-critical specifications, is used for lighting, heating, solvents, and blending into diesel fuel.

3. Liquified Petroleum Gas (LPG). LPG, which consists principally of propane and butane, is produced for use as fuel and is an intermediate material in the manufacture of petrochemicals. The important specifications for proper performance include vapor pressure and control of contaminants.

4. Distillate Fuels. Diesel fuels and domestic heating oils have boiling ranges of about 400°-700° F. The desirable qualities required for distillate fuels include controlled flash and pour points, clean burning, no deposit formation in storage tanks, and a proper diesel fuel cetane rating for good starting and combustion.

5. Residual Fuels. Many marine vessels, power plants, commercial buildings and industrial facilities use residual fuels or combinations of residual and distillate fuels for heating and processing. The two most critical specifications of residual fuels are viscosity and low sulfur content for environmental control.

6. Coke and Asphalt. Coke is almost pure carbon with a variety of uses from electrodes to charcoal briquets. Asphalt, used for roads and roofing materials, must be inert to most chemicals and weather conditions.

7. Solvents. A variety of products, whose boiling points and hydrocarbon composition are closely controlled, are produced for use as solvents. These include benzene, toluene, and xylene.

8. Petrochemicals. Many products derived from crude oil refining, such as ethylene, propylene, butylene, and isobutylene, are primarily intended for use as petrochemical feedstock in the production of plastics, synthetic fibers, synthetic rubbers, and other products.

9. Lubricants. Special refining processes produce lubricating oil base stocks. Additives such as demulsifiers, antioxidants, and viscosity improvers are blended into the base stocks to provide the characteristics required for motor oils, industrial greases, lubricants, and cutting oils. The most critical quality for lubricating-oil base stock is a high viscosity index, which provides for greater consistency under varying temperatures.

E. COMMON REFINERY CHEMICALS.

1. Leaded Gasoline Additives. Tetraethyl lead (TEL) and tetramethyl lead (TML) are additives formerly used to improve gasoline octane ratings but are no longer in common use except in aviation gasoline.

2. Oxygenates. Ethyl tertiary butyl ether (ETBE), methyl tertiary butyl ether (MTBE), tertiary amyl methyl ether (TAME), and other oxygenates improve gasoline octane ratings and reduce carbon monoxide emissions.

3. Caustics. Caustics are added to desalting water to neutralize acids and reduce corrosion. They are also added to desalted crude in order to reduce the amount of corrosive chlorides in the tower overheads. They are used in some refinery treating processes to remove contaminants from hydrocarbon streams.

4. Sulfuric Acid and Hydrofluoric Acid. Sulfuric acid and hydrofluoric acid are used primarily as catalysts in alkylation processes. Sulfuric acid is also used in some treatment processes.

III. PETROLEUM REFINING OPERATIONS.

A. INTRODUCTION. Petroleum refining begins with the distillation, or fractionation, of crude oils into separate hydrocarbon groups. The resultant products are directly related to the characteristics of the crude processed. Most distillation products are further converted into more usable products by changing the size and structure of the hydrocarbon molecules through cracking, reforming, and other conversion processes as discussed in this chapter. These converted products are then subjected to various treatment and separation processes such as extraction, hydrotreating, and sweetening to remove undesirable constituents and improve product quality. Integrated refineries incorporate fractionation, conversion, treatment, and blending operations and may also include petrochemical processing.

B. REFINING OPERATIONS. Petroleum refining processes and operations can be separated into five basic areas:

1. Fractionation (distillation) is the separation of crude oil in atmospheric and vacuum distillation towers into groups of hydrocarbon compounds of differing boiling-point ranges called "fractions" or "cuts."

2. Conversion processes change the size and/or structure of hydrocarbon molecules. These processes include:

    - Decomposition (dividing) by thermal and catalytic cracking;

- Unification (combining) through alkylation and polymerization; and
- Alteration (rearranging) with isomerization and catalytic reforming.

3. Treatment processes are intended to prepare hydrocarbon streams for additional processing and to prepare finished products. Treatment may include the removal or separation of aromatics and naphthenes as well as impurities and undesirable contaminants. Treatment may involve chemical or physical separation such as dissolving, absorption, or precipitation using a variety and combination of processes including desalting, drying, hydrodesulfurizing, solvent refining, sweetening, solvent extraction, and solvent dewaxing.

4. Formulating and Blending is the process of mixing and combining hydrocarbon fractions, additives, and other components to produce finished products with specific performance properties.

5. Other Refining Operations include: light-ends recovery; sour-water stripping; solid waste and wastewater treatment; process-water treatment and cooling; storage and handling; product movement; hydrogen production; acid and tail-gas treatment; and sulfur recovery.

   Auxiliary operations and facilities include: steam and power generation; process and fire water systems; flares and relief systems; furnaces and heaters; pumps and valves; supply of steam, air, nitrogen, and other plant gases; alarms and sensors; noise and pollution controls; sampling, testing, and inspecting; and laboratory, control room, maintenance, and administrative facilities.

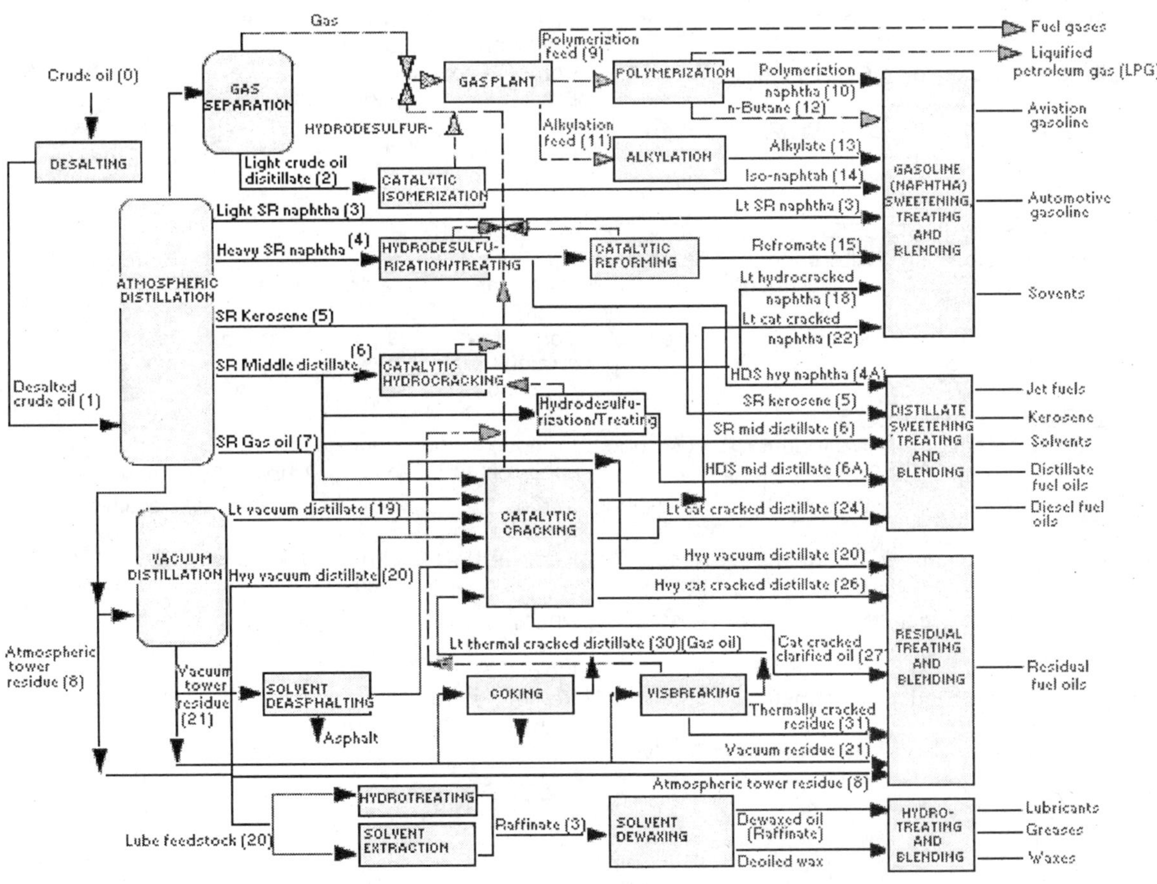

FIGURE IV: 2-6. REFINERY PROCESS CHART.

TABLE IV: 2-3. OVERVIEW OF PETROLEUM REFINING PROCESSES.

Process name	Action	Method	Purpose	Feedstock(s)	Product(s)
FRACTIONATION PROCESSES					
Atmospheric distillation	Separation	Thermal	Separate fractions	Desalted crude oil	Gas, gas oil, distillate, residual
Vacuum distillation	Separation	Thermal	Separate w/o cracking	Atmospheric tower residual	Gas oil, lube stock, residual
CONVERSION PROCESSED--DECOMPOSITION					
Catalytic cracking	Alteration	Catalytic	Upgrade gasoline	Gas oil, coke distillate	Gasoline, petrochemical feedstock
Coking	Polymerize	Thermal	Convert vacuum residuals	Gas oil, coke distillate	Gasoline, petrochemical feedstock
Hydro-cracking	Hydrogenate	Catalytic	Convert to lighter HC's	Gas oil, cracked oil, residual	Lighter, higher-quality products
*Hydrogen steam reforming	Decompose	Thermal/ catalytic	Produce hydrogen	Desulfurized gas, O_2, steam	Hydrogen, CO, CO_2
*Steam cracking	Decompose	Thermal	Crack large molecules	Atm tower hvy fuel/ distillate	Cracked naphtha, coke, residual
Visbreaking	Decompose	Thermal	Reduce viscosity	Atmospheric tower residual	Distillate, tar
CONVERSION PROCESSES--UNIFICATION					
Alkylation	Combining	Catalytic	Unite olefins & isoparaffins	Tower isobutane/ cracker olefin	Iso-octane (alkylate)
Grease compounding	Combining	Thermal	Combine soaps & oils	Lube oil, fatty acid, alky metal	Lubricating grease
Polymerizing	Polymerize	Catalytic	Unite 2 or more olefins	Cracker olefins	High-octane naphtha, petrochemical stocks
CONVERSION PROCESSES--ALTERATION OR REARRANGEMENT					
Catalytic reforming	Alteration/ dehydration	Catalytic	Upgrade low-octane naphtha	Coker/ hydro-cracker naphtha	High oct. Reformate/ aromatic
Isomerization	Rearrange	Catalytic	Convert straight chain to branch	Butane, pentane, hexane	Isobutane/ pentane/ hexane
TREATMENT PROCESSES					
*Amine treating	Treatment	Absorption	Remove acidic contaminants	Sour gas, HCs w/CO_2 & H_2S	Acid free gases & liquid HCs
Desalting	Dehydration	Absorption	Remove contaminants	Crude oil	Desalted crude oil
Drying & sweetening	Treatment	Abspt/ therm	Remove H_2O & sulfur cmpds	Liq Hcs, LPG, alky feedstk	Sweet & dry hydrocarbons
*Furfural extraction	Solvent extr.	Absorption	Upgrade mid distillate & lubes	Cycle oils & lube feed-stocks	High quality diesel & lube oil
Hydrodesulfurization	Treatment	Catalytic	Remove sulfur, contaminants	High-sulfur residual/ gas oil	Desulfurized olefins
Hydrotreating	Hydrogenation	Catalytic	Remove impurities, saturate HC's	Residuals, cracked HC's	Cracker feed, distillate, lube

*Phenol extraction	Solvent extr.	Abspt/ therm	Improve visc. index, color	Lube oil base stocks	High quality lube oils
Solvent deasphalting	Treatment	Absorption	Remove asphalt	Vac. tower residual, propane	Heavy lube oil, asphalt
Solvent dewaxing	Treatment	Cool/ filter	Remove wax from lube stocks	Vac. tower lube oils	Dewaxed lube basestock
Solvent extraction	Solvent extr.	Abspt/ precip.	Separate unsat. oils	Gas oil, reformate, distillate	High-octane gasoline
Sweetening	Treatment	Catalytic	Remv H_2S, convert mercaptan	Untreated distillate/gasoline	High-quality distillate/gasoline

* Note: These processes are not depicted in the refinery process flow chart.

IV. DESCRIPTION OF PETROLEUM REFINING PROCESSES AND RELATED HEALTH AND SAFETY CONSIDERATIONS.

A. CRUDE OIL PRETREATMENT (DESALTING).

1. Description.

a. Crude oil often contains water, inorganic salts, suspended solids, and water-soluble trace metals. As a first step in the refining process, to reduce corrosion, plugging, and fouling of equipment and to prevent poisoning the catalysts in processing units, these contaminants must be removed by desalting (dehydration).

b. The two most typical methods of crude-oil desalting, chemical and electrostatic separation, use hot water as the extraction agent. In chemical desalting, water and chemical surfactant (demulsifiers) are added to the crude, heated so that salts and other impurities dissolve into the water or attach to the water, and then held in a tank where they settle out. Electrical desalting is the application of high-voltage electrostatic charges to concentrate suspended water globules in the bottom of the settling tank. Surfactants are added only when the crude has a large amount of suspended solids. Both methods of desalting are continuous. A third and less-common process involves filtering heated crude using diatomaceous earth.

c. The feedstock crude oil is heated to between 150° and 350°F to reduce viscosity and surface tension for easier mixing and separation of the water. The temperature is limited by the vapor pressure of the crude-oil feedstock. In both methods other chemicals may be added. Ammonia is often used to reduce corrosion. Caustic or acid may be added to adjust the pH of the water wash. Wastewater and contaminants are discharged from the bottom of the settling tank to the wastewater treatment facility. The desalted crude is continuously drawn from the top of the settling tanks and sent to the crude distillation (fractionating) tower.

TABLE IV:2-4. DESALTING PROCESS.

Feedstock	*From*	*Process*	*Typical products . . . To*
Crude	Storage	Treating	Desalted crude . . . Atmospheric distillation tower Waste water . . . . . Treatment

2. Health and Safety Considerations.

a. Fire Prevention and Protection. The potential exists for a fire due to a leak or release of crude from heaters in the crude desalting unit. Low boiling point components of crude may also be released if a leak occurs.

b. Safety. Inadequate desalting can cause fouling of heater tubes and heat exchangers throughout the refinery. Fouling restricts product flow and heat transfer and leads to failures due to increased pressures and temperatures. Corrosion, which occurs due to the presence of hydrogen sulfide, hydrogen chloride, naphthenic (organic) acids, and other contaminants in the crude oil, also causes equipment failure. Neutralized salts (ammonium chlorides and sulfides), when moistened by condensed water, can cause corrosion. Overpressuring the unit is another potential hazard that causes failures.

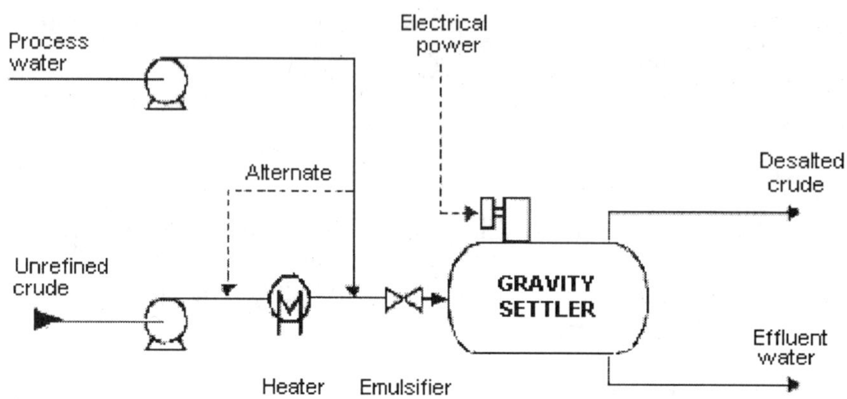

FIGURE IV: 2-7. ELECTROSTAITC DESALTING.

c. Health. Because this is a closed process, there is little potential for exposure to crude oil unless a leak or release occurs. Where elevated operating temperatures are used when desalting sour crudes, hydrogen sulfide will be present. There is the possibility of exposure to ammonia, dry chemical demulsifiers, caustics, and/or acids during this operation. Safe work practices and/or the use of appropriate personal protective equipment may be needed for exposures to chemicals and other hazards such as heat, and during process sampling, inspection, maintenance, and turnaround activities.

Depending on the crude feedstock and the treatment chemicals used, the wastewater will contain varying amounts of chlorides, sulfides, bicarbonates, ammonia, hydrocarbons, phenol, and suspended solids. If diatomaceous earth is used in filtration, exposures should be minimized or controlled. Diatomaceous earth can contain silica in very fine particle size, making this a potential respiratory hazard.

B. CRUDE OIL DISTILLATION (FRACTIONATION).

1. Description. The first step in the refining process is the separation of crude oil into various fractions or straight-run cuts by distillation in atmospheric and vacuum towers. The main fractions or "cuts" obtained have specific boiling-point ranges and can be classified in order of decreasing volatility into gases, light distillates, middle distillates, gas oils, and residuum.

2. Atmospheric Distillation Tower.

a. At the refinery, the desalted crude feedstock is preheated using recovered process heat. The feedstock then flows to a direct-fired crude charge heater where it is fed into the vertical distillation column just above the bottom, at pressures slightly above atmospheric and at temperatures ranging from 650° to 700° F (heating crude oil above these temperatures may cause undesirable thermal cracking). All but the heaviest fractions flash into vapor. As the hot vapor rises in the tower, its temperature is reduced. Heavy fuel oil or asphalt residue is taken from the bottom. At successively higher points on the tower, the various major products including lubricating oil, heating oil, kerosene, gasoline, and uncondensed gases (which condense at lower temperatures) are drawn off.

b. The fractionating tower, a steel cylinder about 120 feet high, contains horizontal steel trays for separating and collecting the liquids. At each tray, vapors from below enter perforations and bubble caps. They permit the vapors to bubble through the liquid on the tray, causing some condensation at the temperature of that tray. An overflow pipe drains the condensed liquids from each tray back to the tray below, where the higher temperature causes re-evaporation. The evaporation, condensing, and scrubbing operation is repeated many times until the desired degree of product purity is reached. Then side streams from certain trays are taken off to obtain the desired fractions. Products ranging from uncondensed fixed gases at the top to heavy fuel oils at the bottom can be taken continuously from a fractionating tower. Steam is often used in towers to lower the vapor pressure and create a partial vacuum. The distillation process separates the major constituents of crude oil into so-called straight-run products. Sometimes crude oil is "topped" by distilling off only the lighter fractions, leaving a heavy residue that is often distilled further under high vacuum.

TABLE IV:2-5. ATMOSPHERIC DISTILLATION PROCESS

Feedstock	*From*	*Process*	*Typical products . . . . . .*	*To*
Crude	Desalting	Separation	Gases	Atmospheric distillation tower
			Naphtas.	Reforming or treating
			Kerosene or distillates . .	Treating
			Gas oil	Catalytic cracking
			Residual	Vacuum tower or visbreaker

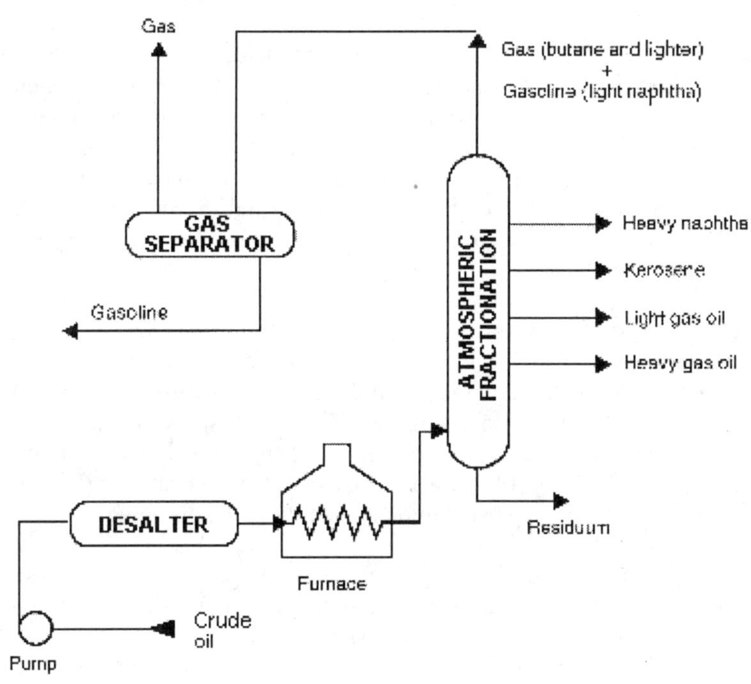

FIGURE IV: 2-8. ATMOSPHERIC DISTILLATION.

3. Vacuum Distillation Tower. In order to further distill the residuum or topped crude from the atmospheric tower at higher temperatures, reduced pressure is required to prevent thermal cracking. The process takes place in one or more vacuum distillation towers. The principles of vacuum distillation resemble those of fractional

distillation and, except that larger-diameter columns are used to maintain comparable vapor velocities at the reduced pressures, the equipment is also similar. The internal designs of some vacuum towers are different from atmospheric towers in that random packing and demister pads are used instead of trays. A typical first-phase vacuum tower may produce gas oils, lubricating-oil base stocks, and heavy residual for propane deasphalting. A second-phase tower operating at lower vacuum may distill surplus residuum from the atmospheric tower, which is not used for lube-stock processing, and surplus residuum from the first vacuum tower not used for deasphalting. Vacuum towers are typically used to separate catalytic cracking feedstock from surplus residuum.

4. Other Distillation Towers (Columns). Within refineries there are numerous other, smaller distillation towers called columns, designed to separate specific and unique products. Columns all work on the same principles as the towers described above. For example, a depropanizer is a small column designed to separate propane and lighter gases from butane and heavier components. Another larger column is used to separate ethyl benzene and xylene. Small "bubble" towers called strippers use steam to remove trace amounts of light products from heavier product streams.

5. Health and Safety Considerations.

   a. Fire Prevention and Protection. Even though these are closed processes, heaters and exchangers in the atmospheric and vacuum distillation units could provide a source of ignition, and the potential for a fire exists should a leak or release occur.

   b. Safety. An excursion in pressure, temperature, or liquid levels may occur if automatic control devices fail. Control of temperature, pressure, and reflux within operating parameters is needed to prevent thermal cracking within the distillation towers. Relief systems should be provided for overpressure and operations monitored to prevent crude from entering the reformer charge.

   The sections of the process susceptible to corrosion include (but may not be limited to) preheat exchanger (HCl and H2S), preheat furnace and bottoms exchanger (H2S and sulfur compounds), atmospheric tower and vacuum furnace (H2S, sulfur compounds, and organic acids), vacuum tower (H2S and organic acids), and overhead (H2S, HCl, and water). Where sour crudes are processed, severe corrosion can occur in furnace tubing and in both atmospheric and vacuum towers where metal temperatures exceed 450° F. Wet H2S also will cause cracks in steel. When processing high-nitrogen crudes, nitrogen oxides can form in the flue gases of furnaces. Nitrogen oxides are corrosive to steel when cooled to low temperatures in the presence of water.

   Chemicals are used to control corrosion by hydrochloric acid produced in distillation units. Ammonia may be injected into the overhead stream prior to initial condensation and/or an alkaline solution may be carefully injected into the hot crude-oil feed. If sufficient wash-water is not injected, deposits of ammonium chloride can form and cause serious corrosion. Crude feedstock may contain appreciable amounts of water in suspension which can separate during startup and, along with water remaining in the tower from steam purging, settle in the bottom of the tower. This water can be heated to the boiling point and create an instantaneous vaporization explosion upon contact with the oil in the unit.

   c. Health. Atmospheric and vacuum distillation are closed processes and exposures are expected to be minimal. When sour (high-sulfur) crudes are processed, there is potential for exposure to hydrogen sulfide in the preheat exchanger and furnace, tower flash zone and overhead system, vacuum furnace and tower, and bottoms exchanger. Hydrogen chloride may be present in the preheat exchanger, tower top zones, and overheads. Wastewater may contain water-soluble sulfides in high concentrations and other water-soluble compounds such as ammonia, chlorides, phenol, mercaptans, etc., depending upon the crude feedstock and the treatment chemicals. Safe work practices and/or the use of appropriate personal protective equipment may be needed for exposures to chemicals and other hazards such as heat and noise, and during sampling, inspection, maintenance, and turnaround activities.

TABLE IV:2-6. VACUUM DISTILLATION PROCESS

Feedstock	*From*	*Process*	*Typical products . . To*	
Residuals	Atmospheric tower	Separation	Gas oils	Catalytic cracker
			Lubricants	Hydrotreating or solvent
			Residual	Deasphalter, visbreaker, or coke

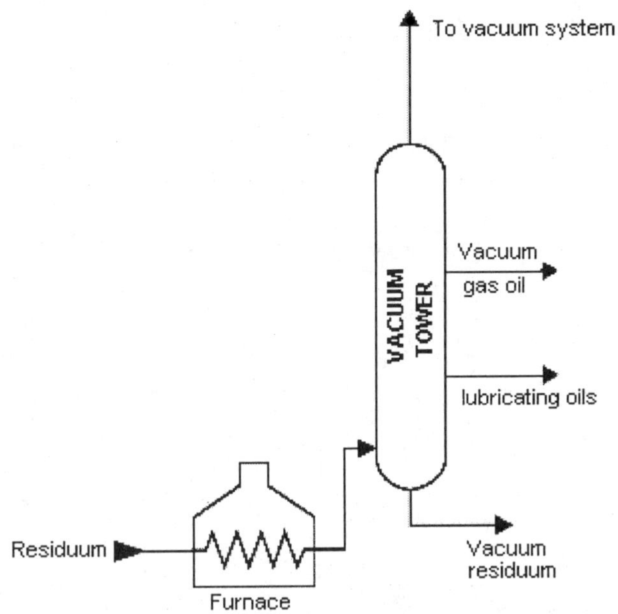

FIGURE IV: 2-9. VACUUM DISTILLATION.

C. SOLVENT EXTRACTION AND DEWAXING.

1. Description. Solvent treating is a widely used method of refining lubricating oils as well as a host of other refinery stocks. Since distillation (fractionation) separates petroleum products into groups only by their boiling-point ranges, impurities may remain. These include organic compounds containing sulfur, nitrogen, and oxygen; inorganic salts and dissolved metals; and soluble salts that were present in the crude feedstock. In addition, kerosene and distillates may have trace amounts of aromatics and naphthenes, and lubricating oil base-stocks may contain wax. Solvent refining processes including solvent extraction and solvent dewaxing usually remove these undesirables at intermediate refining stages or just before sending the product to storage.

2. Solvent Extraction.

   a. The purpose of solvent extraction is to prevent corrosion, protect catalyst in subsequent processes, and improve finished products by removing unsaturated, aromatic hydrocarbons from lubricant and grease stocks. The solvent extraction process separates aromatics, naphthenes, and impurities from the product stream by dissolving or precipitation. The feedstock is first dried and then treated using a continuous countercurrent solvent treatment operation. In one type of process, the feedstock is washed with a liquid in which the substances to be removed are more soluble than in the desired resultant product. In another process, selected solvents are added to cause impurities to precipitate out of the product. In the adsorption process, highly porous solid materials collect liquid molecules on their surfaces.

b. The solvent is separated from the product stream by heating, evaporation, or fractionation, and residual trace amounts are subsequently removed from the raffinate by steam stripping or vacuum flashing. Electric precipitation may be used for separation of inorganic compounds. The solvent is then regenerated to be used again in the process.

c. The most widely used extraction solvents are phenol, furfural, and cresylic acid. Other solvents less frequently used are liquid sulfur dioxide, nitrobenzene, and 2,2' dichloroethyl ether. The selection of specific processes and chemical agents depends on the nature of the feedstock being treated, the contaminants present, and the finished product requirements.

TABLE IV:2-7. SOLVENT EXTRACTION PROCESS

Feedstock	*From*	*Process*	*Typical products . . . To*
Naphthas, distillates, kerosene	Atm. tower	Treating/ blending	High octane gasoline . . Storage Refined fuels . . . . Treating and blending Spent agents . . . . Treatment and blending

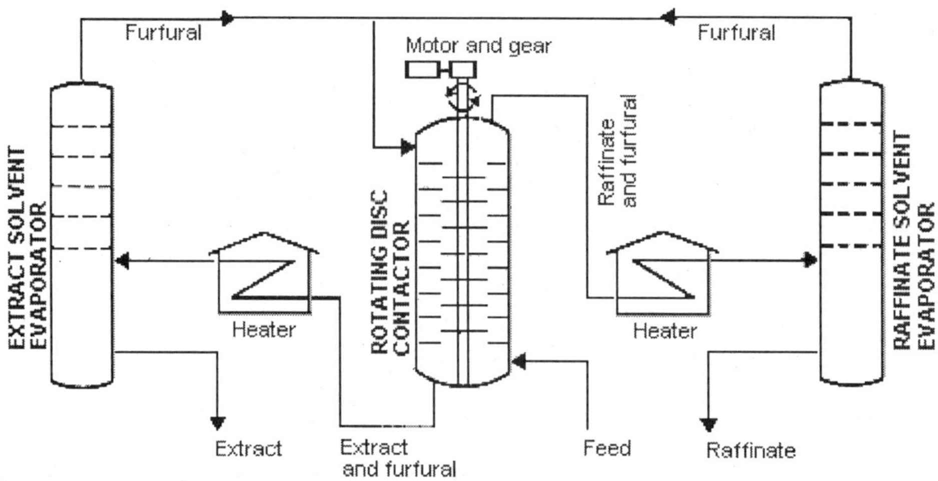

FIGURE IV: 2-10. AROMATICS EXTRACTION.

3. Solvent Dewaxing. Solvent dewaxing is used to remove wax from either distillate or residual basestocks at any stage in the refining process. There are several processes in use for solvent dewaxing, but all have the same general steps, which are: (1) mixing the feedstock with a solvent, (2) precipitating the wax from the mixture by chilling, and (3) recovering the solvent from the wax and dewaxed oil for recycling by distillation and steam stripping. Usually two solvents are used: toluene, which dissolves the oil and maintains fluidity at low temperatures, and methyl ethyl ketone (MEK), which dissolves little wax at low temperatures and acts as a wax precipitating agent. Other solvents that are sometimes used include benzene, methyl isobutyl ketone, propane, petroleum naphtha, ethylene dichloride, methylene chloride, and sulfur dioxide. In addition, there is a catalytic process used as an alternate to solvent dewaxing.

TABLE IV:2-8. SOLVENT DEWAXING PROCESS

Feedstock	*From*	*Process*	*Typical products . .*	*To*
Lube basestock	Vacuum tower	Treating	Dewaxed lubes	Hydrotreating
			Wax	Hydrotreating
			Spent agents	Treatment or recycle

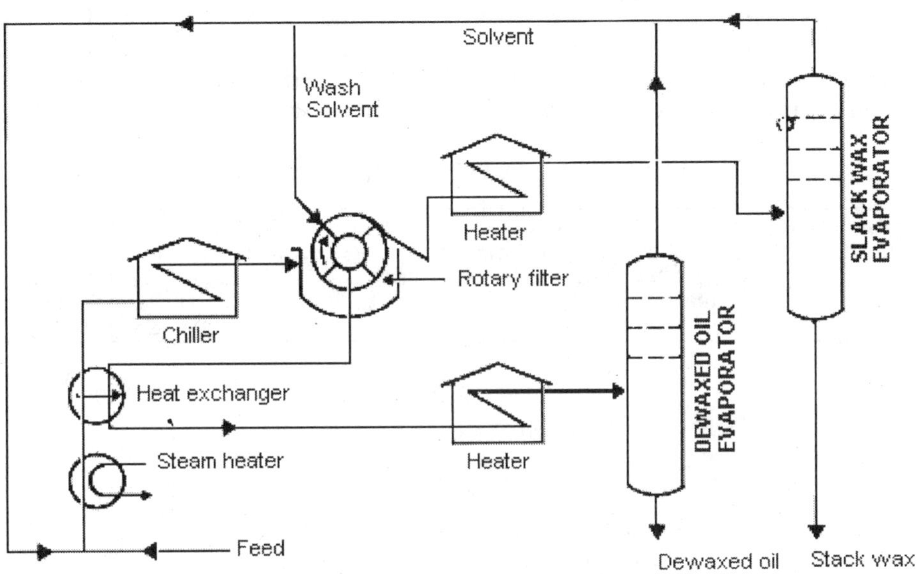

FIGURE IV: 2-11. SOLVENT DEWAXING

Note: Diagrams in Figures IV:2-10, 11, 12, 13, 15, and 20 reproduced with permission from Shell International Petroleum Company.

4. Health and Safety Considerations.

   a. Fire Prevention and Protection. Solvent treatment is essentially a closed process and, although operating pressures are relatively low, the potential exists for fire from a leak or spill contacting a source of ignition such as the drier or extraction heater. In solvent dewaxing, disruption of the vacuum will create a potential fire hazard by allowing air to enter the unit.

   b. Health. Because solvent extraction is a closed process, exposures are expected to be minimal under normal operating conditions. However, there is a potential for exposure to extraction solvents such as phenol, furfural, glycols, methyl ethyl ketone, amines, and other process chemicals. Safe work practices and/or the use of appropriate personal protective equipment may be needed for exposures to chemicals and other hazards such as noise and heat, and during repair, inspection, maintenance, and turnaround activities.

D. THERMAL CRACKING.

1. Description.

   a. Because the simple distillation of crude oil produces amounts and types of products that are not consistent with those required by the marketplace, subsequent refinery processes change the product mix by altering the molecular structure of the hydrocarbons. One of the ways of accomplishing this change is through "cracking," a process that breaks or cracks the heavier, higher boiling-point petroleum fractions into more valuable products such as gasoline, fuel oil, and gas oils. The two basic types of cracking are thermal cracking, using heat and pressure, and catalytic cracking.

b. The first thermal cracking process was developed around 1913. Distillate fuels and heavy oils were heated under pressure in large drums until they cracked into smaller molecules with better antiknock characteristics. However, this method produced large amounts of solid, unwanted coke. This early process has evolved into the following applications of thermal cracking: visbreaking, steam cracking, and coking.

2. Visbreaking Process. Visbreaking, a mild form of thermal cracking, significantly lowers the viscosity of heavy crude-oil residue without affecting the boiling point range. Residual from the atmospheric distillation tower is heated (800°-950° F) at atmospheric pressure and mildly cracked in a heater. It is then quenched with cool gas oil to control overcracking, and flashed in a distillation tower. Visbreaking is used to reduce the pour point of waxy residues and reduce the viscosity of residues used for blending with lighter fuel oils. Middle distillates may also be produced, depending on product demand. The thermally cracked residue tar, which accumulates in the bottom of the fractionation tower, is vacuum flashed in a stripper and the distillate recycled.

TABLE IV:2-9. VISBREAKING PROCESS.

Feedstock	*From*	*Process*	*Typical products . . . . . To*
Residual	Atmospheric tower & Vacuum tower	Decompose	Gasoline or distillate . . Hydrotreating Vapor . . . . Hydrotreater Residue . . . . Stripper or recycle Gases . . . . Gas plant

3. Steam Cracking Process. Steam cracking is a petrochemical process sometimes used in refineries to produce olefinic raw materials (e.g., ethylene) from various feedstock for petrochemicals manufacture. The feedstock range from ethane to vacuum gas oil, with heavier feeds giving higher yields of by-products such as naphtha. The most common feeds are ethane, butane, and naphtha. Steam cracking is carried out at temperatures of 1,500°-1,600° F, and at pressures slightly above atmospheric. Naphtha produced from steam cracking contains benzene, which is extracted prior to hydrotreating. Residual from steam cracking is sometimes blended into heavy fuels.

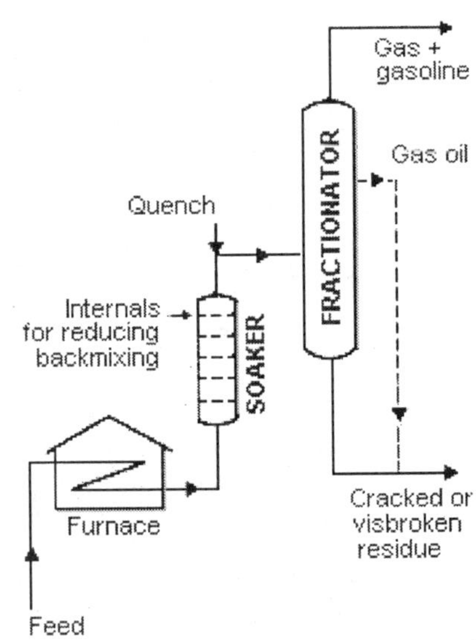

FIGURE IV: 2-12. VISBREAKING.

4. Coking Processes. Coking is a severe method of thermal cracking used to upgrade heavy residuals into lighter products or distillates. Coking produces straight-run gasoline (coker naphtha) and various middle-distillate fractions used as catalytic cracking feedstock. The process so completely reduces hydrogen that the residue is a form of carbon called "coke." The two most common processes are delayed coking and continuous (contact or fluid) coking. Three typical types of coke are obtained (sponge coke, honeycomb coke, and needle coke) depending upon the reaction mechanism, time, temperature, and the crude feedstock.

a. Delayed Coking. In delayed coking the heated charge (typically residuum from atmospheric distillation towers) is transferred to large coke drums which provide the long residence time needed to allow the cracking reactions to proceed to completion. Initially the heavy feedstock is fed to a furnace which heats the residuum to high temperatures (900°-950° F) at low pressures (25-30 psi) and is designed and controlled to prevent premature coking in the heater tubes. The mixture is passed from the heater to one or more coker drums where the hot material is held approximately 24 hours (delayed) at pressures of 25-75 psi, until it cracks into lighter products. Vapors from the drums are returned to a fractionator where gas, naphtha, and gas oils are separated out. The heavier hydrocarbons produced in the fractionator are recycled through the furnace.

After the coke reaches a predetermined level in one drum, the flow is diverted to another drum to maintain continuous operation. The full drum is steamed to strip out uncracked hydrocarbons, cooled by water injection, and decoked by mechanical or hydraulic methods. The coke is mechanically removed by an auger rising from the bottom of the drum. Hydraulic decoking consists of fracturing the coke bed with high-pressure water ejected from a rotating cutter.

b. Continuous Coking. Continuous (contact or fluid) coking is a moving-bed process that operates at temperatures higher than delayed coking. In continuous coking, thermal cracking occurs by using heat transferred from hot, recycled coke particles to feedstock in a radial mixer, called a reactor, at a pressure of 50 psi. Gases and vapors are taken from the reactor, quenched to stop any further reaction, and fractionated. The reacted coke enters a surge drum and is lifted to a feeder and classifier where the larger coke particles are removed as product. The remaining coke is dropped into the preheater for recycling with feedstock. Coking occurs both in the reactor and in the surge drum. The process is automatic in that there is a continuous flow of coke and feedstock.

TABLE IV: 2-10. COKING PROCESSES.

Feedstock	*From*	*Process*	*Typical products . . .*	*To*
Residual	Atmospheric & vacuum catalytic cracker	Decompo-sition	Naphtha, gasoline . .	Distillation column, blending
Clarified oil . .	Catalytic cracker		Coke	Shipping, recycle
Tars	Various units		Gas oil	Catalytic cracking
Wasteater (sour)	Treatment			
Gases	Gas plant			

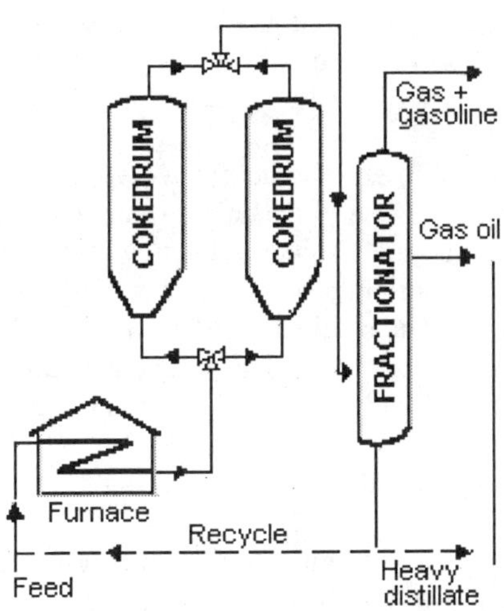

FIGURE IV: 2-13. DELAYED COKING.

5. Health and Safety Considerations.

a. Fire Protection and Prevention. Because thermal cracking is a closed process, the primary potential for fire is from leaks or releases of liquids, gases, or vapors reaching an ignition source such as a heater. The potential for fire is present in coking operations due to vapor or product leaks. Should coking temperatures get out of control, an exothermic reaction could occur within the coker.

b. Safety. In thermal cracking when sour crudes are processed, corrosion can occur where metal temperatures are between 450° and 900° F. Above 900° F coke forms a protective layer on the metal. The furnace, soaking drums, lower part of the tower, and high-temperature exchangers are usually subject to corrosion. Hydrogen sulfide corrosion in coking can also occur when temperatures are not properly controlled above 900° F.

Continuous thermal changes can lead to bulging and cracking of coke drum shells. In coking, temperature control must often be held within a 10°-20° F range, as high temperatures will produce coke that is too hard to cut out of the drum. Conversely, temperatures that are too low will result in a high asphaltic-content slurry. Water or steam injection may be used to prevent buildup of coke in delayed coker furnace tubes. Water must be completely drained from the coker, so as not to cause an explosion upon recharging with hot coke. Provisions for alternate means of egress from the working platform on top of coke drums are important in the event of an emergency.

c. Health. The potential exists for exposure to hazardous gases such as hydrogen sulfide and carbon monoxide, and trace polynuclear aromatics (PNA's) associated with coking operations. When coke is moved as a slurry, oxygen depletion may occur within confined spaces such as storage silos, since wet carbon will adsorb oxygen. Wastewater may be highly alkaline and contain oil, sulfides, ammonia, and/or phenol. The potential exists in the coking process for exposure to burns when handling hot coke or in the event of a steam-line leak, or from steam, hot water, hot coke, or hot slurry that may be expelled when opening cokers. Safe work practices and/or the use of appropriate personal protective equipment may be needed for exposures to chemicals and other hazards such as heat and noise, and during process sampling, inspection, maintenance, and turnaround activities. (Note: coke produced from petroleum is a different product from that generated in the steel-industry coking process.)

E. CATALYTIC CRACKING.

1. Description.

a. Catalytic cracking breaks complex hydrocarbons into simpler molecules in order to increase the quality and quantity of lighter, more desirable products and decrease the amount of residuals. This process rearranges the molecular structure of hydrocarbon compounds to convert heavy hydrocarbon feedstock into lighter fractions such as kerosene, gasoline, LPG, heating oil, and petrochemical feedstock.

b. Catalytic cracking is similar to thermal cracking except that catalysts facilitate the conversion of the heavier molecules into lighter products. Use of a catalyst (a material that assists a chemical reaction but does not take part in it) in the cracking reaction increases the yield of improved-quality products under much less severe operating conditions than in thermal cracking. Typical temperatures are from 850°-950° F at much lower pressures of 10-20 psi. The catalysts used in refinery cracking units are typically solid materials (zeolite, aluminum hydrosilicate, treated bentonite clay, fuller's earth, bauxite, and silica-alumina) that come in the form of powders, beads, pellets or shaped materials called extrudites.

c. There are three basic functions in the catalytic racking process:

- Reaction: Feedstock reacts with catalyst and cracks into different hydrocarbons;
- Regeneration: Catalyst is reactivated by burning off coke; and
- Fractionation: Cracked hydrocarbon stream is separated into various products.

d. The three types of catalytic cracking processes are fluid catalytic cracking (FCC), moving-bed catalytic cracking, and Thermofor catalytic cracking (TCC). The catalytic cracking process is very flexible, and operating parameters can be adjusted to meet changing product demand. In addition to cracking, catalytic activities include dehydrogenation, hydrogenation, and isomerization.

TABLE IV: 2-11. CATALYTIC CRACKING PROCESS

Feedstock	From	Process	Typical products . . . . To
Gas oils	Towers, coker visbreaker	Decomposition, alteration	Gasoline . . . . Treater or blend Gases . . . . Gas plant
Deasphalted oils	Deashpalter		Middle distillates . . . . . Hydrotreat, blend, or recycle Petrochem feedstock . . Petrochem or other Residue . . . . Residual fuel blend

F. FLUID CATALYTIC CRACKING.

1. Description.

a. The most common process is FCC, in which the oil is cracked in the presence of a finely divided catalyst which is maintained in an aerated or fluidized state by the oil vapors. The fluid cracker consists of a catalyst section and a fractionating section that operate together as an integrated processing unit. The catalyst section contains the reactor and regenerator, which, with the standpipe and riser, forms the catalyst circulation unit. The fluid catalyst is continuously circulated between the reactor and the regenerator using air, oil vapors, and steam as the conveying media.

b. A typical FCC process involves mixing a preheated hydrocarbon charge with hot, regenerated catalyst as it enters the riser leading to the reactor. The charge is combined with a recycle stream within the riser, vaporized, and raised to reactor temperature (900°-1,000° F) by the hot catalyst. As the mixture travels up the riser, the charge is cracked at 10-30 psi. In the more modern FCC units, all cracking takes place in the riser. The "reactor" no longer functions as a reactor; it merely serves as a holding vessel for the cyclones. This cracking continues until the oil vapors are separated from the catalyst in the reactor cyclones. The resultant product stream (cracked product) is then charged to a fractionating column where it is separated into fractions, and some of the heavy oil is recycled to the riser.

c. Spent catalyst is regenerated to get rid of coke that collects on the catalyst during the process. Spent catalyst flows through the catalyst stripper to the regenerator, where most of the coke deposits burn off at the bottom where preheated air and spent catalyst are mixed. Fresh catalyst is added and worn-out catalyst removed to optimize the cracking process.

2. Moving Bed Catalytic Cracking. The moving-bed catalytic cracking process is similar to the FCC process. The catalyst is in the form of pellets that are moved continuously to the top of the unit by conveyor or pneumatic lift tubes to a storage hopper, then flow downward by gravity through the reactor, and finally to a regenerator. The regenerator and hopper are isolated from the reactor by steam seals. The cracked product is separated into recycle gas, oil, clarified oil, distillate, naphtha, and wet gas.

3. Thermofor Catalytic Cracking. In a typical thermofor catalytic cracking unit, the preheated feedstock flows by gravity through the catalytic reactor bed. The vapors are separated from the catalyst and sent to a fractionating tower. The spent catalyst is regenerated, cooled, and recycled. The flue gas from regeneration is sent to a carbon-monoxide boiler for heat recovery.

4. Health and Safety Considerations.

a. Fire Prevention and Protection. Liquid hydrocarbons in the catalyst or entering the heated combustion air stream should be controlled to avoid exothermic reactions. Because of the presence of heaters in catalytic cracking units, the possibility exists for fire due to a leak or vapor release. Fire protection including concrete or other insulation on columns and supports, or fixed water spray or fog systems where insulation is not feasible and in areas where firewater hose streams cannot reach, should be considered.

In some processes, caution must be taken to prevent explosive concentrations of catalyst dust during recharge or disposal. When unloading any coked catalyst, the possibility exists for iron sulfide fires. Iron sulfide will ignite spontaneously when exposed to air and therefore must be wetted with water to prevent it from igniting vapors. Coked catalyst may be either cooled below 120° F before it is dumped from the reactor, or dumped into containers that have been purged and inerted with nitrogen and then cooled before further handling.

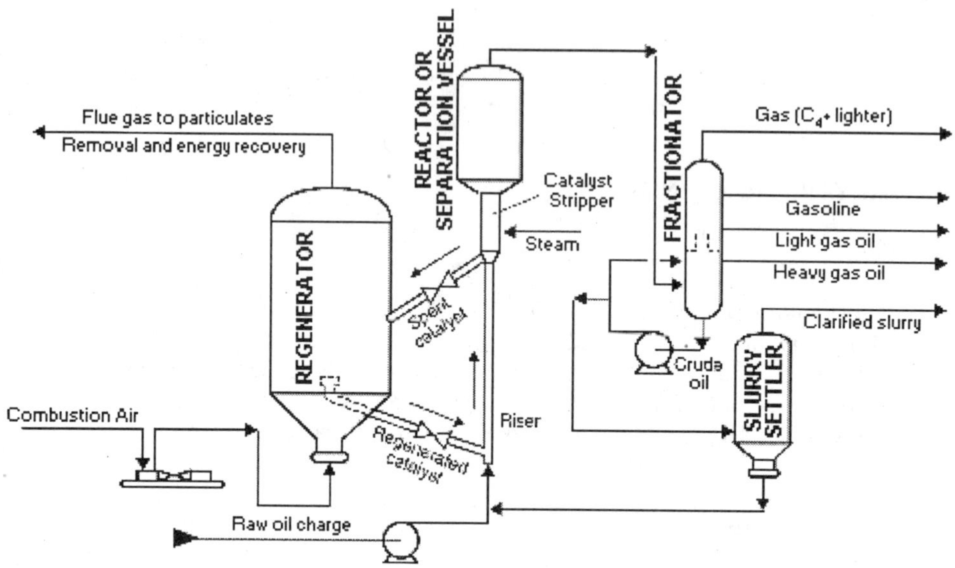

FIGURE IV: 2-14. FLUID CATALYTIC CRACKING.

b. Safety. Regular sampling and testing of the feedstock, product, and recycle streams should be performed to assure that the cracking process is working as intended and that no contaminants have entered the process stream. Corrosives or deposits in the feedstock can foul gas compressors. Inspections of critical equipment including pumps, compressors, furnaces, and heat exchangers should be conducted as needed. When processing sour crude, corrosion may be expected where temperatures are below 900° F. Corrosion takes place where both liquid and vapor phases exist, and at areas subject to local cooling such as nozzles and platform supports.

When processing high-nitrogen feedstock, exposure to ammonia and cyanide may occur, subjecting carbon steel equipment in the FCC overhead system to corrosion, cracking, or hydrogen blistering. These effects may be minimized by water wash or corrosion inhibitors. Water wash may also be used to protect overhead condensers in the main column subjected to fouling from ammonium hydrosulfide. Inspections should include checking for leaks due to erosion or other malfunctions such as catalyst buildup on the expanders, coking in the overhead feeder lines from feedstock residues, and other unusual operating conditions.

c. Health. Because the catalytic cracker is a closed system, there is normally little opportunity for exposure to hazardous substances during normal operations. The possibility exists of exposure to extremely hot (700° F) hydrocarbon liquids or vapors during process sampling or if a leak or release occurs. In addition, exposure to hydrogen sulfide and/or carbon monoxide gas may occur during a release of product or vapor.

Catalyst regeneration involves steam stripping and decoking, and produces fluid waste streams that may contain varying amounts of hydrocarbon, phenol, ammonia, hydrogen sulfide, mercaptan, and other materials depending upon the feedstock, crudes, and processes. Inadvertent formation of nickel carbonyl may occur in cracking processes using nickel catalysts, with resultant potential for hazardous exposures.

Safe work practices and/or the use of appropriate personal protective equipment may be needed for exposures to chemicals and other hazards such as noise and heat; during process sampling, inspection, maintenance and turnaround activities; and when handling spent catalyst, recharging catalyst, or if leaks or releases occur.

G. HYDROCRACKING.

1. Description.

a. Hydrocracking is a two-stage process combining catalytic cracking and hydrogenation, wherein heavier feedstock are cracked in the presence of hydrogen to produce more desirable products. The process employs high pressure, high temperature, a catalyst, and hydrogen. Hydrocracking is used for feedstock that are difficult to process by either catalytic cracking or reforming, since these feedstock are characterized usually by a high polycyclic aromatic content and/or high concentrations of the two principal catalyst poisons, sulfur and nitrogen compounds.

b. The hydrocracking process largely depends on the nature of the feedstock and the relative rates of the two competing reactions, hydrogenation and cracking. Heavy aromatic feedstock is converted into lighter products under a wide range of very high pressures (1,000-2,000 psi) and fairly high temperatures (750°-1,500° F), in the presence of hydrogen and special catalysts. When the feedstock has a high paraffinic content, the primary function of hydrogen is to prevent the formation of polycyclic aromatic compounds. Another important role of hydrogen in the hydrocracking process is to reduce tar formation and prevent buildup of coke on the catalyst. Hydrogenation also serves to convert sulfur and nitrogen compounds present in the feedstock to hydrogen sulfide and ammonia.

c. Hydrocracking produces relatively large amounts of isobutane for alkylation feedstock. Hydrocracking also performs isomerization for pour-point control and smoke-point control, both of which are important in high-quality jet fuel.

2. Hydrocracking Process.

a. In the first stage, preheated feedstock is mixed with recycled hydrogen and sent to the first-stage reactor, where catalysts convert sulfur and nitrogen compounds to hydrogen sulfide and ammonia. Limited hydrocracking also occurs.

b. After the hydrocarbon leaves the first stage, it is cooled and liquefied and run through a hydrocarbon separator. The hydrogen is recycled to the feedstock. The liquid is charged to a fractionator. Depending on the products desired (gasoline components, jet fuel, and gas oil), the fractionator is run to cut out some portion of the first stage reactor out-turn. Kerosene-range material can be taken as a separate side-draw product or included in the fractionator bottoms with the gas oil.

c. The fractionator bottoms are again mixed with a hydrogen stream and charged to the second stage. Since this material has already been subjected to some hydrogenation, cracking, and reforming in the first stage, the operations of the second stage are more severe (higher temperatures and pressures). Like the outturn of the first stage, the second stage product is separated from the hydrogen and charged to the fractionator.

TABLE IV: 2-12. HYDROCRACKING PROCESS.

Feedstock	*From*	*Process*	*Typical products . . . . To*
High pour point	Catalytic cracker, atmospheric & vacuum tower	Decomposition, hydrogenation	Kerosen, jet fuel . . . .Blending
Gas oil	Vacuum tower, coker		Gasoline, distillates . . Blending
Hydrogen	Reformer		Recycle, reformer gas . . Gas plant

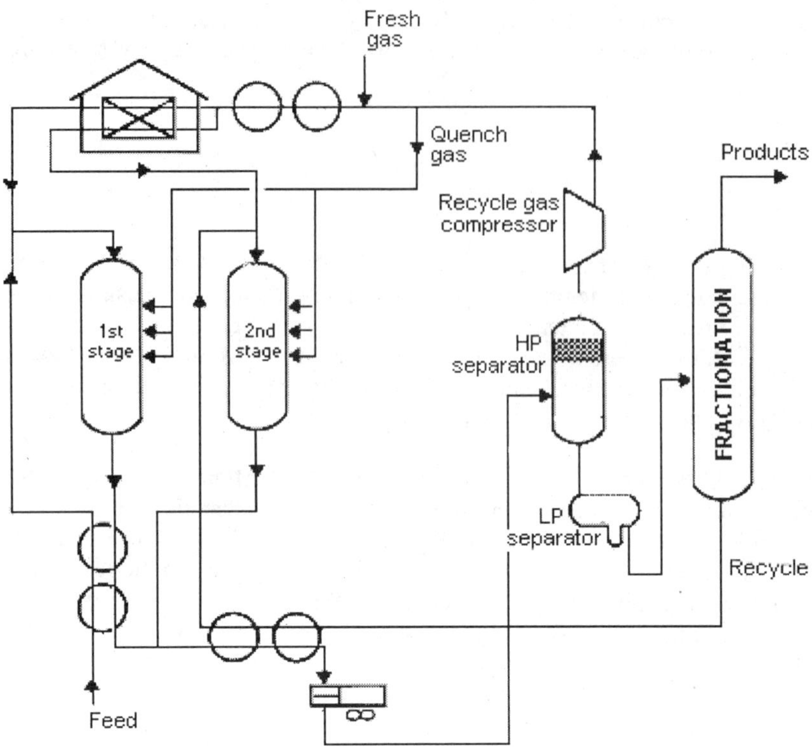

FIGURE IV: 2-15. TWO-STAGE HYDROCRACKING.

3. Health and Safety Considerations.

a. Fire Prevention and Protection. Because this unit operates at very high pressures and temperatures, control of both hydrocarbon leaks and hydrogen releases is important to prevent fires. In some processes, care is needed to ensure that explosive concentrations of catalytic dust do not form during recharging.

b. Safety. Inspection and testing of safety relief devices are important due to the very high pressures in this unit. Proper process control is needed to protect against plugging reactor beds. Unloading coked catalyst requires special precautions to prevent iron sulfide-induced fires. The coked catalyst should either be cooled to below 120° F before dumping, or be placed in nitrogen-inerted containers until cooled.

Because of the operating temperatures and presence of hydrogen, the hydrogen-sulfide content of the feedstock must be strictly controlled to a minimum to reduce the possibility of severe corrosion. Corrosion by wet carbon dioxide in areas of condensation also must be considered. When processing high-nitrogen feedstock, the ammonia and hydrogen sulfide form ammonium hydrosulfide, which causes serious corrosion at temperatures below the water dew point. Ammonium hydrosulfide is also present in sour water stripping.

c. Health. Because this is a closed process, exposures are expected to be minimal under normal operating conditions. There is a potential for exposure to hydrocarbon gas and vapor emissions, hydrogen and hydrogen sulfide gas due to high-pressure leaks. Large quantities of carbon monoxide may be released during catalyst regeneration and changeover. Catalyst steam stripping and regeneration create waste streams containing sour water and ammonia. Safe work practices and/or the use of appropriate personal protective equipment may be needed for exposure to chemicals and other hazards such as noise and heat, during process sampling, inspection, maintenance, and turnaround activities, and when handling spent catalyst.

H. CATALYTIC REFORMING.

1. Description.

a. Catalytic reforming is an important process used to convert low-octane naphthas into high-octane gasoline blending components called reformate. Reforming represents the total effect of numerous reactions such as cracking, polymerization, dehydrogenation, and isomerization taking place simultaneously. Depending on the properties of the naphtha feedstock (as measured by the paraffin, olefin, naphthene, and aromatic content) and catalysts used, reformates can be produced with very high concentrations of toluene, benzene, xylene, and other aromatics useful in gasoline blending and petrochemical processing. Hydrogen, a significant by-product, is separated from the reformate for recycling and use in other processes.

b. A catalytic reformer comprises a reactor section and a product-recovery section. More or less standard is a feed preparation section in which, by combination of hydrotreatment and distillation, the feedstock is prepared to specification. Most processes use platinum as the active catalyst. Sometimes platinum is combined with a second catalyst (bimetallic catalyst) such as rhenium or another noble metal.

c. There are many different commercial catalytic reforming processes including platforming, powerforming, ultraforming, and Thermofor catalytic reforming. In the platforming process, the first step is preparation of the naphtha feed to remove impurities from the naphtha and reduce catalyst degradation. The naphtha feedstock is then mixed with hydrogen, vaporized, and passed through a series of alternating furnace and fixed-bed reactors containing a platinum catalyst. The effluent from the last reactor is cooled and sent to a separator to permit removal of the hydrogen-rich gas stream from the top of the separator for recycling. The liquid product from the bottom of the separator is sent to a fractionator called a stabilizer (butanizer). It makes a bottom product called reformate; butanes and lighter go overhead and are sent to the saturated gas plant.

d. Some catalytic reformers operate at low pressure (50-200 psi), and others operate at high pressures (up to 1,000 psi). Some catalytic reforming systems continuously regenerate the catalyst in other systems. One reactor at a time is taken off-stream for catalyst regeneration, and some facilities regenerate all of the reactors during turnarounds.

2. Health and Safety Considerations.

a. Fire Prevention and Protection. This is a closed system; however, the potential for fire exists should a leak or release of reformate gas or hydrogen occur.

b. Safety. Operating procedures should be developed to ensure control of hot spots during start-up. Safe catalyst handling is very important. Care must be taken not to break or crush the catalyst when loading the beds, as the small fines will plug up the reformer screens. Precautions against dust when regenerating or replacing catalyst should also be considered. Also, water wash should be considered where stabilizer fouling has occurred due to the formation of ammonium chloride and iron salts. Ammonium chloride may form in pretreater exchangers and cause corrosion and fouling. Hydrogen chloride from the hydrogenation of chlorine compounds may form acid or ammonium chloride salt.

c. Health. Because this is a closed process, exposures are expected to be minimal under normal operating conditions. There is potential for exposure to hydrogen sulfide and benzene should a leak or release occur.

Small emissions of carbon monoxide and hydrogen sulfide may occur during regeneration of catalyst. Safe work practices and/or appropriate personal protective equipment may be needed for exposures to chemicals and other hazards such as noise and heat; during testing, inspecting, maintenance and turnaround activities; and when handling regenerated or spent catalyst.

TABLE IV: 2-13. CATALYTIC REFORMING PROCESS

Feedstock	*From*	*Process*	*Typical products . . . . To*
Desulfurized naphtha	Coker	Rearrange, dehydrogenate	High octane gasoline . . Blending
			Aromatics . . . . Petrochemical
Naphthene-rich fractions	hydrocracker, hydrodesulfur		Hydrogen . . . . Recycle, hydrotreat, etc
Straight-run naphtha	Atmospheric fractionator		Gas . . . . Gas plant

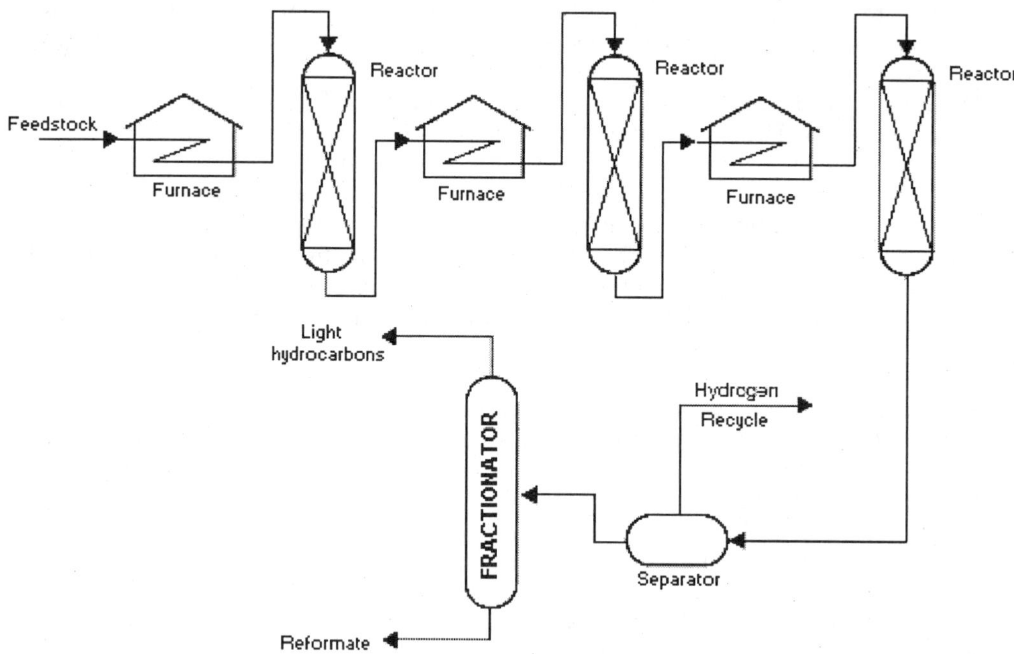

FIGURE IV: 2-16. PLATFORMING PROCESS.

I. CATALYTIC HYDROTREATING.

1. Description. Catalytic hydrotreating is a hydrogenation process used to remove about 90% of contaminants such as nitrogen, sulfur, oxygen, and metals from liquid petroleum fractions. These contaminants, if not removed from the petroleum fractions as they travel through the refinery processing units, can have detrimental effects on the equipment, the catalysts, and the quality of the finished product. Typically, hydrotreating is done prior to processes such as catalytic reforming so that the catalyst is not contaminated by untreated feedstock. Hydrotreating is also used prior to catalytic cracking to reduce sulfur and improve product yields, and to upgrade middle-distillate petroleum fractions into finished kerosene, diesel fuel, and heating fuel oils. In addition, hydrotreating converts olefins and aromatics to saturated compounds.

2. Catalytic Hydrodesulfurization Process. Hydrotreating for sulfur removal is called hydrodesulfurization. In a typical catalytic hydrodesulfurization unit, the feedstock is deaerated and mixed with hydrogen, preheated in a fired heater (600°-800° F) and then charged under pressure (up to 1,000 psi) through a fixed-bed catalytic reactor. In the reactor, the sulfur and nitrogen compounds in the feedstock are converted into H2S and NH3. The reaction products leave the reactor and after cooling to a low temperature enter a liquid/gas separator. The hydrogen-rich gas from the high-pressure separation is recycled to combine with the feedstock, and the low-pressure gas stream rich in H2S is sent to a gas treating unit where H2S is removed. The clean gas is then suitable as fuel for the refinery furnaces. The liquid stream is the product from hydrotreating and is normally sent to a stripping column for removal of H2S and other undesirable components. In cases where steam is used for stripping, the product is sent to a vacuum drier for removal of water. Hydrodesulfurized products are blended or used as catalytic reforming feedstock.

3. Other Hydrotreating Processes.

   a. Hydrotreating processes differ depending upon the feedstock available and catalysts used. Hydrotreating can be used to improve the burning characteristics of distillates such as kerosene. Hydrotreatment of a kerosene fraction can convert aromatics into naphthenes, which are cleaner-burning compounds.

   b. Lube-oil hydrotreating uses catalytic treatment of the oil with hydrogen to improve product quality. The objectives in mild lube hydrotreating include saturation of olefins and improvements in color, odor, and acid nature of the oil. Mild lube hydrotreating also may be used following solvent processing. Operating temperatures are usually below 600° F and operating pressures below 800 psi. Severe lube hydrotreating, at temperatures in the 600°-750° F range and hydrogen pressures up to 3,000 psi, is capable of saturating aromatic rings, along with sulfur and nitrogen removal, to impart specific properties not achieved at mild conditions.

   c. Hydrotreating also can be employed to improve the quality of pyrolysis gasoline (pygas), a by-product from the manufacture of ethylene. Traditionally, the outlet for pygas has been motor gasoline blending, a suitable route in view of its high octane number. However, only small portions can be blended untreated owing to the unacceptable odor, color, and gum-forming tendencies of this material. The quality of pygas, which is high in diolefin content, can be satisfactorily improved by hydrotreating, whereby conversion of diolefins into mono-olefins provides an acceptable product for motor gas blending.

TABLE IV: 2-14 HYDRODESULFURIZATION PROCESS

Feedstock	*From*	*Process*	*Typical products . . To*
Naphthas, distillates sour gas oil, residuals	Atmospheric & vacuum tower, catalytic & thermal cracker	Treating, hydrogenation	Naphtha . . . . Blending Hydrogen . . . . . . Recycle Distillates . . . . . . Blending H_2S_2, ammonia . . Sulfure plant, treater Gas . . . . Gas plant

4. Health and Safety Considerations.

   a. Fire Prevention and Protection. The potential exists for fire in the event of a leak or release of product or hydrogen gas.

   b. Safety. Many processes require hydrogen generation to provide for a continuous supply. Because of the operating temperatures and presence of hydrogen, the hydrogen sulfide content of the feedstock must be strictly controlled to a minimum to reduce corrosion. Hydrogen chloride may form and condense as hydrochloric acid in the lower-temperature parts of the unit. Ammonium hydrosulfide may form in high-temperature, high-pressure units. Excessive contact time and/or temperature will create coking. Precautions need to be taken when unloading coked catalyst from the unit to prevent iron sulfide fires. The coked catalyst should be cooled to below 120° F before removal, or dumped into nitrogen-inerted bins where it can be cooled before further handling. Special antifoam additives may be used to prevent catalyst poisoning from silicone carryover in the coker feedstock.

c. Health. Because this is a closed process, exposures are expected to be minimal under normal operating conditions. There is a potential for exposure to hydrogen sulfide or hydrogen gas in the event of a release, or to ammonia should a sour-water leak or spill occur. Phenol also may be present if high boiling-point feedstock are processed. Safe work practices and/or appropriate personal protective equipment may be needed for exposures to chemicals and other hazards such as noise and heat; during process sampling, inspection, maintenance, and turnaround activities; and when handling amine or exposed to catalyst.

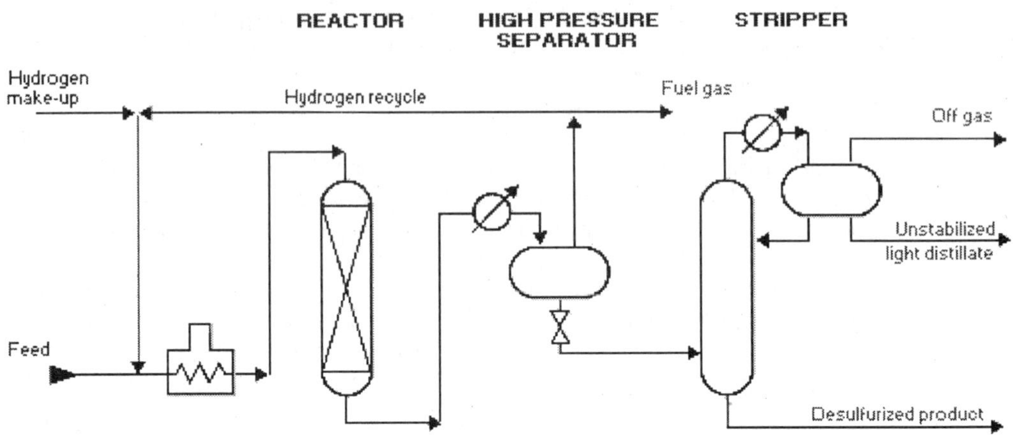

FIGURE IV: 2-17. DISTILLATE HYDRODESULFURIZATION

J. ISOMERIZATION.

1. Description.

a. Isomerization converts n-butane, n-pentane and n-hexane into their respective isoparaffins of substantially higher octane number. The straight-chain paraffins are converted to their branched-chain counterparts whose component atoms are the same but are arranged in a different geometric structure. Isomerization is important for the conversion of n-butane into isobutane, to provide additional feedstock for alkylation units, and the conversion of normal pentanes and hexanes into higher branched isomers for gasoline blending. Isomerization is similar to catalytic reforming in that the hydrocarbon molecules are rearranged, but unlike catalytic reforming, isomerization just converts normal paraffins to isoparaffins.

b. There are two distinct isomerization processes, butane (C4) and pentane/hexane (C5/C6). Butane isomerization produces feedstock for alkylation. Aluminum chloride catalyst plus hydrogen chloride are universally used for the low-temperature processes. Platinum or another metal catalyst is used for the higher-temperature processes. In a typical low-temperature process, the feed to the isomerization plant is n-butane or mixed butanes mixed with hydrogen (to inhibit olefin formation) and passed to the reactor at 230°-340° F and 200-300 psi. Hydrogen is flashed off in a high-pressure separator and the hydrogen chloride removed in a stripper column. The resultant butane mixture is sent to a fractionator (deisobutanizer) to separate n-butane from the isobutane product.

c. Pentane/hexane isomerization increases the octane number of the light gasoline components n-pentane and n-hexane, which are found in abundance in straight-run gasoline. In a typical C5/C6 isomerization process, dried and desulfurized feedstock is mixed with a small amount of organic chloride and recycled hydrogen, and then heated to reactor temperature. It is then passed over supported-metal catalyst in the first reactor where benzene and olefins are hydrogenated. The feed next goes to the isomerization reactor where the paraffins are catalytically isomerized to isoparaffins. The reactor effluent is then cooled and subsequently separated in the product separator into two streams: a liquid product (isomerate) and a recycle hydrogen-gas stream. The isomerate is washed (caustic and water), acid stripped, and stabilized before going to storage.

TABLE IV: 2-15. ISOMERIZATION PROCESSES

Feedstock	*From*	*Process*	*Typical products . . . To*
n-Butane	Various	Rearrangement	Isobutane . . . . Alkylation
n-Butane	Processes		Isopentane . . . . . . Blending
n-Hexane			Isohexene . . . . Blending
			Gas . . . . Gas Plant

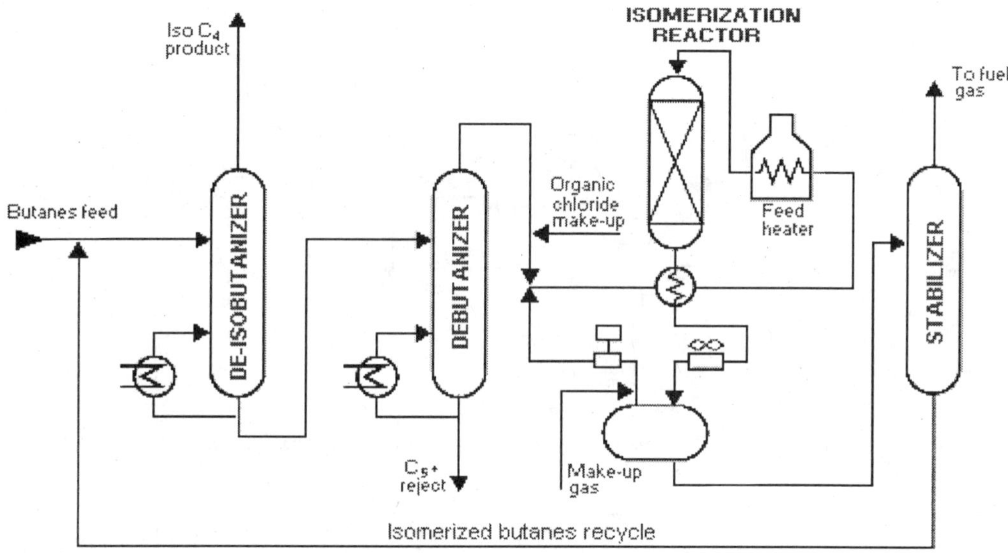

FIGURE IV: 2-18. C4 ISOMERIZATION.

2. Safety and Health Considerations.

   a. Fire Protection and Prevention. Although this is a closed process, the potential for a fire exists should a release or leak contact a source of ignition such as the heater.

   b. Safety. If the feedstock is not completely dried and desulfurized, the potential exists for acid formation leading to catalyst poisoning and metal corrosion. Water or steam must not be allowed to enter areas where hydrogen chloride is present. Precautions are needed to prevent HCl from entering sewers and drains.

   c. Health. Because this is a closed process, exposures are expected to be minimal during normal operating conditions. There is a potential for exposure to hydrogen gas, hydrochloric acid, and hydrogen chloride and to dust when solid catalyst is used. Safe work practices and/or appropriate personal protective equipment may be needed for exposures to chemicals and other hazards such as heat and noise, and during process sampling, inspection, maintenance, and turnaround activities.

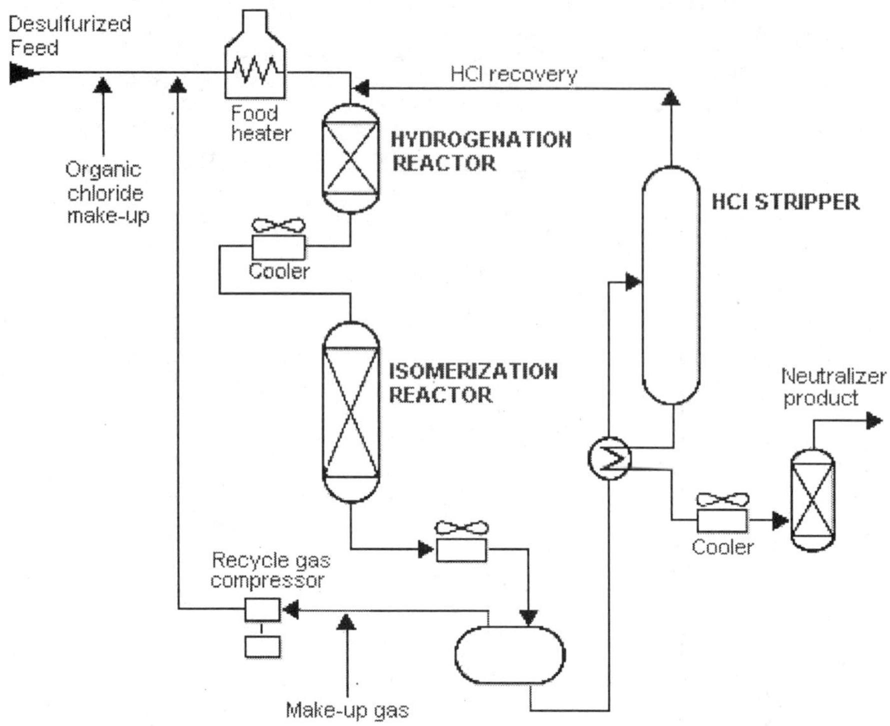

FIGURE IV: 2-19. C5 AND C6 ISOMERIZATION.

K. POLYMERIZATION.

1. Description.

a. Polymerization in the petroleum industry is the process of converting light olefin gases including ethylene, propylene, and butylene into hydrocarbons of higher molecular weight and higher octane number that can be used as gasoline blending stocks. Polymerization combines two or more identical olefin molecules to form a single molecule with the same elements in the same proportions as the original molecules. Polymerization may be accomplished thermally or in the presence of a catalyst at lower temperatures.

b. The olefin feedstock is pretreated to remove sulfur and other undesirable compounds. In the catalytic process the feedstock is either passed over a solid phosphoric acid catalyst or comes in contact with liquid phosphoric acid, where an exothermic polymeric reaction occurs. This reaction requires cooling water and the injection of cold feedstock into the reactor to control temperatures between 300° and 450° F at pressures from 200 psi to 1,200 psi. The reaction products leaving the reactor are sent to stabilization and/or fractionator systems to separate saturated and unreacted gases from the polymer gasoline product.

NOTE: In the petroleum industry, polymerization is used to indicate the production of gasoline components, hence the term "polymer" gasoline. Furthermore, it is not essential that only one type of monomer be involved. If unlike olefin molecules are combined, the process is referred to as "copolymerization." Polymerization in the true sense of the word is normally prevented, and all attempts are made to terminate the reaction at the dimer or trimer (three monomers joined together) stage. However, in the petrochemical section of a refinery, polymerization, which results in the production of, for instance, polyethylene, is allowed to proceed until materials of the required high molecular weight have been produced.

2. Safety and Health Considerations.

   a. Fire prevention and Protection. Polymerization is a closed process where the potential for a fire exists due to leaks or releases reaching a source of ignition.

   b. Safety. The potential for an uncontrolled exothermic reaction exists should loss of cooling water occur. Severe corrosion leading to equipment failure will occur should water make contact with the phosphoric acid, such as during water washing at shutdowns. Corrosion may also occur in piping manifolds, reboilers, exchangers, and other locations where acid may settle out.

   c. Health. Because this is a closed system, exposures are expected to be minimal under normal operating conditions. There is a potential for exposure to caustic wash (sodium hydroxide), to phosphoric acid used in the process or washed out during turnarounds, and to catalyst dust. Safe work practices and/or appropriate personal protective equipment may be needed for exposures to chemicals and other hazards such as noise and heat, and during process sampling, inspection, maintenance, and turnaround activities.

TABLE IV: 2-16. POLYMERIZATION PROCESS

Feedstock	*From*	*Process*	*Typical products . . . . . . To*
Olefins	Cracking processes	Unification	High octqane naphta . . Gasoline blending Petrochem. feedstock . . Petrochemical Liquefied petro. gas . . . Storage

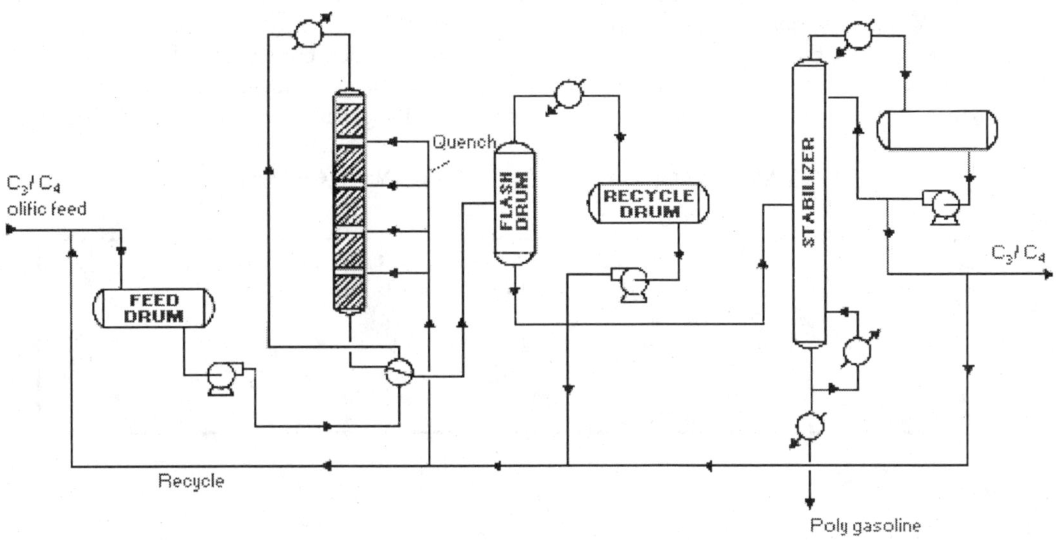

POLYMERIZATION PROCESS.

L. ALKYLATION.

1. Description. Alkylation combines low-molecular-weight olefins (primarily a mixture of propylene and butylene) with isobutene in the presence of a catalyst, either sulfuric acid or hydrofluoric acid. The product is called alkylate and is composed of a mixture of high-octane, branched-chain paraffinic hydrocarbons. Alkylate is a premium blending stock because it has exceptional antiknock properties and is clean burning. The octane number of the alkylate depends mainly upon the kind of olefins used and upon operating conditions.

2. Sulfuric Acid Alkylation Process.

   a. In cascade type sulfuric acid (H2SO4) alkylation units, the feedstock (propylene, butylene, amylene, and fresh isobutane) enters the reactor and contacts the concentrated sulfuric acid catalyst (in concentrations of 85% to 95% for good operation and to minimize corrosion). The reactor is divided into zones, with olefins fed through distributors to each zone, and the sulfuric acid and isobutanes flowing over baffles from zone to zone.

   b. The reactor effluent is separated into hydrocarbon and acid phases in a settler, and the acid is returned to the reactor. The hydrocarbon phase is hot-water washed with caustic for pH control before being successively depropanized, deisobutanized, and debutanized. The alkylate obtained from the deisobutanizer can then go directly to motor-fuel blending or be rerun to produce aviation-grade blending stock. The isobutane is recycled to the feed.

3. Hydrofluoric Acid Alylation Process. Phillips and UOP are the two common types of hydrofluoric acid alkylation processes in use. In the Phillips process, olefin and isobutane feedstock are dried and fed to a combination reactor/settler system. Upon leaving the reaction zone, the reactor effluent flows to a settler (separating vessel) where the acid separates from the hydrocarbons. The acid layer at the bottom of the separating vessel is recycled. The top layer of hydrocarbons (hydrocarbon phase), consisting of propane, normal butane, alkylate, and excess (recycle) isobutane, is charged to the main fractionator, the bottom product of which is motor alkylate. The main fractionator overhead, consisting mainly of propane, isobutane, and HF, goes to a depropanizer. Propane with trace amount of HF goes to an HF stripper for HF removal and is then catalytically defluorinated, treated, and sent to storage. Isobutane is withdrawn from the main fractionator and recycled to the reactor/settler, and alkylate from the bottom of the main fractionator is sent to product blending.

4. The UOP process uses two reactors with separate settlers. Half of the dried feedstock is charged to the first reactor, along with recycle and makeup isobutane. The reactor effluent then goes to its settler, where the acid is recycled and the hydrocarbon charged to the second reactor. The other half of the feedstock also goes to the second reactor, with the settler acid being recycled and the hydrocarbons charged to the main fractionator. Subsequent processing is similar to the Phillips process. Overhead from the main fractionator goes to a depropanizer. Isobutane is recycled to the reaction zone and alkylate is sent to product blending.

TABLE IV: 2-17. ALKYLATION PROCESS

Feedstock	*From*	*Process*	*Typical products . . . . To*
Petroleum gas	Distillation or cracking	Unification	High octane gasoline . . Blending
Olefins	Cat. or hydro cracking		n-Butane & propane . . . Stripper or blender
Isobutane	Isomerization		

5. Health and Safety Considerations.

   a. Fire Protection and Prevention. Alkylation units are closed processes; however, the potential exists for fire should a leak or release occur that allows product or vapor to reach a source of ignition.

   b. Safety. Sulfuric acid and hydrofluoric acid are potentially hazardous chemicals. Loss of coolant water, which is needed to maintain process temperatures, could result in an upset. Precautions are necessary to ensure that equipment and materials that have been in contact with acid are handled carefully and are thoroughly cleaned before they leave the process area or refinery. Immersion wash vats are often provided for neutralization of equipment that has come into contact with hydrofluoric acid. Hydrofluoric acid units should be thoroughly drained and chemically cleaned prior to turnarounds and entry to remove all traces of iron fluoride and hydro-fluoric acid. Following shutdown, where water has been used the unit should be thoroughly dried before hydrofluoric acid is introduced.

Leaks, spills, or releases involving hydrofluoric acid or hydrocarbons containing hydrofluoric acid can be extremely hazardous. Care during delivery and unloading of acid is essential. Process unit containment by curbs, drainage, and isolation so that effluent can be neutralized before release to the sewer system is considered. Vents can be routed to soda-ash scrubbers to neutralize hydrogen fluoride gas or hydrofluoric acid vapors before release. Pressure on the cooling water and steam side of exchangers should be kept below the minimum pressure on the acid service side to prevent water contamination.

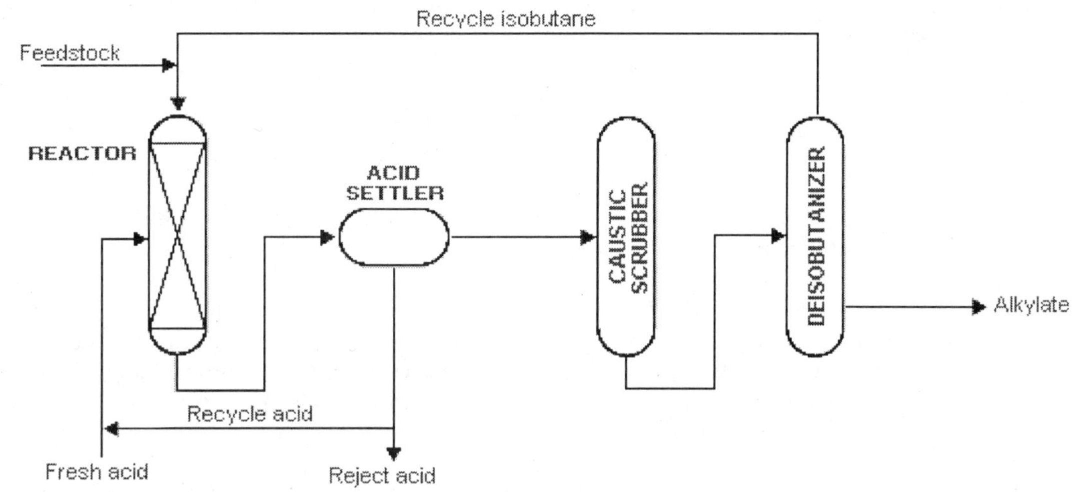

FIGURE IV: 2-21. SULFURIC ACID ALKYLATION

Some corrosion and fouling in sulfuric acid units may occur from the breakdown of sulfuric acid esters or where caustic is added for neutralization. These esters can be removed by fresh acid treating and hot-water washing. To prevent corrosion from hydrofluoric acid, the acid concentration inside the process unit should be maintained above 65% and moisture below 4%.

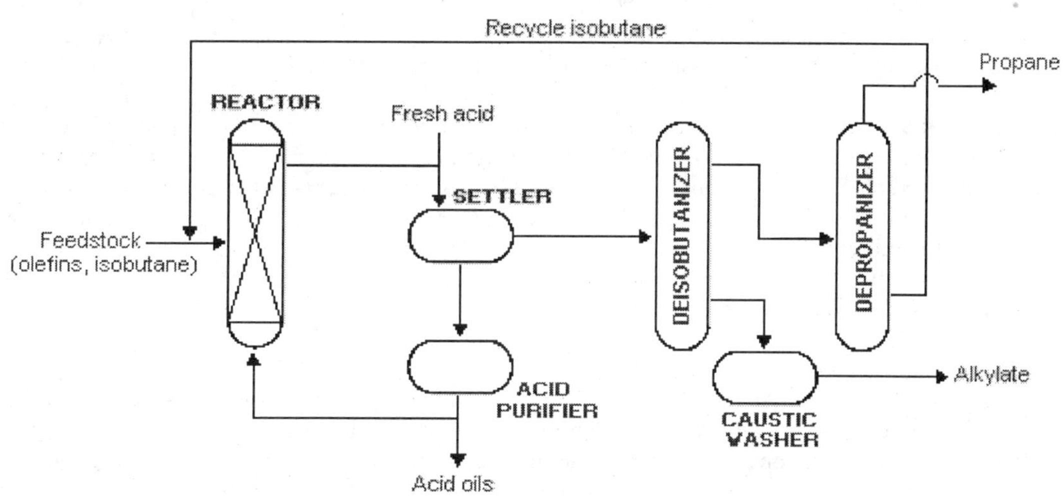

FIGURE IV: 2-22. HYDROGEN FLUORIDE ALKYLATION.

c. Health. Because this is a closed process, exposures are expected to be minimal during normal operations. There is a potential for exposure should leaks, spills, or releases occur. Sulfuric acid and (particularly) hydrofluoric acid are potentially hazardous chemicals. Special precautionary emergency preparedness measures and protection appropriate to the potential hazard and areas possibly affected need to be provided. Safe work practices and appropriate skin and respiratory personal protective equipment are needed for potential exposures to hydrofluoric and sulfuric acids during normal operations such as reading gauges, inspecting, and process sampling, as well as during emergency response, maintenance, and turnaround activities. Procedures should be in place to ensure that protective equipment and clothing worn in hydrofluoric acid activities are decontaminated and inspected before reissue. Appropriate personal protection for exposure to heat and noise also may be required.

M. SWEETENING AND TREATING PROCESSES.

1. Description.

a. Treating is a means by which contaminants such as organic compounds containing sulfur, nitrogen, and oxygen; dissolved metals and inorganic salts; and soluble salts dissolved in emulsified water are removed from petroleum fractions or streams. Petroleum refiners have a choice of several different treating processes, but the primary purpose of the majority of them is the elimination of unwanted sulfur compounds. A variety of intermediate and finished products, including middle distillates, gasoline, kerosene, jet fuel, and sour gases are dried and sweetened. Sweetening, a major refinery treatment of gasoline, treats sulfur compounds (hydrogen sulfide, thiophene and mercaptan) to improve color, odor, and oxidation stability. Sweetening also reduces concentrations of carbon dioxide.

b. Treating can be accomplished at an intermediate stage in the refining process, or just before sending the finished product to storage. Choices of a treating method depend on the nature of the petroleum fractions, amount and type of impurities in the fractions to be treated, the extent to which the process removes the impurities, and end-product specifications. Treating materials include acids, solvents, alkalis, oxidizing, and adsorption agents.

2. Acid, Caustic, or Clay Treating. Sulfuric acid is the most commonly used acid treating process. Sulfuric acid treating results in partial or complete removal of unsaturated hydrocarbons, sulfur, nitrogen, and oxygen compounds, and resinous and asphaltic compounds. It is used to improve the odor, color, stability, carbon residue, and other properties of the oil. Clay/lime treatment of acid-refined oil removes traces of asphaltic materials and other compounds improving product color, odor, and stability. Caustic treating with sodium (or potassium) hydroxide is used to improve odor and color by removing organic acids (naphthenic acids, phenols) and sulfur compounds (mercaptans, H2S) by a caustic wash. By combining caustic soda solution with various solubility promoters (e.g., methyl alcohol and cresols), up to 99% of all mercaptans as well as oxygen and nitrogen compounds can be dissolved from petroleum fractions.

3. Drying and Sweetening. Feedstock from various refinery units are sent to gas treating plants where butanes and butenes are removed for use as alkylation feedstock, heavier components are sent to gasoline blending, propane is recovered for LPG, and propylene is removed for use in petrochemicals. Some mercaptans are removed by water-soluble chemicals that react with the mercaptans. Caustic liquid (sodium hydroxide), amine compounds (diethanolamine) or fixed-bed catalyst sweetening also may be used. Drying is accomplished by the use of water absorption or adsorption agents to remove water from the products. Some processes simultaneously dry and sweeten by adsorption on molecular sieves.

4. Sulfur Recovery. Sulfur recovery converts hydrogen sulfide in sour gases and hydrocarbon streams to elemental sulfur. The most widely used recovery system is the Claus process, which uses both thermal and catalytic-conversion reactions. A typical process produces elemental sulfur by burning hydrogen sulfide under controlled conditions. Knockout pots are used to remove water and hydrocarbons from feed gas streams. The gases are then exposed to a catalyst to recover additional sulfur. Sulfur vapor from burning and conversion is condensed and recovered.

5. Hydrogen Sulfide Scrubbing. Hydrogen sulfide scrubbing is a common treating process in which the hydrocarbon feedstock is first scrubbed to prevent catalyst poisoning. Depending on the feedstock and the

nature of contaminants, desulfurization methods vary from ambient temperature-activated charcoal absorption to high-temperature catalytic hydrogenation followed by zinc oxide treating.

TABLE IV: 2-18. SWEETENING AND TREATING PROCESSES.

Feedstock	*From*	*Process*	*Typical products . . . . . . To*
Gases, finished products, intermediates	Various	Treatment	Butane & butene . . . . Alkylation Propane, distillates . . . Storage Gasoline . . . . Blending Propylene . . . . Petrochemical

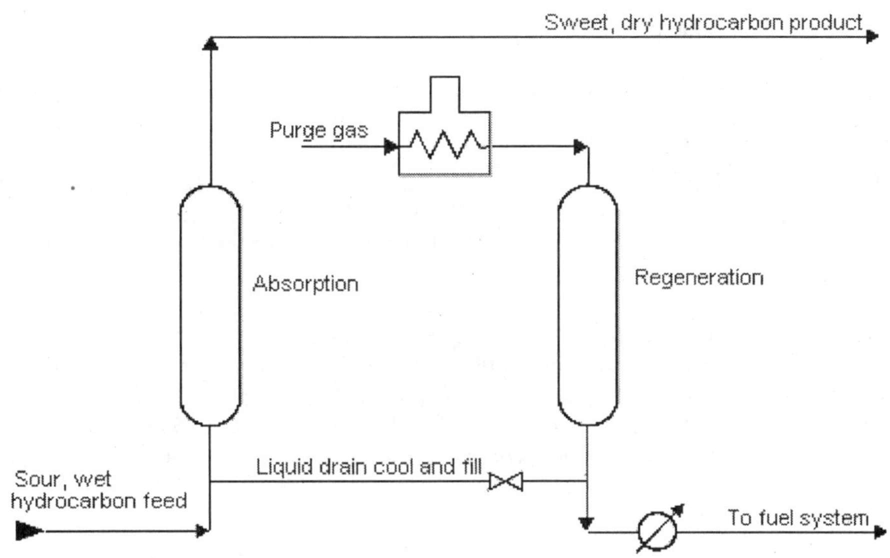

FIGURE IV: 2-23. MOLECULAR SIEVE DRYING AND SWEETENING

6. Health and Safety Considerations.

   a. Fire Protection and Prevention. The potential exists for fire from a leak or release of feedstock or product. Sweetening processes use air or oxygen. If excess oxygen enters these processes, it is possible for a fire to occur in the settler due to the generation of static electricity, which acts as the ignition source.

   b. Health. Because these are closed processes, exposures are expected to be minimal under normal operating conditions. There is a potential for exposure to hydrogen sulfide, caustic (sodium hydroxide), spent caustic, spent catalyst (Merox), catalyst dust and sweetening agents (sodium carbonate and sodium bicarbonate). Safe work practices and/or appropriate personal protective equipment may be needed for exposures to chemicals and other hazards such as noise and heat, and during process sampling, inspection, maintenance, and turnaround activities.

N. UNSATURATED GAS PLANTS.

1. Description. Unsaturated (unsat) gas plants recover light hydrocarbons (C3 and C4 olefins) from wet gas streams from the FCC, TCC, and delayed coker overhead accumulators or fractionation receivers. In a typical unsat gas plant, the gases are compressed and treated with amine to remove hydrogen sulfide either before or

after they are sent to a fractionating absorber where they are mixed into a concurrent flow of debutanized gasoline. The light fractions are separated by heat in a reboiler, the offgas is sent to a sponge absorber, and the bottoms are sent to a debutanizer. A portion of the debutanized hydrocarbon is recycled, with the balance sent to the splitter for separation. The overhead gases go to a depropanizer for use as alkylation unit feedstock.

TABLE IV: 2-19. UNSAT GAS PLANT PROCESS.

Feedstock	*From*	*Process*	*Typical products . . . To*
Gas Oils	FCC, TCC, delayed coker	Treatment	Gasoline . . . . Recycle or treating Gases . . . . Alkylation

2. Health and Safety Considerations.

   a. Fire Prevention and Protection. The potential of a fire exists should spills, releases, or vapors reach a source of ignition.

   b. Safety. In unsat gas plants handling FCC feedstock, the potential exists for corrosion from moist hydrogen sulfide and cyanides. When feedstock are from the delayed coker or the TCC, corrosion from hydrogen sulfide and deposits in the high pressure sections of gas compressors from ammonium compounds is possible.

   c. Health. Because these are closed processes, exposures are expected to be minimal under normal operating conditions. There is a potential for exposures to amine compounds such as monoethanolamine (MEA), diethanolamine (DEA) and methyldiethanolamine (MDEA) and hydrocarbons. Safe work practices and/or appropriate personal protective equipment may be needed for exposures to chemicals and other hazards such as noise and heat, and during process sampling, inspection, maintenance, and turnaround activities.

O. AMINE PLANTS.

1. Description. Amine plants remove acid contaminants from sour gas and hydrocarbon streams. In amine plants, gas and liquid hydrocarbon streams containing carbon dioxide and/or hydrogen sulfide are charged to a gas absorption tower or liquid contactor where the acid contaminants are absorbed by counterflowing amine solutions (i.e., MEA, DEA, MDEA). The stripped gas or liquid is removed overhead, and the amine is sent to a regenerator. In the regenerator, the acidic components are stripped by heat and reboiling action and disposed of, and the amine is recycled.

2. Health and Safety Considerations.

   a. Fire Protection and Prevention. The potential for fire exists where a spill or leak could reach a source of ignition.

   b. Safety. To minimize corrosion, proper operating practices should be established and regenerator bottom and reboiler temperatures controlled. Oxygen should be kept out of the system to prevent amine oxidation.

   c. Health. Because this is a closed process, exposures are expected to be minimal during normal operations. There is potential for exposure to amine compounds (i.e. monoethanolamine, diethanolamine, methyldiethanolamine), hydrogen sulfide and carbon dioxide. Safe work practices and/or appropriate personal protective equipment may be needed for exposures to chemicals and other hazards such as noise and heat, and during process sampling, inspection, maintenance and turnaround activities.

P. SATURATE GAS PLANTS.

1. Description. Saturate (sat) gas plants separate refinery gas components including butanes for alkylation, pentanes for gasoline blending, LPG's for fuel, and ethane for petrochemicals. Because sat gas processes depend

on the feedstock and product demand, each refinery uses different systems, usually absorption-fractionation or straight fractionation. In absorption-fractionation, gases and liquids from various refinery units are fed to an absorber-deethanizer where C2 and lighter fractions are separated from heavier fractions by lean oil absorption and removed for use as fuel gas or petrochemical feed. The heavier fractions are stripped and sent to a debutanizer, and the lean oil is recycled back to the absorber-deethanizer. C3/C4 is separated from pentanes in the debutanizer, scrubbed to remove hydrogen sulfide, and fed to a splitter where propane and butane are separated. In fractionation sat gas plants, the absorption stage is eliminated.

2. Health and Safety Considerations.

a. Fire Protection and Prevention. There is potential for fire if a leak or release reaches a source of ignition such as the unit reboiler.

b. Safety. Corrosion could occur from the presence of hydrogen sulfide, carbon dioxide, and other compounds as a result of prior treating. Streams containing ammonia should be dried before processing. Antifouling additives may be used in absorption oil to protect heat exchangers. Corrosion inhibitors may be used to control corrosion in overhead systems.

c. Health. Because this is a closed process, exposures are expected to be minimal during normal operations. There is potential for exposure to hydrogen sulfide, carbon dioxide, and other products such as diethanolamine or sodium hydroxide carried over from prior treating. Safe work practices and/or appropriate personal protective equipment may be needed for exposures to chemicals and other hazards such as noise and heat, and during process sampling, inspection, maintenance, and turnaround activities.

Q. ASPHALT PRODUCTION.

1. Description.

a. Asphalt is a portion of the residual fraction that remains after primary distillation operations. It is further processed to impart characteristics required by its final use. In vacuum distillation, generally used to produce road-tar asphalt, the residual is heated to about 750° F and charged to a column where vacuum is applied to prevent cracking.

b. Asphalt for roofing materials is produced by air blowing. Residual is heated in a pipe still almost to its flash point and charged to a blowing tower where hot air is injected for a predetermined time. The dehydrogenization of the asphalt forms hydrogen sulfide, and the oxidation creates sulfur dioxide. Steam, used to blanket the top of the tower to entrain the various contaminants, is then passed through a scrubber to condense the hydrocarbons.

c. A third process used to produce asphalt is solvent deasphalting. In this extraction process, which uses propane (or hexane) as a solvent, heavy oil fractions are separated to produce heavy lubricating oil, catalytic cracking feedstock, and asphalt. Feedstock and liquid propane are pumped to an extraction tower at precisely controlled mixtures, temperatures (150°-250° F), and pressures of 350-600 psi. Separation occurs in a rotating disc contactor, based on differences in solubility. The products are then evaporated and steam stripped to recover the propane, which is recycled. Deasphalting also removes some sulfur and nitrogen compounds, metals, carbon residues, and paraffins from the feedstock.

TABLE IV: 2-20. SOLVENT DEASPHALTING PROCESS.

Feedstock	*From*	*Process*	*Typical products . . To*
Residual, reduced crude	Atmospheric tower & Vacuum tower	Treatment	Heavy lube oil . . . Treating or lube blending Asphalt . . . . Storage of shipping Deasphalted oil . . Hydrotreat & catalytic cracker Propane . . . . Recycle

2. Safety and Health Considerations.

a. Fire Protection and Prevention. The potential for a fire exists if a product leak or release contacts a source of ignition such as the process heater. Condensed steam from the various asphalt and deasphalting processes will contain trace amounts of hydrocarbons. Any disruption of the vacuum can result in the entry of atmospheric air and subsequent fire. In addition, raising the temperature of the vacuum tower bottom to improve efficiency can generate methane by thermal cracking. This can create vapors in asphalt storage tanks that are not detectable by flash testing but are high enough to be flammable.

b. Safety. Deasphalting requires exact temperature and pressure control. In addition, moisture, excess solvent, or a drop in operating temperature may cause foaming, which affects the product temperature control and may create an upset.

c. Health. Because these are closed processes, exposures are expected to be minimal during normal operations. Should a spill or release occur, there is a potential for exposure to residuals and asphalt. Air blowing can create some polynuclear aromatics. Condensed steam from the air-blowing asphalt process may also contain contaminants. The potential for exposure to hydrogen sulfide and sulfur dioxide exists in the production of asphalt. Safe work practices and/or appropriate personal protective equipment may be needed for exposures to chemicals and other hazards such as noise and heat, and during process sampling, inspection, maintenance, and turnaround activities.

R. HYDROGEN PRODUCTION.

1. Description.

a. High-purity hydrogen (95%-99%) is required for hydrodesulfurization, hydrogenation, hydrocracking, and petrochemical processes. Hydrogen, produced as a by-product of refinery processes (principally hydrogen recovery from catalytic reformer product gases), often is not enough to meet the total refinery requirements, necessitating the manufacturing of additional hydrogen or obtaining supply from external sources.

b. In steam-methane reforming, desulfurized gases are mixed with superheated steam (1,100°-1,600° F) and reformed in tubes containing a nickel base catalyst. The reformed gas, which consists of steam, hydrogen, carbon monoxide, and carbon dioxide, is cooled and passed through converters containing an iron catalyst where the carbon monoxide reacts with steam to form carbon dioxide and more hydrogen. The carbon dioxide is removed by amine washing. Any remaining carbon monoxide in the product stream is converted to methane.

c. Steam-naphtha reforming is a continuous process for the production of hydrogen from liquid hydrocarbons and is, in fact, similar to steam-methane reforming. A variety of naphthas in the gasoline boiling range may be employed, including fuel containing up to 35% aromatics. Following pretreatment to remove sulfur compounds, the feedstock is mixed with steam and taken to the reforming furnace (1,250°-1,500° F) where hydrogen is produced.

TABLE IV: 2-21. STEAM REFORMING PROCESS.

Feedstock	***From***	***Process***	***Typical products . . . To***
Desufurized refinery gas	Various treatment units	Decomposition	Hydrogen . . . . Processing Carbon dioxide . . . Atmosphere Carbon monoxide . . Methane

2. Health and Safety Considerations.

a. Fire Protection and Prevention. The possibility of fire exists should a leak or release occur and reach an ignition source.

b. Safety. The potential exists for burns from hot gases and superheated steam should a release occur. Inspections and testing should be considered where the possibility exists for valve failure due to contaminants in the hydrogen. Carryover from caustic scrubbers should be controlled to prevent corrosion in preheaters. Chlorides from the feedstock or steam system should be prevented from entering reformer tubes and contaminating the catalyst.

c. Health. Because these are closed processes, exposures are expected to be minimal during normal operating conditions. There is a potential for exposure to excess hydrogen, carbon monoxide, and/or carbon dioxide. Condensate can be contaminated by process materials such as caustics and amine compounds, with resultant exposures. Depending on the specific process used, safe work practices and/or appropriate personal protective equipment may be needed for exposures to chemicals and other hazards such as noise and heat, and during process sampling, inspection, maintenance, and turnaround activities.

S. BLENDING.

1. Description. Blending is the physical mixture of a number of different liquid hydrocarbons to produce a finished product with certain desired characteristics. Products can be blended in-line through a manifold system, or batch blended in tanks and vessels. In-line blending of gasoline, distillates, jet fuel, and kerosene is accomplished by injecting proportionate amounts of each component into the main stream where turbulence promotes thorough mixing. Additives including octane enhancers, metal deactivators, anti-oxidants, anti-knock agents, gum and rust inhibitors, detergents, etc. are added during and/or after blending to provide specific properties not inherent in hydrocarbons.

2. Health and Safety Considerations.

a. Fire Prevention and Protection. Ignition sources in the area need to be controlled in the event of a leak or release.

b. Health. Safe work practices and/or appropriate personal protective equipment may be needed for exposures to chemicals and other hazards such as noise and heat; when handling additives; and during inspection, maintenance, and turnaround activities.

T. LUBRICANT, WAX, AND GREASE MANUFACTURING PROCESSES.

1. Description. Lubricating oils and waxes are refined from the residual fractions of atmospheric and vacuum distillation. The primary objective of the various lubricating oil refinery processes is to remove asphalts, sulfonated aromatics, and paraffinic and isoparaffinic waxes from residual fractions. Reduced crude from the vacuum unit is deasphalted and combined with straight-run lubricating oil feedstock, preheated, and solvent-extracted (usually with phenol or furfural) to produce raffinate.

2. Wax Manufacturing Process. Raffinate from the extraction unit contains a considerable amount of wax that must be removed by solvent extraction and crystallization. The raffinate is mixed with a solvent (propane) and precooled in heat exchangers. The crystallization temperature is attained by the evaporation of propane in the chiller and filter feed tanks. The wax is continuously removed by filters and cold solvent-washed to recover retained oil. The solvent is recovered from the oil by flashing and steam stripping. The wax is then heated with hot solvent, chilled, filtered, and given a final wash to remove all oil.

3. Lubricating Oil Process. The dewaxed raffinate is blended with other distillate fractions and further treated for viscosity index, color, stability, carbon residue, sulfur, additive response, and oxidation stability in extremely selective extraction processes using solvents (furfural, phenol, etc.). In a typical phenol unit, the raffinate is mixed with phenol in the treating section at temperatures below 400° F. Phenol is then separated from the treated oil and recycled. The treated lube-oil base stocks are then mixed and/or compounded with additives to meet the required physical and chemical characteristics of motor oils, industrial lubricants, and metal working oils.

TABLE IV: 2-22. LUBRICATING OIL AND WAX MANUFACTURING PROCESSES.

Feedstock	*From*	*Process*	*Typical products . . . .*	*To*
Lube feedstock and additives	Vacuum tower, solvent dewaxing, hydrotreating solvent extraction, etc.	Treatment	Dewaxed raffinate . .	Lube blend or compound, grease compounding
			Wax	Storage or shipping

4. Grease Compounding. Grease is made by blending metallic soaps (salts of long-chained fatty acids) and additives into a lubricating oil medium at temperatures of 400°-600° F. Grease may be either batch-produced or continuously compounded. The characteristics of the grease depend to a great extent on the metallic element (calcium, sodium, aluminum, lithium, etc.) in the soap and the additives used.

5. Safety and Health Considerations.

   a. Fire Protection and Prevention. The potential for fire exists if a product or vapor leak or release in the lube blending and wax processing areas reaches a source of ignition. Storage of finished products, both bulk and packaged, should be in accordance with recognized practices.

      While the potential for fire is reduced in lube oil blending, care must be taken when making metal-working oils and compounding greases due to the use of higher blending and compounding temperatures and lower flash point products.

   b. Safety. Control of treater temperature is important as phenol can cause corrosion above 400° F. Batch and in-line blending operations require strict controls to maintain desired product quality. Spills should be cleaned and leaks repaired to avoid slips and falls. Additives in drums and bags need to be handled properly to avoid strain. Wax can clog sewer or oil drainage systems and interfere with wastewater treatment.

   c. Health. When blending, sampling, and compounding, personal protection from steam, dusts, mists, vapors, metallic salts, and other additives is appropriate. Skin contact with any formulated grease or lubricant should be avoided. Safe work practices and/or appropriate personal protection may be needed for exposures to chemicals and other hazards such as noise and heat; during inspection, maintenance, and turnaround activities; and while sampling and handling hydrocarbons and chemicals during the production of lubricating oil and wax.

V. OTHER REFINERY OPERATIONS.

A. HEAT EXCHANGERS, COOLERS, AND PROCESS HEATERS.

1. Heating Operations. Process heaters and heat exchangers preheat feedstock in distillation towers and in refinery processes to reaction temperatures. Heat exchangers use either steam or hot hydrocarbon transferred from some other section of the process for heat input. The heaters are usually designed for specific process operations, and most are of cylindrical vertical or box-type designs. The major portion of heat provided to process units comes from fired heaters fueled by refinery or natural gas, distillate, and residual oils. Fired heaters are found on crude and reformer preheaters, coker heaters, and large-column reboilers.

2. Cooling Operations. Heat also may be removed from some processes by air and water exchangers, fin fans, gas and liquid coolers, and overhead condensers, or by transferring heat to other systems. The basic mechanical vapor-compression refrigeration system, which may serve one or more process units, includes an evaporator, compressor, condenser, controls, and piping. Common coolants are water, alcohol/water mixtures, or various glycol solutions.

3. Health and Safety Considerations.

   a. Fire Protection and Prevention. A means of providing adequate draft or steam purging is required to reduce the chance of explosions when lighting fires in heater furnaces. Specific start-up and emergency procedures are required for each type of unit. If fire impinges on fin fans, failure could occur due to overheating. If flammable product escapes from a heat exchanger or cooler due to a leak, fire could occur.

   b. Safety. Care must be taken to ensure that all pressure is removed from heater tubes before removing header or fitting plugs. Consideration should be given to providing for pressure relief in heat-exchanger piping systems in the event they are blocked off while full of liquid. If controls fail, variations of temperature and pressure could occur on either side of the heat exchanger. If heat exchanger tubes fail and process pressure is greater than heater pressure, product could enter the heater with downstream consequences. If the process pressure is less than heater pressure, the heater stream could enter into the process fluid. If loss of circulation occurs in liquid or gas coolers, increased product temperature could affect downstream operations and require pressure relief.

   c. Health. Because these are closed systems, exposures under normal operating conditions are expected to be minimal. Depending on the fuel, process operation, and unit design, there is a potential for exposure to hydrogen sulfide, carbon monoxide, hydrocarbons, steam boiler feed-water sludge, and water-treatment chemicals. Skin contact should be avoided with boiler blowdown, which may contain phenolic compounds. Safe work practices and/or appropriate personal protective equipment against hazards may be needed during process maintenance, inspection, and turnaround activities and for protection from radiant heat, superheated steam, hot hydrocarbon, and noise exposures.

B. STEAM GENERATION.

1. Heater and Boiler Operations. Steam is generated in main generation plants, and/or at various process units using heat from flue gas or other sources. Heaters (furnaces) include burners and a combustion air system, the boiler enclosure in which heat transfer takes place, a draft or pressure system to remove flue gas from the furnace, soot blowers, and compressed-air systems that seal openings to prevent the escape of flue gas. Boilers consist of a number of tubes that carry the water-steam mixture through the furnace for maximum heat transfer. These tubes run between steam-distribution drums at the top of the boiler and water-collecting drums at the bottom of the boiler. Steam flows from the steam drum to the superheater before entering the steam distribution system.

2. Heater Fuel.

   a. Heaters may use any one or combination of fuels including refinery gas, natural gas, fuel oil, and powdered coal. Refinery off-gas is collected from process units and combined with natural gas and LPG in a fuel-gas balance drum. The balance drum provides constant system pressure, fairly stable Btu-content fuel, and automatic separation of suspended liquids in gas vapors, and it prevents carryover of large slugs of condensate into the distribution system. Fuel oil is typically a mix of refinery crude oil with straight-run and cracked residues and other products. The fuel-oil system delivers fuel to process-unit heaters and steam generators at required temperatures and pressures. The fuel oil is heated to pumping temperature, sucked through a coarse suction strainer, pumped to a temperature-control heater, and then pumped through a fine-mesh strainer before being burned.

   b. In one example of process-unit heat generation, carbon monoxide boilers recover heat in catalytic cracking units as carbon monoxide in flue gas is burned to complete combustion. In other processes, waste-heat recovery units use heat from the flue gas to make steam.

3. Steam Distribution. The distribution system consists of valves, fittings, piping, and connections suitable for the pressure of the steam transported. Steam leaves the boilers at the highest pressure required by the process units or electrical generation. The steam pressure is then reduced in turbines that drive process pumps and compressors. Most steam used in the refinery is condensed to water in various types of heat exchangers. The condensate is reused as boiler feedwater or discharged to wastewater treatment. When refinery steam is also

used to drive steam turbine generators to produce electricity, the steam must be produced at much higher pressure than required for process steam. Steam typically is generated by heaters (furnaces) and boilers combined in one unit.

4. Feedwater.

   a. Feedwater supply is an important part of steam generation. There must always be as many pounds of water entering the system as there are pounds of steam leaving it. Water used in steam generation must be free of contaminants including minerals and dissolved impurities that can damage the system or affect its operation. Suspended materials such as silt, sewage, and oil, which form scale and sludge, must be coagulated or filtered out of the water. Dissolved gases, particularly carbon dioxide and oxygen, cause boiler corrosion and are removed by deaeration and treatment. Dissolved minerals including metallic salts, calcium, carbonates, etc., that cause scale, corrosion, and turbine blade deposits are treated with lime or soda ash to precipitate them from the water. Recirculated cooling water must also be treated for hydrocarbons and other contaminants.

   b. Depending on the characteristics of raw boiler feedwater, some or all of the following six stages of treatment will be applicable:

      - Clarification;
      - Sedimentation;
      - Filtration;
      - Ion exchange;
      - Deaeration; and
      - Internal treatment.

5. Health and Safety Considerations.

   a. Fire Protection and Prevention. The most potentially hazardous operation in steam generation is heater startup. A flammable mixture of gas and air can build up as a result of loss of flame at one or more burners during lightoff. Each type of unit requires specific startup and emergency procedures including purging before lightoff and in the event of misfire or loss of burner flamc.

   b. Safety. If feedwater runs low and boilers are dry, the tubes will overheat and fail. Conversely, excess water will be carried over into the steam distribution system and damage the turbines. Feedwater must be free of contaminants that could affect operations. Boilers should have continuous or intermittent blowdown systems to remove water from steam drums and limit buildup of scale on turbine blades and superheater tubes. Care must be taken not to overheat the superheater during startup and shut-down. Alternate fuel sources should be provided in the event of loss of gas due to refinery unit shutdown or emergency. Knockout pots provided at process units remove liquids from fuel gas before burning.

   c. Health. Safe work practices and/or appropriate personal protective equipment may be needed for potential exposures to feedwater chemicals, steam, hot water, radiant heat, and noise, and during process sampling, inspection, maintenance, and turnaround activities.

C. PRESSURE-RELIEF AND FLARE SYSTEMS.

1. Pressure-Relief Systems. Pressure-relief systems control vapors and liquids that are released by pressure-relieving devices and blow-downs. Pressure relief is an automatic, planned release when operating pressure reaches a predetermined level. Blowdown normally refers to the intentional release of material, such as blowdowns from process unit startups, furnace blowdowns, shutdowns, and emergencies. Vapor depressuring is the rapid removal of vapors from pressure vessels in case of fire. This may be accomplished by the use of a rupture disc, usually set at a higher pressure than the relief valve.

2. Safety Relief Valve Operations. Safety relief valves, used for air, steam, and gas as well as for vapor and liquid, allow the valve to open in proportion to the increase in pressure over the normal operating pressure. Safety valves designed primarily to release high volumes of steam usually pop open to full capacity. The overpressure

needed to open liquid-relief valves where large-volume discharge is not required increases as the valve lifts due to increased spring resistance. Pilot-operated safety relief valves, with up to six times the capacity of normal relief valves, are used where tighter sealing and larger volume discharges are required. Nonvolatile liquids are usually pumped to oil-water separation and recovery systems, and volatile liquids are sent to units operating at a lower pressure.

3. Flare Systems. A typical closed pressure release and flare system includes relief valves and lines from process units for collection of discharges, knockout drums to separate vapors and liquids, seals, and/or purge gas for flashback protection, and a flare and igniter system which combusts vapors when discharging directly to the atmosphere is not permitted. Steam may be injected into the flare tip to reduce visible smoke.

4. Pressure Relief Health and Safety Considerations.

    a. Fire Protection and Prevention. Vapors and gases must not discharge where sources of ignition could be present.

    b. Safety. Liquids should not be discharged directly to a vapor disposal system. Flare knockout drums and flares need to be large enough to handle emergency blowdowns. Drums should be provided with relief in the event of over pressure.

    Pressure relief valves must be provided where the potential exists for overpressure in refinery processes due to the ollowing causes:

    - Loss of cooling water, which may greatly reduce pressure in condensers and increase the pressure in the process unit.
    - Loss of reflux volume, which may cause a pressure drop in condensers and a pressure rise in distillation towers because the quantity of reflux affects the volume of vapors leaving the distillation tower.
    - Rapid vaporization and pressure increase from injection of a lower boiling-point liquid including water into a process vessel operating at higher temperatures.
    - Expansion of vapor and resultant over-pressure due to overheated process steam, malfunctioning heaters, or fire.
    - Failure of automatic controls, closed outlets, heat exchanger failure, etc.
    - Internal explosion, chemical reaction, thermal expansion, or accumulated gases.

    Maintenance is important because valves are required to function properly. The most common operating problems are listed below.

    - Failure to open at set pressure, because of plugging of the valve inlet or outlet, or because corrosion prevents proper operation of the disc holder and guides.
    - Failure to reseat after popping open due to fouling, corrosion, or deposits on the seat or moving parts, or because solids in the gas stream have cut the valve disc.
    - Chattering and premature opening, because operating pressure is too close to the set point.

    c. Health. Safe work practices and/or appropriate personal protective equipment may be needed to protect against hazards during inspection, maintenance, and turnaround activities.

D. WASTEWATER TREATMENT.

1. Description. Wastewater treatment is used for process, runoff, and sewerage water prior to discharge or recycling. Wastewater typically contains hydrocarbons, dissolved materials, suspended solids, phenols, ammonia, sulfides, and other compounds. Wastewater includes condensed steam, stripping water, spent caustic solutions, cooling tower and boiler blowdown, wash water, alkaline and acid waste neutralization water, and other process-associated water.

2. Pretreatment Operations. Pretreatment is the separation of hydrocarbons and solids from wastewater. API separators, interceptor plates, and settling ponds remove suspended hydrocarbons, oily sludge, and solids by

gravity separation, skimming, and filtration. Some oil-in-water emulsions must be heated to assist in separating the oil and water. Gravity separation depends on the specific gravity differences between water and immiscible oil globules and allows free oil to be skimmed off the surface of the wastewater. Acidic wastewater is neutralized using ammonia, lime, or soda ash. Alkaline wastewater is treated with sulfuric acid, hydrochloric acid, carbon dioxide-rich flue gas, or sulfur.

3. Secondary Treatment Operations. After pretreatment, suspended solids are removed by sedimentation or air flotation. Wastewater with low levels of solids may be screened or filtered. Flocculation agents are sometimes added to help separation. Secondary treatment processes biologically degrade and oxidize soluble organic matter by the use of activated sludge, unaerated or aerated lagoons, trickling filter methods, or anaerobic treatments. Materials with high adsorption characteristics are used in fixed-bed filters or added to the wastewater to form a slurry which is removed by sedimentation or filtration. Additional treatment methods are used to remove oils and chemicals from wastewater. Stripping is used on wastewater containing sulfides and/or ammonia, and solvent extraction is used to remove phenols.

4. Tertiary Treatment Operations. Tertiary treatments remove specific pollutants to meet regulatory discharge requirements. These treatments include chlorination, ozonation, ion exchange, reverse osmosis, activated carbon adsorption, etc. Compressed oxygen is diffused into wastewater streams to oxidize certain chemicals or to satisfy regulatory oxygen-content requirements. Wastewater that is to be recycled may require cooling to remove heat and/or oxidation by spraying or air stripping to remove any remaining phenols, nitrates, and ammonia.

5. Health and Safety Considerations.

   a. Fire Protection and Prevention. The potential for fire exists if vapors from wastewater containing hydrocarbons reach a source of ignition during treatment.

   b. Health. Safe work practices and/or appropriate personal protective equipment may be needed for exposures to chemicals and waste products during process sampling, inspection, maintenance, and turnaround activities as well as to noise, gases, and heat.

E. COOLING TOWERS.

1. Description. Cooling towers remove heat from process water by evaporation and latent heat transfer between hot water and air. The two types of towers are crossflow and counterflow. Crossflow towers introduce the airflow at right angles to the water flow throughout the structure. In counterflow cooling towers, hot process water is pumped to the uppermost plenum and allowed to fall through the tower. Numerous slats or spray nozzles located throughout the length of the tower disperse the water and help in cooling. Air enters at the tower bottom and flows upward against the water. When the fans or blowers are at the air inlet, the air is considered to be forced draft. Induced draft is when the fans are at the air outlet.

2. Cooling Water. Recirculated cooling water must be treated to remove impurities and dissolved hydrocarbons. Because the water is saturated with oxygen from being cooled with air, the chances for corrosion are increased. One means of corrosion prevention is the addition of a material to the cooling water that forms a protective film on pipes and other metal surfaces.

3. Health and Safety Considerations.

   a. Fire Prevention and Protection. When cooling water is contaminated by hydrocarbons, flammable vapors can be evaporated into the discharge air. If a source of ignition is present, or if lightning occurs, a fire may start. A potential fire hazard also exists where there are relatively dry areas in induced-draft cooling towers of combustible construction.

   b. Safety. Loss of power to cooling tower fans or water pumps could have serious consequences in the operation of the refinery. Impurities in cooling water can corrode and foul pipes and heat exchangers, scale from dissolved salts can deposit on pipes, and wooden cooling towers can be damaged by microorganisms.

c. Health. Cooling-tower water can be contaminated by process materials and by-products including sulfur dioxide, hydrogen sulfide, and carbon dioxide, with resultant exposures. Safe work practices and/or appropriate personal protective equipment may be needed during process sampling, inspection, maintenance, and turnaround activities; and for exposure to hazards such as those related to noise, water-treatment chemicals, and hydrogen sulfide when wastewater is treated in conjunction with cooling towers.

F. ELECTRIC POWER.

1. Description. Refineries may receive electricity from outside sources or produce their own power with generators driven by steam turbines or gas engines. Electrical substations receive power from the utility or power plant for distribution throughout the facility. They are usually located in nonclassified areas, away from sources of vapor or cooling-tower water spray. Transformers, circuit breakers, and feed-circuit switches are usually located in substations. Substations feed power to distribution stations within the process unit areas. Distribution stations can be located in classified areas, providing that classification requirements are met. Distribution stations usually have a liquid-filled transformer and an oil-filled or air-break disconnect device.

2. Health and Safety Considerations.

a. Fire Protection and Prevention. Generators that are not properly classified and are located too close to process units may be a source of ignition should a spill or release occur.

b. Safety. Normal electrical safety precautions including dry footing, high-voltage warning signs, and guarding must be taken to protect against electrocution. Lockout/tagout and other appropriate safe work practices must be established to prevent energization while work is being performed on high-voltage electrical equipment.

c. Health. Safe work practices and/or the use of appropriate personal protective equipment may be needed for exposures to noise, for exposure to hazards during inspection and maintenance activities, and when working around transformers and switches that may contain a dielectric fluid which requires special handling precautions.

G. GAS AND AIR COMPRESSORS.

1. Description. Both reciprocating and centrifugal compressors are used throughout the refinery for gas and compressed air. Air compressor systems include compressors, coolers, air receivers, air dryers, controls, and distribution piping. Blowers are used to provide air to certain processes. Plant air is provided for the operation of air-powered tools, catalyst regeneration, process heaters, steam-air decoking, sour-water oxidation, gasoline sweetening, asphalt blowing, and other uses. Instrument air is provided for use in pneumatic instruments and controls, air motors and purge connections.

2. Health and Safety Considerations.

a. Fire Protection and Prevention. Air compressors should be located so that the suction does not take in flammable vapors or corrosive gases. There is a potential for fire should a leak occur in gas compressors.

b. Safety. Knockout drums are needed to prevent liquid surges from entering gas compressors. If gases are contaminated with solid materials, strainers are needed. Failure of automatic compressor controls will affect processes. If maximum pressure could potentially be greater than compressor or process-equipment design pressure, pressure relief should be provided. Guarding is needed for exposed moving parts on compressors. Compressor buildings should be properly electrically classified, and provisions should be made for proper ventilation.

Where plant air is used to back up instrument air, interconnections must be upstream of the instrument air drying system to prevent contamination of instruments with moisture. Alternate sources of instrument air supply, such as use of nitrogen, may be needed in the event of power outages or compressor failure.

c. Health. Safe work practices and/or appropriate personal protective equipment may be needed for exposure to hazards such as noise and during inspection and maintenance activities. The use of appropriate safeguards must be considered so that plant and instrument air is not used for breathing or pressuring potable water systems.

H. MARINE, TANK CAR, AND TANK TRUCK LOADING AND UNLOADING.

1. Description. Facilities for loading liquid hydrocarbons into tank cars, tank trucks, and marine vessels and barges are usually part of the refinery operations. Product characteristics, distribution needs, shipping requirements, and operating criteria are important when designing loading facilities. Tank trucks and rail tank cars are either top- or bottom-loaded, and vapor-recovery systems may be provided where required. Loading and unloading liquefied petroleum gas (LPG) require special considerations in addition to those for liquid hydrocarbons.

2. Health and Safety Considerations.

a. Fire Protection and Prevention. The potential for fire exists where flammable vapors from spills or releases can reach a source of ignition. Where switch-loading is permitted, safe practices need to be established and followed. Bonding is used to equalize the electrical charge between the loading rack and the tank truck or tank car. Grounding is used at truck and rail loading facilities to prevent flow of stray currents. Insulating flanges are used on marine dock piping connections to prevent static electricity buildup and discharge. Flame arrestors should be installed in loading rack and marine vapor-recovery lines to prevent flashback.

b. Safety. Automatic or manual shutoff systems at supply headers are needed for top and bottom loading in the event of leaks or overfills. Fall protection such as railings are needed for top-loading racks where employees are exposed to falls. Drainage and recovery systems may be provided for storm drainage and to handle spills and leaks. Precautions must be taken at LPG loading facilities not to overload or overpressurize tank cars and trucks.

c. Health. The nature of the health hazards at loading and unloading facilities depends upon the products being loaded and the products previously transported in the tank cars, tank trucks, or marine vessels. Safe work practices and/or appropriate personal protective equipment may be needed to protect against hazardous exposures when loading or unloading, cleaning up spills or leaks, or when gauging, inspecting, sampling, or performing maintenance activities on loading facilities or vapor-recovery systems.

I. TURBINES.

1. Description. Turbines are usually gas- or steam-powered and are typically used to drive pumps, compressors, blowers, and other refinery process equipment. Steam enters turbines at high temperatures and pressures, expands across and drives rotating blades while directed by fixed blades.

2. Health and Safety Considerations.

a. Safety. Steam turbines used for exhaust operating under vacuum should have safety relief valves on the discharge side, both for protection and to maintain steam in the event of vacuum failure. Where maximum operating pressure could be greater than design pressure, steam turbines should be provided with relief devices. Consideration should be given to providing governors and overspeed control devices on turbines.

b. Health. Safe work practices and/or appropriate personal protective equipment may be needed for noise, steam and heat exposures, and during inspection and maintenance activities.

J. PUMPS, PIPING AND VALVES.

1. Description.

a. Centrifugal and positive-displacement (i.e., reciprocating) pumps are used to move hydrocarbons, process water, fire water, and wastewater through piping within the refinery. Pumps are driven by electric motors, steam turbines, or internal combustion engines. The pump type, capacity, and construction materials depend on the service for which it is used.

b. Process and utility piping distribute hydrocarbons, steam, water, and other products throughout the facility. Their size and construction depend on the type of service, pressure, temperature, and nature of the products. Vent, drain, and sample connections are provided on piping, as well as provisions for blanking.

c. Different types of valves are used depending on their operating purpose. These include gate valves, bypass valves, globe and ball valves, plug valves, block and bleed valves, and check valves. Valves can be manually or automatically operated.

2. Health and Safety Considerations.

a. Fire Protection and Prevention. The potential for fire exists should hydrocarbon pumps, valves, or lines develop leaks that could allow vapors to reach sources of ignition. Remote sensors, control valves, fire valves, and isolation valves should be used to limit the release of hydrocarbons at pump suction lines in the event of leakage and/or fire.

b. Safety. Depending on the product and service, backflow prevention from the discharge line may be needed. The failure of automatic pump controls could cause a deviation in process pressure and temperature. Pumps operated with reduced or no flow can overheat and rupture. Pressure relief in the discharge piping should be provided where pumps can be overpressured. Provisions may be made for pipeline expansion, movement, and temperature changes to avoid rupture. Valves and instruments that require servicing or other work should be accessible at grade level or from an operating platform. Operating vent and drain connections should be provided with double-block valves, a block valve and plug, or blind flange for protection against releases.

c. Health. Safe work practices and/or appropriate personal protective equipment may be needed for exposure to hazards such as those related to liquids and vapors when opening or draining pumps, valves, and/or lines, and during product sampling, inspection, and maintenance activities.

K. TANK STORAGE.

1. Description. Atmospheric storage tanks and pressure storage tanks are used throughout the refinery for storage of crudes, intermediate hydrocarbons (during the process), and finished products. Tanks are also provided for fire water, process and treatment water, acids, additives, and other chemicals. The type, construction, capacity and location of tanks depends on their use and materials stored.

2. Health and Safety Considerations.

a. Fire Prevention and Protection. The potential for fire exists should hydrocarbon storage tanks be overfilled or develop leaks that allow vapors to escape and reach sources of ignition. Remote sensors, control valves, isolation valves, and fire valves may be provided at tanks for pump-out or closure in the event of a fire in the tank, or in the tank dike or storage area.

b. Safety. Tanks may be provided with automatic overflow control and alarm systems, or manual gauging and checking procedures may be established to control overfills.

c. Health. Safe work practices and/or appropriate personal protective equipment may be needed for exposure to hazards related to product sampling, manual gauging, inspection, and maintenance activities including confined space entry where applicable.

VI. BIBLIOGRAPHY.

American Petroleum Institute. 1971. *Chemistry and Petroleum for Classroom Use in Chemistry Courses.* Washington, D.C.: American Petroleum Institute.

__________. 1973. *Industrial Hygiene Monitoring Manual for Petroleum Refineries and Selected Petrochemical Operations*. Manual 2700-1/79-1M. Washington, D.C.: American Petroleum Institute.

__________. 1980. *Facts About Oil*. Manual 4200-10/80-25M. Washington, D.C.: American Petroleum Institute.

__________. 1990. *Management of Process Hazards*. RP 750. Washington, D.C.: American Petroleum Institute.

__________. 1990. *Inspection of Piping, Tubing, Valves and Fittings*. RP 574. Washington, D.C.: American Petroleum Institute.

__________. 1991. *Inspection of Fired Boilers and Heaters*. RP 573. Washington, D.C.: American Petroleum Institute.

__________. 1992. *Inspection of Pressure Vessels*. RP 572. Washington, D.C.: American Petroleum Institute.

__________. 1992. *Inspection of Pressure Relieving Devices*. RP 576. Washington, D.C.: American Petroleum Institute.

__________. 1994. *Fire Protection in Refineries*. Sixth Edition. RP 2001. Washington, D.C.: American Petroleum Institute.

Armistead, George, Jr. 1950. *Safety in Petroleum Refining and Related Industries*. New York: John G. Simmons & Co., Inc.

Exxon Company, USA. 1987. *Encyclopedia for the User of Petroleum Products*. Lubetext D400. Houston: Exxon Company, USA.

Hydrocarbon Processing. 1988. *Refining Handbook*. Houston: Gulf Publishing Co.

__________. 1992. *Refining Handbook*. Houston: Gulf Publishing Co.

IARC. [No date given.] *Occupational Exposures in Petroleum Refining*. IARC Monographs, Volume 45.

Kutler, A. A. 1969. "Crude distillation." *Petro/Chem Engineering*. New York: John G. Simmonds & Co., Inc.

Mobil Oil Corporation. 1972. "Light Products Refining, Fuels Manufacture." *Mobil Technical Bulletin*, 1972. Fairfax, Virginia: Mobil Oil Corporation.

Parmeggiani, Luigi, Technical Editor. 1983. *Encyclopaedia of Occupational Health and Safety*. Third Edition. Geneva: International Labour Organization.

Shell International Petroleum Company Limited. 1983. *The Petroleum Handbook*. Sixth Edition. Amsterdam: Elsevier Science Publishers B.V.

Speight, James G. 1980. *The Chemistry and Terminology of Petroleum*. New York: Marcel Dekker, Inc.

Vervalin, Charles H., Editor. 1985. *Fire Protection Manual for Hydrocarbon Processing Plants*. Volume 1, Third edition. Houston: Gulf Publishing Co.

APPENDIX IV: 2-1. GLOSSARY.

ABSORPTION The disappearance of one substance into another so that the absorbed substance loses its identifying characteristics, while the absorbing substance retains most of its original physical aspects. Used in refining to selectively remove specific components from process streams.

ACID TREATMENT A process in which unfinished petroleum products such as gasoline, kerosene, and lubricating oil stocks are treated with sulfuric acid to improve color, odor, and other properties.

ADDITIVE Chemicals added to petroleum products in small amounts to improve quality or add special characteristics.

ADSORPTION Adhesion of the molecules of gases or liquids to the surface of solid materials.

AIR FIN COOLERS A radiator-like device used to cool or condense hot hydrocarbons; also called fin fans.

ALICYCLIC HYDROCARBONS Cyclic (ringed) hydrocarbons in which the rings are made up only of carbon atoms.

ALIPHATIC HYDROCARBONS Hydrocarbons characterized by open-chain structures: ethane, butane, butene, acetylene, etc.

ALKYLATION A process using sulfuric or hydro-fluoric acid as a catalyst to combine olefins (usually butylene) and isobutane to produce a high-octane product known as alkylate.

API GRAVITY An arbitrary scale expressing the density of petroleum products.

AROMATIC Organic compounds with one or more benzene rings.

ASPHALTENES The asphalt compounds soluble in carbon disulfide but insoluble in paraffin naphthas.

ATMOSPHERIC TOWER A distillation unit operated at atmospheric pressure.

BENZENE An unsaturated, six-carbon ring, basic aromatic compound.

BLEEDER VALVE A small-flow valve connected to a fluid process vessel or line for the purpose of bleeding off small quantities of contained fluid. It is installed with a block valve to determine if the block valve is closed tightly.

BLENDING The process of mixing two or more petroleum products with different properties to produce a finished product with desired characteristics.

BLOCK VALVE A valve used to isolate equipment.

BLOWDOWN The removal of hydrocarbons from a process unit, vessel, or line on a scheduled or emergency basis by the use of pressure through special piping and drums provided for this purpose.

BLOWER Equipment for moving large volumes of gas against low-pressure heads.

BOILING RANGE The range of temperature (usually at atmospheric pressure) at which the boiling (or distillation) of a hydrocarbon liquid commences, proceeds, and finishes.

BOTTOMS Tower bottoms are residue remaining in a distillation unit after the highest boiling-point material to be distilled has been removed. Tank bottoms are the heavy materials that accumulate in the bottom of storage tanks, usually comprised of oil, water, and foreign matter.

BUBBLE TOWER A fractionating (distillation) tower in which the rising vapors pass through layers of condensate, bubbling under caps on a series of plates.

CATALYST A material that aids or promotes a chemical reaction between other substances but does not react itself. Catalysts increase reaction speeds and can provide control by increasing desirable reactions and decreasing undesirable reactions.

CATALYTIC CRACKING The process of breaking up heavier hydrocarbon molecules into lighter hydrocarbon fractions by use of heat and catalysts.

CAUSTIC WASH A process in which distillate is treated with sodium hydroxide to remove acidic contaminants that contribute to poor odor and stability.

CHD UNIT See Hydrodesulfurization.

COKE A high carbon-content residue remaining from the destructive distillation of petroleum residue.

COKING A process for thermally converting and upgrading heavy residual into lighter products and by-product petroleum coke. Coking also is the removal of all lighter distillable hydrocarbons that leaves a residue of carbon in the bottom of units or as buildup or deposits on equipment and catalysts.

CONDENSATE The liquid hydrocarbon resulting from cooling vapors.

CONDENSER A heat-transfer device that cools and condenses vapor by removing heat via a cooler medium such as water or lower-temperature hydrocarbon streams.

CONDENSER REFLUX Condensate that is returned to the original unit to assist in giving increased conversion or recovery.

COOLER A heat exchanger in which hot liquid hydrocarbon is passed through pipes immersed in cool water to lower its temperature.

CRACKING The breaking up of heavy molecular weight hydrocarbons into lighter hydrocarbon molecules by the application of heat and pressure, with or without the use of catalysts.

CRUDE ASSAY A procedure for determining the general distillation and quality characteristics of crude oil.

CRUDE OIL A naturally occurring mixture of hydrocarbons that usually includes small quantities of sulfur, nitrogen, and oxygen derivatives of hydrocarbons as well as trace metals.

CYCLE GAS OIL Cracked gas oil returned to a cracking unit.

DEASPHALTING Process of removing asphaltic materials from reduced crude using liquid propane to dissolve nonasphaltic compounds.

DEBUTANIZER A fractionating column used to remove butane and lighter components from liquid streams.

DE-ETHANIZER A fractionating column designed to remove ethane and gases from heavier hydrocarbons.

DEHYDROGENATION A reaction in which hydrogen atoms are eliminated from a molecule. Dehydrogenation is used to convert ethane, propane, and butane into olefins (ethylene, propylene, and butenes).

DEPENTANIZER A fractionating column used to remove pentane and lighter fractions from hydrocarbon streams.

DEPROPANIZER A fractionating column for removing propane and lighter components from liquid streams.

DESALTING Removal of mineral salts (most chlorides, e.g., magnesium chloride and sodium chloride) from crude oil.

DESULFURIZATION A chemical treatment to remove sulfur or sulfur compounds from hydrocarbons.

DEWAXING The removal of wax from petroleum products (usually lubricating oils and distillate fuels) by solvent absorption, chilling, and filtering.

DIETHANOLAMINE A chemical $C4H11O2N$) used to remove $H2S$ from gas streams.

DISTILLATE The products of distillation formed by condensing vapors.

DOWNFLOW Process in which the hydrocarbon stream flows from top to bottom.

DRY GAS Natural gas with so little natural gas liquids that it is nearly all methane with some ethane.

FEEDSTOCK Stock from which material is taken to be fed (charged) into a processing unit.

FLASHING The process in which a heated oil under pressure is suddenly vaporized in a tower by reducing pressure.

FLASH POINT Lowest temperature at which a petroleum product will give off sufficient vapor so that the vapor-air mixture above the surface of the liquid will propagate a flame away from the source of ignition.

FLUX Lighter petroleum used to fluidize heavier residual so that it can be pumped.

FOULING Accumulation of deposits in condensers, exchangers, etc.

FRACTION One of the portions of fractional distillation having a restricted boiling range.

FRACTIONATING COLUMN Process unit that separates various fractions of petroleum by simple distillation, with the column tapped at various levels to separate and remove fractions according to their boiling ranges.

FUEL GAS Refinery gas used for heating.

GAS OIL Middle-distillate petroleum fraction with a boiling range of about 350°-750° F, usually includes diesel fuel, kerosene, heating oil, and light fuel oil.

GASOLINE A blend of naphthas and other refinery products with sufficiently high octane and other desirable characteristics to be suitable for use as fuel in internal combustion engines.

HEADER A manifold that distributes fluid from a series of smaller pipes or conduits.

HEAT As used in the Health Considerations paragraphs of this document, heat refers to thermal burns for contact with hot surfaces, hot liquids and vapors, steam, etc.

HEAT EXCHANGER Equipment to transfer heat between two flowing streams of different temperatures. Heat is transferred between liquids or liquids and gases through a tubular wall.

HIGH-LINE OR HIGH-PRESSURE GAS High-pressure (100 psi) gas from cracking unit distillate drums that is compressed and combined with low-line gas as gas absorption feedstock.

HYDROCRACKING A process used to convert heavier feedstock into lower-boiling, higher-value products. The process employs high pressure, high temperature, a catalyst, and hydrogen.

HYDRODESULFURIZATION A catalytic process in which the principal purpose is to remove sulfur from petroleum fractions in the presence of hydrogen.

HYDROFINISHING A catalytic treating process carried out in the presence of hydrogen to improve the properties of low viscosity-index naphthenic and medium viscosity-index naphthenic oils. It is also applied to paraffin waxes and microcrystalline waxes for the removal of undesirable components. This process consumes hydrogen and is used in lieu of acid treating.

HYDROFORMING Catalytic reforming of naphtha at elevated temperatures and moderate pressures in the presence of hydrogen to form high-octane BTX aromatics for motor fuel or chemical manufacture. This process results in a net production of hydrogen and has rendered thermal reforming somewhat obsolete. It represents the total effect of numerous simultaneous reactions such as cracking, polymerization, dehydrogenation, and isomerization.

HYDROGENATION The chemical addition of hydrogen to a material in the presence of a catalyst.

INHIBITOR Additive used to prevent or retard undesirable changes in the quality of the product, or in the condition of the equipment in which the product is used.

ISOMERIZATION A reaction that catalytically converts straight-chain hydrocarbon molecules into branched-chain molecules of substantially higher octane number. The reaction rearranges the carbon skeleton of a molecule without adding or removing anything from the original material.

ISO-OCTANE A hydrocarbon molecule (2,2,4-trimethylpentane) with excellent antiknock characteristics on which the octane number of 100 is based.

KNOCKOUT DRUM A vessel wherein suspended liquid is separated from gas or vapor.

LEAN OIL Absorbent oil fed to absorption towers in which gas is to be stripped. After absorbing the heavy ends from the gas, it becomes fat oil. When the heavy ends are subsequently stripped, the solvent again becomes lean oil.

LOW-LINE or LOW-PRESSURE GAS Low-pressure (5 psi) gas from atmospheric and vacuum distillation recovery systems that is collected in the gas plant for compression to higher pressures.

NAPHTHA A general term used for low boiling hydrocarbon fractions that are a major component of gasoline. Aliphatic naphtha refers to those naphthas containing less than 0.1% benzene and with carbon numbers from C3 through C16. Aromatic naphthas have carbon numbers from C6 through C16 and contain significant quantities of aromatic hydrocarbons such as benzene (>0.1%), toluene, and xylene.

NAPHTHENES Hydrocarbons (cycloalkanes) with the general formula CnH2n, in which the carbon atoms are arranged to form a ring.

OCTANE NUMBER A number indicating the relative antiknock characteristics of gasoline.

OLEFINS A family of unsaturated hydrocarbons with one carbon-carbon double bond and the general formula CnH2n.

PARAFFINS A family of saturated aliphatic hydrocarbons (alkanes) with the general formula CnH2n+2.

POLYFORMING The thermal conversion of naphtha and gas oils into high-quality gasoline at high temperatures and pressure in the presence of recirculated hydrocarbon gases.

POLYMERIZATION The process of combining two or more unsaturated organic molecules to form a single (heavier) molecule with the same elements in the same proportions as in the original molecule.

PREHEATER Exchanger used to heat hydrocarbons before they are fed to a unit.

PRESSURE-REGULATING VALVE A valve that releases or holds process-system pressure (that is, opens or closes) either by preset spring tension or by actuation by a valve controller to assume any desired position between fully open and fully closed.

PYROLYSIS GASOLINE A by-product from the manufacture of ethylene by steam cracking of hydrocarbon fractions such as naphtha or gas oil.

PYROPHORIC IRON SULFIDE A substance typically formed inside tanks and processing units by the corrosive interaction of sulfur compounds in the hydrocarbons and the iron and steel in the equipment. On exposure to air (oxygen) it ignites spontaneously.

QUENCH OIL Oil injected into a product leaving a cracking or reforming heater to lower the temperature and stop the cracking process.

RAFFINATE The product resulting from a solvent extraction process and consisting mainly of those components that are least soluble in the solvents. The product recovered from an extraction process is relatively free of aromatics, naphthenes, and other constituents that adversely affects physical parameters.

REACTOR The vessel in which chemical reactions take place during a chemical conversion type of process.

REBOILER An auxiliary unit of a fractionating tower designed to supply additional heat to the lower portion of the tower.

RECYCLE GAS High hydrogen-content gas returned to a unit for reprocessing.

REDUCED CRUDE A residual product remaining after the removal by distillation of an appreciable quantity of the more volatile components of crude oil.

REFLUX The portion of the distillate returned to the fractionating column to assist in attaining better separation into desired fractions.

REFORMATE An upgraded naphtha resulting from catalytic or thermal reforming.

REFORMING The thermal or catalytic conversion of petroleum naphtha into more volatile products of higher octane number. It represents the total effect of numerous simultaneous reactions such as cracking, polymerization, dehydrogenation, and isomerization.

REGENERATION In a catalytic process the reactivation of the catalyst, sometimes done by burning off the coke deposits under carefully controlled conditions of temperature and oxygen content of the regeneration gas stream.

SCRUBBING Purification of a gas or liquid by washing it in a tower.

SOLVENT EXTRACTION The separation of materials of different chemical types and solubilities by selective solvent action.

SOUR GAS Natural gas that contains corrosive, sulfur-bearing compounds such as hydrogen sulfide and mercaptans.

STABILIZATION A process for separating the gaseous and more volatile liquid hydrocarbons from crude petroleum or gasoline and leaving a stable (less-volatile) liquid so that it can be handled or stored with less change in composition.

STRAIGHT-RUN GASOLINE Gasoline produced by the primary distillation of crude oil. It contains no cracked, polymerized, alkylated, reformed, or visbroken stock.

STRIPPING The removal (by steam-induced vaporization or flash evaporation) of the more volatile components from a cut or fraction.

SULFURIC ACID TREATING A refining process in which unfinished petroleum products such as gasoline, kerosene, and lubricating oil stocks are treated with sulfuric acid to improve their color, odor, and other characteristics.

SULFURIZATION Combining sulfur compounds with petroleum lubricants.

SWEETENING Processes that either remove obnoxious sulfur compounds (primarily hydrogen sulfide, mercaptans, and thiophens) from petroleum fractions or streams, or convert them, as in the case of mercaptans, to odorless disulfides to improve odor, color, and oxidation stability.

SWITCH LOADING The loading of a high static-charge retaining hydrocarbon (i.e., diesel fuel) into a tank truck, tank car, or other vessel that has previously contained a low-flash hydrocarbon (gasoline) and may contain a flammable mixture of vapor and air.

TAIL GAS The lightest hydrocarbon gas released from a refining process.

THERMAL CRACKING The breaking up of heavy oil molecules into lighter fractions by the use of high temperature without the aid of catalysts.

TURNAROUND A planned complete shutdown of an entire process or section of a refinery, or of an entire refinery to perform major maintenance, overhaul, and repair operations and to inspect, test, and replace process materials and equipment.

VACUUM DISTILLATION The distillation of petroleum under vacuum which reduces the boiling temperature sufficiently to prevent cracking or decomposition of the feedstock.

VAPOR The gaseous phase of a substance that is a liquid at normal temperature and pressure.

VISBREAKING Viscosity breaking is a low-temperature cracking process used to reduce the viscosity or pour point of straight-run residuum.

WET GAS A gas containing a relatively high proportion of hydrocarbons that are recoverable as liquids.

SECTION IV: CHAPTER 3

PRESSURE VESSEL GUIDELINES

Contents:

I. INTRODUCTION.

A. Recent inspection programs for metallic pressure containment vessels and tanks have revealed cracking and damage in a considerable number of the vessels inspected. Safety and hazard evaluations of pressure vessels, as also presented in OSHA Instruction PUB 8-1.5, need to consider the consequences of a leakage or a rupture failure of a vessel.

B. Two consequences result from a complete rupture:

- Blast effects due to sudden expansion of the pressurized fluid.
- Fragmentation damage and injury, if vessel rupture occurs.

C. For a leakage failure, the hazard consequences can range from no effect to very serious effects:

- Suffocation or poisoning, depending on the nature of the contained fluid, if the leakage occurs into a closed space.
- Fire and explosion (physical hazards for a flammable fluid).
- Chemical and thermal burns from contact with process liquids.

D. Only pressure vessels and low pressure storage tanks widely used in process, pulp and paper, petroleum refining, and petrochemical industries and for water treatment systems of boilers and steam generation equipment are covered in this chapter. Excluded are vessels and tanks used in many other applications. Also excluded are other parts of a pressure containment system such as piping and valves.

E. The types and applications of pressure vessels included and excluded in this chapter are summarized in Table IV:3-1. An illustration of a schematic pressure vessel is presented in Figure IV:3-1.

> **NOTE:** Though this review of pressure vessels excludes inspection or evaluation of safety release valves, the compliance officer should be axare that *NO* valves or T-fittings should be present between the vessel and the safety relief valve.

F. Most of the pressure or storage vessels in service in the United States will have been designed and constructed in accordance with one of the following two pressure vessel design codes:

- The ASME Code, or Section VIII of the ASME (American Society of Mechanical Engineers) "Boiler and Pressure Vessel Code."
- The API Standard 620 or the American Petroleum Institute Code which provides rules for lower pressure vessels not covered by the ASME Code.

In addition, some vessels designed and constructed between 1934 to 1956 may have used the rules in the "API-ASME Code for Unfired Pressure Vessels for Petroleum Liquids and Gases." This code was discontinued in 1956.

G. Vessels certification can only be performed by trained inspectors qualified for each code. Written tests and practical experience are required for certification. Usually, the compliance office is not equipped for this task, but is able to obtain the necessary contract services.

TABLE IV:3-1. VESSEL TYPES

Vessels included:	***Vessel types specifically excluded:***
Stationary and unfired	Vessels used as fired boilers
Used for pressure containment of gases and liquids	Vessels used in high-temperature processes (above 315° C, 600° F) or at very low and cryogenic temperatures
Constructed of carbon steel or low alloy steel	Vessels and containers used in transportable systems
Operated at temperatures between -75° and 315° C (-100° and 600° F)	Storage tanks that operate at nominally atmospheric pressure
	Piping and pipelines
	Safety and pressure-relief valves
	Special-purpose vessels, such as those for human occupancy

II. RECENT CRACKING EXPERIENCE IN PRESSURE VESSELS.

A. DEAERATOR SERVICE.

1. Deaeration refers to the removal of noncondensible gases, primarily oxygen, from the water used in a steam generation system. Deaerators are widely used in many industrial applications including power generation, pulp and paper, chemical, and petroleum refining and in many public facilities such as hospitals and schools where steam generation is required. In actual practice, the deaerator vessel can be separate from the storage vessel or combined with a storage vessel into one unit.

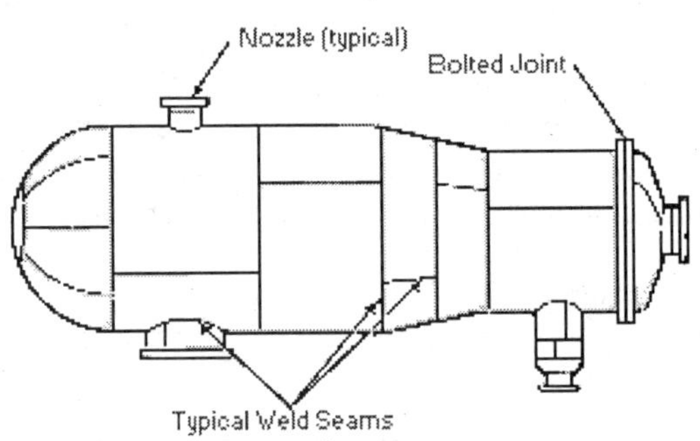

FIGURE IV: 3-1. SOME MAJOR PARTS OF A PRESSURE VESSEL.

2. Typical operational conditions for deaerator vessels range up to about 300 psi and up to about 150° C (300° F). Nearly all of the vessels are designed to ASME Code, resulting in vessel wall thicknesses up to but generally less than 25 mm (1 in). The vessel material is almost universally one of the carbon steel grades.

3. Analysis of incident survey data and other investigations has determined the following features about the deaerator vessel cracking.

    - Water hammer is the only design or operational factor that correlates with cracking.
    - Cracking is generally limited to weld regions of vessels that had not been postweld heat treated.
    - Corrosion fatigue appears to be the predominant mechanism of crack formation and growth.

4. The failures and the survey results have prompted TAPPI (Technical Association of Pulp and Paper Industry), the National Board of Boiler and Pressure Vessel Inspectors, and NACE (National Association of Corrosion Engineers) to prepare inspection, operation and repair recommendations. For inspection, all recommendations suggest:

    - Special attention to the internal surface of all welds and heat-affected zones (HAZ).
    - Use of the wet fluorescent magnetic particle (WFMT) method for inspection.

5. The TAPPI and the NACE recommendations also contain additional items, such as:

    - Inspection by personnel certified to American Society for Nondestructive Testing's SNT-TC-1A minimum Level I and interpretation of the results by minimum Level II.
    - Reinspection within one year for repaired vessels, 1-2 years for vessels with discontinuities but unrepaired, and 3-5 years for vessels found free of discontinuities.

B. AMINE SERVICE.

1. The amine process is used to remove hydrogen sulfide (H2S) from petroleum gases such as propane and butane. It is also used for carbon dioxide (CO2) removal in some processes. Amine is a generic term and includes monoethanolamine (MEA), diethanolamine (DEA) and others in the amine group. These units are used in petroleum refinery, gas treatment and chemical plants.

2. The operating temperatures of the amine process are generally in the 38 to 93° C (100° to 200° F) range and therefore the plant equipment is usually constructed from one of the carbon steel grades. The wall thickness of the pressure vessels in amine plants is typically about 25 mm (1 in.).

3. Although the possibility of cracking of carbon steels in an amine environment has been known for some years, real concern about safety implications was highlighted by a 1984 failure of the amine process pressure vessel. Overall, the survey found about 40% cracking incidence in a total of 294 plants. Cracking had occurred in the absorber/contactor, the regenerator and the heat exchanger vessels, and in the piping and other auxiliary equipment. Several of the significant findings of the survey were:

    - All cracks were in or near welds.
    - Cracking occurred predominantly in stressed or unrelieved (not PWHT) welds.
    - Cracking occurred in all amine vessel processes but was most prevalent in MEA units.
    - WFMT and UT (ultrasonic test) were the predominant detection methods for cracks; internal examination by WFMT was the preferred method.

4. Information from laboratory studies indicate that pure amine does not cause cracking of carbon steels but amine with carbon dioxide in the gas phase causes severe cracking. The presence or absence of chlorides, cyanides, or hydrogen sulfide may also be factors but their full role in the cracking mechanism is not completely known at present.

C. WET HYDROGEN SULFIDE.

1. Wet Hydrogen Sulfide refers to any fluid containing water and hydrogen sulfide (H2S). Hydrogen is generated when steel is exposed to this mixture and the hydrogen can enter into the steel. Dissolved hydrogen can cause cracking, blistering, and embrittlement.

2. The harmful effects of hydrogen generating environments on steel have been known and recognized for a long time in the petroleum and petrochemical industries. In particular, sensitivity to damage by hydrogen increases with the hardness and strength of the steel; and damage and cracking are more apt to occur in high strength steels, as follows:

    - Significant cracks can start from very small hard zones associated with welds; these hard zones are not detected by conventional hardness tests.
    - Initially small cracks can grow by a stepwise form of hydrogen blistering to form through thickness cracks.
    - NACE/API limits on weld hardness may not be completely effective in preventing cracking.
    - Thermal stress relief (postweld heat treatment, PWHT) appears to reduce the sensitivity to and the severity of cracking.

3. Wet hydrogen sulfide has also been found to cause service cracking in liquified petroleum gas (LPG) storage vessels. The service cracking in the LPG vessels occurs predominantly in the weld heat affected zone (HAZ). The vessels are usually spherical with wall thickness in the 20 mm to 75 mm (0.8 in. to 3 in.) range.

4. Recommendations for new and existing wet hydrogen-sulfide vessels to minimize the risk of a major failure include:

    - Use lower-strength steels for new vessels.
    - Schedule an early inspection for vessels more than five years in service.
    - Improve monitoring to minimize breakthrough of hydrogen sulfide.
    - Replace unsafe vessels or downgrade to less severe, usually lower-pressure, service.

D. AMMONIA SERVICE.

1. Commercial refrigeration systems, certain chemical processes, and formulators of agricultural chemicals will be sites of ammonia service tanks. Careful inspections of vessels used for storage of ammonia (in either vapor or liquid form) in recent years have produced evidence of serious stress corrosion cracking problems. The vessels for this service are usually constructed as spheres from one of the carbon steel grades, and they operate in the ambient temperature range.

2. The water and oxygen content in the ammonia has a strong influence on the propensity of carbon steels to crack in this environment. Cracks have a tendency to be found to be in or near the welds in as-welded vessels. Cracks occur both transverse and parallel to the weld direction. Thermal stress relieving seems to be a mitigating procedure for new vessels, but its efficacy for older vessels after a period of operation is dubious partly because small, undetected cracks may be present.

E. PULP DIGESTER SERVICE.

1. The kraft pulping process is used in the pulp and paper industry to digest the pulp in the papermaking process. The operation is done in a relatively weak (a few percent) water solution of sodium hydroxide and sodium sulfide typically in the 110° to 140° C (230° to 285° F) temperature range. Since the early 1950's, a continuous version of this process has been widely used. Nearly all of the vessels are ASME Code vessels made using one of the carbon steel grades with typical design conditions of 175° to 180° C (350° to 360° F) and 150 psig.

2. These vessels had a very good service record with only isolated reports of cracking problems until the occurrence of a sudden rupture failure in 1980. The inspection survey has revealed that about 65% of the properly inspected vessels had some cracking. Some of the cracks were fabrication flaws revealed by the use of more sensitive inspection techniques but most of the cracking was service-induced. The inspection survey and analysis indicates the following features about the cracking.

    - All cracking was associated with welds.
    - Wet fluorescent magnetic particle (WFMT) testing with proper surface preparation was the most effective method of detecting the cracking.
    - Fully stress-relieved vessels were less susceptible.

- No clear correlation of cracking and noncracking could be found with vessel age and manufacture or with process variables and practices.
- Analysis and research indicate that the cracking is due to a caustic stress corrosion cracking mechanism although its occurrence at the relatively low caustic concentrations of the digester process was unexpected.

3. Currently, preventive measures such as weld cladding, spray coatings, and anodic protection are being studied, and considerable information has been obtained. In the meantime, the recommended guideline is to perform an annual examination.

F. SUMMARY OF SERVICE CRACKING EXPERIENCE.

1. The preceding discussion shows a strong influence of chemical environment on cracking incidence. This is a factor that is not explicitly treated in most design codes. Service experience is the best and often the only guide to in-service safety assessment.

2. For vessels and tanks within the scope of this document, the service experience indicates that the emphasis of the inspection and safety assessment should be on:

   - Vessels in deaerator, amine, wet H2S, ammonia and pulp digesting service.
   - Welds and adjacent regions.
   - Vessels that have not been thermally stress relieved (no PWHT of fabrication welds).
   - Repaired vessels, especially those without PWHT after repair.

3. The evaluation of the severity of the detected cracks can be done by fracture mechanics methods. This requires specific information about stresses, material properties, and flaw indications. Generalized assessment guidelines are not easy to formulate. However, fortunately, many vessels in the susceptible applications listed above operate at relatively low stresses, and therefore, cracks have a relatively smaller effect on structural integrity and continued safe operation.

III. NONDESTRUCTIVE EXAMINATION METHODS.

Of the various conventional and advanced nondestructive examination (NDE) methods, five are widely used for the examination of pressure vessels and tanks by certified pressure vessel inspectors. The names and acronyms of these common five methods are:

- Visual Examination (VT)
- Liquid Penetrant Test (PT)
- Magnetic Particle Test (MT)
- Gamma and X-ray Radiography (RT)
- Ultrasonic Test (UT)

VT, PT, and MT can detect only those discontinuities and defects that are open to the surface or are very near the surface. In contrast, RT and UT can detect conditions that are located within the part. For these reasons, the first three are often referred to as "surface" examination methods and the last two as "volumetric" methods. Table II of PUB 8-1.5 summarizes the main features of these five methods.

A. VISUAL EXAMINATION (VT) is easy to conduct and can cover a large area in a short time. It is very useful for assessing the general condition of the equipment and for detecting some specific problems such as severe instances of corrosion, erosion, and hydrogen blistering. The obvious requirements for a meaningful visual examination are a clean surface and good illumination.

B. LIQUID PENETRANT TEST (PT) depends on allowing a specially formulated liquid (penetrant) to seep into an open discontinuity and then detecting the entrapped liquid by a developing agent. When the penetrant is removed from the surface, some of it remains entrapped in the discontinuities. Application of a developer draws out the entrapped penetrant and magnifies the discontinuity. Chemicals which fluoresce under black (ultraviolet) light can be added to the penetrant to aid the detectability and visibility of the developed indications. The essential feature of PT is that the discontinuity must be "open," which means a clean, undisturbed surface.

The PT method is independent of the type and composition of the metal alloy so it can be used for the examination of austenitic stainless steels and nonferrous alloys where the magnetic particle test is not applicable.

C. MAGNETIC PARTICLE TEST (MT).

1. This method depends on the fact that discontinuities in or near the surface perturb magnetic flux lines induced into a ferromagnetic material. For a component such as a pressure vessel where access is generally limited to one surface at a time, the "prod" technique is widely used. The magnetic field is produced in the region around and between the prods (contact probes) by an electric current (either AC or DC) flowing between the prods. The ferromagnetic material requirement basically limits the applicability of MT to carbon and low- alloy steels.

2. The perturbations of the magnetic lines are revealed by applying fine particles of a ferromagnetic material to the surface. The particles can be either a dry powder or a wet suspension in a liquid. The particles can also be treated to fluoresce under black light. These options lead to variations such as the "wet fluorescent magnetic particle test" (WFMT). MT has some capability for detecting subsurface defects. However, there is no easy way to determine the limiting depth of sensitivity since it is highly dependent on magnetizing current, material, and geometry and size of the defect. A very crude approximation would be a depth no more than 1.5 mm to 3 mm (1/16 in. to 1/8 in.).

3. A very important precaution in performing MT is that corners and surface irregularities also perturb the magnetic field. Therefore, examining for defects in corners and near or in welds must be performed with extra care. Another precaution is that MT is most sensitive to discontinuities which are oriented transverse to the magnetic flux lines and this characteristic needs to be taken into account in determining the procedure for inducing the magnetic field.

D. RADIOGRAPHY (RT).

1. The basic principle of radiographic examination of metallic objects is the same as in any other form of radiography such as medical radiography. Holes, voids, and discontinuities decrease the attenuation of the X-ray and produce greater exposure on the film (darker areas on the negative film).

2. Because RT depends on density differences, cracks with tightly closed surfaces are much more difficult to detect than open voids. Also, defects located in an area of an abrupt dimensional change are difficult to detect due to the superimposed density difference. RT is effective in showing defect dimensions on a plane normal to the beam direction but determination of the depth dimension and location requires specialized techniques. Since ionizing radiation is involved, field application of RT requires careful implementation to prevent health hazards.

E. ULTRASONIC TESTING (UT).

1. The fundamental principles of ultrasonic testing of metallic materials are similar to radar and related methods of using electromagnetic and acoustic waves for detection of foreign objects. The distinctive aspect of UT for the inspection of metallic parts is that the waves are mechanical, so the test equipment requires three basic components:

   - Electronic system for generating electrical signal.
   - Transducer system to convert the electrical signal into mechanical vibrations and vice versa and to inject the vibrations into and extract them from the material.
   - Electronic system for amplifying, processing, and displaying the return signal.

   Very short signal pulses are induced into the material and waves reflected back from discontinuities are detected during the "receive" mode. The transmitting and detection can be done with one transducer or with two separate transducers (the tandem technique).

2. Unlike radiography, UT in its basic form does not produce a permanent record of the examination. However, more recent versions of UT equipment include automated operation and electronic recording of the signals.

3. Ultrasonic techniques can also be used for the detection and measurement of general material loss such as by corrosion and erosion. Since wave velocity is constant for a specific material, the transit time between the initial pulse and the back reflection is a measure of the travel distance and the thickness.

F. DETECTION PROBABILITIES AND FLAW SIZING.

1. The implementation of NDE (nondestructive examination) results for structural integrity and safety assessment involves a detailed consideration of two separate but interrelated factors.

    - Detecting the discontinuity.
    - Identifying the nature of the discontinuity and determining its size.

2. Much of the available information on detection and sizing capabilities has been developed for aircraft and nuclear power applications. This kind of information is very specific to the nature of the flaw, the material, and the details of the test technique, and direct transference to other situations is not always warranted.

3. The overall reliability of NDE is obviously an important factor in a safety and hazard assessment. Failing to detect or undersizing existing discontinuities reduces the safety margin while oversizing errors can result in unnecessary and expensive outages. High reliability is achieved through a combination of factors, as follows:

    - Validated procedures, equipment and test personnel.
    - Utilization of diverse methods and techniques.
    - Application of redundancy by repetitive and independent tests.

Finally, it is useful to note that safety assessment depends on evaluating the "largest flaw that may be missed, not the smallest one that can be found."

IV. INFORMATION FOR SAFETY ASSESSMENT.

This chapter and PUB 8-1.5 has a large amount of information on the design rules, inspection requirements, and service experience, relevant to pressure vessels and low pressure storage tanks used in general industrial applications. Though the compliance officer is not usually qualified as a pressure vessel inspector, as a summary and a reminder, Appendix IV:3-1 outlines the information, data, and recordkeeping that are necessary, useful, or indicative of safe management of operating vessels and tanks.

These records, in addition to the construction and maintenance logs, usually are kept by the plant engineer, maintenance supervisor, or facility manager, and will be indicative of the surveillance activities around safe operation of pressure vessels.

V. BIBLIOGRAPHY.

Chuse, R. 1984. *Pressure Vessels: The ASME Code Simplified.* 6th ed. McGraw-Hill: New York.

Forman, B. Fred. 1981. *Local Stresses in Pressure Vessels.* Pressure Vessel Handbook Publishing, Inc.: Tulsa.

Hammer, W. 1981. *Pressure Hazards in Occupational Safety Management and Engineering.* 2nd ed. Prentice-Hall: New York.

McMaster, R.C. and McIntire, P. (eds.) 1982-1987. *Nondestructive Testing Handbook.* 2nd ed., Vols. 1-3. American Society for Metals/American Society of Nondestructive Testing: Columbus.

Megyesy, E.F. 1986. *Pressure Vessel Handbook.* 7th ed. Pressure Vessel Handbook Publishing Inc.: Tulsa.

OSHA Instruction Pub 8-1.5. 1989. *Guidelines for Pressure Vessel Safety Assessment.* Occupational Safety and Health Administration: Washington, D.C.

Thielsch, H. 1975. *Defects and Failures in Pressure Vessels and Piping.* 2nd ed., Chaps. 16 and 17. Reinhold: New York.

Yokell, S. 1986. *Understanding the Pressure Vessel Code.* Chemical Engineering 93(9):75-85.

APPENDIX IV:3-1. RECORDKEEPING DATA FOR STEEL VESSELS AND LOW PRESSURE STORAGE TANKS

INTRODUCTION AND SCOPE. This outline summarizes information and data that will be helpful in assessing the safety of steel pressure vessels and low pressure storage tanks that operate at temperatures between -75° and 315 ° C (-100° and 600° F).

VESSEL IDENTIFICATION AND DOCUMENTATION. Information that identifies the specific vessel being assessed and provides general information about it include the following items:

- Current owner of the vessel
- Vessel location- Original location and current location if it has been moved
- Vessel identification: - Manufacturer's serial number; - National Board number if registered with NB
- Manufacturer identification: - Name and address of manufacturer;- Authorization or identification number of the manufacturer
- Date of manufacture of the vessel
- Data report for the vessel: - ASME U-1 or U-2; API 620 form or other applicable report
- Date vessel was placed in service
- Interruption dates if not in continuous service.

DESIGN AND CONSTRUCTION INFORMATION.

Information that will identify the code or standard used for the design and construction of the vessel or tank and the specific design values, materials, fabrication methods, and inspection methods used include the following items:

- Design code: - ASME Code Section and Division, API Standard or other design code used
- Type of construction: - Shop or field fabricated or other fabrication method
- VIII, division 1 or 2 vessels: - Maximum allowable pressure and temperature; - Minimum design temperature
- API 620 vessels: - Design pressure at top and maximum fill
- Additional requirements included, such as: - Appendix Q (Low-Pressure Storage Tanks For Liquified Hydrocarbon Gases) and; - Appendix R (Low-Pressure Storage Tanks for Refrigerated Products)
- Other design code vessels: - Maximum design and allowable pressures; - Maximum and minimum operating temperatures
- Vessel materials: - ASME, ASTM, or other specification names and numbers for the major parts
- Design corrosion allowance
- Thermal stress relief (PWHT, postweld heat treatment): - Design code requirements; - Type, extent, and conditions of PWHT performed
- Nondestructive examination (NDE) of welds: - Type and extent of examination performed; - Time when NDE was performed (before or after PWHT or hydrotest)

SERVICE HISTORY.

Information on the conditions of operating history of the vessel or tank that will be helpful in safety assessment include the following items:

- Fluids handled: - Type and composition, temperature and pressures
- Type of service: - Continuous, intermittent or irregular

- Significant changes in service conditions: - Changes in pressures, temperatures, and fluid compositions and the dates of the changes
- Vessel history: - Alterations, reratings, and repairs performed; - Date(s) of changes or repairs

IN-SERVICE INSPECTION.

Information about inspections performed on the vessel or tank and the results obtained that will assist in the safety assessment include the following items:

- Inspection(s) performed: - Type, extent, and dates
- Examination methods: - Preparation of surfaces and welds; - Techniques used (visual, magnetic particle, penetrant test, radiography, ultrasonic)
- Qualifications of personnel: - ASNT (American Society for Nondestructive Testing) levels or equivalent of examining and supervisory personnel
- Inspection results and report: - Report form used (NBIC NB-7, API 510 or other); - Summary of type and extent of damage or cracking; - Disposition (no action, delayed action or repaired)

SPECIFIC APPLICATIONS.

Survey results indicate that a relatively high proportion of vessels in operations in several specific applications have experienced in service-related damage and cracking. Information on the following items can assist in assessing the safety of vessels in these applications:

- Service application: - Deaerator, amine, wet hydrogen sulfide, ammonia, or pulp digesting
- Industry bulletins and guidelines for this application: - Owner/operator awareness of information
- Type, extent, and results of examinations: - Procedures, guidelines and recommendations used; - Amount of damage and cracking; - Next examination schedule
- Participation in industry survey for this application
- Problem mitigation: - Written plans and actions

EVALUATION OF INFORMATION.

The information acquired for the above items is not adaptable to any kind of numerical ranking for quantitative safety assessment purposes. However, the information can reveal the owner or user's apparent attention to good practice, careful operation, regular maintenance, and adherence to the recommendations and guidelines developed for susceptible applications. If the assessment indicated cracking and other serious damage problems, it is important that the inspector obtain qualified technical advice and opinion.

SECTION IV: CHAPTER 4

INDUSTRIAL ROBOTS AND ROBOT SYSTEM SAFETY

Contents:

I. INTRODUCTION.

Industrial robots are programmable multifunctional mechanical devices designed to move material, parts, tools, or specialized devices through variable programmed motions to perform a variety of tasks. An industrial robot system includes not only industrial robots but also any devices and/or sensors required for the robot to perform its tasks as well as sequencing or monitoring communication interfaces.

Robots are generally used to perform unsafe, hazardous, highly repetitive, and unpleasant tasks. They have many different functions such as material handling, assembly, arc welding, resistance welding, machine tool load and unload functions, painting, spraying, etc. See Appendix IV:4-1 for common definitions. Most robots are set up for an operation by the teach-and-repeat technique. In this mode, a trained operator (programmer) typically uses a portable control device (a teach pendant) to teach a robot its task manually. Robot speeds during these programming sessions are slow.

This instruction includes safety considerations necessary to operate the robot properly and use it automatically in conjunction with other peripheral equipment. This instruction applies to fixed industrial robots and robot systems only. See Appendix IV:4-2 for the systems that are excluded.

A. ACCIDENTS: PAST STUDIES.

1. Studies in Sweden and Japan indicate that many robot accidents do not occur under normal operating conditions but, instead during programming, program touch-up or refinement, maintenance, repair, testing, setup, or adjustment. During many of these operations the operator, programmer, or corrective maintenance worker may temporarily be within the robot's working envelope where unintended operations could result in injuries.

2. Typical accidents have included the following:

   - A robot's arm functioned erratically during a programming sequence and struck the operator.
   - A materials handling robot operator entered a robot's work envelope during operations and was pinned between the back end of the robot and a safety pole.
   - A fellow employee accidentally tripped the power switch while a maintenance worker was servicing an assembly robot. The robot's arm struck the maintenance worker's hand.

B. ROBOT SAFEGUARDING.

1. The proper selection of an effective robotic safeguarding system should be based upon a hazard analysis of the robot system's use, programming, and maintenance operations. Among the factors to be considered are the tasks a robot will be programmed to perform, start-up and command or programming procedures, environmental conditions, location and installation requirements, possible human errors, scheduled and unscheduled

maintenance, possible robot and system malfunctions, normal mode of operation, and all personnel functions and duties.

2. An effective safeguarding system protects not only operators but also engineers, programmers, maintenance personnel, and any others who work on or with robot systems and could be exposed to hazards associated with a robot's operation. A combination of safeguarding methods may be used. Redundancy and backup systems are especially recommended, particularly if a robot or robot system is operating in hazardous conditions or handling hazardous materials. The safeguarding devices employed should not themselves constitute or act as a hazard or curtail necessary vision or viewing by attending human operators.

II. TYPES AND CLASSIFICATION OF ROBOTS.

Industrial robots are available commercially in a wide range of sizes, shapes, and configurations. They are designed and fabricated with different design configurations and a different number of axes or degrees of freedom. These factors of a robot's design influence its working envelope (the volume of working or reaching space). Diagrams of the different robot design configurations are shown in Figure IV: 4-1.

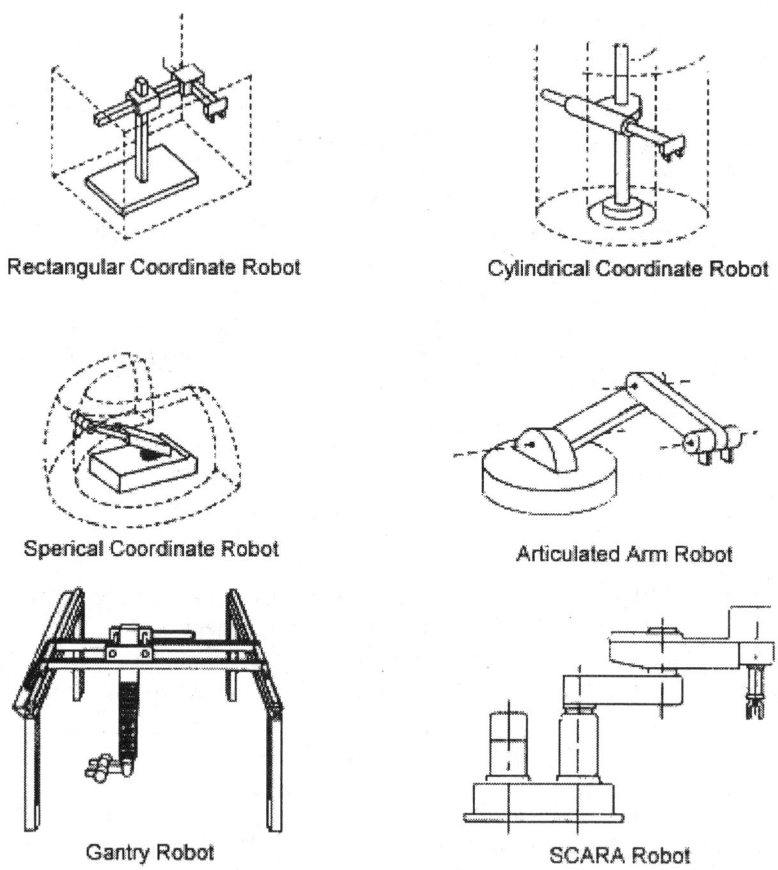

FIGURE IV: 4-1. ROBOT ARM DESIGN CONFIGURATIONS

A. SERVO AND NONSERVO.

All industrial robots are either servo or nonservo controlled. Servo robots are controlled through the use of sensors that continually monitor the robot's axes and associated components for position and velocity. This feedback is compared to pretaught information which has been programmed and stored in the robot's memory. Nonservo robots do not have the feedback capability, and their axes are controlled through a system of mechanical stops and limit switches.

B. TYPE OF PATH GENERATED. Industrial robots can be programmed from a distance to perform their required and preprogrammed operations with different types of paths generated through different control techniques. The three different types of paths generated are Point-to-Point Path, Controlled Path, and Continuous Path.

1. Point-to-Point Path. Robots programmed and controlled in this manner are programmed to move from one discrete point to another within the robot's working envelope. In the automatic mode of operation, the exact path taken by the robot will vary slightly due to variations in velocity, joint geometries, and point spatial locations. This difference in paths is difficult to predict and therefore can create a potential safety hazard to personnel and equipment.

2. Controlled Path. The path or mode of movement ensures that the end of the robot's arm will follow a predictable (controlled) path and orientation as the robot travels from point to point. The coordinate transformations required for this hardware management are calculated by the robot's control system computer. Observations that result from this type of programming are less likely to present a hazard to personnel and equipment.

3. Continuous Path. A robot whose path is controlled by storing a large number or close succession of spatial points in memory during a teaching sequence is a continuous path controlled robot. During this time, and while the robot is being moved, the coordinate points in space of each axis are continually monitored on a fixed time base, e.g., 60 or more times per second, and placed into the control system's computer memory. When the robot is placed in the automatic mode of operation, the program is replayed from memory and a duplicate path is generated.

C. ROBOT COMPONENTS. Industrial robots have four major components: the mechanical unit, power source, control system, and tooling (Figure IV: 4-2).

1. Mechanical Unit. The robot's manipulative arm is the mechanical unit. This mechanical unit is also comprised of a fabricated structural frame with provisions for supporting mechanical linkage and joints, guides, actuators (linear or rotary), control valves, and sensors. The physical dimensions, design, and weight-carrying ability depend on application requirements.

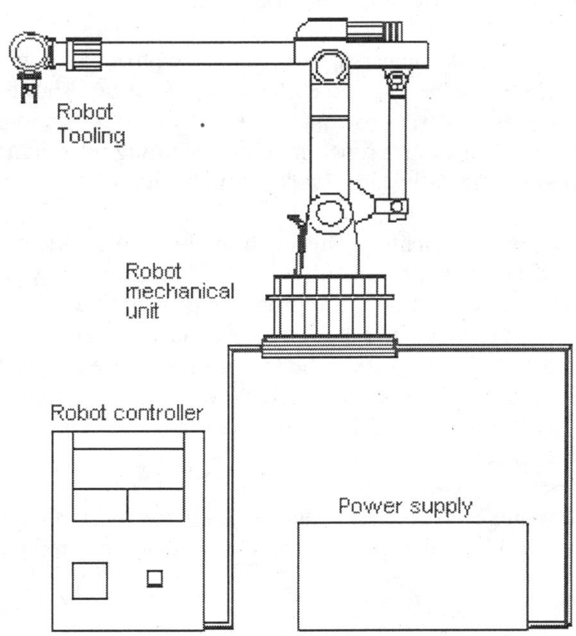

FIGURE IV: 4-2. INDUSTRIAL ROBOTS: MAJOR COMPONENTS.

2. Power Sources.

   a. Energy is provided to various robot actuators and their controllers as pneumatic, hydraulic, or electrical power. The robot's drives are usually mechanical combinations powered by these types of energy, and the selection is usually based upon application requirements. For example, pneumatic power (low-pressure air) is used generally for low weight carrying robots.

   b. Hydraulic power transmission (high-pressure oil) is usually used for medium to high force or weight applications, or where smoother motion control can be achieved than with pneumatics. Consideration should be given to potential hazards of fires from leaks if petroleum-based oils are used.

   c. Electrically powered robots are the most prevalent in industry. Either AC or DC electrical power is used to supply energy to electromechanical motor-driven actuating mechanisms and their respective control systems. Motion control is much better, and in an emergency an electrically powered robot can be stopped or powered down more safely and faster than those with either pneumatic or hydraulic power.

D. CONTROL SYSTEMS.

1. Either auxiliary computers or embedded microprocessors are used for practically all control of industrial robots today. These perform all of the required computational functions as well as interface with and control associated sensors, grippers, tooling, and other associated peripheral equipment. The control system performs the necessary sequencing and memory functions for on-line sensing, branching, and integration of other equipment. Programming of the controllers can be done on-line or at remote off-line control stations with electronic data transfer of programs by cassette, floppy disc, or telephone modem.

2. Self-diagnostic capability for troubleshooting and maintenance greatly reduces robot system downtime. Some robot controllers have sufficient capacity, in terms of computational ability, memory capacity, and input-output capability to serve also as system controllers and handle many other machines and processes. Programming of robot controllers and systems has not been standardized by the robotics industry; therefore, the manufacturers use their own proprietary programming languages which require special training of personnel.

E. ROBOT PROGRAMMING BY TEACHING METHODS. A program consists of individual command steps which state either the position or function to be performed, along with other informational data such as speed, dwell or delay times, sample input device, activate output device, execute, etc.

When establishing a robot program, it is necessary to establish a physical or geometrical relationship between the robot and other equipment or work to be serviced by the robot. To establish these coordinate points precisely within the robot's working envelope, it is necessary to control the robot manually and physically teach the coordinate points. To do this as well as determine other functional programming information, three different teaching or programming techniques are used: lead-through, walk-through, and off-line.

1. Lead-Through Programming or Teaching. This method of teaching uses a proprietary teach pendant (the robot's control is placed in a "teach" mode), which allows trained personnel physically to lead the robot through the desired sequence of events by activating the appropriate pendant button or switch. Position data and functional information are "taught" to the robot, and a new program is written (Figure IV:4-3). The teach pendant can be the sole source by which a program is established, or it may be used in conjunction with an additional programming console and/or the robot's controller. When using this technique of teaching or programming, the person performing the teach function can be within the robot's working envelope, with operational safeguarding devices deactivated or inoperative.

2. Walk-Through Programming or Teaching. A person doing the teaching has physical contact with the robot arm and actually gains control and walks the robot's arm through the desired positions within the working envelope (Figure IV:4-4).

FIGURE IV: 4-3. ROBOT LEAD-THROUGH PROGRAMMING OR TEACHING.

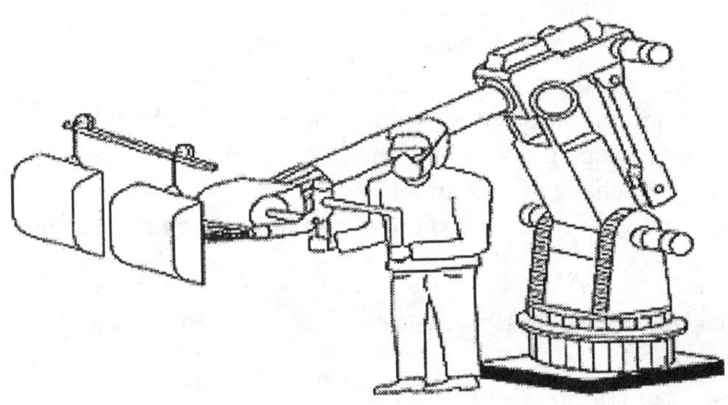

FIGURE IV: 4-4. WALK-THROUGH PROGRAMMING OR TEACHING.

During this time, the robot's controller is scanning and storing coordinate values on a fixed time basis. When the robot is later placed in the automatic mode of operation, these values and other functional information are replayed and the program run as it was taught. With the walk-through method of programming, the person doing the teaching is in a potentially hazardous position because the operational safeguarding devices are deactivated or inoperative.

3. Off-Line Programming. The programming establishing the required sequence of functional and required positional steps is written on a remote computer console (Figure IV:4-5). Since the console is distant from the robot and its controller, the written program has to be transferred to the robot's controller and precise positional data established to achieve the actual coordinate information for the robot and other equipment. The program can be transferred directly or by cassette or floppy discs. After the program has been completely transferred to the robot's controller, either the lead-through or walk-through technique can be used for obtaining actual positional coordinate information for the robot's axes.

When programming robots with any of the three techniques discussed above, it is generally required that the program be verified and slight modifications in positional information made. This procedure is called program touch-up and is normally carried out in the teach mode of operation. The teacher manually leads or walks the robot through the programmed steps. Again, there are potential hazards if safeguarding devices are deactivated or inoperative.

4. DEGREES OF FREEDOM. Regardless of the configuration of a robot, movement along each axis will result in either a rotational or a translational movement. The number of axes of movement (degrees of freedom) and their arrangement, along with their sequence of operation and structure, will permit movement of the robot to any point within its envelope. Robots have three arm movements (up-down, in-out, side-to-side). In addition, they can have as many as three additional wrist movements on the end of the robot's arm: yaw (side to side), pitch (up and down), and rotational (clockwise and counterclockwise).

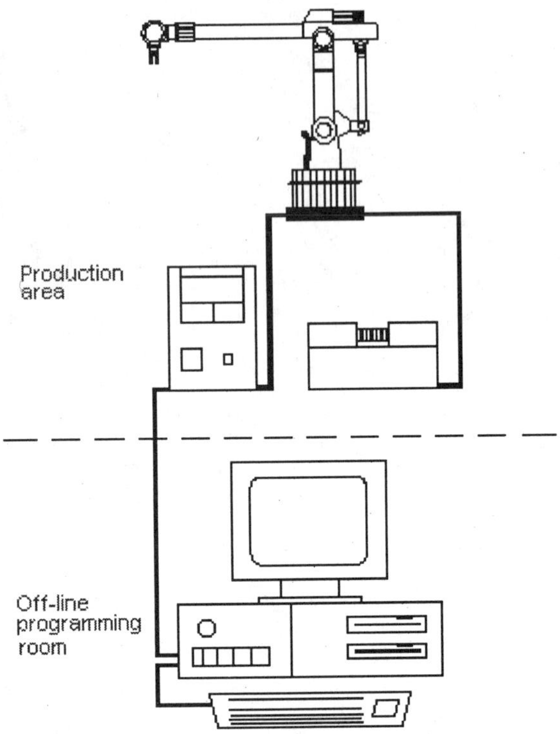

FIGURE IV: 4-5. OFF-LINE PROGRAMMING OR TEACHING.

III. HAZARDS.

The operational characteristics of robots can be significantly different from other machines and equipment. Robots are capable of high-energy (fast or powerful) movements through a large volume of space even beyond the base dimensions of the robot (see Figure IV:4-6). The pattern and initiation of movement of the robot is predictable if the item being "worked" and the environment are held constant. Any change to the object being worked (i.e., a physical model change) or the environment can affect the programmed movements.

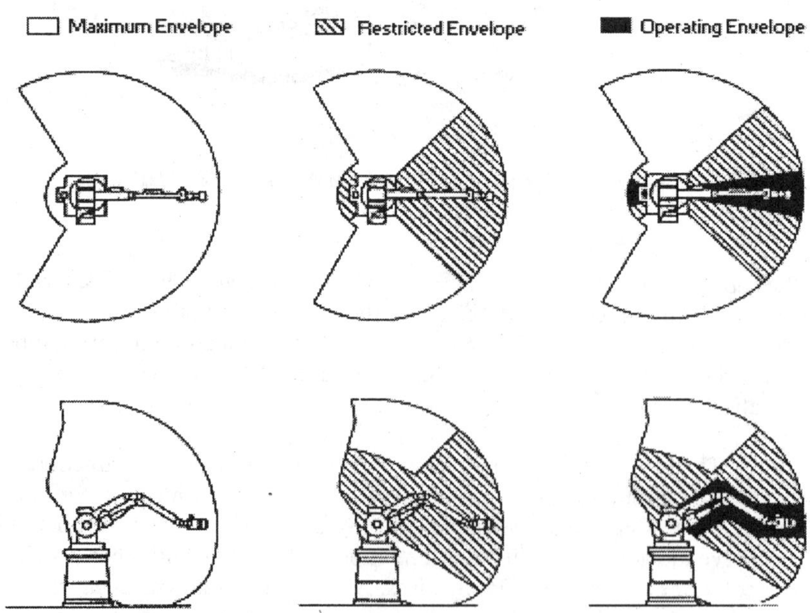

FIGURE IV: 4-6. A ROBOT'S WORK ENVELOPE.

Some maintenance and programming personnel may be required to be within the restricted envelope while power is available to actuators. The restricted envelope of the robot can overlap a portion of the restricted envelope of other robots or work zones of other industrial machines and related equipment. Thus, a worker can be hit by one robot while working on another, trapped between them or peripheral equipment, or hit by flying objects released by the gripper.

A robot with two or more resident programs can find the current operating program erroneously calling another existing program with different operating parameters such as velocity, acceleration, or deceleration, or position within the robot's restricted envelope. The occurrence of this might not be predictable by maintenance or programming personnel working with the robot. A component malfunction could also cause an unpredictable movement and/or robot arm velocity.

Additional hazards can also result from the malfunction of, or errors in, interfacing or programming of other process or peripheral equipment. The operating changes with the process being performed or the breakdown of conveyors, clamping mechanisms, or process sensors could cause the robot to react in a different manner.

A. TYPES OF ACCIDENTS. Robotic incidents can be grouped into four categories: a robotic arm or controlled tool causes the accident, places an individual in a risk circumstance, an accessory of the robot's mechanical parts fails, or the power supplies to the robot are uncontrolled.

1. Impact or Collision Accidents. Unpredicted movements, component malfunctions, or unpredicted program changes related to the robot's arm or peripheral equipment can result in contact accidents.

2. Crushing and Trapping Accidents. A worker's limb or other body part can be trapped between a robot's arm and other peripheral equipment, or the individual may be physically driven into and crushed by other peripheral equipment.

3. Mechanical Part Accidents. The breakdown of the robot's drive components, tooling or end-effector, peripheral equipment, or its power source is a mechanical accident. The release of parts, failure of gripper mechanism, or the failure of end-effector power tools (e.g., grinding wheels, buffing wheels, deburring tools, power screwdrivers, and nut runners) are a few types of mechanical failures.

4. Other Accidents. Other accidents can result from working with robots. Equipment that supplies robot power and control represents potential electrical and pressurized fluid hazards. Ruptured hydraulic lines could create dangerous high-pressure cutting streams or whipping hose hazards. Environmental accidents from arc flash, metal spatter, dust, electromagnetic, or radio-frequency interference can also occur. In addition, equipment and power cables on the floor present tripping hazards.

B. SOURCES OF HAZARDS. The expected hazards of machine to humans can be expected with several additional variations, as follows.

1. Human Errors. Inherent prior programming, interfacing activated peripheral equipment, or connecting live input-output sensors to the microprocessor or a peripheral can cause dangerous, unpredicted movement or action by the robot from human error. The incorrect activation of the "teach pendant" or control panel is a frequent human error. The greatest problem, however, is overfamiliarity with the robot's redundant motions so that an individual places himself in a hazardous position while programming the robot or performing maintenance on it.

2. Control Errors. Intrinsic faults within the control system of the robot, errors in software, electromagnetic interference, and radio frequency interference are control errors. In addition, these errors can occur due to faults in the hydraulic, pneumatic, or electrical subcontrols associated with the robot or robot system.

3. Unauthorized Access. Entry into a robot's safeguarded area is hazardous because the person involved may not be familiar with the safeguards in place or their activation status.

4. Mechanical Failures. Operating programs may not account for cumulative mechanical part failure, and faulty or unexpected operation may occur.

5. Environmental Sources. Electromagnetic or radio-frequency interference (transient signals) should be considered to exert an undesirable influence on robotic operation and increase the potential for injury to any person working in the area. Solutions to environmental hazards should be documented prior to equipment start-up.

6. Power Systems. Pneumatic, hydraulic, or electrical power sources that have malfunctioning control or transmission elements in the robot power system can disrupt electrical signals to the control and/or power-supply lines. Fire risks are increased by electrical overloads or by use of flammable hydraulic oil. Electrical shock and release of stored energy from accumulating devices also can be hazardous to personnel.

7. Improper Installation. The design, requirements, and layout of equipment, utilities, and facilities of a robot or robot system, if inadequately done, can lead to inherent hazards.

IV. INVESTIGATION GUIDELINES.

A. MANUFACTURED, REMANUFACTURED, AND REBUILT ROBOTS.

1. All robots should meet minimum design requirements to ensure safe operation by the user. Consideration needs to be given to a number of factors in designing and building the robots to industry standards. If older or obsolete robots are rebuilt or remanufactured, they should be upgraded to conform to current industry standards.

2. Every robot should be designed, manufactured, remanufactured, or rebuilt with safe design and manufacturing considerations. Improper design and manufacture can result in hazards to personnel if minimum industry standards are not conformed to on mechanical components, controls, methods of operation, and other required information necessary to insure safe and proper operating procedures. To ensure that robots are designed, manufactured, remanufactured, and rebuilt to ensure safe operation, it is recommended that they comply with Section 4 of the ANSI/RIA R15.06-1992 standard for Manufacturing, Remanufacture, and Rebuild of Robots.

B. INSTALLATION.

1. A robot or robot system should be installed by the users in accordance with the manufacturer's recommendations and in conformance to acceptable industry standards. Temporary safeguarding devices and practices should be used to minimize the hazards associated with the installation of new equipment. The facilities, peripheral equipment, and operating conditions which should be considered are:

   - Installation specifications;
   - Physical facilities;
   - Electrical facilities;
   - Action of peripheral equipment integrated with the robot;
   - Identification requirements;
   - Control and emergency stop requirements; and
   - Special robot operating procedures or conditions.

2. To ensure safe operating practices and safe installation of robots and robot systems, it is recommended that the minimum requirements of Section 5 of the ANSI/RIA R15.06-1992, Installation of Robots and Robot Systems be followed. In addition, OSHA's Lockout/Tagout standards (29 CFR 1910.147 and 1910.333) must be followed for servicing and maintenance.

V. CONTROL AND SAFEGUARDING PERSONNEL.

For the planning stage, installation, and subsequent operation of a robot or robot system, one should consider the following.

A. RISK ASSESSMENT. At each stage of development of the robot and robot system a risk assessment should be performed. There are different system and personnel safeguarding requirements at each stage. The appropriate level of safeguarding determined by the risk assessment should be applied. In addition, the risk assessments for each stage of development should be documented for future reference.

B. SAFEGUARDING DEVICES. Personnel should be safeguarded from hazards associated with the restricted envelope (space) through the use of one or more safeguarding devices:

- Mechanical limiting devices;
- Nonmechanical limiting devices;
- Presence-sensing safeguarding devices;
- Fixed barriers (which prevent contact with moving parts); and
- Interlocked barrier guards.

C. AWARENESS DEVICES. Typical awareness devices include chain or rope barriers with supporting stanchions or flashing lights, signs, whistles, and horns. They are usually used in conjunction with other safeguarding devices.

D. SAFEGUARDING THE TEACHER. Special consideration must be given to the teacher or person who is programming the robot. During the teach mode of operation, the person performing the teaching has control of the robot and associated equipment and should be familiar with the operations to be programmed, system interfacing, and control functions of the robot and other equipment. When systems are large and complex, it can be easy to activate improper functions or sequence functions improperly. Since the person doing the training can be within the robot's restricted envelope, such mistakes can result in accidents. Mistakes in programming can result in unintended movement or actions with similar results. For this reason, a restricted speed of 250 mm/sec. or 10 in/sec. should be placed on any part of the robot during training to minimize potential injuries to teaching personnel.

Several other safeguards are suggested in the ANSI/RIA R15.06-1992 standard to reduce the hazards associated with teaching a robotic system.

E. OPERATOR SAFEGUARDS. The system operator should be protected from all hazards during operations performed by the robot. When the robot is operating automatically, all safeguarding devices should be activated, and at no time should any part of the operator's body be within the robot's safeguarded area.

For additional operator safeguarding suggestions, see the ANSI/RIA R15.06-1992 standard, Section 6.6.

F. ATTENDED CONTINUOUS OPERATION. When a person is permitted to be in or near the robots restricted envelope to evaluate or check the robots motion or other operations, all continuous operation safeguards must be in force. During this operation, the robot should be at slow speed, and the operator would have the robot in the teach mode and be fully in control of all operations.

Other safeguarding requirements are suggested in the ANSI/RIA R15.06-1992 standard, Section 6.7.

G. MAINTENANCE AND REPAIR PERSONNEL. Safeguarding maintenance and repair personnel is very difficult because their job functions are so varied. Troubleshooting faults or problems with the robot, controller, tooling, or other associated equipment is just part of their job. Program touchup is another of their jobs as is scheduled maintenance, and adjustments of tooling, gages, recalibration, and many other types of functions.

While maintenance and repair is being performed, the robot should be placed in the manual or teach mode, and the maintenance personnel perform their work within the safeguarded area and within the robots restricted envelope. Additional hazards are present during this mode of operation because the robot system safeguards are not operative.

To protect maintenance and repair personnel, safeguarding techniques and procedures as stated in the ANSI/RIA R15.06-1992 standard, Section 6.8, are recommended.

H. MAINTENANCE. Maintenance should occur during the regular and periodic inspection program for a robot or robot system. An inspection program should include, but not be limited to, the recommendations of the robot manufacturer and manufacturer of other associated robot system equipment such as conveyor mechanisms, parts feeders, tooling, gages, sensors, and the like.

These recommended inspection and maintenance programs are essential for minimizing the hazards from component malfunction, breakage, and unpredicted movements or actions by the robot or other system equipment. To ensure proper maintenance, it is recommended that periodic maintenance and inspections be documented along with the identity of personnel performing these tasks.

I. SAFETY TRAINING. Personnel who program, operate, maintain, or repair robots or robot systems should receive adequate safety training, and they should be able to demonstrate their competence to perform their jobs safely. Employers can refer to OSHA's publication 2254 (Revised), "Training Requirements in OSHA Standards and Training Guidelines."

J. GENERAL REQUIREMENTS. To ensure minimum safe operating practices and safeguards for robots and robot systems covered by this instruction, the following sections of the ANSI/RIA R15.06-1992 must also be considered:

- Section 6 - Safeguarding Personnel;
- Section 7 - Maintenance of Robots and Robot Systems;
- Section 8 - Testing and Start-up of Robots and Robot Systems; and
- Section 9 - Safety Training of Personnel.

Robots or robotic systems must comply with the following regulations: Occupational Safety and Health Administration, OSHA 29 CFR 1910.333, Selection and Use of Work Practices, and OSHA 29 CFR Part 1910.147, The Control of Hazardous Energy (Lockout/Tagout).

VI. BIBLIOGRAPHY.

American National Standards Institute (ANSI) American National Safety Standard ANSI/ RIA R15.06-1992. *Industrial Robots and Robot Systems - Safety Requirements.* American National Standards Institute, Inc., 1430 Broadway, New York, New York 10018

National Institute for Occupational Safety and Health (NIOSH) Alert Publication No. 85103. *Request for Assistance in Preventing the Injury of Workers by Robots.* National Institute for Occupational Safety and Health, Division of Safety Research, 944 Chestnut Ridge Road, Morgantown, West Virginia 26505

National Institute for Occupational Safety and Health (NIOSH) Technical Report Publication No. 880108. *Safe Maintenance Guidelines for Robotic Workstations.* National Institute for Occupational Safety and Health, Division of Safety Research, 944 Chestnut Ridge Road, Morgantown, West Virginia 26505

National Safety Council Data Sheet 1-717-85. *Robots.* National Safety Council, 444 N. Michigan Avenue, Chicago, Illinois 60611

Occupational Safety and Health Administration Publication No. 2254 (Revised). *Training Requirements in OSHA Standards and Training Guidelines.* Superintendent of Documents, U.S. Government Printing Office, Washington, D.C. 20210

Occupational Safety and Health Administration Publication No. 3067. *Concepts and Techniques of Machine Safeguarding.* U.S. Department of Labor, 1980 (reprinted 1983). Superintendent of Documents, U.S. Government Printing Office, Washington, D.C. 20210

OSHA Instruction Publication No. 8-1.3. 1987. *Guideline for Robotics Safety.* Occupational Safety and Health Administration, Washington, D.C.

Robotic Industries Association, 900 Victors Way, P.O. Box 3724, Ann Arbor, Michigan 48106

Underwriters Laboratories, Inc. 1996. *Standard for Industrial Robots and Robotic Equipment.* ANSI/UL1740. Northbrook, Illinois

APPENDIX IV:4-1. GLOSSARY FOR ROBOTICS AND ROBOTIC SYSTEMS.

Actuator A power mechanism used to effect motion of the robot; a device that converts electrical, hydraulic, or pneumatic energy into robot motion.

Application Program The set of instructions that defines the specific intended tasks of robots and robot systems. This program may be originated and modified by the robot user.

Attended Continuous Operation The time when robots are performing (production) tasks at a speed no greater than slow speed through attended program execution.

Attended Program Verification The time when a person within the restricted envelope (space) verifies the robot's programmed tasks at programmed speed.

Automatic Mode The robot state in which automatic operation can be initiated.

Automatic Operation The time during which robots are performing programmed tasks through unattended program execution.

Awareness Barrier Physical and/or visual means that warns a person of an approaching or present hazard.

Awareness Signal A device that warns a person of an approaching or present hazard by means of audible sound or visible light.

Axis The line about which a rotating body (such as a tool) turns.

Barrier A physical means of separating persons from the restricted envelope (space).

Control Device Any piece of control hardware providing a means for human intervention in the control of a robot or robot system, such as an emergency-stop button, a start button, or a selector switch.

Control Program The inherent set of control instructions that defines the capabilities, actions and responses of the robot system. This program is usually not intended to be modified by the user.

Coordinated Straight Line Motion Control wherein the axes of the robot arrive at their respective end points simultaneously, giving a smooth appearance to the motion. Control wherein the motions of the axes are such that the Tool Center Point (TCP) moves along a prespecified type of path (line, circle, etc.)

Device Any piece of control hardware such as an emergency-stop button, selector switch, control pendant, relay, solenoid valve, sensor, etc.

Drive Power The energy source or sources for the robot actuators.

Emergency Stop The operation of a circuit using hardware-based components that overrides all other robot controls, removes drive power from the robot actuators, and causes all moving parts to stop.

Enabling Device A manually operated device that permits motion when continuously activated. Releasing the device stops robot motion and motion of associated equipment that may present a hazard.

End-effector An accessory device or tool specifically designed for attachment to the robot wrist or tool mounting plate to enable the robot to perform its intended task. (Examples may include gripper, spot-weld gun, arc-weld gun, spray-paint gun, or anyother application tools.)

Energy Source Any electrical, mechanical, hydraulic, pneumatic, chemical, thermal, or other source.

Envelope (Space), Maximum The volume of space encompassing the maximum designed movements of all robot parts including the end-effector, workpiece, and attachments.

Restricted Envelope (Space) That portion of the maximum envelope to which a robot is restricted by limiting devices. The maximum distance that the robot can travel after the limiting device is actuated defines the boundaries of the restricted envelope (space) of the robot.

NOTE: The safeguarding interlocking logic and robot program may redefine the restricted envelope (space) as the robot performs its application program. (See Appendix D of the ANSI/RIA R15.06-1992 Specification).

Operating Envelope (Space) That portion of the restricted envelope (space) that is actually used by the robot while performing its programmed motions.

Hazard A situation that is likely to cause physical harm.

Hazardous Motion Any motion that is likely to cause personal physical harm.

Industrial Equipment Physical apparatus used to perform industrial tasks, such as welders, conveyors, machine tools, fork trucks, turn tables, positioning tables, or robots.

Industrial Robot A reprogrammable, multifunctional manipulator designed to move material, parts, tools, or specialized devices through variable programmed motions for the performance of a variety of tasks.

Industrial Robot System A system that includes industrial robots, the end-effectors, and the devices and sensors required for the robots to be taught or programmed, or for the robots to perform the intended automatic operations, as well as the communication interfaces required for interlocking, sequencing, or monitoring the robots.

Interlock An arrangement whereby the operation of one control or mechanism brings about or prevents the operation of another.

Joint Motion A method for coordinating the movement of the joints such that all joints arrive at the desired location simultaneously.

Limiting Device A device that restricts the maximum envelope (space) by stopping or causing to stop all robot motion and is independent of the control program and the application programs.

Maintenance The act of keeping the robots and robot systems in their proper operating condition.

Mobile Robot A self-propelled and self-contained robot that is capable of moving over a mechanically unconstrained course.

Muting The deactivation of a presence-sensing safeguarding device during a portion of the robot cycle.

Operator The person designated to start, monitor, and stop the intended productive operation of a robot or robot system. An operator may also interface with a robot for productive purposes.

Pendant Any portable control device, including teach pendants, that permits an operator to control the robot from within the restricted envelope (space) of the robot.

Presence-Sensing Safeguarding Device A device designed, constructed, and installed to create a sensing field or area to detect an intrusion into the field or area by personnel, robots, or other objects.

Program

1.(noun) A sequence of instructions to be executed by the computer or robot controller to control a robot or robot system. 2.(verb) to furnish (a computer) with a code of instruction. 3.(verb) to teach a robot system a specific set of movements and instructions to accomplish a task.

Rebuild To restore the robot to the original specifications of the manufacturer, to the extent possible.

Remanufacture To upgrade or modify robots to the revised specifications of the manufacturer and applicable industry standards.

Repair To restore robots and robot systems to operating condition after damage, malfunction, or wear.

Robot Manufacturer A company or business involved in either the design, fabrication, or sale of robots, robot tooling, robotic peripheral equipment or controls, and associated process ancillary equipment.

Robot System Integrator A company or business who either directly or through a subcontractor will assume responsibility for the design, fabrication, and integration of the required robot, robotic peripheral equipment, and other required ancillary equipment for a particular robotic application.

Safeguard A barrier guard, device, or safety procedure designed for the protection of personnel.

Safety Procedure An instruction designed for the protection of personnel.

Sensor A device that responds to physical stimuli (such as heat, light, sound, pressure, magnetism, motion, etc.) and transmits the resulting signal or data for providing a measurement, operating a control, or both.

Service To adjust, repair, maintain, and make fit for use.

Single Point of Control The ability to operate the robot such that initiation or robot motion from one source of control is possible only from that source and cannot be overridden from another source.

Slow Speed Control A mode of robot motion control where the velocity of the robot is limited to allow persons sufficient time either to withdraw the hazardous motion or stop the robot.

Start-up Routine application of drive power to the robot or robot system.

Start-up, Initial Initial drive power application to the robot or robot syst Teach The generation and storage of a series of positional data points effected by moving the robot arm through a path of intended motions.

Teach Mode The control state that allows the generation and storage of positional data points effected by moving the robot arm through a path of intended motions.

Teacher A person who provides the robot with a specific set of instructions to perform a task.

Tool Center Point (TCP) The origin of the tool coordinate system.

User A company, business, or person who uses robots and who contracts, hires, or is responsible for the personnel associated with robot operation.

APPENDIX IV: 4-2. OTHER ROBOTIC SYSTEMS NOT COVERED BY THIS CHAPTER.

Service robots are machines that extend human capabilities.

Automatic guided-vehicle systems are advanced material-handling or conveying systems that involve a driverless vehicle which follows a guide-path.

Undersea and space robots include in addition to the manipulator or tool that actually accomplishes a task, the vehicles or platforms that transport the tools to the site. These vehicles are called remotely operated vehicles (ROV's) or autonomous undersea vehicles (AUV's); the feature that distinguishes them is, respectively, the presence or absence of an electronics tether that connects the vehicle and surface control station.

Automatic storage and retrieval systems are storage racks linked through automatically controlled conveyors and an automatic storage and retrieval machine or machines that ride on floor-mounted guide rails and power-driven wheels.

Automatic conveyor and shuttle systems are comprised of various types of conveying systems linked together with various shuttle mechanisms for the prime purpose of conveying materials or parts to prepositioned and predetermined locations automatically.

Teleoperators are robotic devices comprised of sensors and actuators for mobility and/or manipulation and are controlled remotely by a human operator.

Mobile robots are freely moving automatic programmable industrial robots.

Prosthetic robots are programmable manipulators or devices for missing human limbs.

Numerically controlled machine tools are operated by a series of coded instructions comprised of numbers, letters of the alphabet, and other symbols. These are translated into pulses of electrical current or other output signals that activate motors and other devices to run the machine. em after one of the following events:

- Manufacture or modification;
- Installation or reinstallation;
- Programming or program editing; and
- Maintenance or repair

SECTION V: CHAPTER 1

DEMOLITION

Contents:

I. PREPARATORY OPERATIONS.

Before the start of every demolition job, the demolition contractor should take a number of steps to safeguard the health and safety of workers at the job site. These preparatory operations involve the overall planning of the demolition job, including the methods to be used to bring the structure down, the equipment necessary to do the job, and the measures to be taken to perform the work safely. Planning for a demolition job is as important as actually doing the work. Therefore, all planning work should be performed by a competent person experienced in all phases of the demolition work to be performed.

The American National Standards Institute (ANSI) in its ANSI A10.6-1983 - Safety Requirements For Demolition Operations states:

"No employee shall be permitted in any area that can be adversely affected when demolition operations are being performed. Only those employees necessary for the performance of the operations shall be permitted in these areas."

A. ENGINEERING SURVEY.

1. Prior to starting all demolition operations, OSHA Standard 1926.850(a) requires that an engineering survey of the structure must be conducted by a competent person. The purpose of this survey is to determine the condition of the framing, floors, and walls so that measures can be taken, if necessary, to prevent the premature collapse of any portion of the structure. When indicated as advisable, any adjacent structure(s) or improvements should also be similarly checked. The demolition contractor must maintain a written copy of this survey. Photographing existing damage in neighboring structures is also advisable.

2. The engineering survey provides the demolition contractor with the opportunity to evaluate the job in its entirety. The contractor should plan for the wrecking of the structure, the equipment to do the work, manpower requirements, and the protection of the public. The safety of all workers on the job site should be a prime consideration. During the preparation of the engineering survey, the contractor should plan for potential hazards such as fires, cave-ins, and injuries.

3. If the structure to be demolished has been damaged by fire, flood, explosion, or some other cause, appropriate measures, including bracing and shoring of walls and floors, shall be taken to protect workers and any adjacent structures. It shall also be determined if any type of hazardous chemicals, gases, explosives, flammable material, or similar dangerous substances have been used or stored on the site. If the nature of a substance cannot be easily determined, samples should be taken and analyzed by a qualified person prior to demolition.

4. During the planning stage of the job, all safety equipment needs should be determined. The required number and type of respirators, lifelines, warning signs, safety nets, special face and eye protection, hearing protection, and other worker protection devices outlined in this manual should be determined during the preparation of the engineering survey. A comprehensive plan is necessary for any confined space entry.

B. UTILITY LOCATION.

1. One of the most important elements of the pre-job planning is the location of all utility services. All electric, gas, water, steam, sewer, and other services lines should be shut off, capped, or otherwise controlled, at or outside the building before demolition work is started. In each case, any utility company that is involved should be notified in advance, and its approval or services, if necessary, shall be obtained.

2. If it is necessary to maintain any power, water, or other utilities during demolition, such lines shall be temporarily relocated as necessary and/or protected. The location of all overhead power sources should also be determined, as they can prove especially hazardous during any machine demolition. All workers should be informed of the location of any existing or relocated utility service.

C. MEDICAL SERVICES AND FIRST AID.

1. Prior to starting work, provisions should be made for prompt medical attention in case of serious injury. The nearest hospital, infirmary, clinic, or physician shall be located as part of the engineering survey. The job supervisor should be provided with instructions for the most direct route to these facilities. Proper equipment for prompt transportation of an injured worker, as well as a communication system to contact any necessary ambulance service, must be available at the job site. The telephone numbers of the hospitals, physicians, or ambulances shall be conspicuously posted.

2. In the absence of an infirmary, clinic, hospital, or physician that is reasonably accessible in terms of time and distance to the work site, a person who has a valid certificate in first aid training from the U.S. Bureau of Mines, the American Red Cross, or equivalent training should be available at the work site to render first aid.

3. A properly stocked first aid kit as determined by an occupational physician, must be available at the job site. The first aid kit should contain approved supplies in a weatherproof container with individual sealed packages for each type of item. It should also include rubber gloves to prevent the transfer of infectious diseases. Provisions should also be made to provide for quick drenching or flushing of the eyes should any person be working around corrosive materials. Eye flushing must be done with water containing no additives. The contents of the kit shall be checked before being sent out on each job and at least weekly to ensure the expended items are replaced.

D. POLICE AND FIRE CONTACT. The telephone numbers of the local police, ambulance, and fire departments should be available at each job site. This information can prove useful to the job supervisor in the event of any traffic problems, such as the movement of equipment to the job, uncontrolled fires, or ther police/fire matters. The police number may also be used to report any vandalism, unlawful entry to the job site, or accidents requiring police assistance.

E. FIRE PREVENTION AND PROTECTION.

1. A "fire plan" should be set up prior to beginning a demolition job. This plan should outline the assignments of key personnel in the event of a fire and provide an evacuation plan for workers on the site. Common sense should be the general rule in all fire prevention planning, as follows:

   - All potential sources of ignition should be evaluated and the necessary corrective measures taken.
   - Electrical wiring and equipment for providing light, heat, or power should be installed by a competent person and inspected regularly.
   - Equipment powered by an internal combustion engine should be located so that the exhausts discharge well away from combustible materials and away from workers.
   - When the exhausts are piped outside the building, a clearance of at least six inches should be maintained between such piping and combustible material.
   - All internal combustion equipment should be shut down prior to refueling. Fuel for this equipment should be stored in a safe location.
   - Sufficient firefighting equipment should be located near any flammable or combustible liquid storage area.
   - Only approved containers and portable tanks should be used for the storage and handling of flammable and combustible liquids.

2. Heating devices should be situated so that they are not likely to overturn and shall be installed in accordance with their listing, including clearance to combustible material or equipment. Temporary heating equipment, when utilized, should be maintained by competent personnel.

3. Smoking should be prohibited at or in the vicinity of hazardous operations or materials. Where smoking is permitted, safe receptacles shall be provided for smoking materials.

4. Roadways between and around combustible storage piles should be at least 15 feet wide and maintained free from accumulation of rubbish, equipment, or other materials. When storing debris or combustible material inside a structure, such storage shall not obstruct or adversely affect the means of exit.

5. A suitable location at the job site should be designated and provided with plans, emergency information, and equipment, as needed. Access for heavy fire-fighting equipment should be provided on the immediate job site at the start of the job and maintained until the job is completed.

6. Free access from the street to fire hydrants and to outside connections for standpipes, sprinklers, or other fire extinguishing equipment, whether permanent or temporary, should be provided and maintained at all times, as follows:

   - Pedestrian walkways should not be so constructed as to impede access to hydrants.
   - No material or construction should interfere with access to hydrants, Siamese connections, or fire-extinguishing equipment.

7. A temporary or permanent water supply of volume, duration, and pressure sufficient to operate the fire-fighting equipment properly should be made available. Standpipes with outlets should be provided on large multi story buildings to provide for fire protection on upper levels. If the water pressure is insufficient, a pump should also be provided.

8. An ample number of fully charged portable fire extinguishers should be provided throughout the operation. All motor-driven mobile equipment should be equipped with an approved fire extinguisher.

9. An alarm system, e.g., telephone system, siren, two-way radio, etc., shall be established in such a way that employees on the site and the local fire department can be alerted in case of an emergency. The alarm code and reporting instructions shall be conspicuously posted and the alarm system should be serviceable at the job site during the demolition. Fire cutoffs shall be retained in the buildings undergoing alterations or demolition until operations necessitate their removal.

II. SPECIAL STRUCTURES DEMOLITION.

A. SAFE WORK PRACTICES WHEN DEMOLISHING A CHIMNEY, STACK, SILO, OR COOLING TOWER.

1. Inspection and Planning. When preparing to demolish any chimney, stack, silo, or cooling tower, the first step must be a careful, detailed inspection of the structure by an experienced person. If possible, architectural/engineering drawings should be consulted. Particular attention should be paid to the condition of the chimney or stack. Workers should be on the lookout for any structural defects such as weak or acid-laden mortar joints, and any cracks or openings. The interior brickwork in some sections of industrial chimney shafts can be extremely weak. If stack has been banded with steel straps, these bands shall be removed only as the work progresses from the top down. Sectioning of the chimney by water, etc. should be considered.

2. Safe Work Practice.

   a. When hand demolition is required, it should be carried out from a working platform.

      - Experienced personnel must install a self-supporting tubular scaffold, suspended platform, or knee-braced scaffolding around the chimney. Particular attention should be paid to the design, support, and tie-in (braces) of the scaffold.
      - A competent person should be present at all times during the erection of the scaffold.

- It is essential that there be adequate working clearance between the chimney and the work platform.
- Access to the top of the scaffold should be provided by means of portable walkways.
- The platforms should be decked solid and the area from the work platform to the wall should be bridged with a minimum of two-inch thick lumber.
- A back rail 42 inches above the platform, with a midrail covered with canvas or mesh, should be installed around the perimeter of the platform to prevent injury to workers below. Debris netting may be installed below the platform.
- Excess canvas or plywood attachments can form a wind-sail that could collapse the scaffold.
- When working on the work platform, all personnel should wear hard hats, long-sleeve shirts, eye and face protection, such as goggles and face shields, respirators, and safety belts, as required.
- Care should be taken to assign the proper number of workers to the task. Too many people on a small work platform can lead to accidents.

b. An alternative to the erection of a self-supporting tubular steel scaffold is to "climb" the structure with a creeping bracket scaffold. Careful inspection of the masonry and a decision as to the safety of this alternative must be made by a competent person. It is essential that the masonry of the chimney be in good enough condition to support the bracket scaffold.

c. The area around the chimney should be roped off or barricaded and secured with appropriate warning signs posted. No unauthorized entry should be permitted to this area. It is also good practice to keep a worker, i.e., a supervisor, operating engineer, another worker, or a "safety person," on the ground with a form of communication to the workers above.

d. Special attention should be paid to weather conditions when working on a chimney. No work should be done during inclement weather such as during lightning or high wind situations. The work site should be wetted down, as needed, to control dust.

3. Debris Clearance. If debris is dropped inside the shaft, it can be removed through an opening in the chimney at grade level.

- The opening at grade must be kept relatively small in order not to weaken the structure. If a larger opening is desired, a professional engineer should be consulted.
- When removing debris by hand, an overhead canopy of adequate strength should be provided. If machines are used for removal of debris, proper overhead protection for the operator should be used.
- Excessive debris should not be allowed to accumulate inside or outside the shaft of the chimney as the excess weight of the debris can impose pressure on the wall of the structure and might cause the shaft to collapse.
- The foreman should determine when debris is to be removed, halt all demolition during debris removal, and make sure the area is clear of cleanup workers before continuing demolition.

4. Demolition by Deliberate Collapse.

a. Another method of demolishing a chimney or stack is by deliberate collapse. Deliberate collapse requires extensive planning and experienced personnel, and should be used only when conditions are favorable. There must be a clear space for the fall of the structure of at least 45 degrees on each side of the intended fall line and 1½ times the total height of the chimney. Considerable vibration may be set up when the chimney falls, so there should be no sewers or underground services on the line of the fall. Lookouts must be posted on the site and warning signals must be arranged. The public and other workers at the job site must be kept well back from the fall area.

b. The use of explosives is one way of setting off deliberate collapse. This type of demolition should be undertaken only by qualified persons. The entire work area shall be cleared of nonessential personnel before any explosives are placed. Though the use of explosives is a convenient method of bringing down a chimney or stack, there is a considerable amount of vibration produced, and caution should be taken if there is any likelihood of damage.

B. DEMOLITION OF PRESTRESSED CONCRETE STRUCTURES.

1. The different forms of construction used in a number of more or less conventional structures built during the last few decades will give rise to a variety of problems when the time comes for them to be demolished. Prestressed concrete structures fall in this general category. The most important aspect of demolishing a prestressed concrete structure takes place during the engineering survey. During the survey, a qualified person should determine if the structure to be demolished contains any prestressed members.

2. It is the responsibility of the demolition contractor to inform all workers on the demolition job site of the presence of prestressed concrete members within the structure. They should also instruct them in the safe work practice which must be followed to safely perform the demolition. Workers should be informed of the hazards of deviating from the prescribed procedures and the importance of following their supervisor's instruction.

3. Pretensioned members usually do not have any end anchors, the wires being embedded or bonded within the length of the member. Simple Pretensioned beams and slabs of spans up to about 7 meters (23 feet) can be demolished in a manner similar to ordinary reinforced concrete. Pretensioned beams and slabs may be lifted and lowered to the ground as complete units after the removal of any composite concrete covering to tops and ends of the units. To facilitate breaking up, the members should be turned on their sides. Lifting from the structure should generally be done from points near the ends of the units or from lifting point positions. Reuse of lifting eyes, if in good condition, is recommended whenever possible. When units are too large to be removed, consideration should be given to temporary supporting arrangements.

There are four main categories of prestressed members. The category or categories should be determined before attempting demolition, bearing in mind that any prestressed structure may contain elements of more than one category.

- **Category 1** Members are prestressed before the application of the superimposed loads, and all cables or tendons are fully bonded in the concrete or grouted within ducts.
- **Category 2** Like Category 1, but the tendons are left ungrouted. This type of construction can sometimes be recognized from the access points that may have been provided for inspection of the cables and anchors. More recently, unbonded tendons have been used in the construction of beams, slabs, and other members; these tendons are protected by grease and surrounded by plastic sheathing, instead of the usual metal duct.
- **Category 3** Members are prestressed progressively as building construction proceeds and the dead load increases, using bonded tendons as in Category 1.
- **Category 4** Like Category 3, but using unbonded tendons as in Category 2.

Examples of construction using members of Categories 3 or 4 are relatively rare. However, they may be found, for example in the podium of a tall building or some types of bridges. They require particular care in demolition.

FIGURE V: 1-1. CATEGORIES OF PRESTRESSED CONSTRUCTION

C. PRECAST UNITS STRESSED SEPARATELY FROM THE MAIN FRAMES OF THE STRUCTURE, WITH END ANCHORS AND GROUTED AND UNGROUTED DUCTS. Before breaking up, units of this type should be lowered to the ground, if possible. It is advisable to seek the counsel of a professional engineer before carrying out this work, especially where there are ungrouted tendons. In general, this is true because grouting is not always 100% efficient. After lowering the units can be turned on their side with the ends up on blocks after any composite concrete is removed. This may suffice to break the unit and release the prestress; if not, a sand bag screen, timbers, or a blast mat as a screen should be erected around the ends and demolition commenced, taking care to clear the area of any personnel. It should be borne in mind that the end blocks may be heavily reinforced and difficult to break up.

1. Monolithic Structures. The advice of the professional engineer experienced in prestressed work should be sought before any attempt is made to expose the tendons or anchorages of structures in which two or more members have been stressed together. It will usually be necessary for temporary supports to be provided so that the tendons and the anchorage can be cautiously exposed. In these circumstances it is essential that indiscriminate attempts to expose and destress the tendons and anchorages not be made.

2. Progressively Prestressed Structures. In the case of progressively prestressed structures, it is essential to obtain the advice of a professional engineer, and to demolish the structure in strict accordance with the engineer's method of demolition. The stored energy in this type of structure is large. In some cases, the inherent properties of the stressed section may delay failure for some time, but the presence of these large prestressing forces may cause sudden and complete collapse with little warning.

D. SAFE WORK PRACTICES WHEN WORKING IN CONFINED SPACES.

1. Demolition contractors often come in contact with confined spaces when demolishing structure at industrial sites. These confined spaces can be generally categorized in two major groups: those with open tops and a depth that restricts the natural movement of air, and enclosed spaces with very limited openings for entry. Examples of these spaces include storage tanks, vessels, degreasers, pits vaults, casing, and silos.

2. The hazards encountered when entering and working in confined spaces are capable of causing bodily injury, illness, and death. Accidents occur among workers because of failure to recognize that a confined space is a potential hazard. It should therefore be considered that the most unfavorable situation exists in every case and that the danger of explosion, poisoning, and asphyxiation will be present at the onset of entry.

III. SAFE BLASTING PROCEDURES.

A. GENERAL SAFE WORK PRACTICES.

1. Blasting Survey and Site Preparation.

   a. Prior to the blasting of any structure or portion thereof, a complete written survey must be made by a qualified person of all adjacent improvements and underground utilities. When there is a possibility of excessive vibration due to blasting operations, seismic or vibration tests should be taken to determine proper safety limits to prevent damage to adjacent or nearby buildings, utilities, or other property.

   b. The preparation of a structure for demolition by explosives may require the removal of structural columns, beams or other building components. This work should be directed by a structural engineer or a competent person qualified to direct the removal of these structural elements. Extreme caution must be taken during this preparatory work to prevent the weakening and premature collapse of the structure.

   c. The use of explosives to demolish smokestacks, silos, cooling towers, or similar structures should be permitted only if there is a minimum of 90 of open space extended for at least 150% of the height of the structure or if the explosives specialist can demonstrate consistent previous performance with tighter constraints at the site.

2. Fire Precautions.

a. The presence of fire near explosives presents a severe danger. Every effort should be made to ensure that fires or sparks do not occur near explosive materials. Smoking, matches, firearms, open flame lamps, and other fires, flame, or heat-producing devices must be prohibited in or near explosive magazines or in areas where explosives are being handled, transported, or used. In fact, persons working near explosives should not even carry matches, lighters, or other sources of sparks or flame. Open fires or flames should be prohibited within 100 feet of any explosive materials. In the vent of a fire which is in imminent danger of contact with explosives, all employees must be removed to a safe area.

b. Electrical detonators can be inadvertently triggered by stray RF (radio frequency) signals from two-way radios. RF signal sources should be restricted from or near to the demolition site, if electrical detonators are used.

3. Personnel Selection.

a. A blaster is a competent person who uses explosives. A blaster must be qualified by reason of training, knowledge, or experience in the field of transporting, storing, handling, and using explosives. In addition, the blaster should have a working knowledge of state and local regulations which pertain to explosives. Training courses are often available from manufacturers of explosives and blasting safety manuals are offered by the Institute of Makers of Explosives (IME) as well as other organizations.

b. Blasters shall be required to furnish satisfactory evidence of competency in handling explosives and in safely performing the type of blasting required. A competent person should always be in charge of explosives and should be held responsible for enforcing all recommended safety precautions in connection with them.

B. TRANSPORTATION OF EXPLOSIVES.

1. Vehicle Safety.

a. Vehicles used for transporting explosives shall be strong enough to carry the load without difficulty, and shall be in good mechanical condition. All vehicles used for the transportation of explosives shall have tight floors, and any exposed spark-producing metal on the inside of the body shall be covered with wood or some other nonsparking material. Vehicles or conveyances transporting explosives shall only be driven by, and shall be under the supervision of, a licensed driver familiar with the local, state, and Federal regulations governing the transportation of explosives. No passengers should be allowed in any vehicle transporting explosives.

b. Explosives, blasting agents, and blasting supplies shall not be transported with other materials or cargoes. Blasting caps shall not be transported in the same vehicle with other explosives. If an open-bodied truck is used, the entire load should be completely covered with a fire and water-resistant tarpaulin to protect it from the elements. Vehicles carrying explosives should not be loaded beyond the manufacturer's safe capacity rating, and in no case should the explosives be piled higher than the closed sides and ends of the body.

c. Every motor vehicle or conveyance used for transporting explosives shall be marked or placarded with warning signs required by OSHA and the DOT. Each vehicle used for transportation of explosives shall be equipped minimally with at least a ten-pound rated, serviceable ABC fire extinguisher. All drivers should be trained in the use of the extinguishers on their vehicle.

d. In transporting explosives, congested traffic and high density population areas should be avoided, where possible, and no unnecessary stops should be made. Vehicles carrying explosives, blasting agents, or blasting supplies shall not be taken inside a garage or shop for repairs or servicing. No motor vehicle transporting explosives shall be left unattended.

C. STORAGE OF EXPLOSIVES.

1. Inventory Handling and Safe Handling.

a. All explosives must be accounted for at all times and all not being used must be kept in a locked magazine. A complete detailed inventory of all explosives received and placed in, removed from, and returned to the magazine should be maintained at all times. Appropriate authorities must be notified of any loss, theft, or unauthorized entry into a magazine.

b. Manufacturers' instructions for the safe handling and storage of explosives are ordinarily enclosed in each case of explosives. The specifics of storage and handling are best referred to these instructions and the aforementioned IME manuals. They should be carefully followed. Packages of explosives should not be handled roughly. Sparking metal tools should not be used to open wooden cases. Metallic slitters may be used for opening fiberboard cases, provided the metallic slitter does not come in contact with the metallic fasteners of the case.

c. The oldest stock should always be used first to minimize the chance of deterioration from long storage. Loose explosives or broken, defective, or leaking packages can be hazardous and should be segregated and properly disposed of in accordance with the specific instructions of the manufacturer. If the explosives are in good condition it may be advisable to repack them. In this case, the explosives supplier should be contacted. Explosives cases should not be opened or explosives packed or repacked while in a magazine.

2. Storage Conditions.

a. Providing a dry, well-ventilated place for the storage of explosives is one of the most important and effective safety measures. Exposure to weather damages most kinds of explosives, especially dynamite and caps. Every precaution should be taken to keep them dry and relatively cool. Dampness or excess humidity may be the cause of misfires resulting in injury or loss of life. Explosives should be stored in properly constructed fire and bullet-resistant structures, located according to the IME American Table of Distances and kept locked at all times except when opened for use by an authorized person. Explosives should not be left, kept, or stored where children, unauthorized persons, or animals have access to them, nor should they be stored in or near a residence.

b. Detonators should be stored in a separate magazine located according to the IME American Table of Distances. DETONATORS SHOULD NEVER BE STORED IN THE SAME MAGAZINE WITH ANY OTHER KIND OF EXPLOSIVES.

c. Ideally, arrangements should be made whereby the supplier delivers the explosives to the job site in quantities which will be used up during the work day. An alternative would be for the supplier to return to pick up unused quantities of explosives. If it is necessary for the contractor to store his explosives, he should be familiar with all local requirements for such storage.

D. PROPER USE OF EXPLOSIVES.

1. Blasting operations shall be conducted between sunup and sundown, whenever possible. Adequate signs should be sounded to alert to the hazard presented by blasting. Blasting mats or other containment should be used where there is danger of rocks or other debris being thrown into the air or where there are buildings or transportation systems nearby. Care should be taken to make sure mats and other protection do not disturb the connections to electrical blasting caps.

2. Radio, television, and radar transmitters create fields of electrical energy that can, under exceptional circumstances, detonate electric blasting caps. Certain precautions must be taken to prevent accidental discharge of electric blasting caps from current induced by radar, radio transmitters, lightning, adjacent power lines, dust storms, or other sources of extraneous or static electricity. These precautions shall include:

- Ensuring that mobile radio transmitters on the job site that are less than 100 feet away from electric blasting caps, in other than original containers, shall be de-energized and effectively locked.
- The prominent display of adequate signs, warning against the use of mobile radio transmitters, on all roads within 1,000 feet of the blasting operations.

- Maintaining the minimum distances recommended by the IME between the nearest transmitter and electric blasting caps.
- The suspension of all blasting operations and removal of persons from the blasting area during the approach and progress of an electric storm.
- After loading is completed, there should be as little delay as possible before firing. Each blast should be fired under the direct supervision of the blaster, who should inspect all connections before firing and who should personally see that all persons are in the clear before giving the order to fire. Standard signals, which indicate that a blast is about to be fired and a later all-clear signal shall have been adopted. It is important that everyone working in the area be familiar with these signals and that they be strictly obeyed.

E. PROCEDURES AFTER BLASTING.

1. Inspection After the Blast. Immediately after the blast has been fired, the firing line shall be disconnected from the blasting machine and short-circuited. Where power switches are used, they shall be locked open or in the off position. Sufficient time shall be allowed for dust, smoke, and fumes to leave the blasted area before returning to the spot. An inspection of the area and the surrounding rubble shall be made by the blaster to determine if all charges have been exploded before employees are allowed to return to the operation. All wires should be traced and the search for unexploded cartridges made by the blaster.

2. Disposal of Explosives.

   a. Explosives, blasting agents, and blasting supplies that are obviously deteriorated or damaged should not be used, they should be properly disposed of. Explosives distributors will usually take back old stock. Local fire marshals or representatives of the United States Bureau of Mines may also arrange for its disposal. Under no circumstances should any explosives be abandoned.

   b. Wood, paper, fiber, or other materials that have contained high explosives should not be used again for any purpose, but should be destroyed by burning. These materials should not be burned in a stove, fireplace, or other confined space. Rather, they should be burned at an isolated outdoor location, at a safe distance from thoroughfares, magazines, and other structures. It is important to check that the containers are entirely empty before burning. During burning, the area should be adequately protected from intruders and all persons kept at least 100 feet from the fire.

IV. BIBLIOGRAPHY.

Malmberg, K.B. 1975. EPA *Demolition and Renovation Inspection Procedures*. U.S.E.P.A.: Washington, D.C.

National Association of Demolition Contractors (NADC). 1981. *Demolition Safety Manual*. NADC: Hillside, IL.

Occupational Safety and Health Administration. *OSHA Safety and Health Standards, Construction*, (29 CFR 1926). 1989. U.S. Government Printing Office: Washington DC.

SECTION V: CHAPTER 2

EXCAVATIONS: HAZARD RECOGNITION IN TRENCHING AND SHORING

Contents:

I. INTRODUCTION.

Excavating is recognized as one of the most hazardous construction operations. OSHA recently revised Subpart P, Excavations, of 29 CFR 1926.650, .651, and .652 to make the standard easier to understand, permit the use of performance criteria where possible, and provide construction employers with options when classifying soil and selecting employee protection methods.

This chapter is intended to assist OSHA Technical Manual users, safety and health consultants, OSHA field staff, and others in the recognition of trenching and shoring hazards and their prevention.

II. DEFINITIONS.

A. ACCEPTED ENGINEERING PRACTICES are procedures compatible with the standards of practice required of a registered professional engineer.

B. ADJACENT STRUCTURE STABILITY refers to the stability of the foundation(s) of adjacent structures whose location may create surcharges, changes in soil conditions, or other disruptions that have the potential to extend into the failure zone of the excavation or trench.

C. COMPETENT PERSON is an individual who is capable of identifying existing and predictable hazards or working conditions that are hazardous, unsanitary, or dangerous to employees, and who has authorization to take prompt corrective measures to eliminate or control these hazards and conditions.

D. CONFINED SPACE is a space that, by design and/or configuration, has limited openings for entry and exit, unfavorable natural ventilation, may contain or produce hazardous substances, and is not intended for continuous employee occupancy.

E. EXCAVATION. An Excavation is any man-made cut, cavity, trench, or depression in an earth surface that is formed by earth removal. A Trench is a narrow excavation (in relation to its length) made below the surface of the ground. In general, the depth of a trench is greater than its width, and the width (measured at the bottom) is not greater than 15 ft (4.6 m). If a form or other structure installed or constructed in an excavation reduces the distance between the form and the side of the excavation to 15 ft (4.6 m) or less (measured at the bottom of the excavation), the excavation is also considered to be a trench.

F. HAZARDOUS ATMOSPHERE is an atmosphere that by reason of being explosive, flammable, poisonous, corrosive, oxidizing, irritating, oxygen-deficient, toxic, or otherwise harmful may cause death, illness, or injury to persons exposed to it.

G. INGRESS AND EGRESS mean "entry" and "exit," respectively. In trenching and excavation operations, they refer to the provision of safe means for employees to enter or exit an excavation or trench.

H. PROTECTIVE SYSTEM refers to a method of protecting employees from cave-ins, from material that could fall or roll from an excavation face or into an excavation, and from the collapse of adjacent structures. Protective systems include support systems, sloping and benching systems, shield systems, and other systems that provide the necessary protection.

I. REGISTERED PROFESSIONAL ENGINEER is a person who is registered as a professional engineer in the state where the work is to be performed. However, a professional engineer who is registered in any state is deemed to be a "registered professional engineer" within the meaning of Subpart P when approving designs for "manufactured protective systems" or "tabulated data" to be used in interstate commerce.

J. SUPPORT SYSTEM refers to structures such as underpinning, bracing, and shoring that provide support to an adjacent structure or underground installation or to the sides of an excavation or trench.

K. SUBSURFACE ENCUMBRANCES include underground utilities, foundations, streams, water tables, transformer vaults, and geological anomalies.

L. SURCHARGE means an excessive vertical load or weight caused by spoil, overburden, vehicles, equipment, or activities that may affect trench stability.

M. TABULATED DATA are tables and charts approved by a registered professional engineer and used to design and construct a protective system.

N. UNDERGROUND INSTALLATIONS include, but are not limited to, utilities (sewer, telephone, fuel, electric, water, and other product lines), tunnels, shafts, vaults, foundations, and other underground fixtures or equipment that may be encountered during excavation or trenching work.

O. UNCONFINED COMPRESSIVE STRENGTH is the load per unit area at which soil will fail in compression. This measure can be determined by laboratory testing, or it can be estimated in the field using a pocket penetrometer, by thumb penetration tests, or by other methods.

P. DEFINITIONS THAT ARE NO LONGER APPLICABLE. For a variety of reasons, several terms commonly used in the past are no longer used in revised Subpart P. These include the following:

1. Angle of Repose Conflicting and inconsistent definitions have led to confusion as to the meaning of this phrase. This term has been replaced by Maximum Allowable Slope.

2. Bank, Sheet Pile, and Walls Previous definitions were unclear or were used inconsistently in the former standard.

3. Hard Compact Soil and Unstable Soil The new soil classification system in revised Subpart P uses different terms for these soil types.

III. OVERVIEW: SOIL MECHANICS.

A number of stresses and deformations can occur in an open cut or trench. For example, increases or decreases in moisture content can adversely affect the stability of a trench or excavation. The following diagrams show some of the more frequently identified causes of trench failure.

A. TENSION CRACKS. Tension cracks usually form at a horizontal distance of 0.5 to 0.75 times the depth of the trench, measured from the top of the vertical face of the trench. See the accompanying drawing for additional details.

FIGURE 5:2-1. TENSION CRACK.

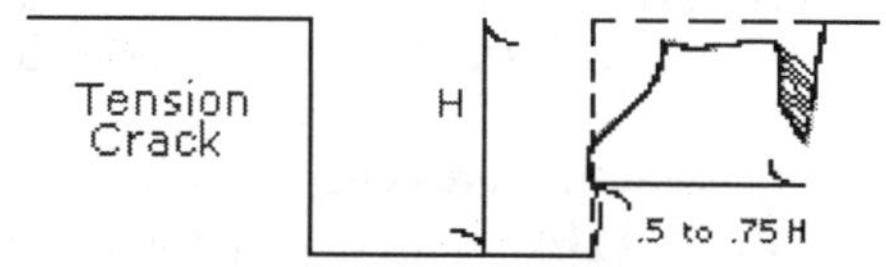

B. SLIDING or sluffing may occur as a result of tension cracks, as illustrated below.

FIGURE 5:2-2. SLIDING.

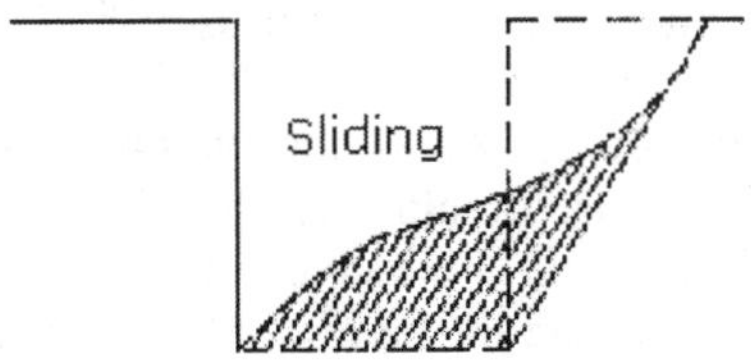

C. TOPPLING. In addition to sliding, tension cracks can cause toppling. Toppling occurs when the trench's vertical face shears along the tension crack line and topples into the excavation.

FIGURE 5:2-3. TOPPLING.

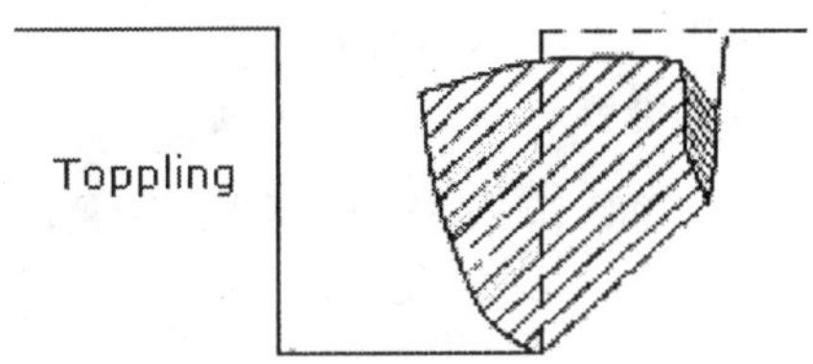

D. SUBSIDENCE AND BULGING. An unsupported excavation can create an unbalanced stress in the soil, which, in turn, causes subsidence at the surface and bulging of the vertical face of the trench. If uncorrected, this condition can cause face failure and entrapment of workers in the trench.

FIGURE 5:2-4. SUBSIDENCE AND BULGING.

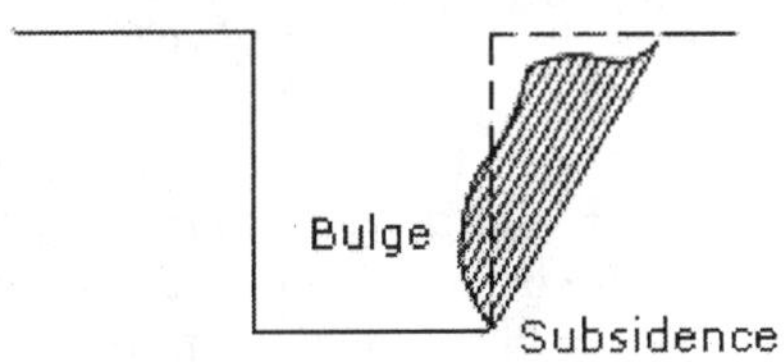

E. HEAVING OR SQUEEZING. Bottom heaving or squeezing is caused by the downward pressure created by the weight of adjoining soil. This pressure causes a bulge in the bottom of the cut, as illustrated in the drawing above. Heaving and squeezing can occur even when shoring or shielding has been properly installed.

FIGURE 5:2-5. HEAVING OR SQUEEZING.

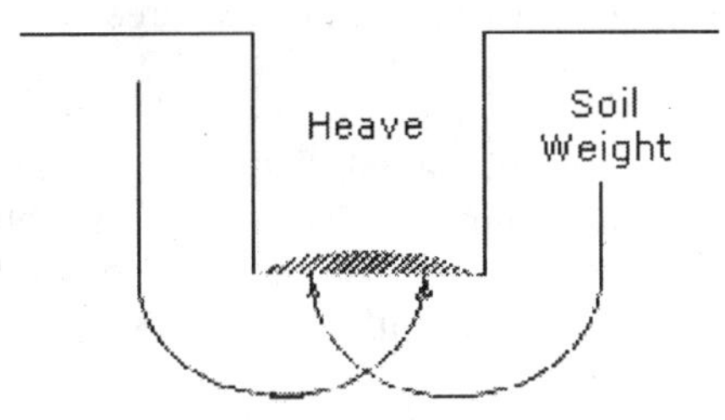

F. BOILING is evidenced by an upward water flow into the bottom of the cut. A high water table is one of the causes of boiling. Boiling produces a "quick" condition in the bottom of the cut, and can occur even when shoring or trench boxes are used.

FIGURE 5:2-6. BOILING.

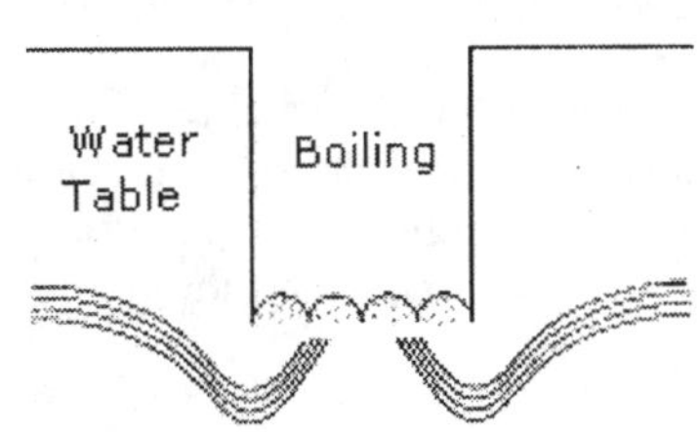

G. UNIT WEIGHT OF SOILS refers to the weight of one unit of a particular soil. The weight of soil varies with type and moisture content. One cubic foot of soil can weigh from 110 pounds to 140 pounds or more, and one cubic meter (35.3 cubic feet) of soil can weigh more than 3,000 pounds.

IV. DETERMINATION OF SOIL TYPE.

OSHA categorizes soil and rock deposits into four types, A through D, as follows:

A. STABLE ROCK is natural solid mineral matter that can be excavated with vertical sides and remain intact while exposed. It is usually identified by a rock name such as granite or sandstone. Determining whether a deposit is of this type may be difficult unless it is known whether cracks exist and whether or not the cracks run into or away from the excavation.

B. TYPE A SOILS are cohesive soils with an unconfined compressive strength of 1.5 tons per square foot (tsf) (144 kPa) or greater. Examples of Type A cohesive soils are often: clay, silty clay, sandy clay, clay loam and, in some cases, silty clay loam and sandy clay loam. (No soil is Type A if it is fissured, is subject to vibration of any type, has previously been disturbed, is part of a sloped, layered system where the layers dip into the excavation on a slope of 4 horizontal to 1 vertical (4H:1V) or greater, or has seeping water.

C. TYPE B SOILS are cohesive soils with an unconfined compressive strength greater than 0.5 tsf (48 kPa) but less than 1.5 tsf (144 kPa). Examples of other Type B soils are: angular gravel; silt; silt loam; previously disturbed soils unless otherwise classified as Type C; soils that meet the unconfined compressive strength or cementation requirements of Type A soils but are fissured or subject to vibration; dry unstable rock; and layered systems sloping into the trench at a slope less than 4H:1V (only if the material would be classified as a Type B soil).

D. TYPE C SOILS are cohesive soils with an unconfined compressive strength of 0.5 tsf (48 kPa) or less. Other Type C soils include granular soils such as gravel, sand and loamy sand, submerged soil, soil from which water is freely seeping, and submerged rock that is not stable. Also included in this classification is material in a sloped, layered system where the layers dip into the excavation or have a slope of four horizontal to one vertical (4H:1V) or greater.

E. LAYERED GEOLOGICAL STRATA. Where soils are configured in layers, i.e., where a layered geologic structure exists, the soil must be classified on the basis of the soil classification of the weakest soil layer. Each layer may be classified individually if a more stable layer lies below a less stable layer, i.e., where a Type C soil rests on top of stable rock.

V. TEST EQUIPMENT AND METHODS FOR EVALUATING SOIL TYPE.

Many kinds of equipment and methods are used to determine the type of soil prevailing in an area, as described below.

A. POCKET PENETROMETER. Penetrometers are direct-reading, spring-operated instruments used to determine the unconfined compressive strength of saturated cohesive soils. Once pushed into the soil, an indicator sleeve displays the reading. The instrument is calibrated in either tons per square foot (tsf) or kilograms per square centimeter (kPa). However, Penetrometers have error rates in the range of ± 20-40%.

1. Shearvane (Torvane). To determine the unconfined compressive strength of the soil with a shearvane, the blades of the vane are pressed into a level section of undisturbed soil, and the torsional knob is slowly turned until soil failure occurs. The direct instrument reading must be multiplied by 2 to provide results in tons per square foot (tsf) or kilograms per square centimeter (kPa).

2. Thumb Penetration Test. The thumb penetration procedure involves an attempt to press the thumb firmly into the soil in question. If the thumb makes an indentation in the soil only with great difficulty, the soil is probably Type A. If the thumb penetrates no further than the length of the thumb nail, it is probably Type B soil, and if the thumb penetrates the full length of the thumb, it is Type C soil. The thumb test is subjective and is therefore the least accurate of the three methods.

3. Dry Strength Test. Dry soil that crumbles freely or with moderate pressure into individual grains is granular. Dry soil that falls into clumps that subsequently break into smaller clumps (and the smaller clumps can be broken only with difficulty) is probably clay in combination with gravel, sand, or silt. If the soil breaks into clumps that do not break into smaller clumps (and the soil can be broken only with difficulty), the soil is considered unfissured unless there is visual indication of fissuring.

B. PLASTICITY OR WET THREAD TEST. This test is conducted by molding a moist sample of the soil into a ball and attempting to roll it into a thin thread approximately 1/8 inch (3 mm) in diameter (thick) by 2 inches (50 mm) in length. The soil sample is held by one end. If the sample does not break or tear, the soil is considered cohesive.

C. VISUAL TEST. A visual test is a qualitative evaluation of conditions around the site. In a visual test, the entire excavation site is observed, including the soil adjacent to the site and the soil being excavated. If the soil remains in clumps, it is cohesive; if it appears to be coarse-grained sand or gravel, it is considered granular. The evaluator also checks for any signs of vibration.

During a visual test, the evaluator should check for crack-line openings along the failure zone that would indicate tension cracks, look for existing utilities that indicate that the soil has previously been disturbed, and observe the open side of the excavation for indications of layered geologic structuring.

The evaluator should also look for signs of bulging, boiling, or sluffing, as well as for signs of surface water seeping from the sides of the excavation or from the water table. If there is standing water in the cut, the evaluator should check for "quick" conditions (see Paragraph III. F. in this chapter). In addition, the area adjacent to the excavation should be checked for signs of foundations or other intrusions into the failure zone, and the evaluator should check for surcharging and the spoil distance from the edge of the excavation.

VI. SHORING TYPES.

Shoring is the provision of a support system for trench faces used to prevent movement of soil, underground utilities, roadways, and foundations. Shoring or shielding is used when the location or depth of the cut makes sloping back to the maximum allowable slope impractical. Shoring systems consist of posts, wales, struts, and sheeting. There are two basic types of shoring, timber and aluminum hydraulic.

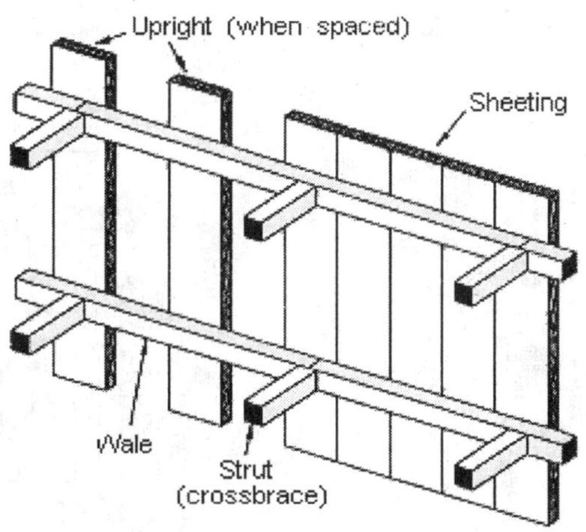

FIGURE V: 2-7. TIMBER SHORING.

A. HYDRAULIC SHORING. The trend today is toward the use of hydraulic shoring, a prefabricated strut and/or wale system manufactured of aluminum or steel. Hydraulic shoring provides a critical safety advantage over timber shoring because workers do not have to enter the trench to install or remove hydraulic shoring. Other advantages of most hydraulic systems are that they:

- Are light enough to be installed by one worker;
- Are gauge-regulated to ensure even distribution of pressure along the trench line;
- Can have their trench faces "preloaded" to use the soil's natural cohesion to prevent movement; and
- Can be adapted easily to various trench depths and widths.

All shoring should be installed from the top down and removed from the bottom up. Hydraulic shoring should be checked at least once per shift for leaking hoses and/or cylinders, broken connections, cracked nipples, bent bases, and any other damaged or defective parts.

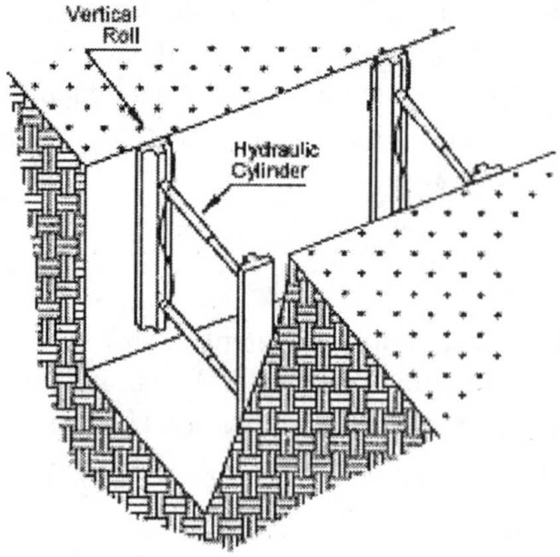

Vertical Aluminum Hydraulic Shoring
(Spot Bracing)

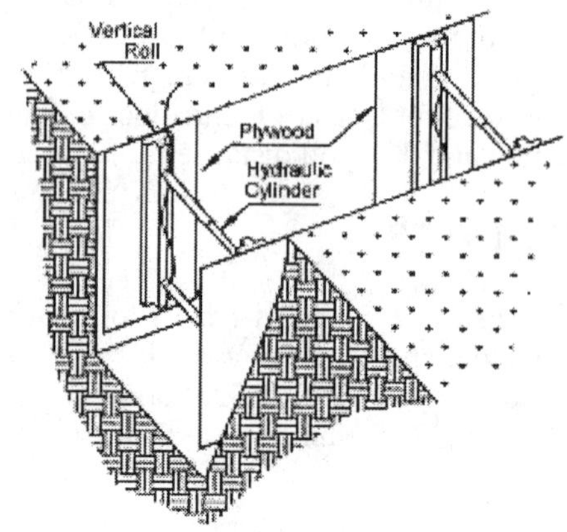

Vertical Aluminum Hydraulic Shoring
(With Plywood)

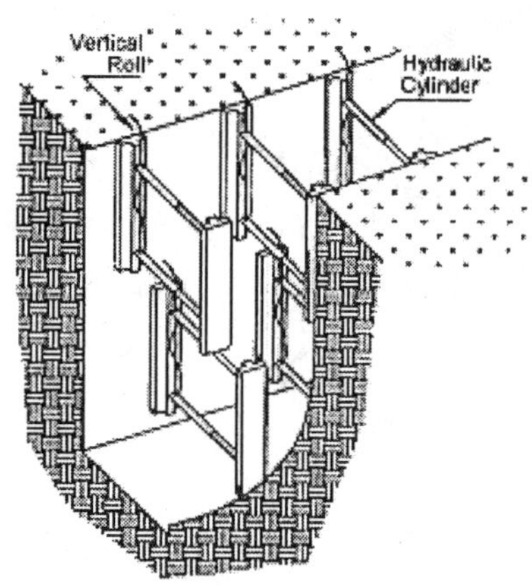

Vertical Aluminum Hydraulic Shoring
(Stacked)

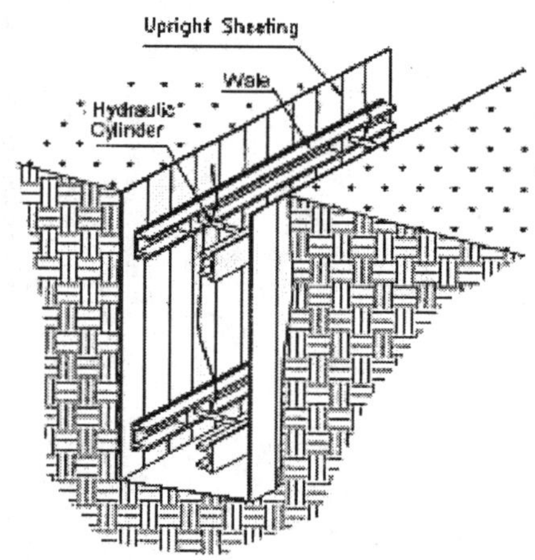

Aluminum Hydraulic Shoring Waler System
(Typical)

FIGURE V: 2-8. SHORING VARIATIONS: TYPICAL ALUMINUM HYDRAULIC SHORING INSTALLATIONS.

B. PNEUMATIC SHORING works in a manner similar to hydraulic shoring. The primary difference is that pneumatic shoring uses air pressure in place of hydraulic pressure. A disadvantage to the use of pneumatic shoring is that an air compressor must be on site.

1. Screw Jacks. Screw jack systems differ from hydraulic and pneumatic systems in that the struts of a screw jack system must be adjusted manually. This creates a hazard because the worker is required to be in the trench in order to adjust the strut. In addition, uniform "preloading" cannot be achieved with screw jacks, and their weight creates handling difficulties.

2. Single-Cylinder Hydraulic Shores. Shores of this type are generally used in a water system, as an assist to timber shoring systems, and in shallow trenches where face stability is required.

3. Underpinning. This process involves stabilizing adjacent structures, foundations, and other intrusions that may have an impact on the excavation. As the term indicates, underpinning is a procedure in which the foundation is physically reinforced. Underpinning should be conducted only under the direction and with the approval of a registered professional engineer.

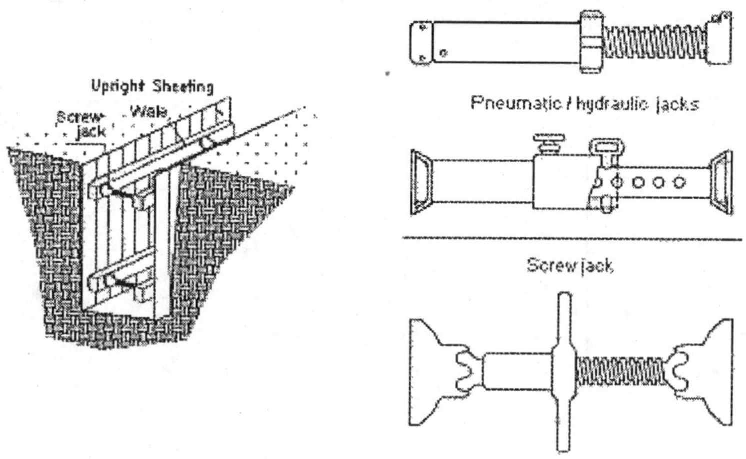

FIGURE V: 2-9. SHORING VARIATIONS.

VII. SHIELDING TYPES.

A. TRENCH BOXES are different from shoring because, instead of shoring up or otherwise supporting the trench face, they are intended primarily to protect workers from cave-ins and similar incidents. The excavated area between the outside of the trench box and the face of the trench should be as small as possible. The space between the trench boxes and the excavation side are backfilled to prevent lateral movement of the box. Shields may not be subjected to loads exceeding those which the system was designed to withstand.

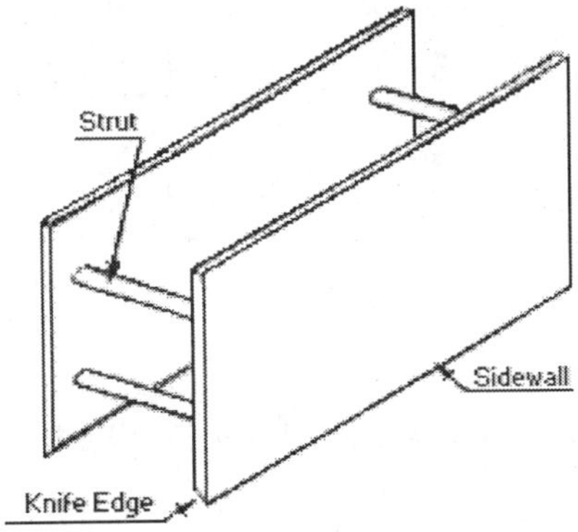

FIGURE V: 2-10. TRENCH SHIELD.

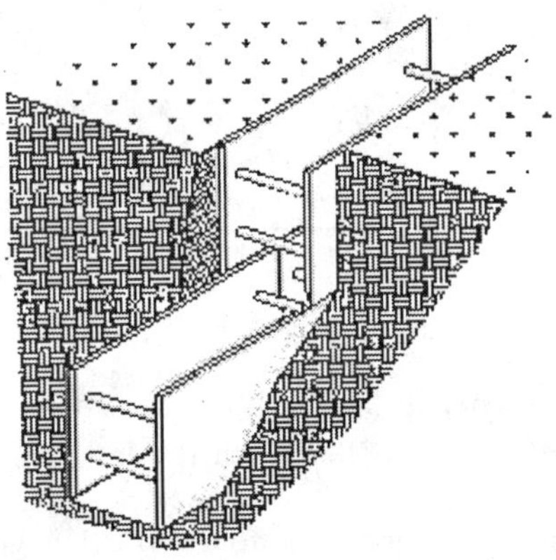

FIGURE V: 2-11. TRENCH SHIELD, STACKED.

B. COMBINED USE. Trench boxes are generally used in open areas, but they also may be used in combination with sloping and benching. The box should extend at least 18 in (0.45 m) above the surrounding area if there is sloping toward excavation. This can be accomplished by providing a benched area adjacent to the box.

Earth excavation to a depth of 2 ft (0.61 m) below the shield is permitted, but only if the shield is designed to resist the forces calculated for the full depth of the trench and there are no indications while the trench is open of possible loss of soil from behind or below the bottom of the support system. Conditions of this type require observation on the effects of bulging, heaving, and boiling as well as surcharging, vibration, adjacent structures, etc., on excavating below the bottom of a shield. Careful visual inspection of the conditions mentioned above is the primary and most prudent approach to hazard identification and control.

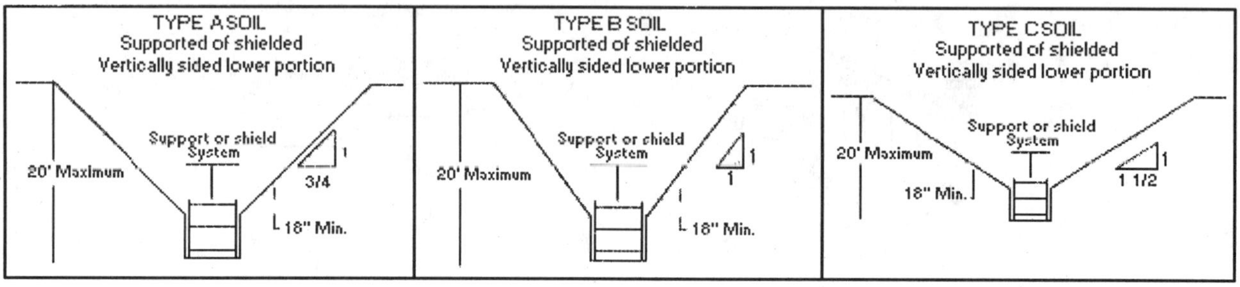

FIGURE V: 2-12. SLOPE AND SHIELD CONFIGURATIONS.

VIII. SLOPING AND BENCHING.

SLOPING. Maximum allowable slopes for excavations less than 20 ft (6.09 m) based on soil type and angle to the horizontal are as follows:

TABLE V:2-1. ALLOWABLE SLOPES.

Soil type	*Height/Depth ratio*	*Slope angle*
Stable Rock	Vertical	90°
Type A	¾:1	53°
Type B	1:1	45°
Type C	1½:1	34°
Type A (short-term)	½:1	63°
(For a maximum excavation depth of 12 ft)		

B. BENCHING. There are two basic types of benching, simple and multiple. The type of soil determines the horizontal to vertical ratio of the benched side.

As a general rule, the bottom vertical height of the trench must not exceed 4 ft (1.2 m) for the first bench. Subsequent benches may be up to a maximum of 5 ft (1.5 m) vertical in Type A soil and 4 ft (1.2 m) in Type B soil to a total trench depth of 20 ft (6.0 m). All subsequent benches must be below the maximum allowable slope for that soil type. For Type B soil the trench excavation is permitted in cohesive soil only.

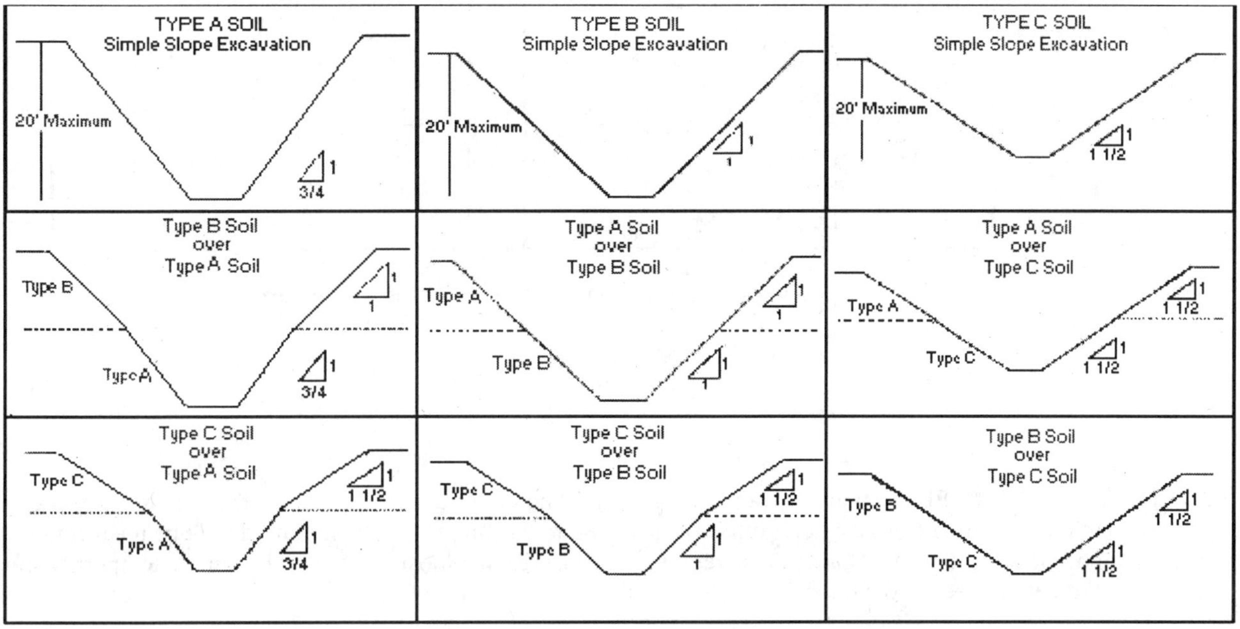

FIGURE V: 2-13. SLOPE CONFIGURATIONS: EXCAVATIONS IN LAYERED SOILS.

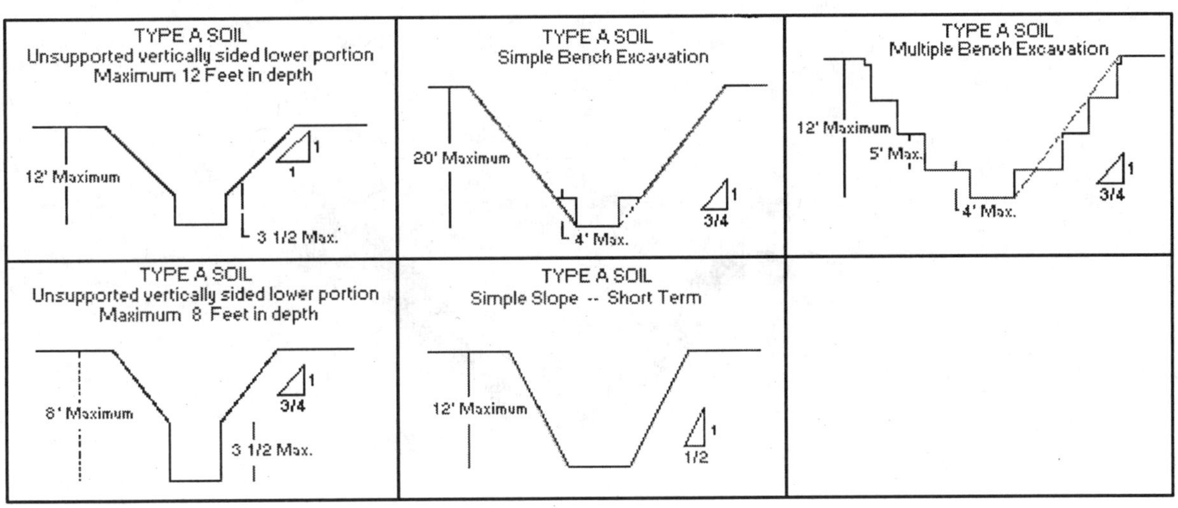

FIGURE V: 2-14. EXCAVATIONS MADE IN TYPE A SOIL.

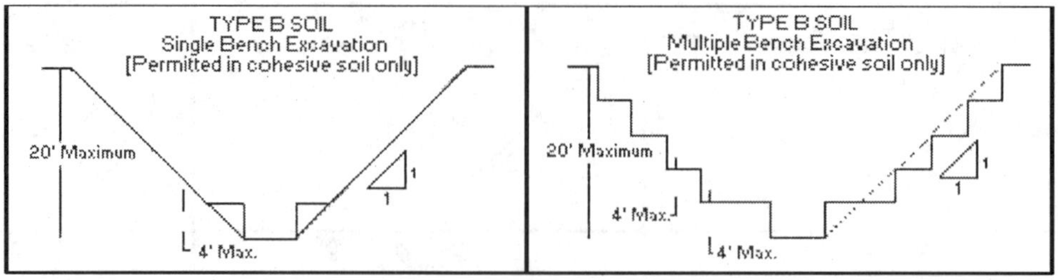

FIGURE V: 2-15. EXCAVATIONS MADE IN TYPE B SOIL.

IX. SPOIL.

A. TEMPORARY SPOIL. Temporary spoil must be placed no closer than 2 ft (0.61 m) from the surface edge of the excavation, measured from the nearest base of the spoil to the cut. This distance should not be measured from the crown of the spoil deposit. This distance requirement ensures that loose rock or soil from the temporary spoil will not fall on employees in the trench.

Spoil should be placed so that it channels rainwater and other run-off water away from the excavation. Spoil should be placed so that it cannot accidentally run, slide, or fall back into the excavation.

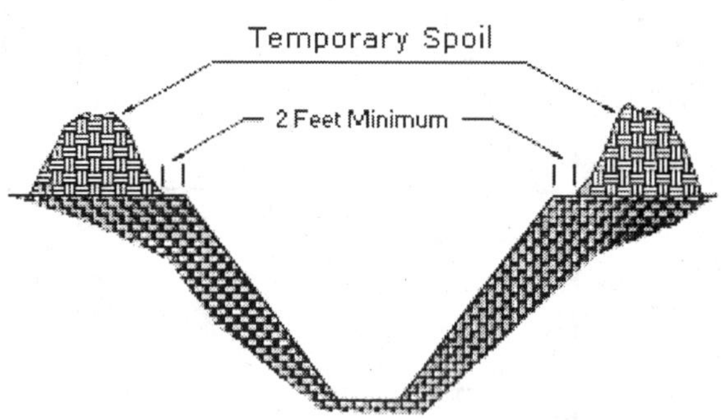

FIGURE V: 2-16. TEMPORARY SPOIL.

B. PERMANENT SPOIL. Permanent spoil should be placed at some distance from the excavation. Permanent spoil is often created where underpasses are built or utilities are buried. The improper placement of permanent spoil, i.e. insufficient distance from the working excavation, can cause an excavation to be out of compliance with the horizontal-to-vertical ratio requirement for a particular excavation. This can usually be determined through visual observation. Permanent spoil can change undisturbed soil to disturbed soil and dramatically alter slope requirements.

X. SPECIAL HEALTH AND SAFETY CONSIDERATIONS.

A. COMPETENT PERSON. The designated competent person should have and be able to demonstrate the following:

- Training, experience, and knowledge of: - soil analysis; - use of protective systems; and; - requirements of 29 CFR Part 1926 Subpart P.
- Ability to detect: - conditions that could result in cave-ins; - failures in protective systems; - hazardous atmospheres; and; - other hazards including those associated with confined spaces.
- Authority to take prompt corrective measures to eliminate existing and predictable hazards and to stop work when required.

B. SURFACE CROSSING OF TRENCHES. Surface crossing of trenches should be discouraged; however, if trenches must be crossed, such crossings are permitted only under the following conditions:

- Vehicle crossings must be designed by and installed under the supervision of a registered professional engineer.
- Walkways or bridges must be provided for foot traffic. These structures shall: - have a safety factor of 4; - have a minimum clear width of 20 in (0.51 m); - be fitted with standard rails; and - extend a minimum of 24 in (.61 m) past the surface edge of the trench.

C. INGRESS AND EGRESS. Access to and exit from the trench require the following conditions:

- Trenches 4 ft or more in depth should be provided with a fixed means of egress.
- Spacing between ladders or other means of egress must be such that a worker will not have to travel more than 25 ft laterally to the nearest means of egress.
- Ladders must be secured and extend a minimum of 36 in (0.9 m) above the landing.
- Metal ladders should be used with caution, particularly when electric utilities are present.

D. EXPOSURE TO VEHICLES. Procedures to protect employees from being injured or killed by vehicle traffic include:

- Providing employees with and requiring them to wear warning vests or other suitable garments marked with or made of reflectorized or high-visibility materials.
- Requiring a designated, trained flagperson along with signs, signals, and barricades when necessary.

E. EXPOSURE TO FALLING LOADS. Employees must be protected from loads or objects falling from lifting or digging equipment. Procedures designed to ensure their protection include:

- Employees are not permitted to work under raised loads.
- Employees are required to stand away from equipment that is being loaded or unloaded.
- Equipment operators or truck drivers may stay in their equipment during loading and unloading if the equipment is properly equipped with a cab shield or adequate canopy.

F. WARNING SYSTEMS FOR MOBILE EQUIPMENT. The following steps should be taken to prevent vehicles from accidentally falling into the trench:

- Barricades must be installed where necessary.
- Hand or mechanical signals must be used as required.
- Stop logs must be installed if there is a danger of vehicles falling into the trench.
- Soil should be graded away from the excavation; this will assist in vehicle control and channeling of run-off water.

G. HAZARDOUS ATMOSPHERES AND CONFINED SPACES. Employees shall not be permitted to work in hazardous and/or toxic atmospheres. Such atmospheres include those with:

- Less than 19.5% or more than 23.5% oxygen;
- A combustible gas concentration greater than 20% of the lower flammable limit; and
- Concentrations of hazardous substances that exceed those specified in the Threshold Limit Values for Airborne Contaminants established by the ACGIH (American Conference of Governmental Industrial Hygienists).

All operations involving such atmospheres must be conducted in accordance with OSHA requirements for occupational health and environmental controls (see Subpart D of 29 CPR 1926) for personal protective equipment and for lifesaving equipment (see Subpart E, 29 CFR 1926). Engineering controls (e.g., ventilation) and respiratory protection may be required.

When testing for atmospheric contaminants, the following should be considered:

- Testing should be conducted before employees enter the trench and should be done regularly to ensure that the trench remains safe.
- The frequency of testing should be increased if equipment is operating in the trench.
- Testing frequency should also be increased if welding, cutting, or burning is done in the trench.

Employees required to wear respiratory protection must be trained, fit-tested, and enrolled in a respiratory protection program. Some trenches qualify as confined spaces. When this occurs, compliance with the Confined Space Standard is also required.

H. EMERGENCY RESCUE EQUIPMENT. Emergency rescue equipment is required when a hazardous atmosphere exists or can reasonably be expected to exist. Requirements are as follows:

- Respirators must be of the type suitable for the exposure. Employees must be trained in their use and a respirator program must be instituted.
- Attended (at all times) lifelines must be provided when employees enter bell-bottom pier holes, deep confined spaces, or other similar hazards.
- Employees who enter confined spaces must be trained.

I. STANDING WATER AND WATER ACCUMULATION. Methods for controlling standing water and water accumulation must be provided and should consist of the following if employees are permitted to work in the excavation:

- Use of special support or shield systems approved by a registered professional engineer.
- Water removal equipment, i.e. well pointing, used and monitored by a competent person.
- Safety harnesses and lifelines used in conformance with 29 CFR 1926.104.
- Surface water diverted away from the trench.
- Employees removed from the trench during rainstorms.
- Trenches carefully inspected by a competent person after each rain and before employees are permitted to re-enter the trench.

J. INSPECTIONS. Inspections shall be made by a competent person and should be documented. The following guide specifies the frequency and conditions requiring inspections:

- Daily and before the start of each shift;
- As dictated by the work being done in the trench;
- After every rainstorm;
- After other events that could increase hazards, e.g. snowstorm, windstorm, thaw, earthquake, etc.;
- When fissures, tension cracks, sloughing, undercutting, water seepage, bulging at the bottom, or other similar conditions occur;
- When there is a change in the size, location, or placement of the spoil pile; and

- When there is any indication of change or movement in adjacent structures.

XI. BIBLIOGRAPHY.

29 CFR 1926, Subpart P. *Excavations*.

Construction Safety Association of Ontario. *Trenching Safety*. 74 Victoria St., Toronto, Ontario, Canada M5C2A5.

International Labour Office (ILO). *Building Work: A Compendium of Occupational Safety and Health Practice.* International Occupational Safety and Health Information Centre (CIS): ILO, Geneva, Switzerland.

National Safety Council. *Accident Prevention Manual for Industrial Operations, Engineering and Technology*, 9th ed., Chicago, IL: National Safety Council.

National Safety Council. *Protecting Worker's Lives: A Safety and Health Guide for Unions*. Chicago, IL: National Safety Council.

National Safety Council. Industrial Data Sheets: I-482, *General Excavation*, and I-254, *Trench Excavation*, Chicago, IL: National Safety Council.

National Utility Contractors Association, *Competent Person Manual*-1991.

NBS/NIOSH, *Development of Draft Construction Safety Standards for Excavations*. Volume I, April 1983. NIOSH 83-103, Pub. No. 84-100-569. Volume II, April 1983. NIOSH 83-2693, Pub. No. 83-233-353.

Scardino, A.J., Jr. 1993. *Hazard Identification and Control--Trench Excavation*. Lagrange, TX: Carlton Press.

APPENDIX V: 2-1. SITE ASSESSMENT QUESTIONS

During first and subsequent visits to a construction or facility maintenance location, the compliance officer (or the site's safety officer or other competent person) may find the following questions useful.

1. Is the cut, cavity, or depression a trench or an excavation?
2. Is the cut, cavity, or depression more than 4 ft (1.2 m) in depth?
3. Is there water in the cut, cavity, or depression?
4. Are there adequate means of access and egress?
5. Are there any surface encumbrances?
6. Is there exposure to vehicular traffic?
7. Are adjacent structures stabilized?
8. Does mobile equipment have a warning system?
9. Is a competent person in charge of the operation?
10. Is equipment operating in or around the cut, cavity, or depression?
11. Are procedures required to monitor, test, and control hazardous atmospheres?
12. Does a competent person determine soil type?
13. Was a soil testing device used to determine soil type?

14. Is the spoil placed 2 ft (0.6 m) or more from the edge of the cut, cavity, or depression?

15. Is the depth 20 ft (6.1 m) or more for the cut, cavity, or depression?

16. Has a registered professional engineer approved the procedure if the depth is more than 20 ft (6.1 m)?

17. Does the procedure require benching or multiple benching? Shoring? Shielding?

18. If provided, do shields extend at least 18 in (0.5 m) above the surrounding area if it is sloped toward the excavation?

19. If shields are used, is the depth of the cut more than 2 ft (0.6 m) below the bottom of the shield?

20. Are any required surface crossings of the cut, cavity, or depression the proper width and fitted with hand rails?

21. Are means of egress from the cut, cavity, or depression no more than 25 ft (7.6m) from the work?

22. Is emergency rescue equipment required?

23. Is there documentation of the minimum daily excavation inspection?

SECTION V: CHAPTER 3

CONTROLLING LEAD EXPOSURES IN THE CONSTRUCTION INDUSTRY: ENGINEERING AND WORK PRACTICE CONTROLS

Contents:

I. INTRODUCTION.

A. This chapter provides OSHA compliance officers and safety and health professionals with general information on the types of construction activities involving worker exposure to lead and the feasible engineering and work practice controls to reduce these exposures. The construction activities identified range from those such as abrasive blasting and welding, cutting, and burning, where exposures to lead are often high, to encapsulating lead-based paint or using lead pots, where exposures are generally low.

B. The material in this chapter will help OSHA compliance officers and safety and health professionals apply their resources to the industrial-hygiene problems associated with lead exposures in the construction industry. General engineering and work practice controls that can be applied to almost any construction activity are addressed under Paragraph II, below.

C. This chapter also describes those lead-related tasks and operations that give rise to lead exposures among construction workers. Recommended/feasible engineering controls (e.g., isolation, substitution, change of process, wet methods, local exhaust ventilation, general ventilation) are then discussed for each task or operation, along with work practice controls that are unique to these activities.

D. The current OSHA standard (29 CFR 1926.62) for lead exposure in construction has a permissible exposure limit (PEL) of 50 micrograms per cubic meter of air (50 μg/m3), measured as an 8-hour time-weighted average (TWA). As with all OSHA health standards, when the PEL is exceeded, the hierarchy of controls requires employers to institute feasible engineering and work practice controls as the primary means to reduce and maintain employee exposures to levels at or below the PEL. When all feasible engineering and work practice controls have been implemented but have proven inadequate to meet the PEL, employers must nonetheless implement these controls and must supplement them with appropriate respiratory protection. The employer also must ensure that employees wear the respiratory protection provided when it is required.

E. Certain lead-related construction tasks commonly produce exposures above the PEL and often orders of magnitude above the PEL. The OSHA lead standard for construction is unique in that it groups tasks (Appendix V:3-1) that are presumed to be associated with employee exposures above the PEL into three lead-exposure ranges. The exposure ranges assigned to the different categories of tasks are based on data collected by OSHA and other sources including two advisory groups.

F. Until an employer performs an employee-exposure assessment and determines the magnitude of the exposures actually occurring during the lead-related activity, the employer must assume that employees performing that task are exposed to the lead concentrations indicated in Appendix V:3-1. For all three groups of tasks, employers are required to provide respiratory protection appropriate to the task's presumed exposure level, protective work clothing and equipment, change areas, hand-washing facilities, training, and the initial medical surveillance prescribed by paragraph (d)(2)(v) of the standard (29 CFR 1926.62). The only difference in the provisions applying to the three categories of tasks is the degree of respiratory protection required.

II. ENGINEERING AND WORK PRACTICE CONTROLS.

A. ENGINEERING CONTROLS. Engineering controls, such as ventilation, and good work practices are the preferred methods of minimizing exposures to airborne lead at the worksite. The engineering control methods that can be used to reduce or eliminate lead exposures can be grouped into three main categories: (1) substitution; (2) isolation; and (3) ventilation. Engineering controls are the first line of defense in protecting workers from hazardous exposures.

1. Substitution.

a. Substitution includes using a material that is less hazardous than lead, changing from one type of process equipment to another, or even, in some cases, changing the process itself to reduce the potential exposure to lead. In other words, material, equipment, or an entire process can be substituted to provide effective control of a lead hazard. However, in choosing alternative methods, a hazard evaluation should be conducted to identify inherent hazards of the method and equipment.

b. Examples of substitution include:

- Use of a less hazardous material: applying a nonleaded paint rather than a coating that contains lead.
- Change in process equipment: using less dusty methods such as vacuum blast cleaning, wet abrasive blast cleaning, shrouded power tool cleaning, or chemical stripping to substitute for open abrasive blast cleaning to reduce exposure to respirable airborne particulates containing lead.
- Change in process: performing demolition work using mobile hydraulic shears instead of a cutting torch to reduce exposure to lead fumes generated by heating lead compounds.

c. Any material that is being considered as a substitute for a lead-based paint should be evaluated to ensure that it does not contain equally or more toxic components (e.g., cadmium or chromates). Because substitute materials can also be hazardous, employers must obtain a Material Safety Data Sheet (MSDS) before a material is used in the workplace. If the MSDS identifies the material as hazardous, as defined by OSHA's hazard communication standard (29 CFR 1926.59), an MSDS must be maintained at the job site and proper protective measures must be implemented prior to usage of the material.

2. Isolation is a method of limiting lead exposure to those employees who are working directly with it. A method which isolates lead contamination and thus protects both nonessential workers, bystanders, and the environment is to erect a sealed containment structure around open abrasive blasting operations. However, this method may substantially increase the lead exposures of the workers doing the blasting inside the structure. The containment structure must therefore be provided with negative-pressure exhaust ventilation to reduce workers' exposure to lead, improve visibility, and reduce emissions from the enclosure.

3. Ventilation.

a. Ventilation, either local or dilution (general), is probably the most important engineering control available to the safety and health professional to maintain airborne concentrations of lead at acceptable levels. Local exhaust ventilation, which includes both portable ventilation systems and shrouded tools supplied with ventilation, is generally the preferred method. If a local exhaust system is properly designed, it will capture and control lead particles at or near the source of generation and transport these particles to a collection system before they can be dispersed into the work environment.

b. Dilution ventilation, on the other hand, allows lead particles generated by work activities to spread throughout the work area and then dilutes the concentration of particles by circulating large quantities of air into and out from the work area. For work operations where the sources of lead dust generation are numerous and widely distributed (e.g., open abrasive blasting conducted in containment structures), dilution ventilation may be the best control.

c. Examples of ventilation controls include the following:

- Power tools that are equipped with dust collection shrouds or other attachments for dust removal and are exhausted through a HEPA vacuum system.
- Vacuum blast nozzles (vacuum blasting is a variation on open abrasive blasting). In this type of blasting, the blast nozzle has local containment (a shroud) at its end, and containment is usually accomplished through brush-lined attachments at the outer periphery and a vacuum inlet between the blast nozzle and the outer brushes.
- Containment structures that are provided with negative-pressure dilution ventilation systems to reduce airborne lead concentrations within the enclosure, increase visibility, and control emissions of particulate matter to the environment.

B. WORK PRACTICE CONTROLS. Work practices involve the way a task is performed. OSHA has found that appropriate work practices can be a vital aid in lowering worker exposures to hazardous substances and in achieving compliance with the PEL. Some fundamental and easily implemented work practices are: (1) good housekeeping, (2) use of appropriate personal hygiene practices, (3) periodic inspection and maintenance of process and control equipment, (4) use of proper procedures to perform a task, (5) provision of supervision to ensure that the proper procedures are followed, and (6) use of administrative controls.

1. Housekeeping.

a. A rigorous housekeeping program is necessary in many jobs to keep airborne lead levels at or below permissible exposure limits. Good housekeeping involves a regular schedule of housekeeping activities to remove accumulations of lead dust and lead-containing debris. The schedule should be adapted to exposure conditions at a particular worksite.

b. All workplace surfaces must be maintained as free as practicable of accumulations of lead dust. Lead dust on overhead ledges, equipment, floors, and other surfaces must be removed to prevent traffic, vibration, or random air currents from re-entraining the lead-laden dust and making it airborne again. Regularly scheduled clean-ups are important because they minimize the re-entrainment of lead dust into the air, which otherwise serves as an additional source of exposure that engineering controls are generally not designed to control.

c. Vacuuming is considered the most reliable method of cleaning dusty surfaces, but any effective method that minimizes the likelihood of re-entrainment may be used (for example, a wet floor scrubber). When vacuuming equipment is used, the vacuums must be equipped with high-efficiency particulate air (HEPA) filters. Blowing with compressed air is generally prohibited as a cleaning method, unless the compressed air is used in conjunction with a ventilation system that is designed to capture the airborne dust created by the compressed air (e.g. dust "blowdown" inside a negative-pressure containment structure). In addition, all persons doing the cleanup should be provided with suitable respiratory protection and personal protective clothing to prevent contact with lead.

d. Where feasible, lead-containing debris and contaminated items accumulated for disposal should be wet-misted before handling. Such materials must be collected and put into sealed impermeable bags or other closed impermeable containers. Bags and containers must be labeled to indicate that they contain lead-containing waste.

2. Personal Hygiene Practices.

a. Personal hygiene is also an important element in any program to protect workers from exposure to lead dust. When employee exposure is above the PEL, the lead standard requires the employer to provide, and ensure that workers use, adequate shower facilities (where feasible), hand-washing facilities, clean change areas, and separate noncontaminated eating areas. Employees must also wash their hands and faces prior to eating, drinking, using tobacco products, or applying cosmetics, and they must not eat, drink, use tobacco products, or apply cosmetics in any work area where the PEL is exceeded. In addition, employees must not enter lunchroom facilities or eating areas while wearing protective work clothing or equipment unless surface lead dust has first been removed from the clothing or equipment by vacuuming or another cleaning method that limits dispersion of lead dust.

b. Workers who do not shower and change into clean clothing before leaving the worksite may contaminate their homes and vehicles with lead dust. Other members of the household may then be exposed to harmful amounts of lead. A recent NIOSH publication (NIOSH 1992) points out the dangers of "take-home" lead contamination. For the same reason, vehicles driven to the worksite should be parked where they will not be contaminated with lead.

c. The personal hygiene measures described above will reduce worker exposure to lead and decrease the likelihood of lead absorption caused by ingestion or inhalation of lead particles. In addition, these measures will minimize employee exposure to lead after the work shift ends, significantly reduce the movement of lead from the worksite, and provide added protection to employees and their families.

3. Change Areas.

   a. When employee airborne exposures to lead are above the PEL, the employer must provide employees with a clean change area that is equipped with storage facilities for street clothes and a separate area with facilities for the removal and storage of lead-contaminated protective work clothing and equipment. Separate clean and dirty change areas are essential in preventing cross-contamination of the employees' street and work clothing.

   b. Clean change areas are used to remove street clothes, to suit up in clean work clothes (protective clothing), and to don respirators prior to beginning work, and to dress in street clothes after work. No lead-contaminated items are permitted to enter the clean change area. Work clothing should be worn only on the job site. Under no circumstances should lead-contaminated work clothes be laundered at home or taken from the worksite, except to be laundered professionally or properly disposed of following applicable Federal, State, and local regulations.

4. Showers. When employee exposures exceed the PEL, the employer must provide employees with suitable shower facilities, where feasible, so that exposed employees can remove accumulated lead dust from their skin and hair prior to leaving the worksite. Where shower facilities are available, employees must shower at the end of the work shift before changing into their street clothes and leaving the worksite. Showers must be equipped with hot and cold water, in accordance with 29 CFR 1926.51(f)(4)(iv).

5. Washing Facilities. Washing facilities must be provided to employees in accordance with the requirements of 29 CFR 1926.51(f). Water, soap, and clean towels are to be provided for this purpose. Where showers are not provided, the employer must ensure that employees wash their hands and faces at the end of the work shift.

6. Eating Facilities. The employer must provide employees who are exposed to lead at levels exceeding the PEL with eating facilities or designated areas that are readily accessible to employees and must ensure that the eating area is free from lead contamination. To further minimize the possibility of food contamination and reduce the likelihood of additional lead absorption from contaminated food, beverages, tobacco, and cosmetic products, the employer must prohibit the storage, use, or consumption of these products in any area where lead dust or fumes may be present.

7. Periodic Inspection and Maintenance. Periodic inspection and maintenance of process equipment and control equipment such as ventilation systems are another important work practice control. At worksites where full containment is used as an environmental control, the failure of the ventilation system for the containment area can result in hazardous exposures to workers within the enclosure. Equipment that is near failure or in disrepair will not perform as intended. Regular inspections can detect abnormal conditions so that timely maintenance can be performed. If process and control equipment is routinely inspected, maintained, and repaired, or is replaced before failure occurs, there is less chance that hazardous employee exposures will occur.

8. Performance of Task.

   a. In addition to the work practice controls previously described in Section B, the employer must provide training and information to employees as required by OSHA's lead in construction (29 CFR 1926.62), hazard communication (29 CFR 1926.59), and safety training and education (29 CFR 1926.21) standards. One important element of this program is training workers to follow the proper work practices and procedures for their jobs. Workers must know the proper way to perform job tasks to minimize their exposure to lead and to maximize the effectiveness of engineering controls. For example, if a worker performs a task away from (rather than close to) an exhaust hood, the control measure will be unable to capture the particulates generated by the task and will thus be ineffective.

b. In certain applications such as abatement in buildings, wet methods can significantly reduce the generation of lead-containing dust in the work area. Wetting of surfaces with water mist prior to sanding, scraping, or sawing, and wetting lead-containing building components prior to removal will minimize airborne dust generation during these activities. Failure to operate engineering controls properly may also contaminate the work area. Workers can be informed of safe operating procedures through fact sheets, discussions at safety meetings, and other educational means.

9. Supervision.

a. Good supervision is another important work practice. It provides needed support for ensuring that proper work practices are followed by workers. By directing a worker to position the exhaust hood properly or to improve work practice, such as standing to the side or upwind of the cutting torch to avoid the smoke plume, a supervisor can do much to minimize unnecessary employee exposure to airborne contaminants.

b. The OSHA construction standard for lead also requires that a competent person perform frequent and regular inspections of job sites, materials, and equipment. A competent person is defined by the standard as one who is capable of identifying existing and predictable lead hazards and who has authorization to take prompt corrective measures to eliminate them.

10. Administrative Controls.

a. Administrative controls are another form of work practice controls that can be used to influence the way a task is performed. Controls of this type generally involve scheduling of the work or the worker. For example, employee exposure can be controlled by scheduling construction activities or workers' tasks in ways that minimize employee exposure levels. One method the employer can use is to schedule the most dust- or fume-producing operations for a time when the fewest number of employees will be present.

b. Another method is worker rotation, which involves rotating employees into and out of contaminated areas in the course of a shift, thereby reducing the full-shift exposure of any given employee. When a worker is rotated out of the job that involves lead exposure, he or she is assigned to an area of the worksite that does not involve lead exposure. If this method is used to control worker exposure to lead, the lead standard requires that the employer implement a job rotation schedule that (1) identifies each affected worker, (2) lists the duration and exposure levels at each job or work station where each affected employee is located, and (3) lists any other information that may be useful in assessing the reliability of administrative controls to reduce exposure to lead.

III. OPERATIONS.

This section describes the job operations that take place in construction worksites and involve worker exposures to lead. Although this list of operations is extensive, it is not necessarily inclusive (i.e, other construction activities not mentioned here may also involve lead exposure). OSHA's lead standard for onstruction applies to any construction activity that potentially exposes workers to airborne concentrations of lead.

A. OPEN ABRASIVE BLAST CLEANING.

1. The most common method of removing lead-based paints is open abrasive blast cleaning. The abrasive medium, generally steel shot/grit, sand, or slag, is propelled through a hose by compressed air. The abrasive material abrades the surface of the structure, exposing the steel substrate underneath. The abrasive also conditions the substrate, forming a "profile" of the metal, which improves the adherence of the new paint. Work is generally organized so that blasting proceeds for approximately one-half day, followed by compressed air cleaning of the steel and application of the prime coat of paint. Prime coat painting must follow blasting immediately to prevent surface rust from forming. Intermediate or finish coats of paint are applied later.

2. Structures that are typically cleaned by open abrasive blasting are bridges, tanks and towers, locks and dams, pipe racks, pressure vessels and process equipment, supporting steel, and metal buildings. Until recently, abrasive blasting work was conducted in unobstructed air. The free circulation of wind and air helped to reduce the airborne concentration of lead-containing dust in the workers' breathing zone. Tarpaulins were generally used only to protect neighboring homes and automobiles from a damaging blast of abrasive or to reduce residents' complaints about overspray, dust, and dirt. Currently, some state and local regulations require the use of enclosures or containment structures to prevent the uncontrolled dispersal of lead dust and debris into the environment. Although containment structures are designed to reduce the dispersion of lead into the environment, they usually increase worker exposure to airborne lead, reduce visibility, and increase the risk of slip and fall injuries due to waste material build-up on the footing surface of the enclosure.

3. Containment structures vary in their design and in their effectiveness in containing debris. Some containment structures consist of tarpaulins made of open mesh fabrics (screens) that are loosely fitted around the blasting area; some use rigid materials such as wood, metal, or plastic to enclose the blasting area; and some use a combination of flexible and rigid materials. Large air-moving devices may be connected to the enclosed containment structure to exhaust dust-laden air and create a negative pressure with respect to the ambient atmosphere.

4. Containment or enclosure structures can be broadly classified as either partial or full. Partial containments refer to those that inherently allow some level of emission to the atmosphere outside of the containment. An example of a partial containment is a structure with loosely hung permeable tarps and partially sealed joints and entryways. Full containment refers to a relatively tight enclosure (with tarps that are generally impermeable and fully sealed joints and entryways) where minimal or no fugitive emissions are expected to reach the outside environment. Partial or full containments can be used to contain entire structures or portions thereof. Examples of the kinds of engineering controls and work practices that can be implemented to protect blasting workers are presented below.

5. Engineering Controls.

   a. Containment/ventilation systems should be designed and operated so as to create a negative pressure within the structure, which reduces the dispersion of lead into the environment. The containment/ventilation system should be designed to optimize the flow of ventilation air past the worker(s), thereby reducing the airborne concentration of lead and increasing visibility. This can be accomplished by employing either a downdraft or crossdraft ventilation system that is properly balanced by a make-up air supply. Designs for the containment structure and ventilation systems should be specific to each task, because conditions can vary substantially from one worksite to another. The dust-laden air must be filtered prior to its release into the atmosphere.

   b. Mini-enclosures, which have smaller cross-sectional areas than conventional enclosures, can be erected. Mini-enclosures have advantages over larger conventional enclosures because the same size fan and dust collector can achieve much higher velocities past the helmets of the workers. Mini-enclosure containment structures are usually light-weight, low wind-loading structures that isolate that area where blasting and surface priming is taking place on a given day.

   c. The risk of silicosis is high among workers exposed to abrasive blasting with silica-containing media, and this hazard is difficult to control. The National Institute for Occupational Safety and Health (NIOSH) has therefore recommended since 1974 that silica sand (or other substances containing more than 1% crystalline silica) be prohibited as abrasive blasting material. A variety of materials such as slags and steel grit are available as alternative blasting media. Because some substitute materials may have their own unique hazards, the MSDS for the substitute material should be consulted before it is used.

   d. Blast cleaning with recyclable abrasive such as steel grit or aluminum oxide requires specialized equipment for vacuuming or collecting the abrasive for reuse, separating the lead dust and fines from the reusable abrasive, and, in the case of steel grit, maintaining clean, dry air to avoid rusting of the abrasive. In addition, the abrasive classifier must be extremely efficient in removing lead dust, to prevent it from being reintroduced into the containment and combining with the paint to increase worker exposures. Recycling equipment must be well-maintained and regularly monitored to ensure it is removing lead effectively.

e. When site conditions warrant, less dusty methods should be used in place of open abrasive blast cleaning. These include:

- Vacuum-blast cleaning;
- Wet abrasive blast cleaning;
- High-pressure water jetting;
- High-pressure water jetting with abrasive injection;
- Ultrahigh-pressure water jetting;
- Sponge jetting;
- Carbon-dioxide (dry-ice) blasting;
- Chemical stripping; and
- Power-tool cleaning.

6. Work Practice Controls. Construction employers engaged in open abrasive blast cleaning operations should implement the following control measures:

- Develop and implement a good respiratory protection program in accordance with OSHA requirements in 29 CFR 1926.103.
- Provide workers with Type CE abrasive-blast respirators; these are the only respirators suitable for use in abrasive-blasting operations. Currently there are only three models of Type CE abrasive blast respirators certified by MSHA/ NIOSH: - A continuous-flow respirator with a loose-fitting hood that has a protection factor of 25; - A continuous-flow respirator with a tight-fitting face-piece that has a protection factor of 50; and - A pressure-demand respirator with a tight-fitting face-piece that has a protection factor of 2000.

The first two models (i.e. the continuous-flow respirators) should be used only for abrasive blast operations where the abrasive materials do not include silica sand and the level of contaminant in the ambient air does not exceed 25 or 50 times the recommended exposure limit, respectively. The third model, which is a pressure-demand respirator, must be worn whenever silica sand is used as an abrasive material (NIOSH 1993). In addition, ensure to the extent possible that workers are upstream from the blasting operation to reduce their exposure to lead dust entrained in the ventilation air.

B. VACUUM BLAST CLEANING.

1. Vacuum blasting is a variation of open abrasive blasting. In this configuration, the blast nozzle has local containment (a shroud) at its end and containment is usually accomplished by brush-lined attachments at the outer periphery and a vacuum inlet between the blast nozzle and the outer brushes (Waagbo and McPhee 1991). The brushes prevent dispersion of the abrasive and debris as they rebound from the steel surface. These particles are removed from the work area by the built-in vacuum system. The abrasive itself can either be disposed of or cleaned and recycled.

2. If used properly, vacuum blast cleaning can achieve cleaning of good quality with minimal dust generation except in areas where access is difficult because of configuration (such as between back-to-back angles). A variety of heads are available to achieve a tight seal for inside corners, outside corners, and flat surfaces. The advantages of vacuum blasting are that most of the waste material and abrasive is collected at the site of generation and is therefore not transported to the breathing zone of the worker, and that there may be little or no need for containment.

3. Vacuum blasting has several disadvantages (Knoy 1990). It is more time-consuming than conventional open abrasive blasting because the abrasive blast nozzle must be smaller to be able to capture the ricocheting abrasive and dust. This restricts the dispersion of the abrasive and thus the size of the area that can be cleaned. Abrasive also may escape the vacuum head if the brush attachments do not seal completely around the substrate; poor seals may be caused by operator fatigue, poor work practices or irregular surfaces and edges. Small areas and areas with gross irregularities cannot be effectively sealed by the shroud. The vacuum system and brushes

obscure the blast surface, and some areas may therefore need to be blasted repeatedly because they are missed on the first or second pass. In addition, some vacuum heads are so heavy that mechanical suspension systems are needed to support them, and even then, the blasters may need to take frequent breaks.

C. WET ABRASIVE BLAST CLEANING.

1. Wet abrasive blast cleaning is a modification of traditional open abrasive blast cleaning. This system uses compressed air to propel the abrasive medium to the surface being cleaned; however, water (which reduces dusting) is injected into the abrasive stream either before or after the abrasive exits the nozzle (Figure V:3-1).

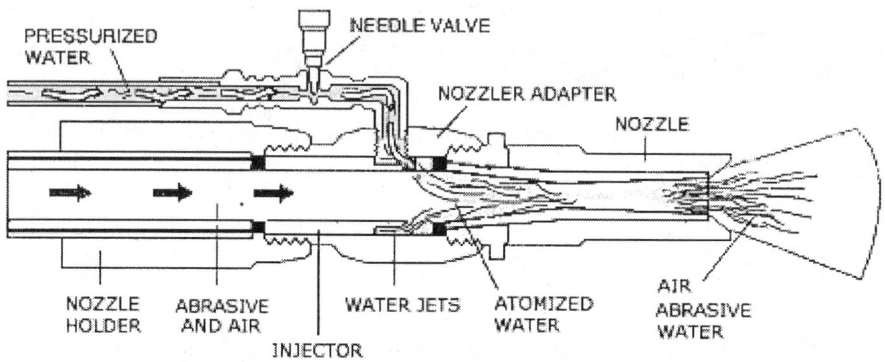

FIGURE V: 3-1. WET ABRASIVE BLAST CLEANING.

2. The disadvantages of using water are that inhibitors may be necessary to avoid flash rusting, the containment must be designed to capture the water and debris generated by the cleaning process, wet abrasive/paint debris is more difficult to handle and transport than dry debris, and, unless the water can be filtered, it may add to the volume of debris generated.

3. Because many corrosion inhibitors (e.g., nitrates, nitrites, and amines) raise industrial hygiene concerns, their use must be considered carefully.

D. HIGH-PRESSURE WATER JETTING.

1. High-pressure water jetting (6,000 to 25,000 psi) utilizes a pressure pump, a large volume of water, a specialized lance and nozzle assembly and, in some cases, a supply of inhibitor to prevent flash rusting. High-pressure water can remove loose paint and rust, but will not efficiently remove tight paint or tight rust, or mill scale. This technique does not create a profile (mechanically induced toothing pattern to enhance the adhesion of high-performance coatings) on its own, but if the original surface was blast cleaned, high-pressure water jetting can be used to remove the old paint and restore the original profile.

2. Because of the water, this kind of jetting generates little dust. The containment must be constructed to collect water rather than to control dust emissions. The debris generated is comprised of the removed paint and rust, along with the water. If the lead debris can be adequately filtered from the water, the volume of debris is low. If not, the volume of debris can be high. Typically, five to ten gallons of water per minute are used.

3. Productivity can be high with this method if the objective is to remove only loose, flaky paint. If the objective is to remove tight paint, productivity may be low. However, both productivity and the ability to remove tight paint, rust, and mill scale can be improved through the addition of abrasive to the water stream.

E. HIGH-PRESSURE WATER JETTING WITH ABRASIVE INJECTION.

1. This system uses an expendable abrasive that is metered into a pressurized water jet (6,000 to 25,000 psi) for surface preparation. Although airborne lead exposures are virtually eliminated with this approach, wet abrasive is more difficult to handle and move than dry abrasive, and the volume of debris also increases. Because the abrasive exposes the bare substrate, inhibitors such as sodium nitrate or amines are often added to the water to prevent flash rusting.

2. Abrasives used for injection include sand and slag materials, as well as soluble abrasives such as sodium bicarbonate. The sodium bicarbonate will not remove paint, rust, and mill scale as efficiently as sand or slag abrasives. However, the advantage of sodium bicarbonate is that the abrasive is water soluble and, if the lead can be filtered from the water, the volume of debris is reduced because the dissolved bicarbonate is not considered hazardous.

F. ULTRAHIGH-PRESSURE WATER JETTING.

1. Ultrahigh-pressure water jetting utilizes pressurized water at pressures in excess of 25,000 psi. Ultrahigh-pressure water jetting is similar to high pressure water jetting except that the ultrahigh variant uses even higher pressures. This means that it cleans more efficiently and removes tight paint and rust more effectively. In addition, the volume of water required is reduced, with less than five gallons per minute typically used. Because of the water, little dust is generated.

2. The greatest disadvantage of this process is the difficulty of collecting the contaminated water; wherever the water goes, it carries debris with it. Dust generation, debris generation, and the type of containment necessary in ultrahigh-pressure water jetting are comparable to those in high-pressure water jetting. Inhibitors are also often required to avoid flash rusting. Mill scale is not removed; however, if the surface was previously blast cleaned, the profile of the original substrate can be restored.

G. SPONGE JETTING.

1. Sponge jetting involves the use of specialized blasting equipment that propels a combination of an abrasive material (e.g., steel, garnet) encased in a soft sponge (foam) medium. The high-density foam cleaning medium is absorptive and can be used either wet or dry. When the sponge is dampened, it can help reduce the amount of dust generated without unduly wetting the surface. The medium provides the impact needed to break the paint coating up into larger particles, and particle rebound is low because of the energy absorbed by the foam.

2. The relatively small volume of dust generated by this method can help to reduce containment requirements, although some screens and tarping are necessary to isolate the work area and to allow the sponge and debris to be collected. Productivity is lower than for open abrasive blasting using more traditional abrasives, according to contractors who have used this product (CONSAD 1993).

H. CARBON-DIOXIDE (DRY-ICE) BLASTING.

1. Cryogenic cleaning by blasting with dry-ice pellets is one of the least-tried methods of surface preparation (Figure V:3-2). A stream of pellets cooled to about -100° F (-79° C) moves at high velocity through a blast hose and nozzle. The pellets impinge on the surface and then sublime, leaving only paint debris to be cleaned up. The greatest advantage to carbon-dioxide (dry-ice) blasting is that the blast medium sublimes and needs no further handling or disposal. However, when used in confined spaces, the potential for creating an oxygen-deficient environment is significant and must always be guarded against.

2. The cost of cryogenic cleaning, however, is still often prohibitive. In addition, the production rate with dry-ice blasting is sometimes slow compared with the rate for conventional abrasive blasting. Finally, because only the paint is removed, the surface may need to be "brush-off" blasted with an abrasive to produce a rough surface to facilitate adhesion of the new coating.

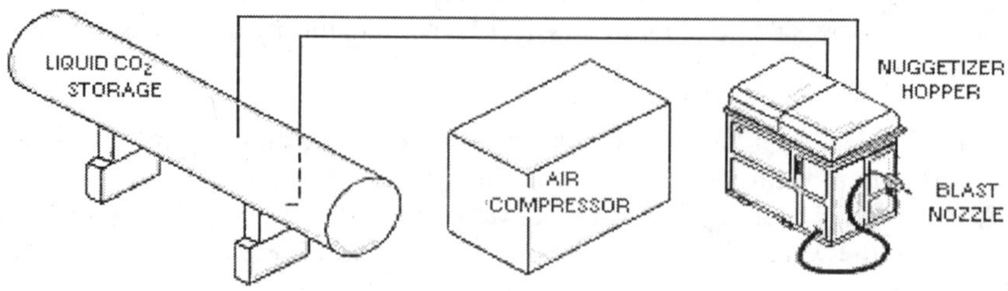

FIGURE V: 3-2. DRY-ICE BLAST CLEANING

I. WELDING, BURNING, AND TORCH CUTTING.

1. Welding and cutting activities that potentially involve exposure to lead can occur as part of a number of construction projects such as highway/railroad bridge rehabilitation (including elevated mass-transit lines), demolition, and indoor and outdoor industrial facility maintenance and renovation. Lead exposures are generated when a piece of lead-based painted steel is heated to its melting point either by an oxyacetylene torch or an arc welder. In this situation, lead becomes airborne as a volatilized component of the coating.

2. The amount of time a worker may spend actually welding or cutting can vary from only a few minutes up to a full shift. In addition, the coating being worked on may consist of several layers of lead-based paint, each of which could contain as much as 50% lead. Taken together, these factors suggest that a worker's exposure to airborne lead during welding or cutting activities can vary widely and may be exceedingly high. Lead burning, a process by which virgin or alloyed lead is melted with a torch or otherwise fused to another lead object, is typically performed in maintenance operations on electrostatic precipitators or during the installation of lead shot, bricks, or sheets in the walls or floors of health-care x-ray units or industrial sites. Lead health hazards in this operation, as in welding and torch cutting, are from lead that is superheated and released into the worker's breathing zone in the form of a fume.

3. Engineering Controls. The engineering controls that can be used, depending on feasibility, are:

   a. Local exhaust ventilation (LEV) that has a flanged hood and is equipped with HEPA filtration may be appropriate where the use of LEV does not create safety hazards. Use of a flexible duct system requires that the welder be instructed to keep the duct close to the emission source and to ensure the duct is not twisted or bent.

   b. A fume-extractor gun that removes fumes from the point of generation (Figure V:3-3) is an alternative to an exhaust hood for gas-shielded arc-welding processes. Such extraction systems can reduce breathing zone concentrations by 70% or more (Hughes and Amendola, 1982). These systems require that the gun and shielding gas flow rates be carefully balanced to maintain weld quality and still provide good exhaust flow.

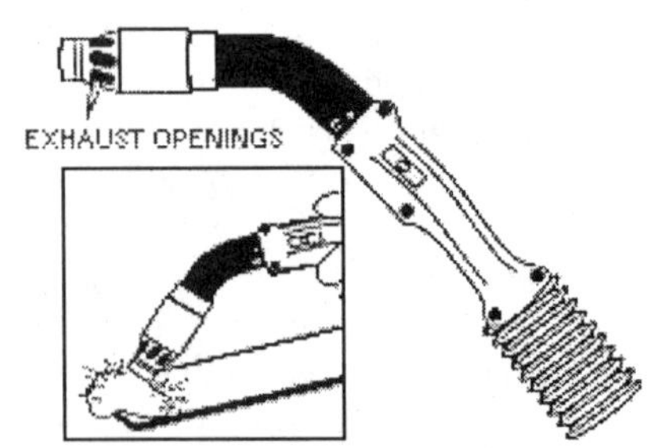

FIGURE V: 3-3. FUME-EXTRACTOR GUN.

   c. A longer cutting torch can be used in some situations to increase the distance from the lead source to the worker's breathing zone.

d. Hydraulic shears can sometimes be used to mechanically cut steel that is coated with lead based-paint. The use of this method is limited by the ability of the shears to reach the cutting area.

e. Whenever possible, pneumatic air tools should be used to remove rivets in lieu of burning and torch cutting.

4. Work Practice Controls. The following work practice controls will help to reduce worker exposures to lead during welding, burning, and torch cutting:

a. Strip back all lead-based paint for a distance of at least 4 inches in all directions from the area of heat application. Chemical stripping, vacuum-shrouded hand tools, vacuum blasting, or other suitable method may be used. However, in enclosed spaces, strip back or protect the workers with air-line respirators in accordance with the requirements of 29 CFR 1926.354(c).

b. Ensure that workers avoid the smoke plume by standing to the side or upwind of the cutting torch whenever the configuration of the job permits.

c. Prohibit burning to remove lead-based paint. Paint should be removed using other methods, such as chemical stripping, power tools (e.g. needle guns) with vacuum attachments, etc.

J. SPRAY PAINTING WITH LEAD-BASED PAINT.

1. In the construction field, the primary source of lead exposure in painting is red lead primers, although many finish coatings continue to contain a small percentage of lead. For most interior or exterior construction painting projects, workers employ conventional compressed-air spray equipment. Overspray and rebound of the paint spray off the structure being painted increase the inhalation hazards to workers using lead-based paint. The magnitude of the painter's particulate exposure to lead is dependent on the product used, its lead content, and the quantity of paint applied.

2. Engineering Controls. The following engineering controls will reduce or eliminate worker exposures to lead during painting:

a. Applying non-lead containing paints and primers.

b. To the extent possible, replacing lead chromate with zinc.

c. Hand-applying lead-based paint by brush or roller coating methods rather than spray methods.

d. Using local exhaust ventilation with proper filtration. (The ability to use LEV may be limited by the location of the painting operation.)

K. MANUAL SCRAPING AND SANDING OF LEAD-BASED PAINTS.

1. Hand scraping of lead-based paints involves the use of a hand-held scraping tool to remove paint from coated surfaces. The health hazards in this activity are caused by the lead dust and paint chips produced in the scraping process. Hand sanding can also produce excessive dust. These activities are typically performed during residential and commercial/institutional lead abatement projects.

2. Engineering and Work Practice Controls. Controls that employers can implement to protect workers performing scraping and sanding of lead-based paints are:

a. Use of wet-sanding and wet-scraping methods in conjunction with HEPA vacuuming or HEPA mechanical ventilation. Wet methods include misting of peeling paint with water before scraping, and sanding and misting of debris prior to sweeping or vacuuming.

b. Use of shrouded power tools with HEPA vacuum attachments. The shroud must be kept flush with the surface.

c. Use of techniques with known low exposure potential, such as encapsulation and removal or replacement instead of hand-scraping and hand-sanding.

L. MANUAL DEMOLITION AND/OR REMOVAL OF PLASTER WALLS OR BUILDING COMPONENTS.

1. The demolition of lead-painted plaster walls or building components is usually performed by striking a wall with a sledge hammer or similar tool. This results in an uncontrolled release of dust. High levels of airborne total dust and lead dust can be generated by breaking lead-painted plaster into small pieces.

2. Removal and replacement is the process of removing components (such as windows, doors, kitchen cabinets, and trim) that have lead-painted surfaces and installing new components that are free of lead-containing paint. Exposures may result from the release of dust and paint-chip particles when these items are removed with a prybar or cut with a saw. Unless the component is seriously deteriorated, occupational exposures during this operation are minimal.

3. Engineering and Work Practice Controls. Controls and practices used to reduce lead exposures during demolition/removal of architectural components are:

   a. Install partitions or other temporary barriers to allow for partial containment of dust to minimize exposures to other workers and building occupants.

   b. Keep surfaces and debris moist when disturbing them.

   c. Remove wallboard by cutting it into large pieces/sections with a carpet knife or shrouded saw with HEPA vacuum attachment.

M. HEAT-GUN REMOVAL OF LEAD-BASED PAINT.

1. In this activity, the worker uses a heat gun, a tool similar in design to a hand-held hair dryer. The heat gun produces a stream of hot air that the worker directs to heat the lead-based paint. This heat separates the substrate, which is subsequently scraped off with a putty knife or similar tool. The health hazards encountered are generated by lead fumes released into the air during the heating process and lead particulates created during the scraping process.

2. Engineering and Work Practice Controls. The controls used to reduce lead exposure during heat gun operations are:

   a. Provide thermostatic control for heat guns to restrict operating temperatures to the lowest temperature that will allow for the effective removal of lead-based paint. [NOTE:HUD places a 700° F limit on the use of heat guns. However, NIOSH (1990) reports that the 700° F restriction for heat-gun nozzle airstream temperatures appears to limit the effectiveness of the guns in removing paint. To compensate for the airstream temperature limitation, workers often hold the gun nozzle close to the surfaces (less than one inch). This reduces the surface area heated and potentially increases the time required for paint removal and prolongs the duration of exposure. On the other hand, commercial heat guns operating at airstream temperatures of 1,000° F can generate and disperse high levels of airborne lead.]

   b. Use techniques with known low exposure potential such as encapsulation and removal/replacement instead of hand scraping with a heat gun.

N. CHEMICAL STRIPPING OF LEAD-BASED PAINT.

1. Chemical stripping of old paint coatings is performed by applying solvent- or caustic-based strippers to the surface, either by hand or spray gun. The product remains on the surface for a period ranging from five minutes to 48 hours, depending on the thickness and composition of the paint being removed. Mechanical scrapers, a

vacuum system, or pressurized water are then used to remove the product and the stripped paint. Finally, a wet-vac system is used to clean the surface, although hard-to-reach areas may not be accessible to the vacuum. This system may employ vibrating brushes to help release the paint from the surface.

2. Some chemical stripping products are toxic (e.g. methylene chloride) when inhaled or absorbed through the skin, and many are skin irritants or skin corrosives. Consequently, appropriate controls must be implemented when using chemical strippers. Although OSHA does not prohibit the use of methylene chloride-based stripping products, some local, State, and other Federal authorities may prohibit its use in residential units.

3. In the industrial arena, some disadvantages of chemical stripping are that containment and collection of the waste materials may be difficult and productivity may be low compared with the rate for conventional abrasive blasting. Because chemical stripping removes only the paint, a final abrasive blast must often be performed to remove rust and mill scale and to provide the metal profile required for adhesion of the new paint. Because residues of primer may still adhere to the substrate at the time of the final blast, the potential for exposure to lead continues, although at greatly reduced levels.

O. ENCAPSULATION OF LEAD-BASED PAINT.

1. Encapsulation refers to processes that makes lead paint inaccessible by covering or sealing the lead-painted surfaces. This may be achieved by installing sheet-rock walls on top of the paint, covering surfaces with fiberglass, or recoating housing components with a nonlead-based paint. Encapsulation is the best strategy if it provides relatively long term protection and does not require routine maintenance to ensure the integrity of the encapsulant. Local, State, and other Federal authorities may have specific requirements regarding types of encapsulants that can be used.

2. If surfaces are peeling or deteriorating and scraping is necessary prior to encapsulation, this method will produce lead dust and paint chips. If encapsulation is used over a surface covered with intact paint, little dust is generated and cleanup and waste disposal problems are therefore minimized.

3. Encapsulation may be a temporary measure because the lead-based paint that remains under the encapsulant may have to be disturbed at a future time and create a new potential for lead exposure. Encapsulation is particularly attractive as a control method when large surfaces such as walls, ceilings, and floors are involved because encapsulation requires little containment or cleanup and does not threaten the environment. Documentation of encapsulation is important because of the potential for exposures to underlying lead-based paint during maintenance, future renovation, and eventual demolition.

4. Engineering Controls. During encapsulation operations, engineering controls may not be necessary to protect workers from lead exposures. Results from HUD air sampling (NIOSH 1990) indicate that 8-hour TWA exposures are well below 50 mg/m3 for this activity.

P. POWER-TOOL CLEANING. Power-tool cleaning involves the use of power-operated impact, grinding, or brushing tools. Power tools available for paint removal include needle guns, disc sanders, grinders, power wire brushes, rotary hammers, rotary peeners, and scarifiers. Each can be used with or without a local exhaust ventilation control. Health hazards in this operation come from lead dust and paint chips created during tool use.

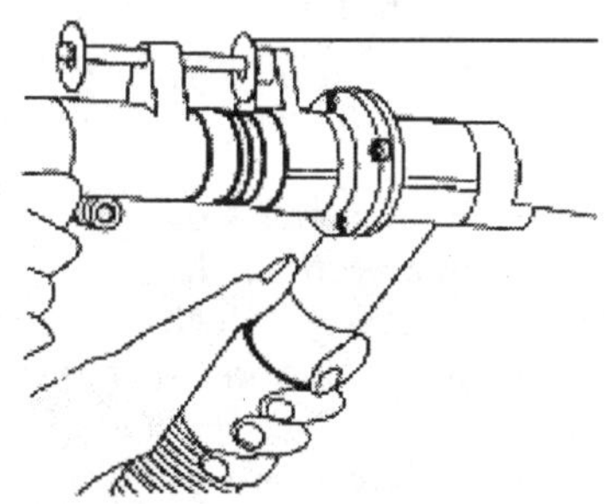

FIGURE V: 3-4. EXAMPLE OF A SHROUDED TOOL.

1. Engineering and Work Practice Controls. The following controls and practices are recommended to reduce worker exposures to lead during power-tool cleaning of lead-painted surfaces.

   a. Use shrouded tools wherever feasible (shrouding can restrict accessibility to the work area) with vacuum attachment to collect dust and debris at the point of generation (Figure V:3-4). Exhaust ventilation must be equipped with an appropriate HEPA filtration/collection system.

b. Keep shroud flush with the surface during cleaning. Dust generation is minimal, but dust can escape when cleaning areas are of difficult configuration because it may not be possible to maintain a seal between the tool and the surface in these areas.

Q. USE OF LEAD POTS. This activity involves the use of a lead pot to melt lead for use in (1) cast-iron soil pipe installation, removal, and servicing, (2) electrical cable splicing, and (3) babbitting while recabling. The health hazard in these operations arises from lead fumes becoming airborne. These operations are discussed below.

1. Cast-Iron Soil Pipe Installation and Removal.

a. Lead caulking is used in commercial construction building applications, most commonly in the joining or sealing of cast iron soil pipes. The lead used for this purpose must be liquefied. The process of heating the lead and applying it as a liquid presents an opportunity for exposure to lead-oxide fumes. The primary exposure to fumes occurs while dipping the ladle into the lead pot, carrying the ladle by hand to the solder area, and pouring solder into the pipe joint. Pot dressing is another source of lead fumes. Additional exposures to lead fumes can occur during repair and maintenance operations in which pipe joints are heated to melt the lead caulking and are then pulled apart.

b. Engineering Controls. The controls used with lead pots include a portable local exhaust ventilation system mounted directly on or near the pot to control lead fumes, or a thermostatic control device installed on the lead pot to prevent overheating to reduce the amount of lead fumes generated.

c. Work Practice Controls. During the repair and removal of cast iron pipes, workers can disconnect the pipe by cutting it (above and below the leaded joints) without creating a lead exposure problem.

2. Electrical Cable Splicing.

a. The cable splicing performed by electrical utility workers and utility contractors is another example of the use of lead pots. In this operation, the lead is typically melted above ground and then lowered by the assistant to the splicer, who is located in the manhole or underground vault. The splicer pours the lead from one ladle over the copper joint and catches the excess in another ladle held below. This process is repeated several times until the metal is too cold to pour. A lead metal sheath is then slipped over the connection and the ends are sealed with molten lead.

b. Engineering Controls. During cable splicing, the following engineering controls are used: (1) a portable local exhaust ventilation system is mounted directly on or near the lead pot; (2) a thermostatic control device can be installed on the lead pot to prevent overheating and reduce the amount of lead fumes; and (3) rubber or plastic connectors can be used instead of molten lead as the sealing method (NIOSH 1993b).

3. Babbitting While Recabling.

a. Elevators receive new wire ropes (i.e., cables) every 10 to 15 years on average. When recabling elevator ropes, it is necessary to secure the ends of the wire ropes in the baskets of thimble rods (sockets) at each end of the cable to keep the multi strand cable from unwinding. This is accomplished by pouring a tin-based babbitt material into the sockets to keep all of the strands in place within the socket.

b. Engineering Controls. The tin-based babbitt material is usually melted in a thermostatically controlled pot that keeps the lead at a temperature below that at which it fumes. This job is done within the confines of the hoist way, which normally has an updraft that will draw lead fumes up through the hoist way and away from the worker and release the fumes over the roof of the building.

c. Where legally permitted, some elevator companies have switched the wire ropes on older elevators from sockets that utilize poured babbitt metal to wedge clamps or to sockets utilizing a thermoplastic/epoxy mixture (Personal communication, E. Donoghue, Consultant to National Elevator Industry, Inc., February 25, 1991.)

R. SOLDERING AND BRAZING.

1. Soldering and brazing are techniques that are used to join metal pieces or parts. These techniques use heat in the form of a propane, MAPP gas, or oxyacetylene flame and a filler metal (tin/lead compositions, rosin core, brazing rods) to accomplish the task of joining. This activity is usually performed by workers in the plumbing trades. The potential exposure source is the filler metal that contains lead.

2. Soldering and brazing operations present similar health hazards (airborne lead fumes) but to a different degree. Most soldering operations occur at temperatures that are less than 800°F. The melting point of the filler metals is usually quite low (<600°F) and the activity does not generate significant concentrations of metal fumes. Brazing operations usually occur at temperatures in excess of 800°F. The temperature of the operation is of major importance because temperature determines the vapor pressure of the metals that are heated and therefore the potential concentration of metal fumes to which the employee may be exposed.

3. Because most field soldering and brazing work is conducted with a torch, it is difficult to regulate operating temperatures to within recommended limits to reduce the amount of metal fumes generated. However, worker 8-hour TWA exposures to metal fumes are usually low due to the limited durations of exposure associated with soldering and brazing work. Electricians soldering electrical connections, plumbers soldering nonpotable water lines, or roofers repairing tin flashing could all experience these short-term and intermittent lead exposures.

4. Engineering Controls. In confined areas, portable local exhaust ventilation can be used to remove metal fumes and gases associated with this type of work.

S. USE OF LEAD-CONTAINING MORTAR IN CHEMICAL (ACID) STORAGE AND PROCESS TANKS.

1. High-pressure acid tanks used in the mining industry (especially during gold refining), as well as tanks (called "accumulators") found in some older paper mills and perhaps in other industries, are often lined with a specialized tile or lead brick. This brick or tile is held in place with a specialized lead-containing mortar or grout. Every three to five years the linings of these tanks must be repointed (i.e. the grout between the tiles/brick must be restored), repaired, or relined with an entirely new lining (either tile or lead brick). Speciality contractors are hired to do this work.

2. After inspection of the lining, the damaged pointing (i.e. mortar between the tiles/brick) must be removed. This is done by a crew of workers simultaneously chipping away the old mortar by hand. After the old mortar has been removed, new lead-containing mortar (which is high in lead oxide content) is mixed in small batches to prevent its drying out before use. This mortar is then used as pointing between the tiles or bricks. Lead must be included in the grout to ensure the structural integrity of the tank, i.e., lead is one of the few materials that can withstand the corrosive effects of the acids in use. The health hazards in these operations arise from lead dust and particulates, with a potential for high airborne levels in the mortar mixing area.

3. The following engineering controls apply:

    a. Portable local exhaust ventilation should be used to remove lead dust and particulates from localized areas.

    b. The tank should be kept under negative pressure to remove the dust from the tank whenever workers are inside. All exhaust systems and vacuum equipment must be equipped with HEPA filters.

T. HANDLING LEAD SHOT, BRICKS, OR SHEETS, AND LEAD-FOIL PANELS.

1. Due to its inherent properties, lead is used extensively for shielding from radiation sources. The three principal projects involving the handling of lead shot, bricks, or sheets and lead-foil panels include the construction of linear accelerator suites, radiology (x-ray) suites, and industrial processing tanks. Installation and cutting of solid lead sheets, lead foil panels, and lead brick as well as the pouring of lead shot into cavities produce lead dust in varying quantities (Personal communication, G. Hyde, Baltimore Lead Burning Corporation, March 5, 1991). Additional exposures to lead can occur when lead sheets and bricks are fused with a welding torch or cut with a power saw when they are made into shielding containers.

2. Engineering Controls. Portable local exhaust ventilation should be used for lead burning (melting/fusing) and sawing operations involving lead sheets and lead bricks, where exposures can easily exceed the PEL. Engineering controls may not be necessary to protect workers from lead exposure when pouring lead shot into cavities or when cutting lead-foil sheets.

U. REINSULATION OVER EXISTING MINERAL WOOL.

1. Mineral wool insulation manufactured before about 1970 has been found to have lead particles. According to industry sources, lead slag is no longer used in the manufacture of mineral wool, although lead can be present as a trace impurity (CONSAD 1993).

2. Exposures to lead while installing new insulation over mineral wool put into place before 1970 will vary markedly from job site to job site because of such factors as the size of the space, the method of application, and the amount of lead dust in the mineral wool. Workers perform this work in relatively confined areas (such as an attic) or in an open bay structure. Moreover, the worker can install insulation manually (e.g., when installing rigid preformed insulation around pipes) or mechanically using, for example, a pneumatic blower (e.g., when blowing fiberglass or mineral wool into place over existing mineral-wool insulation). Exposures are likely to be highest when insulation is blown into place in a confined space. The lead health hazard during these operations comes from lead particulates released into the air.

3. Engineering Controls. No feasible engineering controls are known to exist for this operation.

V. REMOVAL AND REPAIR OF STAINED-GLASS WINDOWS.

1. The removal and repair of stained-glass windows includes several distinct activities:

   - Removing the glass from the building and tracing the location of the pieces of glass;
   - Disassembling the lead strips ("came") and removing the lead putty seals;
   - Cleaning or replacing of the individual pieces of glass; and
   - Reassembling and soldering the "came."

2. Only the first activity, removal, takes place at the construction site. Health hazards arise from lead dust released into the air during glass removal. The other activities typically take place in a workshop and are therefore covered under the general industry lead standard (29 CFR 1910.1025).

3. Engineering and Work Practice Controls. The control that applies is precleaning the stained-glass window with HEPA vacuums or damp wiping to remove loose dust before removal could lower exposures during removal. Whenever feasible, use portable local exhaust ventilation.

W. INDUSTRIAL VACUUMING.

1. Industrial vacuuming involves the use of vehicle-mounted vacuum systems to clean various areas of industrial facilities. The companies providing this industrial cleaning service are hired to clean such areas as catwalks, structural beams, and floors to reduce worker exposure to accumulated dust and debris that results from normal operations. The kinds of facilities where vacuuming could result in lead exposure include nonferrous metal plants (excluding aluminum) or steel plants melting lead-containing scrap in an electric-arc furnace.

2. Engineering and Work Practice Controls. Conventional cartridge dust collection systems capable of meeting the EPA ambient air quality requirements of 1.5 μg/m3 for discharge air should be used.

X. MISCELLANEOUS ACTIVITIES. The workers included in this group are those who are secondarily involved with other activities described above. These workers are not directly involved in the lead-generating activity but may nonetheless be exposed to lead in the course of their work. The lead exposure of this group of workers can be caused either by the activities of their co-workers or their own work-related activities.

Three separate miscellaneous activities representative of the jobs performed by this group of workers are discussed below. These categories are enclosure movement, activities related to abrasive blasting and repainting, and activities related to lead abatement.

1. Enclosure Movement.

    a. Moving containment enclosures involves the setting up, tearing down, and handling of flexible nylon, plastic, and cotton tarpaulins as well as framing members which together form the sides of the enclosure. Projects such as the abrasive blasting of bridges, elevated highways, water towers, and storage tanks are likely to involve some type of containment structure.

    b. Generally, the sequence of events in setting up, tearing down, and moving of a containment structure on a bridge project is as follows: First, a section of the bridge is blast-cleaned and primed within the enclosure. After completion, the sides of the containment (tarpaulins and framing) are dismantled so that the structure can be moved forward to the next section of the bridge to be worked on. After relocating the containment structure, the framing and plastic nylon coverings for the sides are reinstalled before blasting and repainting operations begin. As the enclosures are torn down and moved, the workers involved are exposed to lead dust generated by the movement of the containment components.

    c. Engineering and Work Practice Controls. The inside of the containment should be cleaned prior to tear-down and movement to remove the lead-contaminated dust that accumulates on the walls and ledges. If compressed air is used to clean, the ventilation system of the containment structure must be operating and workers must wear appropriate respirators.

2. Abrasive Blasting and Repainting Activities.

    a. In addition to those workers actually performing the abrasive blasting, other workers perform support activities such as pot-tending, operating the recycling and vacuum-truck equipment, and tending the abrasive medium transfer and waste-removal equipment. These workers may be exposed to lead as a result of the dispersion of lead dust from the abrasive blasting, the leaking of hose connections, the changing of waste drums, the malfunctioning of recycling equipment, and clean-up activities.

    b. Engineering Controls. Full containment enclosures provided with exhaust ventilation and conventional cartridge filtration units should be used to reduce lead-dust emissions from the abrasive blasting area and prevent environmental contamination. In addition, small HEPA-filtered vacuum systems should be used to clean lead dust from work clothing and to clean up small spills.

    c. Work Practice Controls. Controls involve regular inspections and timely maintenance of process equipment and control equipment help to eliminate dust leaks and prevent equipment malfunction or failure.

Y. LEAD ABATEMENT ACTIVITIES (COMMERCIAL/INSTITUTIONAL AND RESIDENTIAL).

1. These miscellaneous activities occur in conjunction with lead abatement or in-place management activities that have been previously addressed (i.e. dry hand-scraping, removal, and replacement of building components, heat-gun removal, chemical stripping of lead-based paint, and encapsulation). These ancillary activities include washing, HEPA vacuuming, enclosure set-up and tear-down, and waste disposal.

2. Engineering Controls. Exposure levels are generally below the PEL during the performance of these activities, and engineering controls are therefore not likely to be necessary.

3. Work Practice Controls. Surfaces and debris should be kept moist when they are being disturbed. Before sweeping or vacuuming, dust and debris should be misted with water to reduce airborne dust. Plastic sheeting should also be misted with water before handling to reduce dust.

4. All retained liquid waste should be poured through a filter cloth to remove paint chips and other debris prior to disposal. Filtered materials as well as other waste and debris should be placed in appropriately labeled, 6-mil

plastic bags or sealed containers suitable for the transport of lead waste, and stored in a secure area pending disposal in accordance with State and/or local requirements.

IV. BIBLIOGRAPHY.

CONSAD Research Corporation. 1993. *Economic Analysis of OSHA's Interim Final Standard for Lead in Construction.* Performed under contract to the Office of Regulatory Analysis, OSHA. Contract No. J-9-F-1-0011.

Hughes, R.T., and Amendola, A.A. 1982. *Recirculating Exhaust Air.* Plant Engineering, March issue.

Knoy, T. 1990. *Vacuum Blasting of an Elevated Water Storage Tank.* Steel Structures Painting Council, Lead Paint Removal From Industrial Structures: Meeting the Challenge, SSPC 90-01-:95-99.

NIOSH. 1993a. *Notice to All Users of Type CE, Abrasive-Blast Supplied-Air Respirators.* NIOSH Division of Safety Research.

NIOSH. 1993b. *Health Hazard Evaluation Report.* Boston Edison Company. DHHS NIOSH Publication HETA 90-075.

NIOSH. 1992. NIOSH Alert: *Request for Assistance in Preventing Lead Poisoning in Construction Workers.* DHHS, April 1992.

NIOSH. 1990. *Health Hazard Evaluation Report.* HUD Lead-Based Paint Abatement Demonstration Project. DHHS NIOSH Publication HETA 90-070-2181.

Waagbo, S. and McPhee, W. 1991. *Vacuum Blasting Lead-Based Paint Structures.* Steel Structures Painting Council, Lead Paint Removal: Meeting the Challenge, SSPC 91-05:117-130.

APPENDIX V: 3-1. LEAD-RELATED CONSTRUCTION TASKS AND THEIR PRESUMED 8-HOUR TWA EXPOSURE LEVELS.

> 50 to 500 $\mu g/m^3$	**> 500 $\mu g/m^3$ to 2,500 $\mu g/m^3$**	**> 2,500 $\mu g/m^3$**
Manual demolition	Using lead-containing mortar	Abrasive blasting
Dry manual scraping	Lead burning	Welding
Dry manual sanding	Rivet busting	Torch cutting
Heat gun use	Power tool cleaning without dust collection systems	Torch burning
Power tool cleaning with dust collection systems	Cleanup of dry expendable abrasive blasting jobs	
Spray painting with lead paint	Abrasive blasting enclosure movement and removal	

The current OSHA lead standard for construction (29 CFR 1926.62) is unique in that it groups tasks presumed to create employee exposures above the PEL of 50 µg/m3 as an 8-hour TWA. Until the employer performs an employee exposure assessment and determines actual employee exposure, the employer must assume that employees performing one of these tasks are exposed to the levels of lead indicated for that task in this Appendix. For all three groups of tasks, employers are required to provide respiratory protection appropriate to the task's presumed exposure level, protective work clothing and equipment, change areas, hand-washing facilities, training, and initial medical surveillance as prescribed by paragraph (d)(2)(v) of the standard. The only difference in the provisions applying to these groups is in the degree of respiratory protection required.

SECTION VI: CHAPTER 1

HOSPITAL INVESTIGATIONS: HEALTH HAZARDS

Contents:

I. INTRODUCTION.

A. INCIDENCE AND CAUSAL FACTORS. As of 1988, 4% of the total U.S. work force was employed by hospitals. The National Safety Council (NSC) reports that hospital employees are 41% more likely to need time off due to injury or illness than employees in other industries.

A survey of 165 clinical laboratories in Minnesota showed that the most frequent type of injuries were needle sticks (63%) followed by cuts and scrapes (21%). Hospital workers frequently report stress as a predisposing factor for accidents. Sprains and strains (often representing low back injury) were the most common type of workers compensation claim in 1983 as reported by the Bureau of Labor Statistics. See Chapter VII:1 for more information.

B. GUIDANCE. In 1988, NIOSH published Guidelines for Protecting the Safety and Health of Health Care Workers; the American Association of Critical-Care Nurses has published a handbook on the occupational hazards encountered in the critical care environment; and the NSC has a Safety Guide for hospital environments.

The hazards of exposure to waste, anesthetic gases, cytotoxic drugs, and blood-borne diseases such as hepatitis and HIV/AIDS are the subject of NIOSH criteria documents and OSHA policy statements.

II. TYPICAL HAZARDS AND HEALTH EFFECTS.

This chapter covers hospital or health care facility-specific employee hazards. Biological, chemical and physical agents presenting potential exposure to health care employees are reviewed in Appendices VI:1-1 through VI:1-3. These lists are not inclusive.

III. INVESTIGATION GUIDELINES.

A. HOSPITAL RECORDS. Hospital's OSHA 200 Log versus employee medical clinic care of employees-is there a possible trend in injuries and illnesses related to typical hazards? If available, check the hospital's safety program records and facility-enabling or operation equipment licenses, e.g., NRC radioisotope and radiation-source license.

B. HOSPITAL SAFETY PROGRAM.

1. Note any previous health and safety inspections by local health departments, fire departments, regulatory or accrediting agencies, such as the Joint Commission on Accreditation of Healthcare Organizations (JCAHO), College of American Pathologists (CAP), and the American Osteopathic Association (AOA).

2. The policies and procedures should outline the training that all employees must receive. General hospital training should include fire and electrical safety, infection control procedures, and the hazard communication

program. The policies and procedures should also delineate appropriate personnel and methods for preparation, mixing, application, storage, removal, and disposal of any hazardous agents. Emergency procedures should include provisions for fires, chemical or radioactive spills, extensive blood or body fluid spills, release of compressed, toxic, and corrosive gases, and power failure.

3. A safety committee and/or infection control committee should be established within the hospital. Periodic inspection and monitoring is the responsibility of the safety committee. Immunizations, other than the mandatory vaccination for Hepatitis B, should be offered to personnel at risk.

4. All electrical equipment used in the hospital must be approved for safety by Underwriters Laboratory (UL) or another OSHA-approved body.

5. Biosafety cabinets should be labeled and certified by the manufacturer and/or a safety officer. The cabinets should be placed in the room at a position where doors, windows, and traffic flow will not create turbulence around the face of the cabinet.

C. WALKAROUND: INFORMAL INTERVIEWS. The worker interviews should concentrate on compliance with appropriate policies and procedures. The employee should be able to verbalize what actions to take in the event of an emergency, i.e., accidental chemical or radioactive spill.

The employee should be aware of the hazards of the products with which he or she works.

Observe the employees' lifting practices. Observe the Walkaround Inspection for Health Hazards. Table VI:1-1 contains a suggested area checklist.

D. SCREENING SAMPLES. All sampling is based on the CSHO's professional judgment.

E. SAMPLING METHODS.

1. When sampling, it is important to ensure that it is a typical day, i.e. normal exposure time.

2. Bioaerosols can be evaluated using the ACGIH Bioaerosol Committee's Guidelines. These guidelines contain information on sampling, analysis, and recommendations for remedial actions. Hospital infection control personnel should assist in bioaerosol determinations, as this is nonroutine sampling and is specific for preidentified organisms. Specialized bioaerosol sampling equipment is available through the OSHA Health Response Team.

3. Some of the most commonly found chemicals, e.g., formaldehyde, xylene, halothane, and acrylamide, can be screened using detector tubes. For nitrous oxide, passive monitors can be used to monitor exposure as stated in OSHA Method No. ID-166. Specific sampling for chemical agents, such as ethylene oxide, methyl methacrylate, ribavirin, nitrous oxide, halothane, and other waste anesthetic gases, can be found in the Chemical Information Manual.

4. Lasers are calibrated by the manufacturer, but the laser system must be checked prior to each procedure and during extended procedures. Classifications of lasers must coincide with actual measurement of output (See Figure VI:1-1). Generally, measurements are required when the manufacturer's information is not available, when the laser system has not been classified or when alterations have been made to the laser system that may have changed its classification. Measurements should only be made by personnel trained in laser technology.

5. Records of alignment and power density can be checked against the manufacturer's equipment specifications. Maximum Permissible Exposure (MPE) values to the eyes and skin are given in tables 5, 6, and 7 of the ANSI standard (Z136.1-1986) as well as the ACGIH standard. Requirements for measurements and criteria for calculating the MPE's are given in sections 8 and 9 of the ANSI standard.

6. With regard to X-ray machines, film badges or their equivalent should be used for long-term monitoring. Ionizing radiation (x-ray)-Screening of radiation levels may be performed by using thermoluminescence detectors, pocket dosimeters, and Geiger-Mueller counters.

TABLE VI:1-1. WALKAROUND INSPECTION FOR HEALTH HAZARDS

Area	To check
Every Area	Floor slipperiness Adequate marking of hazards and chemical labeling Handling of infectious and chemical wastes Spill and emergency procedures Use of appropriate personal protective equipment Adequate hand-washing facilities Presence of impervious containers for needles and other sharp objects Where equipped, the aerator and local exhaust ventilation for ethylene oxide sterilizers, along with any sampling or vapor badge records Where equipped, the steam autoclave drain should be free of debris Electrical equipment and wiring must meet electrical standards
Pharmacy	Availability of a class II type A or B biological safety cabinet for mixing chemotherapeutic drugs Accurate, clear labels on all drugs, chemicals, and biologicals
Laboratory	Uncluttered work areas, clear ventilation slots, and properly labeled ductwork in laboratory hoods and biological safety cabinets Specimen handling Use of pipettes (no mouth pipetting) Gas cylinder placement and storage Maintenance records for laboratory hoods and other equipment Centrifuge tubes with caps Food should never be stored in refrigerators with lab specimens Readily detectable vapors, fumes, or dust Laser or radiation hazards
Operating room	Handling of waste anesthetic gases Air conditioning and humidity (should be about 50%) Static electricity control
Radiation area	Level of radiation Maintenance and radiation logs

8. Electrical equipment used in the hospital must follow Hospital Grades under Underwriter's Laboratory (UL) Standards 498 and 544. Resistance measurement and leakage measurements can be determined using the criteria in the National Fire Protection Association's Health Care Facilities Handbook, Chapter 7.

IV. CONTROLS AND PREVENTION.

A. ENGINEERING. All rooms should have adequate ventilation to remove contaminants. If air recirculation is required, then adequate filtering must be installed.

NOTE: Intensive care units (ICU's), particularly neonatal ICU's, may be designed without walls between patient spaces. This may allow aerosolized chemicals and x-ray radiation to escape to neighboring areas.

Class 1	The least-hazardous class. Considered incapable of providing damaging levels of laser emissions.
Class 2	Applies only to visible laser emissions and may be viewed directly for time periods of less than or equal to 0.25 seconds, which is the aversion response time.
Class 3a	Dangerous under direct or reflected vision. These lasers are restricted to the visible electromagnetic spectrum.
Class 3b	May extend across the whole electromagnetic spectrum and are hazardous when viewed intrabeam.
Class 4	The highest-energy class of lasers, also extending across the electromagnetic spectrum. This class of laser presents significant fire, skin, and eye hazards.

FIGURE VI: 1-1. LASER CLASSIFICATIONS

B. LOCAL VENTILATION.

1. General Points.

a. Hoods should be used for specific procedures, such as mixing antineoplastic drugs. A scavenging system that contains a proper gas disposal system must be in place and operable.

b. Portable suction devices may be used for direct removal of contaminants. Portable ventilation should be used for smoke plume removal during laser surgery.

c. Ethylene oxide should be ventilated through a nonrecycled or dedicated ventilation system. For a discussion of ventilation of aeration units, sterilizer door areas, sterilizer relief valves, and ventilation during cylinder changes, see the appendix of 29 CFR 1910.1047 (Ethylene Oxide). Alarms for inadequate ventilation and automatic shutdown should be in place. Air pressure in laboratories and isolation rooms should be negative so that contaminated air is drawn through the exhaust vents rather than circulating throughout the rest of the building.

d. Biological safety cabinets are primary containment devices used by workers when handling moderate and high risk organisms. There are three types of biological safety cabinets:

- Class I: Open fronted, negative pressure, ventilated cabinet.
- Class II: HEPA-filtered, recirculated air cabinet with an open front face.
- Class III: A totally enclosed HEPA-filtered cabinet of gas-tight construction. A class-III biological safety cabinet provides the highest protection to a worker. The effectiveness of the biological safety cabinets is dependent on air flow; therefore, the front intake grill and rear exhaust grill should not be blocked.

e. All windows must be covered or blacked out in laser surgical areas for protection of employees outside the surgical area.

f. Installation of automatic fire and explosion detection and protection equipment is recommended. The type should be specific to the hazard in the area.

g. In the morgue, local vacuum systems should be in place for power saws. Shields should be in place when significant splash hazards are anticipated.

h. There should be a separate storage area for radioactive sources. This area should be adequately shielded.

i. Laser systems, especially ones with high voltage capacitance, should be adequately covered. Also, bleeders and proper grounding should be attached to the system. All operating room doors to rooms that house lasers should contain safety interlocks, which shutdown the laser system if anyone enters the room.

j. Mixing of Methyl Methacrylate should be done in a closed system.

k. Ultraviolet lamps have been used to prevent tuberculosis transmission.

2. Administrative.

a. Workers should receive health and safety training.

b. Vaccination for rubella, measles, mumps, and influenza is recommended, especially for women of child-bearing age.

c. Work-related stressors, such as inadequate work space, unreasonable work load, lack of readily available resources, inadequate and unsafe equipment, should be considered.

d. Appropriate emergency equipment (i.e., fire extinguishers, showers, eye wash) should be readily available.

e. Perform periodic environmental sampling when indicated.

f. Replace hazardous substances with less hazardous substances whenever possible (i.e., plastic for glass, small packets of chemicals, pre-poured formalin containers). Provide appropriate containers for disposal of sharps, hazardous waste, personal protective equipment.

g. Provide conveniently located and supplied hand washing facilities.

h. Document and retain inventories of radioactive materials. Only authorized personnel should have access to storage areas.

3. Maintenance Schedules.

a. Hospital-grade electrical equipment including anesthesia machines, portable x-ray machines and laser systems, biological safety cabinets, and exhaust ventilation systems should have a preventive maintenance schedule. Testing intervals of electric equipment shall be set by the institution.

b. A specific person should have the responsibility for assuring proper maintenance of the portable x-ray machines. Preventive and corrective maintenance programs for x-ray machines are detailed in 21 CFR 1000, Radiological Health.

c. The anesthesia machine should be inspected and maintained at least every four months. This should be done by factory service representatives or other qualified personnel. Leakage of gas should be less than 100 ml/min during normal operation.

d. The entire laser system should be properly maintained and serviced according to the manufacturer's instructions. Only qualified personnel from the manufacturer or in-house shall maintain the system. Maintenance may only be done according to written standard operating procedures.

e. A written log is recommended for any detected leak and any service done on an ethylene oxide chamber. Sterilizer/aerator door gaskets, valves, and fittings must be replaced when necessary.

4. Training.

a. All hospital staff members should have training on electrical and fire safety, hazard communication, and infection control by qualified personnel. Some educators recommend hands on training with pre- and post-tests.

b. In the hospital, specific training regarding hazardous substances should be given. Only qualified personnel may handle the hazardous substances or operate the specified machines.

5. Warning Signs.

a. Specific requirements regarding the warning signs to be used on electrical equipment are outlined in UL No. 544. This should include a Hospital Grade warning. Warning signs should be placed in areas where exposure to ribavirin, antineoplastic agent spills, ethylene oxide, or lasers is likely to occur.

b. Contract employees should not endanger hospital employees and can be controlled sometimes through use of privileges contracts.

C. WORK PRACTICES.

1. Hands should be washed frequently and thoroughly. Workers should wash immediately after direct contact with any chemical, drug, blood, or other body fluid.

2. No eating, drinking, smoking or application of cosmetics should take place in the lab.

3. Needles and other sharp objects should be disposed of promptly in impervious containers. Needles should not be clipped or recapped by hand. See CPL 2-2.44B.

4. There should be immediate and proper disposal of biohazardous waste.

5. Mouth pipetting is to be prohibited.

6. Care should be taken not to create aerosols.

7. Appropriate personal dosimetry devices should be worn when working with radioactive materials.

8. Electrical equipment that appears to be damaged or in poor repair should not be used. Any shocks from electrical equipment should be reported promptly to the maintenance department.

9. Cylinders of compressed gases should be kept secured. They should never be dropped or allowed to strike each other with force.

10. Large pieces of broken glass should be removed with brooms and disposed of in a separate container. Small pieces can be removed with tongs. Glass should never be removed with fingers.

11. Vaporizers of anesthesia machines should be turned off when not in use. Also, proper face masks, sufficiently inflated endotracheal tubes, and prevention of anesthetic spills will decrease the amount of waste anesthetic gases in the operating room.

12. Antineoplastic drug contact requires the use of an isotonic wash to the body or eyes.

D. PERSONAL PROTECTIVE EQUIPMENT.

1. Lab coats should be worn in the laboratory area and removed before leaving. Plastic or rubber aprons should be worn when there is a potential for splashing.

2. Gloves should be worn when performing tasks such as handling hazardous chemicals, specimens, or hot materials. The type of glove should be selected according to the task being performed, as follows:

- Latex or vinyl type gloves should be changed frequently and inspected for punctures before putting them on.
- Double gloving to decrease the risk of exposure by penetration is recommended if it does not interfere with the task.
- Less permeable surgical latex gloves are recommended over polyvinyl gloves.
- Lead-lined gloves are to be worn in the direct x-ray field.

3. Rubber-soled shoes should be worn to prevent slips and falls. Rubber-lined shoe coverings may also be used to protect against spills or dropped objects. Fluid-proof shoes must be worn if there is a possibility of leakage to the skin.

4. Protective eyewear or shields should be used if splashes of a hazardous substance are likely to occur. Goggles that are tight-fitting may prevent irritation of the eyes if aerosolized chemicals are present. Goggles that protect the cornea, conjunctive and other ocular tissue are required for all personnel in the operating room during laser surgery. The wavelength of the laser output is the most important factor in determining the type of eye protection to be used. Opaque goggles are to be worn if in the direct x-ray field.

5. Impervious or low permeability gowns should be worn when in contact with antineoplastic drugs, ribavirin and blood/body fluids. These gowns should be properly stored in the area of use if contaminated. Soiled gowns should be washed or discarded. Lead-lined aprons are to be worn if in the x-ray field.

6. Respirators may be required in case of emergencies, such as accidental spills and/or exposure to specific chemicals, e.g., formaldehyde and ethylene oxide. Check for a respirator program.

V. BIBLIOGRAPHY.

American Hospital Association (AHA). 1966. *Housekeeping Manual for Health Care Facilities*. AHA: Chicago.

American National Standards Institute (ANSI). 1988. *Standard for the Safe Use of Lasers in Health Care Facilities*. ANSI Z136.3-1988.

American National Standards Institute (ANSI). 1982. *Standard for Anesthetic Equipment-Scavenging Systems for Excess Anesthetic Gases*. ANSI Z79.11-1982.

Chaff, L.F. 1989. *Safety Guide for Healthcare Institutions*. 4th ed. American Hospital Publishing: Chicago.

Charney, W. and Schirmer, J. 1990. *Essentials of Modern Hospital Safety*. Lewis Publishers, Inc.: Chelsea, MI.

ECRI. 1989. *Ethylene Oxide: Protecting Your Employees*. Hospital Hazardous Management. Vol. 1, Nos. 1 & 2, Contact ECRI at 5200 Butler Pike, Plymouth Meeting, PA 19462, (215) 825-6000.

Hoeltge, G.A. 1986. *Documentation of Safe Work Practices in the Clinical Laboratory*. Clinics in Laboratory Medicine 6(4):787-798.

Joint Commission on the Accreditation of Hospitals. 1986. *Accreditation Manual for Hospitals*. JCAH: Chicago.

Miller, B.M. et al. 1986. *Laboratory Safety Principles and Practices*. American Society for Microbiology: Washington D.C.

National Council for Radiation Protection and Measurements (NCRP). 1978. *Radiation Protection for Medical and Allied Health Personnel.* NCRP Report #48. NCRP: Washington, DC.

National Fire Protection Association (NFPA). 1987. *Health Care Facilities Handbook.* 2nd ed. NFPA: Quincy, MA.

National Institute for Occupational Safety and Health (NIOSH). 1977. *Criteria for a Recommended Standard: Occupational Exposure to Waste Anesthetic Gases and Vapors*. DHEW: NIOSH Publication No. 77-140. NIOSH: Cincinnati, OH.

National Institute for Occupational Safety and Health (NIOSH). 1988. *Guidelines for Protecting the Safety and Health of Health Care Workers.* DHHS: NIOSH Publication No. 88-119. NIOSH: Cincinnati, OH.

National Institute for Occupational Safety and Health (NIOSH). 1989. *Ethylene Oxide Sterilizers in Health Care Facilities: Engineering Controls and Work Practices.* DHHS: NIOSH Publication No. 89-116. July 13, 1989. NIOSH: Cincinnati, OH.

Occupational Safety and Health Administration (OSHA). 1987. *Chemical Information Manual.* U.S. Government Printing Office: Washington, DC.

Occupational Safety and Health Administration (OSHA). 1989. *Enforcement Procedures for Occupational Exposure to Hepatitis B Virus (HBV) and Human Immunodeficiency Virus (HIV).* OSHA: Washington, DC.

Occupational Safety and Health Administration (OSHA). 1986. *Guidelines for Antineoplastic (Cytotoxic) Drugs.* OSHA: Washington, DC.

Patterson, W. et al. 1985. *Occupational Hazards to Hospital Personnel.* Annals of Internal Medicine. 102:658-680.

Richmond, J. 1988. *Safe Practices and Procedures for Working with Human Specimens in Biomedical Research Laboratories.* Journal of Clinical Immunoassay 11(3): 115-119.

Rose, S.L. 1984. *Clinical Laboratory Safety.* J.B. Lippincott Co.: Philadelphia, PA.

Sommargren, C.E. (ed.). 1989. *AACN Handbook on Occupational Hazards for the Critical Care Nurse.* American Association of Critical-Care Nurses, Occupational Hazards Task Force. AACN: Chicago, IL.

Texas Hospital Association (THA) and Shared Hospital Electrical Safety Services. 1973. *Electrical Safety Guide.* THA: Austin.

United States Department of Health and Human Services. 1988. *Biosafety in Microbiological and Biomedical Laboratories.* HHS: Washington, DC.

Vesley, D., Hartman, M.S., and Heidi, M. 1988. *Laboratory Acquired Infections and Injuries in Clinical Laboratories*: A 1986 Survey.
American Journal of Public Health 78(9):1213-1215.

APPENDIX VI:1-1. BIOLOGICAL AGENTS--BLOOD AND BODY FLUIDS.

Use or exposure

Contact with blood and body fluids may occur as a result of medical and surgical procedures, such as labor and delivery, blood or body fluid collection and analysis, the handling of contaminated waste (e.g., gloves, linens, bandages, protective clothing, etc.) or the suctioning of airways. Exposure usually occurs because inadequate infection control procedures are in use.

Health effects

Acute: The severity of infection depends on:

- The number of pathogens encountered;
- The worker's resistance, which is affected by such things as: state of health, predisposing diseases, age, sex, and hereditary factors;
- Portal of entry (via inhalation, ingestion, mucous membrane or skin contact, or direct inoculation); and
- Virulence of the organism.

Chronic: Reproductive consequences ranging from congenital anomalies to death of the fetus and other chronic diseases, such as cirrhosis of the liver and primary liver cancer, may result from some viruses including hepatitis B, rubella, cytomegalovirus, herpes, and human immunodeficiency virus (HIV).

Biological agents--Refer to OSHA Instruction CPL 2-2.44B: *Enforcement Procedures for Occupational Exposure to Hepatitis B Virus (HBV) and Human Immunodeficiency Virus (HIV).*

APPENDIX VI:1-2. CHEMICAL AGENTS.

The following are specifically mentioned because of the severity of their health effects. This list is by no means all-inclusive.

Use or exposure	***Health effects***
Ethylene oxide (EtO)	
A disinfectant and sterilant, it is usually used in the central supply area. Exposure usually occurs from improper aeration of the ethylene oxide chamber after the sterilizing process. It can also occur in	**Acute:** Respiratory and eye irritation, vomiting, and diarrhea. **Chronic:** Altered behavior, anemia, secondary respiratory infections, skin

outpatient surgery clinics and in the cardiac catheterization laboratory (29 CFR 1910.1047).

sensitization, miscarriages, and reproductive problems. Carcinogen.

Waste anesthetic gases such as nitrous oxide, Halothane, Enfluorane

Waste gases result from poor work practices during the anesthetization of patients, improper or inadequate maintenance of the machine, and/or patient exhalation after the surgical procedure (recovery).

Acute: Drowsiness, irritability, depression, headaches, nausea, and problems with coordination and judgement.

Chronic: Embryotoxicity, liver and kidney disease, and cancer.

Antineoplastic (cancer) drugs, such as Vincristine, Dacarbazine, Mitomycin, Cytosine Arabinoside, and Fluorouracil

Antineoplastic drugs, used in the treatment of cancer and other tumors, are usually given as intravenous fluids. Mixing usually occurs in the pharmacy area of the hospital in a biological safety cabinet. Exposure may occur during preparation, administration or disposal of the drug and equipment.

Acute: Severe soft-tissue damage, fetotoxicity, headaches, lightheadedness, dizziness, and nausea.

Chronic: Chromosomal damage, teratogenesis, and carcinogenesis.

Methyl methacrylate (MMA)

An acrylic cement-like substance used to secure prostheses to bone during orthopedic surgery. Exposure usually occurs during mixing, preparation, and in the operating room.

Acute: Eye, skin and mucous membrane irritant. Acute effects have varied from a decrease in blood pressure to (rarely) cardiac arrest.

Chronic: Degeneration of the liver, mutagenesis, and teratogenesis.

Ribavirin

An antiviral drug used to treat some infants and young children with lower respiratory syncytial virus (RSV) infections. This drug is aerosolized to a respirable size of approximately 1.3 microns and is usually administered to the patient in an oxygen tent or face mask. This is when exposure can occur.

Acute: Headaches, coughing, dry upper respiratory tract and dry-burning eyes.

Chronic: Carcinogenesis, fertility impairment and fetotoxicity.

Used as a fixative and is commonly found in most laboratories and the morgue (29 CFR 1910.1048).

Acute: Eye and respiratory irritation from the liquid and vapor forms. Severe abdominal pains, nausea, vomiting and possible loss of consciousness could occur, if ingested in large amounts.

Chronic: High concentration of vapor inhaled for long periods can cause laryngitis, bronchitis or bronchial pneumonia. Prolonged exposure may cause conjunctivitis. Nasal tumors have been reported in animals. Suspected carcinogen.

Toluene or Xylene

Solvents used to fix tissue specimens and rinse stains. They are primarily found in the histology, hematology, microbiology, and cytology laboratories (29 CFR 1910.1000 Subpart Z).

Acute: Eye and mucous membrane irritation from vapor and liquid forms. Dizziness, headache, and mental confusion from inhalation of vapor. Ingestion or absorption through the skin can cause poisoning. There is a potential for thermal burns as it is extremely flammable.

Chronic: If the xylene or toluene contains benzene as an impurity, repeated breathing of the vapor over long periods may cause leukemia. Prolonged skin contact may cause dermatitis. Tolune has been implicated in reproductive disorders.

Acrylamide

The resin, usually found in research labs, is used to make gels for biochemical separations (29 CFR 1910.1000 Subpart Z).

Acute: Eye and skin irritation.

Chronic: Central nervous system disorders, i.e., polyneuropathy. Suspected carcinogen. Mutagen.

APPENDIX VI:1-3. PHYSICAL AGENTS.

Use or exposure	***Health effects***
Laser	
Used in the operating rooms for excision and cauterization of tissue. Class 3b and 4 lasers are most often used. Exposure usually occurs from unintentional operation and/or when proper controls are not in effect. The high electrical energy used to generate the beam is a potential shock hazard. The smoke plume during a surgical procedure and the laser's reaction to certain explosive or flammable agents also present hazards in the operating room.	**Acute:** From direct beam exposure, burns to skin and eyes possibly resulting in blindness. Chemical by-products in the smoke plume may cause irritation to the eyes, nose and throat, and nausea (see OSHA Hazard Information Bulletins). Biological and inert particulates can also be found in the smoke plume but these have not been well studied for their effects. **Chronic:** Unknown.
Ionizing radiation	
Portable and fixed X-ray machines are used for diagnostic procedures. Exposure occurs when unprotected employees are near a machine in operation. The degree of exposure depends on the amount of radiation, the duration of exposure, the distance from the source and the type of shielding in place.	Effects of radiation exposure are somatic (body) and/or genetic (offspring) in nature. **Acute:** Erythema and dermatitis. Large whole-body exposures cause nausea, vomiting, diarrhea, weakness, and death.

Kits containing radioactive isotopes or specimens and excreta of humans and animals who have received radionucleotides may pose a hazard. Exposure may also result from handling of radioactive spills (29 CFR 1910.1096).

Chronic: Skin cancer and bone marrow suppression. Genetic effects may lead to congenital defects in the employee's offspring.

Magnetic radiation

Magnetic resonance instrumentation.

No conclusive effects are documented.

Electrical hazards

Exposure may occur when there is lack of maintenance to any electrical equipment, abuse, and lack of understanding of the equipment and/or its controls. Oxygen-enriched atmospheres and water may contribute to hazardous conditions.

Acute: Painful shocks, respiratory inhibition, deep burns (electric and thermal) heart rate irregularities, death.

Chronic: No documented effects.

Ultraviolet radiation

Ultraviolet lamps are sometimes used in biological safety cabinets.

Acute: Skin burns, damage to the eye.

Chronic: No documented effects other than cataracts.

Compressed gases

Compressed gases are used in many clinical laboratories. They are found in varying sizes and in pure or mixed states. Examples: ammonia, carbon dioxide, and nitrogen.

Compressed gases can be toxic, radioactive, flammable, and explosive. These effects arise from the compression of the gas and the health effects of the chemical itself.

Glassware is used as bottles, beakers, flasks, test tubes, pipettes, and tubing. Chipped, cracked, badly etched glassware and sharp edges present hazards, as does broken glass.

Cuts, scratches, abrasions, are potential locations for infection.

SECTION VI: CHAPTER 2

CONTROLLING OCCUPATIONAL EXPOSURE TO HAZARDOUS DRUGS

Contents:

I. INTRODUCTION.

A. In response to numerous inquiries,[1] OSHA published guidelines for the management of Cytotoxic (antineoplastic) drugs in the work place in 1986.[106]. At that time, surveys indicated little standardization in the use of engineering controls and personal protective equipment (P.E.).[56],[73] Although practices improved in subsequent years, problems still exist.[111] In addition, the occupational management of these chemicals has been further clarified. These trends, in conjunction with many information requests, have prompted OSHA to revise its recommendations for hazardous drug handling. Furthermore, some of these agents are covered under the Hazard Communication Standard (HCS) [29 CFR 1910.1200].[107]

B. In order to provide recommendations consistent with current scientific knowledge, this informational guidance document has been expanded to cover hazardous drugs (HD), in addition to the cytotoxic drugs (CD) that were covered in the 1986 guidelines. The recommendations apply to all settings where employees are occupationally exposed to HD's, such as hospitals, physicians' offices and home health care agencies. It is recognized that sections dealing with work areas and prevention of employee exposure refer to workplaces where pharmaceuticals are used in concentrations appropriate for patient therapy. In those settings where employees work with drugs in a more potentially hazardous form, (e.g., a more concentrated form, encountered in certain components of pharmaceutical manufacturing), measures that afford employees a greater degree of protection from exposure are commonly employed and should be used.

C. This review will:

- Provide criteria for classifying drugs as hazardous.
- Summarize the evidence supporting the management of HD's as an occupational hazard.
- Discuss the equipment and worker education recommended as well as the legal requirements of standards for the protection of workers exposed and potentially exposed to HD's.
- Update the important aspects of medical surveillance.
- List some common HD's currently in use.

Anesthetic agents have not been considered in this review. However, exposure to some of these agents is a recognized health hazard,[104] and a separate document on this topic is in the planning stage.

II. CATEGORIZATION OF DRUGS AS HAZARDOUS.

The purpose of this Paragraph is to describe the biological effects of those pharmaceuticals that are considered hazardous. A number of pharmaceuticals in the health care setting may pose occupational risk to employees through acute and chronic workplace exposure. Past attention had mainly focused on drugs used to treat cancer. However, it is clear that many other agents also have toxicity profiles of concern. This recognition prompted the American Society of Hospital Pharmacists (ASHP) to define a class of agents as "hazardous drugs".[3] That report specified concerns about antineoplastic and non-antineoplastic hazardous drugs in use in most institutions throughout the country. OSHA shares this concern.

A. CHARACTERISTICS. The ASHP Technical Assistance Bulletin (TAB) described four drug characteristics, each of which could be considered hazardous:

- genotoxicity;
- carcinogenicity;
- teratogenicity or fertility impairment; and
- serious organ or other toxic manifestation at low doses in experimental animals or treated patients.

Appendix VI:2-1 of this review lists some common drugs that are considered hazardous by the above criteria. There is no standardized reference for this information nor is there complete consensus on all agents listed.

B. HAZARD DEFINITION BASED ON PHARMACOLOGY/TOXICOLOGY. Professional judgment by personnel trained in pharmacology/toxicology is essential in designating drugs as hazardous, and reference [65] provides information regarding the development of such a list at one institution. Some drugs, which have a long history of safe use in humans despite in vitro or animal evidence of toxicity, may be excluded by the institution's experts by considerations such as those used to formulate GRAS (Generally Regarded As Safe) lists by the FDA under the Food, Drug, and Cosmetics Act. In contrast, investigational drugs are new chemicals for which there is often little information on potential toxicity. Structure or activity relationships with similar chemicals and in vitro data can be considered in determining potential toxic effects. Investigational drugs should be prudently handled as HD's unless adequate information becomes available to exclude them.

Some major considerations by professionals trained in pharmacology/toxicology[65] in designating a drug as hazardous are:

- Is the drug designated as Therapeutic Category 10:00 (Antineoplastic Agent) in the American Hospital Formulary Service Drug Information?[68]
- Does the manufacturer suggest the use of special isolation techniques in its handling, administration, or disposal?
- Is the drug known to be a human mutagen, carcinogen, teratogen or reproductive toxicant?
- Is the drug known to be carcinogenic or teratogenic in animals (drugs known to be mutagenic in multiple bacterial systems or animals should also be considered hazardous)?
- Is the drug known to be acutely toxic to an organ system?

Some of the abbreviations used in this review are listed in Table VI:2-1 below.

III. BACKGROUND: HAZARDOUS DRUGS AS OCCUPATIONAL RISKS.

Preparation, administration, and disposal of HD's may expose pharmacists, nurses, physicians, and other health care workers to potentially significant workplace levels of these chemicals. The literature establishing these agents as occupational hazards deals primarily with CD's; however, documentation of adverse exposure effects from other HD's is rapidly accumulating. [15, 40-43,59] The degree of absorption that takes place during work and the significance of secondary early biological effects on each individual encounter are difficult to assess and may vary depending on the HD. As a result, it is difficult to set safe levels of exposure on the basis of current scientific information. However, there are several lines of evidence supporting the toxic potential of these drugs if handled improperly. Therefore, it is essential to minimize exposure to all HD's. Summary tables of much of the data presented below can be found in Sorsa[95] and Rogers.[84]

TABLE VI: 2-1. LIST OF ABBREVIATIONS

ANSI	American National Standards Institute
ASHP	American Society of Hospital Pharmacists
BSC	Biological Safety Cabinet
CD	Cytotoxic Drug
EPA	Environmental Protection Agency
HD	Hazardous Drug
HCS	Hazard Communication Standard
HEPA	High Efficiency Particulate Air
IARC	International Agency for Research on Cancer
MSDS	Material Safety Data Sheet
NIOSH	National Institute for Occupational Safety and Health
NTP	National Toxicology Program
OSHA	Occupational Safety and Health Administration
PPE	Personal Protective Equipment

A. MECHANISM OF ACTION. Most HD's either bind directly to genetic material in the cell nucleus or affect cellular protein synthesis. Cytotoxic drugs may not distinguish between normal and cancerous cells. The growth and reproduction of the normal cells are often affected during treatment of cancerous cells.

B. ANIMAL DATA. Numerous studies document the carcinogenic, mutagenic, and teratogenic effects of HD exposure in animals. They are well summarized in the pertinent IARC publications.[37-43] Alkylating agents present the strongest evidence of carcinogenicity (e.g., cyclophosphamide, mechlorethamine hydrochloride [nitrogen mustard]). However, other classes, such as some antibiotics, have been implicated as well. Extensive evidence for mutagenic and reproductive effects can be found in all antineoplastic classes. The antiviral agent ribavirin has additionally been shown to be teratogenic in all rodent species tested.[31,49] The ASHP recommends that all pharmaceutical agents that are animal carcinogens be handled as human carcinogens.

C. HUMAN DATA AT THERAPEUTIC LEVELS.

1. Many HD's are known human carcinogens, for which there is no safe level of exposure. The development of secondary malignancies is a well-documented side effect of chemotherapy treatment.[52,86,90,115] Leukemia has been most frequently observed. However, other secondary malignancies, such as bladder cancer and lymphoma, have been documented in patients treated for other, usually solid, primary malignancies.[52,114]

2. Chromosomal aberrations can result from chemotherapy treatment as well. One study, on chlorambucil, reveals chromosomal damage in recipients to be cumulative and related to both dose and duration of therapy.[77] Numerous case reports have linked chemotherapeutic treatment to adverse reproductive outcomes.[7,88,91,98] Testicular and ovarian dysfunction, including permanent sterility, have occurred in male and female patients who have received CD's either singly or in combination.[14] In addition, some ntineoplastic agents are known or suspected to be transmitted to infants through breast milk.[79]

3. The literature also documents the effects of these drugs on other organ systems. Extravasation of some agents can cause severe soft-tissue injury, consisting of necrosis and sloughing of exposed areas.[23,78,87] Other HD's, such as pentamidine and zidovudine (formerly AZT), are known to have significant side effects (i.e., hematologic abnormalities), in treated patients.[4,33] Serum transaminase elevation has also been reported in treated patients.[4,33]

D. OCCUPATIONAL EXPOSURE: AIRBORNE LEVELS.

1. Monitoring efforts for cytotoxic drugs have detected measurable air levels when exhaust biological safety cabinets (BSC) were not used for preparation or when monitoring was performed inside the BSC.[50,73] Concentrations of fluorouracil ranging from 0.12 to 82.26 ng/m3 have been found during monitoring of drug preparation without a BSC, implying an opportunity for respiratory exposure.[73] Elevated concentrations of cyclophosphamide were found by these authors as well. Cyclophosphamide has also been detected on the HEPA filters of flow hoods used in HD preparation, demonstrating aerosolization of the drug and an exposure opportunity mitigated by effective engineering controls.[81]

2. A recent study has reported wipe samples of cyclophosphamide, one of the class I IARC carcinogens, on surfaces of work stations in an oncology pharmacy and outpatient treatment areas (sinks and countertops). Concentrations ranged from 0.005 to 0.035 mcg/cm2, documenting opportunity for dermal exposure[60]

3. Administration of drugs via aerosolization can lead to measurable air concentrations in the breathing zone of workers providing treatment. Concentrations of up to 18 mcg/m3 have been found by personal air sampling of workers administering pentamidine.[67] Similar monitoring for ribavirin has found concentrations as high as 316 mcg/m3.31

E. OCCUPATIONAL EXPOSURE: BIOLOGICAL EVIDENCE OF ABSORPTION.

1. Urinary Mutagenicity.

   a. Falck et al. were the first to note evidence of mutagenicity in the urine of nurses who handled cytotoxic drugs.[26] The extent of this effect increased over the course of the work week. With improved handling practices, a decrease in mutagenic activity was seen. [27] Researchers have also studied pharmacy personnel who reconstitute antineoplastic drugs. These employees showed increasingly mutagenic urine over the period of exposure; when they stopped handling the drugs, activity fell within 2 days to the level of unexposed controls.[5,76] They also found mutagenicity in workers using horizontal laminar flow BSC's that decreased to control levels with the use of vertical flow containment BSC's.[76]

   b. Other studies have failed to find a relationship between exposure and urine mutagenicity.[25] Sorsa[99] summarizes this information and discusses the factors, such as differences in urine collection timing and variations in the use of PPE, which could lead to disparate results. Differences may also be related to smoking status; smokers exposed to CD's exhibit greater urine mutagenicity than exposed nonsmokers or control smokers -- suggesting contamination of the work area by CD's and some contribution of smoking to their mutagenic profile.[9]

2. Urinary Thioethers are glutathione conjugated metabolites of alkylating agents, which have been evaluated as an indirect means of measuring exposure. Workers who handle cytotoxic drugs have been reported to have increased levels compared to controls and also have increasing thioether levels over a 5-day work week.[44,48] Other studies of nurses who handle CD's and of treated patients have yielded variable results, which could be due to confounding by smoking, PPE, and glutathione-S-transferase activity.[11]

3. Urinary Metabolites. Venitt assayed the urine of pharmacy and nursing personnel handling cisplatin and found platinum concentrations at or below the limit of detection for both workers and controls.[112] Hirst found cyclophosphamide in the urine of two nurses who handled the drug, documenting worker absorption[35] (Hirst also documented skin absorption in human volunteers by using gas chromatography after topical application of the drug.) Urinary pentamidine recovery has also been reported in exposed health care workers.[94]

F. OCCUPATIONAL EXPOSURE: HUMAN EFFECTS.

1. Cytogenetic Effects.

   a. A number of studies have examined the relationship of exposure to CD's in the workplace to chromosomal aberrations. These studies have looked at a variety of markers for damage, including sister chromatid exchanges (SCE), structural aberrations (e.g., gaps, breaks, translocations), and micronuclei in peripheral blood lymphocytes. The results have been somewhat conflicting. Several authors found increases in one or more markers[74,75,80,113] Increased mutation frequency has been reported as well.[17]

b. Other studies have failed to find a significant difference between workers and controls.[99,101] Some researchers have found higher individual elevations[28] or a relationship between the number of drugs handled and SCE's.[8] These disparate results are not unexpected. The difficulties in quantitating exposure have resulted in different exposure magnitudes between studies; workers in several negative studies appear to have a lower overall exposure.[101] In addition, differences in the use of PPE and work technique will alter absorption of CD's and resultant biologic effects.

c. Finally, techniques for SCE measurement may not be optimal. A recent study that looked at correlation of phosphoramide-induced SCE levels with duration of anticancer drug handling found a statistically significant correlation coefficient of zero.[63,66] Taken together, the evidence indicates an excess of markers of mutagenic exposure in unprotected workers.

2. Reproductive Effects.

a. Reproductive effects associated with occupational exposure to CD's have been well documented. Hemminki et al.[32] found no difference in exposure between nurses who had spontaneous abortions and those who had normal pregnancies. However, the study group consisted of nurses who were employed in surgical or medical floors of a general hospital. When the relationship between CD exposure and congenital malformations was explored, the study group was expanded to include oncology nurses, among others, and an odds ratio of 4.7 was found for exposures of more than once per week. This observed odds ratio is statistically significant.

b. Selevan et al.[89] found a relationship between CD exposure and spontaneous abortion in a case-control study of Finnish nurses. This well-designed study reviewed the reproductive histories of 568 women (167 cases) and found a statistically significant odds ratio of 2.3. Similar results were obtained in another large case-control study of French nurses,[102] and a study of Baltimore-area nurses found a significantly higher proportion of adverse pregnancy outcomes when exposure to antineoplastic agents occurred during the pregnancy.[85] The nurses involved in these studies usually prepared and administered the drugs. Therefore, workplace exposure of these groups of professionals to such products has been associated with adverse reproductive outcomes in several investigations.

3. Other Effects.

a. Hepatocellular damage has been reported in nurses working in an oncology ward; the injury appeared to be related to intensity and duration of work exposure to CD's[96] Symptoms such as lightheadedness, dizziness, nausea, headache, and allergic reactions have also been described in employees after the preparation and administration of antineoplastic drugs in unventilated areas.[22,86] In occupational settings, these agents are known to be toxic to the skin and mucous membranes, including the cornea.[69,82]

b. Pentamidine has been associated with respiratory damage in one worker who administered the aerosol. The injury consisted of a decrease in diffusing capacity that improved after exposure ceased.[29] The onset of bronchospasm in a pentamidine-exposed worker has also been reported.[22] Employees involved in the aerosol administration of ribavirin have noted symptoms of respiratory tract irritation.[55] A number of medications including psyllium and various antibiotics are known respiratory and dermal sensitizers. Exposure in susceptible individuals can lead to asthma or allergic contact dermatitis.

IV. WORK AREAS.

Risks to personnel working with HD's are a function of the drugs' inherent toxicity and the extent of exposure. The main routes of exposure are: inhalation of dusts or aerosols, dermal absorption, and ingestion. Contact with contaminated food or cigarettes represents the primary means of ingestion. Opportunity for exposure to HD's may occur at many points in the handling of these drugs.

A. PHARMACY OR OTHER PREPARATION AREAS.

1. In large oncology centers, HD's are usually prepared in the pharmacy. However, in small hospitals, outpatient treatment areas, and physicians' offices they have been prepared by physicians or nurses without appropriate engineering controls and protective apparel.[16,20] Many HD's must be reconstituted, transferred from one container to another, or manipulated before administration to patients. Even if care is taken, opportunity for absorption through inhalation or direct skin contact can occur.[35,36,73,116]

2. Examples of manipulations that can cause splattering, spraying, and aerosolization include:

    - withdrawal of needles from drug vials;
    - drug transfer using syringes and needles or filter straws;
    - breaking open of ampules; and
    - expulsion of air from a drug-filled syringe.

    Evaluation of these preparation techniques, using fluorescent dye solutions, has shown contamination of gloves and the sleeves and chest of gowns.[97]

3. Horizontal airflow work benches provide an aseptic environment for the preparation of injectable drugs. However, these units provide a flow of filtered air originating at the back of the work space and exiting toward the employee using the unit. Thus, they increase the likelihood of drug exposure to both the preparer and other personnel in the room. As a result, the use of horizontal BSC's is contra-indicated in the preparation of HD's. Smoking, drinking, applying cosmetics, and eating where these drugs are prepared, stored, or used also increase the chance of exposure.

B. ADMINISTRATION OF DRUGS TO PATIENTS.

1. Administration of drugs to patients is generally performed by nurses or physicians. Drug injection into the IV line, clearing of air from the syringe or infusion line, and leakage at the tubing, syringe, or stopcock connection present opportunities for skin contact and aerosol generation. Clipping used needles and crushing used syringes can produce considerable aerosolization as well.

2. Such techniques where needles and syringes are contaminated with blood or other potentially infectious material are prohibited by the Bloodborne Pathogens Standard.[109] Prohibition of clipping or crushing of any needle or syringe is sound practice.

3. Excreta from patients who have received certain antineoplastic drugs may contain high concentrations of the drug or its hazardous metabolites. For example, patients receiving cyclophosphamide excrete large amounts of the drug and its mutagenic metabolites.[46,92] Patients treated with cisplatin have been shown to excrete potentially hazardous amounts of the drug. [112] Unprotected handling of urine or urine-soaked sheets by nursing or housekeeping personnel poses a source of exposure.

C. DISPOSAL OF DRUGS AND CONTAMINATED MATERIALS. Contaminated materials used in the preparation and administration of HD's, such as gloves, gowns, syringes and vials, present a hazard to support and housekeeping staff. The use of properly labeled, sealed and covered disposal containers, handled by trained and protected personnel, should be routine, and is required under the Bloodborne Pathogens Standard[109] if such items are contaminated with blood or other potentially infectious materials. HD's and contaminated materials should be disposed of in accordance with federal, state, and local laws. Disposal of some of these drugs is regulated by the EPA. Drugs that are unused commercial chemical products and are considered by the EPA to be toxic wastes must be disposed of in accordance with 40 CFR part 261.[33,24] Spills can also represent a hazard; the employer should ensure that all employees are familiar with appropriate spill procedures.

D. SURVEY OF CURRENT WORK PRACTICES. Surveys of U.S. cancer centers and oncology clinics reveal wide variation in work practices, equipment or training for personnel preparing CD's.[56,73] This lack of standardization results in a high prevalence of potential occupational exposure to CD's. One survey found that 40% of hospital pharmacists reported a skin exposure to CD's at least once a month, and only 28% had medical surveillance programs in their workplaces.[16] Nurses, particularly those in outpatient settings, were found to be even less well protected than pharmacists.[111] Such findings emphasize current lack of protection for all personnel who risk potential exposure to HD's.

V. PREVENTION OF EMPLOYEE EXPOSURE.

A. HAZARDOUS DRUG SAFETY AND HEALTH PLAN.

1. Where hazardous drugs, as defined in this review, are used in the workplace, sound practice dictates that a written Hazardous Drug Safety and Health Plan be developed. Such a plan assists in:

   - Protecting employees from health hazards associated with HD's.
   - Keeping exposures as low as reasonably achievable.

2. When a Hazardous Drug Safety and Health Plan is developed, it should be readily available and accessible to all employees, including temporary employees, contractors, and trainees. The ASHP recommends that the Plan include each of the following elements and indicate specific measures that the employer is taking to ensure employee protection:[3]

   - Standard operating procedures relevant to safety and health considerations to be followed when health care workers are exposed to hazardous drugs.
   - Criteria that the employer uses to determine and implement control measures to reduce employee exposure to hazardous drugs, including engineering controls, the use of personal protective equipment, and hygiene practices.
   - A requirement that ventilation systems and other protective equipment function properly, and specific measures to ensure proper and adequate performance of such equipment.
   - Provision for information and training.
   - The circumstances under which the use of specific HD's (that is, FDA investigational drugs) require prior approval from the employer before implementation.
   - Provision for medical examinations of potentially exposed personnel.
   - Designation of personnel responsible for implementation of the Hazardous Drug Safety and Health Plan including the assignment of a Hazardous Drug Officer (who is an industrial hygienist, nurse, or pharmacist health and safety representative) and, if appropriate, establishment of a Hazardous Drug Committee or a joint Hazardous Drug Committee/Chemical Committee.

3. The ASHP further recommends that specific consideration of the following provisions be included where appropriate:

   - Establishment of a designated HD handling area.
   - Use of containment devices such as biological safety cabinets.
   - Procedures for safe removal of contaminated waste.
   - Decontamination procedures.

   The ASHP recommends that the Hazardous Drug Safety and Health Plan be reviewed and its effectiveness reevaluated at least annually and updated as necessary.

4. A comparison of OSHA 200 log entries to employee medical clinic appointment or visit rosters can be made[3] to establish if there is evidence of disorders that could be related to hazardous drugs. Previous health and safety inspections by local health departments, fire departments, regulatory, or accrediting agencies may be helpful for the facility's planning purposes as well as any OSHA review of hazards and programs in the facility. Joint Commission on Accreditation of Healthcare Organizations (JCAHO), or College of American Pathologists (CAP) review of facilities may contain information on hazardous drugs used in the facility.

B. DRUG PREPARATION PRECAUTIONS.

1. Work Area. The ASHP recommends that HD preparation be performed in a restricted, preferably, centralized area. Signs restricting the access of unauthorized personnel are to be prominently displayed. Eating, drinking, smoking, chewing gum, applying cosmetics, and storing food in the preparation area should be prohibited.[71] The ASHP recommends that procedures for spills and emergencies, such as skin or eye contact, be available to workers, preferably posted in the area.[3]

2. Biological Safety Cabinets. Class II or III Biological Safety Cabinets (BSC) that meet the current National Sanitation Foundation Standard [49,70,72] should minimize exposure to HD's during preparation. Although these cabinets are designed for biohazards, several studies have documented reduced urine mutagenicity in CD-exposed workers or reduced environmental levels after the institution of BSC's.[5,51,61] If a BSC is unavailable, for example in private practice office, accepted practice is the sharing of a cabinet (e.g., several medical offices share a cabinet) or sending the patient to a center where HD's can be prepared in a BSC. Alternatively, preparation can be performed in a facility with a BSC and the drugs transported to the area of administration. Use of a dedicated BSC, where only HD's are prepared, is prudent practice.

3. Types Of BSC's.

    a. Four main types of Class II BSC's are available. They all have downward airflow and HEPA filters. They are differentiated by the amount of air recirculated within the cabinet, whether this air is vented to the room or the outside, and whether contaminated ducts are under positive or negative pressure. These four types are described below:

        - Type A cabinets recirculate approximately 70% of cabinet air through HEPA filters back into the cabinet; the rest is discharged through a HEPA filter into the preparation room. Contaminated ducts are under positive pressure.
        - Type B1 cabinets have higher velocity air inflow, recirculate 30% of the cabinet air, and exhaust the rest to the outside through HEPA filters. They have negative-pressure contaminated ducts and plenums.
        - Type B2 systems are similar to Type B1 except that no air is recirculated.
        - Type B3 cabinets are similar to Type A in that they recirculate approximately 70% of cabinet air. However, the other 30% is vented to the outside and the ducts are under negative pressure.

    b. Class III cabinets are totally enclosed with gas tight construction. The entire cabinet is under negative pressure, and operations are performed through attached gloves. All air is HEPA filtered.

    c. Class II, type B, or Class III BSC's are recommended since they vent to the outside.[3] Those without air recirculation are the most protective. If the BSC has an outside exhaust, it should be vented away from air intake units.

    d. The exhaust fan or blower on the vertical airflow hood should be on at all times, except when the hood is being mechanically repaired or moved. If the blower is turned off, the hood should be decontaminated before reuse.[3,72] Each BSC should be equipped with a continuous monitoring device to allow confirmation of adequate air flow and cabinet performance. The cabinet should be in an area with minimal air turbulence; this will reduce leakage to the environment.[6,70] Additional information on design and performance testing of BSC's can be found in papers by Avis and Levchuck,[6] Bryan and Marback,[10] and the National Sanitation Foundation.[70] Practical information regarding space needs and use of BSC's is contained in the ASHP's 1990 technical assistance bulletin.[3]

    e. Ventilation and biosafety cabinets installed should be maintained and evaluated for proper performance in accordance with the manufacturer's instructions.

4. Decontamination.

    a. The cabinet should be cleaned according to the manufacturer's instructions. Some manufacturers have recommended weekly decontamination, as well as whenever spills occur, or when the cabinet requires moving, service or certification.

    b. Decontamination should consist of surface cleaning with water and detergent followed by thorough rinsing. The use of detergent is recommended because there is no single accepted method of chemical deactivation for all agents involved.[13,45] Quaternary ammonium cleaners should be avoided due to the possibility of vapor build-up in recirculated air.[3] Ethyl alcohol or 70% isopropyl alcohol may be used with the cleaner if the contamination is soluble only in alcohol.[3] Alcohol vapor build-up has also been a concern, so the use of alcohol should be avoided in BSC's where air is recirculated.[3] Spray cleaners should also be avoided due to the risk of spraying the HEPA filter. Ordinary decontamination procedures, which include fumigation with a germicidal agent, are inappropriate in a BSC used for HD's because such procedures do not remove or deactivate the drugs.

c. Removable work trays, if present, should be lifted in the BSC so the back and any sump below can be cleaned. During cleaning, the worker should wear PPE similar to that used for spills. Ideally, the sash should remain down during cleaning, however, a NIOSH-approved respirator appropriate for the hazard must be worn by the worker if the sash will be lifted during the process. The exhaust fan/blower should be left on. Cleaning should proceed from least to most contaminated areas. The drain spillage trough area should be cleaned at least twice since it can be heavily contaminated. All materials from the decontamination process should be handled as HD's and disposed of in accordance with federal, state, and local laws.

5. Service and Certification.

a. The ASHP recommends that BSC's be serviced and certified by a qualified technician every six months or any time the cabinet is moved or repaired.[3,71] Technicians servicing these cabinets or changing the HEPA filters should be aware of HD risks through hazard communication training from their employers and should use the same personal protective equipment as recommended for large spills.

b. Certification of the BSC includes performance testing as outlined in the procedures of the National Sanitation Foundation's Standard Number [49.70] Helpful information on such testing can be found in the ASHP 1990 technical assistance bulletins,[3] the BSC manufacturer's equipment manuals, and Bryan and Marback's paper.[10] HEPA filters should be changed when they restrict airflow or if they are contaminated by an accidental spill. They should be bagged in plastic and disposed of as HD's. Any time the cabinet is turned off or transported it should be sealed with plastic.

6. Personal Protective Equipment.

a. Gloves. Research indicates that the thickness of the gloves used in handling HD's is more important than the type of material, since all materials tested have been found to be permeable to some HD's.[3,19,53] The best results are seen with latex gloves. Therefore, latex gloves should be used for the preparation of HD's unless the drug-product manufacturer specifically stipulates that some other glove provides better protection.[19,53,72,93,100] Thicker, longer latex gloves that cover the gown cuff are recommended for use with HD's. Gloves with minimal or no powder are preferred since the powder may absorb contamination.[3,104]

The above referenced sources have noted great variability in permeability within and between glove lots. Therefore, double gloving is recommended if it does not interfere with an individual's technique.[3] Because all gloves are permeable to some extent and their permeability increases with time, they should be changed regularly (hourly) or immediately if they are torn, punctured, or contaminated with a spill. Hands should always be washed before gloves are put on and after they are removed. Employees need thorough training in proper methods for contaminated glove removal.

b. Gowns. A protective disposable gown made of lint-free, low-permeability fabric with a closed front, long sleeves, and elastic or knit closed cuffs should be worn. The cuffs should be tucked under the gloves. If double gloves are worn, the outer glove should be over the gown cuff and the inner glove should be under the gown cuff. When the gown is removed, the inner glove should be removed last. Gowns and gloves in use in the HD preparation area should not be worn outside the HD preparation area.[3]

As with gloves, there is no ideal material. Research has found nonporous Tyvek and Kaycel to be more permeable than Saranex-laminated Tyvek and polyethylene-coated Tyvek after four hours of exposure to the CD's tested.[54] However, little airflow is allowed with the latter materials. As a result, manufacturers have produced gowns with Saranex or polyethylene reinforced sleeves and front in an effort to decrease permeability in the most exposure prone areas, but little data exists on decreasing exposure.

c. Respiratory Protection. A BSC is essential for the preparation of HD's. Where a BSC is not currently available, a NIOSH-approved respirator appropriate for the hazard must be worn to afford protection until the BSC is installed. [NIOSH recommendation at the time of this publication is for a respirator with a high-efficiency filter, preferably a powered, air-purifying respirator.] The use of respirators must comply with OSHA's Respiratory Protection Standard,[105] which outlines the aspects of a respirator program, including selection, fit testing, and worker training. Surgical masks are not appropriate since they do not prevent aerosol inhalation. Permanent respirator use, in lieu of BSC's, is imprudent practice and should not be a substitute for engineering controls.

d. Eye and Face Protection. Whenever splashes, sprays, or aerosols of HD's may be generated that can result in eye, nose, or mouth contamination, chemical-barrier face and eye protection must be provided and used in accordance with 29 CFR 1910.133. Eyeglasses with temporary side shields are inadequate protection.

When a respirator is used to provide temporary protection as described above, and splashes, sprays, or aerosols are possible, employee protection should be:

- a respirator with a full face piece; or
- a plastic face shield or splash goggles complying with ANSI standards[2] when using a respirator of less than full-face piece design.

Eyewash facilities should also be made available.

e. PPE Disposal and Decontamination. All gowns, gloves, and disposable materials used in preparation should be disposed of according to the hospital's hazardous drug waste procedures and as described under this review's section on Waste Disposal. Goggles, face shields and respirators may be cleaned with mild detergent and water for reuse.

C. WORK EQUIPMENT. NIH has recommended that work with HD's be carried out in a BSC on a disposable, plastic-backed paper liner. The liner should be changed after preparation is completed for the day or after a shift, whichever comes first. Liners should also be changed after a spill.[103] Syringes and IV sets with Luer-lock fittings should be used for HD's. Syringe size should be large enough so that they are not full when the entire drug dose is present.

A covered disposable container should be used to contain excess solution. A covered sharps container should be in the BSC. The ASHP recommends that HD-labeled plastic bags be available for all contaminated materials (including gloves, gowns, and paper liners), so that contaminated material can be immediately placed in them and disposed of in accordance with ASHP recommendations.[3]

1. Work Practices. Correct work practices are essential to worker protection. Aseptic technique is assumed as a standard practice in drug preparation. The general principles of aseptic technique, therefore, will not be detailed here. It should be noted, however, that BSC benches differ from horizontal flow units in several ways that require special precautions. Manipulations should not be performed close to the work surface of a BSC. Unsterilized items, including liners and hands, should be kept downstream from the working area.

Entry and exit of the cabinet should be perpendicular to the front. Rapid lateral hand movements should be avoided. Additional information can be found in the National Sanitation Foundation Standard 49 for Class II (Laminar Flow) Biohazard Cabinetry[70] and Avis and Levchuck's paper.[6] All operators should be trained in these containment area protocols. All PPE should be donned before work is started in the BSC. All items necessary for drug preparation should be placed within the BSC before work is begun. Extraneous items should be kept out of the work area.

a. Labeling. In addition to standard pharmacy labeling practices, all syringes and IV bags containing HD's should be labeled with a distinctive warning label, such as:

SPECIAL HANDLING/DISPOSAL PRECAUTIONS: Those HD's covered under HCS must also have labels in accordance with section (f) of the standard to warn employees handling the drug(s) of the hazards.

b. Needles. The ASHP recommends that all syringes and needles used in the course of preparation be placed in "sharps" containers for disposal without being crushed, clipped or capped.[3,103]

c. Priming. Prudent practice dictates that drug administration sets be attached and primed within the BSC, prior to addition of the drug. This eliminates the need to prime the set in a less well-controlled environment and ensures that any fluid that escapes during priming contains no drug. If priming must occur at the site of administration, the intravenous line should be primed with non-drug containing fluid or a back-flow closed system should be used.[3]

d. Handling Vials. Extremes of positive and negative pressure in medication vials should be avoided, e.g., attempting to withdraw 10 cc of fluid from a 10-cc vial or placing 10 cc of a fluid into an air-filled 10-cc vial. The use of large-bore needles, #18 or #20, avoids high-pressure syringing of solutions. However, some experienced personnel believe that large-bore needles are more likely to drip. Multiuse dispensing pins are recommended to avoid these problems.

   Venting devices such as filter needles or dispensing pins permit outside air to replace the withdrawn liquid. Proper worker education is essential before using these devices.[3] Although venting devices are recommended, another technique is to add diluent slowly to the vial by alternately injecting small amounts and allowing displaced air to escape into the syringe. When all diluent has been added, a small amount of additional air may be withdrawn to create a slight negative pressure in the vial. This should not be expelled into room air because it may contain drug residue. It should either be injected into a vacuum vial or remain in the syringe to be discarded.

   If any negative pressure must be applied to withdraw a dosage from a stoppered vial and handling safety is compromised, an air-filled syringe should be used to equalize pressure in the stoppered vial. The volume of drug to be withdrawn can be replaced by injecting small amounts of air into the vial and withdrawing equal amounts of liquid until the required volume is withdrawn. The drug should be cleared from the needle and hub (neck) of the syringe before separating to reduce spraying on separation.

e. Handling Ampules. Prudent practice requires that ampules with dry material should be "gently tapped down" before opening to move any material in the top of the ampule to the bottom quantity. A sterile gauze pad should be wrapped around the ampule neck before breaking the top.[3] This can protect against cuts and catch airborne powder or aerosol. If diluent is to be added, it should be injected slowly down the inside wall of the ampule. The ampule should be tilted gently to ensure that all the powder is wet before agitating it to dissolve the contents.

   After the solution is withdrawn from the ampule with a syringe, the needle should be cleared of solution by holding it vertically with the point upwards; the syringe should be tapped to remove air bubbles. Any bubbles should be expelled into a closed container.

f. Packaging HD's for Transport. The outside of bags or bottles containing the prepared drug should be wiped with moist gauze. Entry ports should be wiped with moist alcohol pads and capped. Transport should occur in sealed plastic bags and in containers designed to avoid breakage. Shipped HD's that are subject to EPA regulation as hazardous waste are also subject to Department of Transportation (DOT) regulations as specified in 49 CFR part 172.[101].

g. Nonliquid HD's. The handling of nonliquid forms of HD's requires special precautions as well. Tablets which may produce dust or potential exposure to the handler should be counted in a BSC. Capsules, i.e., gel-caps or coated tablets, are unlikely to produce dust unless broken in handling. These are counted in a BSC on equipment designated for HD's only, because even manual counting devices may be covered with dust from the drugs handled. Automated counting machines should not be used unless an enclosed process isolates the hazard from the employee(s).

   Compounding should also occur in a BSC. A gown and gloves should be worn. (If a BSC is unavailable, an appropriate NIOSH-approved respirator must be worn.)

2. Drug Administration.

a. Personal Protective Equipment. The National Study Commission on Cytotoxic Exposure has recommended that personnel administering HD's wear gowns, latex gloves, and chemical splash goggles or equivalent safety glasses as described under the PPE section, preparation.[71] NIOSH-approved respirators should be worn when administering aerosolized drugs.

b. Administration Kit. Protective and administration equipment may be packaged together and labeled as a HD administration kit. Such a kit could include:

- personal protective equipment;
- gauze (4" × 4") for cleanup;
- alcohol wipes;
- disposable plastic-backed absorbent liner;
- puncture-resistant container for needles and syringes;
- a thick sealable plastic bag (with warning label); and
- accessory warning labels.

c. Work Practices. Safe work practices when handling HD's should include the following:

- Hands should be washed before donning and after removing gloves. Gowns or gloves that become contaminated should be changed immediately. Employees should be trained in proper methods to remove contaminated gloves and gowns. After use, gloves and gowns should be disposed of in accordance with ASHP recommendations.
- Infusion sets and pumps, which should have Luer-lock fittings, should be observed for leakage during use. A plastic-backed absorbent pad should be placed under the tubing during administration to catch any leakage. Sterile gauze should be placed around any push sites; IV tubing connection sites should be taped.
- Priming IV sets or expelling air from syringes should be carried out in a BSC. If done at the administration site, ASHP recommends that the line be primed with non-drug containing solution or that a back-flow closed system be used. IV containers with venting tubes should not be used.[3]
- Syringes, IV bottles and bags, and pumps should be wiped clean of any drug contamination with sterile gauze. Needles and syringes should not be crushed or clipped. They should be placed in a puncture-resistant container, then into the HD disposal bag with all other HD-contaminated materials.
- Administration sets should be disposed of intact. Disposal of the waste bag should follow HD disposal requirements. Unused drugs should be returned to the pharmacy.
- Protective goggles should be cleaned with detergent and properly rinsed. All protective equipment should be disposed of upon leaving the patient care area.
- Nursing stations where these drugs will be administered should have spill and emergency skin and eye decontamination kits available and relevant MSDS's for guidance. The HCS requires MSDS'S to be readily available in the workplace to all employees working with hazardous chemicals.
- PPE should be used during the administration of oral HD's if splashing is possible.

A large number of investigational HD's are under clinical study in health care facilities. Personnel not directly involved in the investigation should not administer these drugs unless they have received adequate instructions regarding safe handling procedures. Literature regarding potential toxic effects of investigational drugs should be evaluated prior to the drug's introduction into the workplace.[65]

The increased use of HD's in the home environment necessitates special precautions. Employees involved in home care delivery should follow the above work practices and employers should make administration and spill kits available. Home health care workers should have emergency protocols with them as well as phone numbers and addresses in the event emergency care becomes necessary.[3] Waste disposal for drugs delivered for home use and other home contaminated material should also be considered by the employer and should follow applicable regulations.

d. Aerosolized Drugs. The administration of aerosolized HD's requires special engineering controls to prevent exposure to health care workers and others in the vicinity. In the case of pentamidine, these controls include treatment booths with local exhaust ventilation designed specifically for its administration. A variety of ventilation methods have also been used for the administration of ribavirin. These include isolation rooms with separate HEPA filtered ventilation systems and administration via endotracheal tube.[30,47] Engineering controls used to manage employee exposure to anesthetic gases is a traditional example of occupational chemical management. Both isolation and ventilation are used for these volatile HD's.

3. Caring for Patients Receiving HD's. In accordance with the Bloodborne Pathogens Standard, universal precautions must be observed to prevent contact with blood or other potentially infectious materials. Under circumstances in which differentiation between body fluid types is difficult or impossible, all body fluids should be considered potentially infectious materials and must be managed as dictated in the Bloodborne Pathogens Standard.[109]

a. Personal Protective Equipment. Personnel dealing with excreta, primarily urine, from patients who have received HD's in the last 48 hours should be provided with and wear latex or other appropriate gloves and disposable gowns, to be discarded after each use or whenever contaminated, as detailed under Waste Disposal. Eye protection should be worn if splashing is possible. Such excreta contaminated with blood, or other potentially infectious materials as well, should be managed according to the Bloodborne Pathogen Standard. Hands should be washed after removal of gloves or after contact with the above substances.

b. Linen contaminated with HD's or excreta from patients who have received HD's in the past 48 hours is a potential source of exposure to employees. Linen soiled with blood or other potentially infectious materials as well as contaminated with excreta must also be managed according to the Bloodborne Pathogens Standard.[109] Linen contaminated with HD's should be placed in specially marked laundry bags and then placed in a labeled impervious bag. The laundry bag and its contents should be prewashed, and then the linens added to other laundry for a second wash. Laundry personnel should wear latex gloves and gowns while handling prewashed material.

c. Reusable Items. Glassware or other contaminated reusable items should be washed twice with detergent by a trained employee wearing double latex gloves and a gown.

4. Waste Disposal.

a. Equipment. Thick, leak-proof plastic bags, colored differently from other hospital trash bags, should be used for routine accumulation and collection of used containers, discarded gloves, gowns, and any other disposable material. Bags containing hazardous chemicals (as defined by Section C of HCS), shall be labeled in accordance with Section F of the Hazard Communication Standard where appropriate. Where the Hazard Communication Standard does not apply, labels should indicate that bags contain HD-related wastes.

Needles, syringes, and breakable items not contaminated with blood or other potentially infectious materials should be placed in a "sharps" container before they are stored in the waste bag. Such items that are contaminated with blood or other potentially infectious material must be placed in a "sharps" container. Similarly, needles should not be clipped or capped nor syringes crushed if contaminated by blood or other potentially infectious material, except for a rare instance where a medical procedure requires recapping. The waste bag should be kept inside a covered waste container clearly labeled "HD WASTE ONLY." At least one such receptacle should be located in every area where the drugs are prepared or administered. Waste should not be moved from one area to another. The bag should be sealed when filled and the covered waste container taped.

b. Handling. Prudent practice dictates that every precaution be taken to prevent contamination of the exterior of the container. Personnel disposing of HD waste should wear gowns and protective gloves when handling waste containers with contaminated exteriors. Prudent practice further dictates that such a container with a contaminated exterior be placed in a second container in a manner that eliminates contamination of the second container. HD waste handlers should also receive hazard communication training as discussed below under Paragraph VIII: D of this chapter.

c. Disposal. Hazardous drug-related wastes should be handled separately from other hospital trash and disposed of in accordance with applicable EPA, state, and local regulations for hazardous waste.[24,110] This disposal can occur at either an incinerator or a licensed sanitary landfill for toxic wastes, as appropriate. Commercial waste disposal is performed by a licensed company. While awaiting removal, the waste should be held in a secure area in covered, labeled drums with plastic liners.

Chemical inactivation traditionally has been a complicated process that requires specialized knowledge and training. The MSDS should be consulted regarding specific advice on cleanup. IARC13 and Lunn et al.[58] have validated inactivation procedures for specific agents that are effective. However, these procedures vary from drug to drug and may be impractical for small amounts. Care must be taken because of unique problems presented by the cleanup of some agents, such as by-product formation.[57] Serious consideration should be given to alternative disposal methods.

5. Spills. Emergency procedures to cover spills or inadvertent release of hazardous drugs should be included in the facility's overall health and safety program. Incidental spills and breakages should be cleaned up immediately by a properly protected person trained in the appropriate procedures. The area should be identified with a warning sign to limit access to the area. Incident reports should be filed to document the spill and persons exposed.

a. Personnel Contamination. Contamination of protective equipment or clothing, or direct skin or eye contact should be treated by:

- Immediately removing the gloves or gown.
- Immediate cleansing of the affected skin with soap and water.
- Flooding an affected eye at an eyewash fountain or with water or isotonic eyewash designated for that purpose for at least 15 minutes, for eye exposure.
- Obtaining medical attention. (Protocols for emergency procedures should be maintained at the designated sites for such medical care. Medical attention should also be sought for inhalation of HD's in powder form.)
- Documenting the exposure in the employee's medical record.

b. Clean-up of Small Spills. The ASHP considers small spills to be those less than 5 ml. The 5-ml volume of material should be used to categorize spills as large or small. Spills of less than 5 ml or 5 gm outside a BSC should be cleaned up immediately by personnel wearing gowns, double latex gloves, and splash goggles. An appropriate NIOSH-approved respirator should be used for either powder or liquid spills where airborne powder or aerosol is or has been generated.

- Liquids should be wiped with absorbent gauze pads; solids should be wiped with wet absorbent gauze. The spill areas should then be cleaned three times using a detergent solution followed by clean water.
- Any broken glass fragments should be picked up using a small scoop (never the hands) and placed in a "sharps" container. The container should then go into a HD disposal bag, along with used absorbent pads and any other contaminated waste.
- Contaminated reusable items, for example glassware and scoops, should be treated as outlined above under Reusable Items.

c. Clean-up of Large Spills. When a large spill occurs, the area should be isolated and aerosol generation avoided. For spills larger than 5 ml, liquid spread is limited by gently covering with absorbent sheets or spill-control pads or pillows. If a powder is involved, damp cloths or towels should be used. Specific individuals should be trained to clean up large spills.

Protective apparel, including respirators, should be used as with small spills when there is any suspicion of airborne powder or that an aerosol has been or will be generated. Most CD's are not volatile; however, this may not be true for all HD's. The volatility of the drug should be assessed in selecting the type of respiratory protection.

As discussed under waste disposal, chemical inactivation should be avoided in this setting. All contaminated surfaces should be thoroughly cleaned three times with detergent and water. All contaminated absorbent sheets and other materials should be placed in the HD disposal bag.

d. Spills in BSC's. Extensive spills within a BSC necessitate decontamination of all interior BSC surfaces after completion of the spill cleanup. The ASHP[3] recommends this action for spills larger than 150 ml or the contents of one vial. If the HEPA filter of a BSC is contaminated, the unit should be labeled and sealed in plastic until the filter can be changed and disposed of properly by trained personnel wearing appropriate protective equipment.

e. Spill Kits. Spill kits, clearly labeled, should be kept in or near preparation and administrative areas. The MSDS'S include sections on emergency procedures, including appropriate personal protective equipment. The ASHP recommends that kits include: chemical splash goggles, two pairs of gloves, utility gloves, a low-permeability gown, 2 sheets (12" × 12") of absorbent material, 250-ml and 1-liter spill control pillows, a "sharps" container, a small scoop to collect glass fragments, and two large HD waste-disposal bags.[3]

Prior to cleanup, appropriate protective equipment should be donned. Absorbent sheets should be incinerable. Protective goggles and respirators should be cleaned with mild detergent and water after use.

6. Storage and Transport.

a. Storage Areas. Access to areas where HD's are stored should be limited to authorized personnel with signs restricting entry.[72] A list of drugs covered by HD policies and information on spill and emergency contact procedures should be posted or easily available to employees. Optimally, facilities used for storing HD's should not be used for other drugs, and such facilities should be designed to prevent containers from falling to the floor, e.g. bins with barrier fronts. Warning labels should be applied to all HD containers, as well as the shelves and bins where these containers are permanently stored.

b. Receiving Damaged HD Packages. Damaged shipping cartons should be opened in an isolated area or a BSC by a designated employee wearing double gloves, a gown, goggles, and appropriate respiratory protection. Individuals must be trained to process damaged packages as well.

The ASHP recommends that broken containers and contaminated packaging mats be placed in a "sharps" container and then into HD disposal bags.[3] The bags should then be closed and placed in receptacles, as described in the section on Waste Disposal. The appropriate protective equipment and waste disposal materials should be kept in the area where shipments are received, and employees should be trained in their use and the risks of exposure to HD's.

c. Transport. HD's should be securely capped or sealed, placed in sealed clear plastic bags, and transported in containers designed to avoid breakage. Personnel involved in transporting HD's should be trained in spill procedures, including sealing off the contaminated area and calling for appropriate assistance.

All HD containers should be labeled as noted in Drug Preparation Work Practices. If transport methods that produce stress on contents (such as pneumatic tubes) are used, guidance from the OSHA clarification of 1910.1030 with respect to transport (M.4.b.(8)(c)) should be followed. This clarification provides for use of packaging material inside the tube to prevent breakage. These recommendations that pertain to the Bloodborne Pathogens Standard are prudent practice for HD's, e.g. padded inserts for carriers.

VI. MEDICAL SURVEILLANCE.

Workers who are potentially exposed to chemical hazards should be monitored in a systematic program of medical surveillance intended to prevent occupational injury and disease.[3,71,72] The purpose of surveillance is to identify the earliest reversible biologic effects so that exposure can be reduced or eliminated before the employee sustains irreversible damage. The occurrence of exposure-related disease or other adverse health effects should prompt immediate re-evaluation of primary preventive measures (e.g., engineering controls, personal protective equipment). In this manner, medical surveillance acts as a check on the appropriateness of controls already in use.[62]

For detection and control of work-related health effects, job-specific medical evaluations should be performed, as follows:

- before job placement;
- periodically during employment;
- following acute exposures; and
- at the time of job termination or transfer (exit examination).

This information should be collected and analyzed in a systematic fashion to allow early detection of disease patterns in individual workers and groups of workers.

A. PRE-PLACEMENT MEDICAL EXAMINATIONS.

1. Sound medical practice dictates that employees who will be working with HD's in the workplace have an initial evaluation consisting of a history, physical exam, and laboratory studies. Information made available, by the employer, to the examining physician should be:

    - A description of the employee's duties as they relate to the employee's exposure.
    - The employee's exposure levels or anticipated exposure levels.
    - A description of any personal protective equipment used or to be used.
    - Information from previous medical examinations of the employee, which is not readily available to the examining physician.

2. The history details the individual's medical and reproductive experience with emphasis on potential risk factors, such as past hematopoietic, malignant, or hepatic disorders. It also includes a complete occupational history with information on extent of past exposures (including environmental sampling data, if possible) and use of protective equipment. Surrogates for worker exposure, in the absence of environmental sampling data, include:

    - records of drugs and quantities handled;
    - hours spent handling these drugs per week; and
    - number of preparations/administrations per week.

3. The physical examination should be complete, but the skin, mucous membranes, cardiopulmonary and lymphatic systems, and liver should be emphasized. An evaluation for respirator use must be performed in accordance with 29 CFR 1910.134, if the employee will wear a respirator. The laboratory assessment may include a complete blood count with differential, liver function tests, blood urea nitrogen, creatinine, and a urine dipstick. Other aspects of the physical and laboratory evaluation should be guided by known toxicities of the HD of exposure. Due to poor reproducibility, interindividual variability, and lack of prognostic value regarding disease development, no biological monitoring tests (e.g., genotoxic markers) are currently recommended for routine use in employee surveillance. Biological marker testing should be performed only within the context of a research protocol.

B. PERIODIC MEDICAL EXAMINATIONS. Recognized occupational medicine experts in the HD area recommend these exams to update the employee's medical, reproductive, and exposure histories. They are recommended on a yearly basis or every two to three years. The interval between exams is a function of the opportunity for exposure, duration of exposure, and possibly the age of the worker at the discretion of the occupational medicine physician, guided by the worker's history. careful documentation of an individual's routine exposure and any acute accidental exposures are made. The physical examination and laboratory studies follow the format outlined in the preplacement examination.[64]

C. POSTEXPOSURE EXAMINATIONS. Post exposure evaluation is tailored to the type of exposure (e.g., spills or needle sticks from syringes containing HD's). An assessment of the extent of exposure is made and included in the confidential database (discussed below) and in an incident report. The physical examination focuses on the involved area as well as other organ systems commonly affected (i.e. for CD's the skin and mucous membranes; for aerosolized HD's the pulmonary system). Treatment and laboratory studies follow as indicated and should be guided by emergency protocols.

D. EXIT EXAMINATIONS. The exit examination completes the information on the employee's medical, reproductive and exposure histories. Examination and laboratory evaluation should be guided by the individual's history of exposures and follow the outline of the periodic evaluation.

E. EXPOSURE-HEALTH OUTCOME LINKAGE. Exposure assessment of all employees who have worked with HD's is important, and the maintenance of records is required by 29 CFR 1910.1020. The use of previously outlined exposure surrogates is acceptable, although actual environmental or employee monitoring data is preferable. An MSDS can serve as an exposure record. Details of the use of personal protective equipment and engineering controls present should be included. A confidential database should be maintained with information regarding the individual's medical and reproductive history, with linkage to exposure information to facilitate epidemiologic review.

F. REPRODUCTIVE ISSUES. The examining physician should consider the reproductive status of employees and inform them regarding relevant reproductive issues. The reproductive toxicity of hazardous drugs should be carefully explained to all workers who will be exposed to these chemicals, and is required for those chemicals covered by the HCS. Unfortunately, no information is available regarding the reproductive risks of HD handling with the current use of BSC's and PPE. However, as discussed earlier, both spontaneous abortion and congenital malformation excesses have been documented among workers handling some of these drugs without currently recommended engineering controls and precautions. The facility should have a policy regarding reproductive toxicity of HD's and worker exposure in male and female employees and should follow that policy.

VII. HAZARD COMMUNICATION.

This paragraph is for informational purposes only and is not a substitute for the requirements of the Hazard Communication Standard.

A. DISCUSSION.

1. The Hazard Communication Standard (HCS)[107] is applicable to some drugs. It defines a hazardous chemical as any chemical that is a physical hazard or a health hazard. Physical hazard refers to characteristics such as combustibility or reactivity. A health hazard is defined as a chemical for which there is statistically significant evidence based on at least one study conducted in accordance with established scientific principles that acute or chronic health effects may occur in exposed employees. Appendixes A and B of the HCS outline the criteria used to determine whether an agent is hazardous.

2. According to HCS Appendix A, agents with any of the following characteristics are considered hazardous:

    - Carcinogens;
    - Corrosives;
    - Toxic or highly toxic (defined on the basis of median lethal doses);
    - Irritants;
    - Sensitizers; and
    - Target organ effectors, including reproductive toxins, hepatotoxins, nephrotoxins, neurotoxins, agents that act on the hematopoietic system, and agents that damage the lungs, skin, eyes, or mucous membranes

    Both human and animal data are to be used in this determination. HCS Appendix C lists sources of toxicity information.

3. As a result of the February 21, 1990, Supreme Court decision,[21] all provisions of the HCS [29 CFR 1910. 1200][107] are now in effect for all industrial segments. This includes the coverage of drugs and pharmaceuticals in the non-manufacturing sector. On February 9, 1994, OSHA issued a revised Hazard Communication Final Rule with technical clarification regarding drugs and pharmaceutical agents.

4. The HCS requires that drugs posing a health hazard (with the exception of those in solid, final form for direct administration to the patient, i.e., tablets or pills) be included on lists of hazardous chemicals to which employees are exposed.[107] Their storage and use locations can be confirmed by reviewing purchasing office

records of currently used and past used agents such as those in Appendix VI:2-1. Employee exposure records, including workplace monitoring, biological monitoring, and MSDS's as well as employee medical records related to drugs posing a health hazard must be maintained and access to them provided to employees in accordance with 29 CFR 1910.1020. Training required under the HCS should include all employees potentially exposed to these agents, not only health care professional staff, but also physical plant, maintenance, or support staff.

5. MSDS's are required to be prepared and transmitted with the initial shipment of all hazardous chemicals including covered drugs and pharmaceutical products. This excludes drugs defined by the Federal Food, Drug, and Cosmetic Act that are in solid, final form for direct administration to the patient (e.g. tablets, pills, or capsules) or that are packaged for sale to consumers in a retail establishment. Package inserts and the Physician's Desk Reference are not acceptable in lieu of requirements of MSDS's under the Standard. Items mandated by the Standard will use the term shall instead of should.

B. WRITTEN HAZARD COMMUNICATION PROGRAM.

1. Employers shall develop, implement, and maintain at the workplace a written hazard communication program for employees handling or otherwise exposed to chemicals, including drugs that represent a health hazard to employees. The written program will describe how the criteria specified in the Standard concerning labels and other forms of warning, MSDS's, and employee information and training will be met. This also includes the following:

- A list of the covered hazardous drugs known to be present using an identity that is referenced on the appropriate MSDS.
- The methods the employer will use to inform employees of the hazards of nonroutine tasks in their work areas.
- The methods the employer will use to inform employees of other employers of hazards at the work site.

2. The employer shall make the written hazard communication program available, upon request, to employees, their designated representatives, and the Assistant Secretary of OSHA in accordance with requirements of the HCS.

C. MSDS's. In accordance with requirements in the Hazard Communication Standard, the employer must maintain MSDS's accessible to employees for all covered HD's used in the hospital. Specifics regarding MSDS content are contained in the Standard. Essential information includes: health hazards, primary exposure routes, carcinogenic evaluations, acute exposure treatment, chemical inactivators, solubility, stability, volatility, PPE required, and spill procedures for each covered HD. MSDS's shall also be made readily available upon request to employees, their designated representatives, or the Assistant Secretary of OSHA.

VIII. TRAINING AND INFORMATION DISSEMINATION.

A. DISCUSSION.

1. In compliance with the Hazard Communication Standard, all personnel involved in any aspect of the handling of covered HD's (physicians, nurses, pharmacists, housekeepers, employees involved in receiving, transport or storage) must receive information and training to appraise them of the hazards of HD's present in the work area.[71] Such information should be provided at the time of an employee's initial assignment to a work area where HD's are present and prior to assignments involving new hazards. The employer should provide annual refresher information and training.

2. The National Study Commission on Cytotoxic Exposure has recommended that knowledge and competence of personnel be evaluated after the first orientation or training session, and then yearly, or more often if a need is perceived.[71] Evaluation may involve direct observation of an individual's performance on the job. In addition, non-HD solutions should be used for evaluation of preparation technique; quinine, which will fluoresce under ultraviolet light, provides an easy mechanism for evaluation of technique.

B. EMPLOYEE INFORMATION. Employees must be informed of the requirements of the Hazard Communication Standard, 29 CFR 1910.1200, as follows:

- Any operation/procedure in their work area where drugs that present a hazard are present.
- The location and availability of the written hazard communication program.
- Any operations or procedure in their work area where other HD's are present.
- The location and availability of any other plan regarding HD's.

C. EMPLOYEE TRAINING.

1. Employee training must include at least the following elements:

   - Methods and observations that may be used to detect the presence or release of an HCS-covered hazardous drug in the work area (such as monitoring conducted by the employer, continuous monitoring devices, visual appearance or odor of covered HD's being released, etc.).
   - The physical and health hazards of the covered HD's in the work area.
   - The measures employees can take to protect themselves from these hazards. This includes specific procedures that the employer has implemented to protect the employees from exposure to such drugs, such as identification of covered drugs and those to be handled as hazardous, appropriate work practices, emergency procedures (for spills or employee exposure).
   - Personal protective equipment, and the details of the hazard communication program developed by the employer, including an explanation of the labeling system and the MSDS, and how employees can obtain and use the appropriate hazard information.

2. It is essential that workers understand the carcinogenic potential and reproductive hazards of these drugs. Both females and males should understand the importance of avoiding exposure, especially early in pregnancy, to the drugs, so that they can make informed decisions about the hazards involved. In addition, the facility's policy regarding reproductive toxicity of HD's should be explained to workers. Updated information should be provided to employees on a regular basis and whenever their jobs involve new hazards. Medical staff and other personnel who are not hospital employees should be informed of hospital policies and of the expectation that they will comply with these policies.

IX. RECORDKEEPING.

Any workplace exposure record created in connection with HD handling shall be kept, transferred, and made available for at least 30 years and medical records shall be kept for the duration of employment plus 30 years in accordance with the Access to Employee Exposure and Medical Records Standard (29 CFR 1910.1020).[108] In addition, sound practice dictates that training records should include the following information:

- Dates of the training sessions;
- Contents or a summary of the training sessions;
- Names and qualifications of the persons conducting the training; and
- Names and job titles of all persons attending the training sessions.

Training records should be maintained for three years from the date on which the training occurred.

X. REFERENCES.

1. American Medical Association Council on Scientific Affairs. 1985. "Guidelines for handling parenteral antineoplastics." J.A.M.A. 253:1590-2.

2. American National Standards Institute. 1968. *Occupational and Educational Eye and Face Protection.* ANSI Z87.1.

3. American Society of Hospital Pharmacists. 1990. ASHP Technical Assistance Bulletin on Handling Cytotoxic and Hazardous Drugs. *Am. J. Hosp. Pharm.* 47:1033-49.

4. Andersen, R., Boedicker, M., Ma, M., and Goldstein, E.J.C. 1986. "Adverse Reactions Associated with Pentamidine Isethionate in AIDS Patients: Recommendations for Monitoring Therapy." *Drug Intell. Clin. Pharm.* 20:862-8.

5. Anderson, R.W., Puckett, W.H., Dana, W.J., et al. 1982. "Risk of handling injectable antineoplastic agents." *Am. J. Hosp. Pharm.* 39:1881-87.

6. Avis, K.E., and Levchuck, J.W. 1984. "Special considerations in the use of vertical laminar flow workbenches." *Am. J. Hosp. Pharm.* 41:81-7.

7. Barber, R.K. 1981. "Fetal and neonatal effects of cytotoxic agents." *Obstet. Gynecol.* 51:41S-47S.

8. Benhamou, S., Pot-Deprun, J., Sancho-Garnier, H., and Chouroulinkov, I. 1988. "Sister chromatid exchanges and chromosomal aberrations in lymphocytes of nurses handling cytostatic drugs." I 41:350-3.

9. Bos, R.P., Leenars, A.O., Theuws, J.L., and Henderson, P.T. 1982. "Mutagenicity of urine from nurses handling cytostatic drugs, influence of smoking." *Int. Arch. Occ. Envir. Health* 50:359-69.

10. Bryan, D., and Marback, R.C. 1984. "Laminar-airflow equipment certification: What the pharmacist needs to know." *Am. J. Hosp. Pharm.* 41:1343-9.

11. Burgaz, S., Ozdamar, Y.N., and Karakaya, A.E. 1988. "A signal assay for the detection of genotoxic compounds: Application on the urine of cancer patients on chemotherapy and of nurses handling cytotoxic drugs." *Human Toxicol.* 7:557-60.

12. California Department of Health Services Occupational Health Surveillance and Evaluation Program. 1986. "Health care worker exposure to ribavirin aerosol: field investigation Fl-86-009." Berkeley: California Department of Health Services.

13. Castegnaro, M., Adams, J., Armour, M.A., et al., eds. 1985. *Laboratory decontamination and destruction of carcinogens in laboratory wastes: Some antineoplastic agents*. International Agency for Research on Cancer. Scientific Publications No. 73. Lyon, France: IARC.

14. Chapman, R.M. 1984. "Effect of cytotoxic therapy on sexuality and gonadal function." *In Toxicity of Chemotherapy*. Perry, M.C. and J.W. Yarbro (eds.). 343-363. Orlando: Grune & Stratton.

15. Chen, C.H., Vazquez-Padua, M., and Cheng, Y.C. 1990. "Effect of anti-human immunodeficiency virus nucleoside analogs on mDNA and its implications for delayed toxicity." *Mol. Pharm.* 39:625-628.

16. Christensen, C.J., Lemasters, G.K., and Wakeman, M.A. 1990. "Work practices and policies of hospital pharmacists preparing antineoplastic agents." *J. Occup. Med.* 32:508-12.

17. Chrysostomou, A., Morley, A.A., and Seshadri, R. 1984. "Mutation frequency in nurses and pharmacists working with cytotoxic drugs." *Aust. N.Z.J. Med.* 14:831-4.

18. Connor, J.D., Hintz, M., and Van Dyke, R. 1984. "Ribavirin pharmacokinetics in children and adults during therapeutic trials." In *Clinical Applications of Ribavirin*, Smith, R.A., Knight, V., and Smith, J.A.D. (eds.) Orlando: Academic Press.

19. Connor, T.H., Laidlaw, J.L., Theiss, J.C., et al. 1984. "Permeability of latex and polyvinyl chloride gloves to carmustine." *Am. J. Hosp. Pharm.* 41:676-9.

20. Crudi, C.B. 1980. "A compounding dilemma: I've kept the drug sterile but have I contaminated myself?" *Nat. Intra. Therapy* J. 3:77-80.

21. Dole v. United Steelworkers. 1990. 494 U.S.26.

22. Doll, D.C. 1989. Aerosolised pentamidine. Lancet ii:1284-5.

23. Duvall, E., and Baumann, B. 1980. "An unusual accident during the administration of chemotherapy." *Cancer Nurs*. 3:305-6.

24. Environmental Protection Agency. 1991. *Discarded commercial chemical products, off specification species, container residues, and spill residues thereof.* 40 CFR 261.33(f).

25. Everson, R.B., Ratcliffe, J.M., Flack, P.M., et al. 1985. "Detection of low levels of urinary mutagen excretion by chemotherapy workers which was not related to occupational drug exposures." *Cancer Research* 45:6487-97.

26. Falck, K., Grohn, P., Sorsa, M., et al. 1979. "Mutagenicity in urine of nurses handling cytostatic drugs." *Lancet* i:1250-1.

27. Falck, K., Sorsa, M., and Vainio, H. 1981. "Use of the bacterial fluctuation test to detect mutagenicity in urine of nurses handling cytostatic drugs." (abstract). *Mutation Res*. 85:236-7.

28. Ferguson, L.R., Everts, R., Robbie, M.A., et al. 1988. "The use within New Zealand of cytogenetic approaches to monitoring of hospital pharmacists for exposure to cytotoxic drugs: Report of a pilot study in Auckland." *Aust J Hosp Pharm*. 18:228-33.

29. Gude, J.K. 1989. "Selective delivery of pentamidine to the lung by aerosol." *Am. Rev. Resp. Dis.* 139:1060.

30. Guglielmo, B.J., Jacobs, R.A., and Locksley, R.M. 1989. "The exposure of health care workers to ribavirin aerosol." *J.A.M.A*. 261:1880-1.

31. Harrison, R., Bellows, J., Rempel, D., et al. 1988. "Assessing exposures of health-care personnel to aerosols of ribavirin -- California." *Morbidity and Mortality Weekly Report* 37:560-3.

32. Hemminki, K., Kyyronen, P., and Lindbohm, M.L. 1985. "Spontaneous abortions and malformations in the offspring of nurses exposed to anaesthetic gases, cytostatic drugs, and other potential hazards in hospitals, based on registered information of outcome." *J. Epidem. Comm. Health* 39:141-7.

33. Henderson, D.K., and Gerberding, J.L. 1989. "Prophylactic zidovudine after occupational exposure to the human immunodeficiency virus: An interim analysis." *J. Infectious Diseases* 160:321-7.

34. Hillyard, I. W. 1980. "The preclinical toxicology and safety of ribavirin." In *Ribavirin: a broad spectrum antiviral agent*. Smith, R.A., and Kirkpatrick, W. (eds.) New York: Academic Press.

35. Hirst, M., Tse, S., Mills, D.G., et al. 1984. "Occupational exposure to cyclophosphamide." *Lancet* 1:186-8.

36. Hoy, R.H. and Stump, L.M. 1984. "Effect of an air-venting filter device on aerosol production from vials." *Am. J. Hosp. Pharm*. 41:324-6.

37. International Agency for Research on Cancer. *IARC Monographs on the Evaluation of the Carcinogenic Risk of Chemicals to Man: Some Aziridines, S-, and O-mustards and selenium*. Lyon, France. 1975; Vol. 9.

38. International Agency for Research on Cancer. 1976. *IARC Monographs on the Evaluation of the Carcinogenic Risk of Chemicals to Man: Some naturally occurring substances*. Vol. 10 Lyon, France: IARC.

39. International Agency for Research on Cancer. 1981. *IARC Monographs on the Evaluation of the Carcinogenic Risk of Chemicals to Humans: Some Antineoplastic and Immunosuppressive Agents.* Vol. 26. Lyon, France: IARC.

40. International Agency for Research on Cancer. 1982. *IARC Monographs on the Evaluation of the Carcinogenic Risk of Chemicals to Humans: Chemicals, Industrial Processes and Industries Associated with Cancer in Humans*. Vol. 1-29 (Suppl. 4). Lyon, France: IARC.

41. International Agency for Research on Cancer. 1987. *IARC Monographs on the Evaluation of the Carcinogenic Risk of Chemicals to Humans; Genetic and related effects: An updating of selected IARC Monographs from Volumes 1-42. Vol. 1-42 (Suppl. 6)*. Lyon, France: IARC.

42. International Agency for Research on Cancer: 1987. *IARC Monographs on the Evaluation of the Carcinogenic Risk of Chemicals to Humans; Overall evaluations of carcinogenicity: An updating of IARC Monographs Volumes 1 to 42. Vol. 1-42 (Suppl. 7)*. Lyon, France: IARC.

43. International Agency for Research on Cancer. 1990. *IARC Monographs on the Evaluation of the Carcinogenic Risk of Chemicals to Humans: Pharmaceutical Drugs*. Vol. 50. Lyon, France: IARC.

44. Jagun, O., Ryan, M., and Waldron, H.A. 1982. "Urinary thioether excretion in nurses handling cytotoxic drugs." *Lancet* i:443-4.

45. Johnson, E.G., and Janosik, J.E. 1989. "Manufacturers' recommendations for handling spilled antineoplastic agents." *Am. J. Hosp. Pharm.* 46:318-9.

46. Juma, F.D., Rogers, H.J., Trounce, J.R., and Bradbrook, I.D. 1978. "Pharmacokinetics of intravenous cyclophosphamide in man, estimated by gas-liquid chromatography." *Cancer Chemother. Pharmacol.* 1:229-31.

47. Kacmarek, R.M. 1990. "Ribavirin and pentamidine aerosols: Caregiver beware!" *Respiratory Care* 35:1034-6.

48. Karakaya, A.E., Burgaz, S., and Bayhan, A. 1989. "The significance of urinary thioethers as indicators of exposure to alkylating agents." *Arch. Toxicol. 13 (suppl.):*117-9.

49. Kilham, L. and Ferm, V.H. 1977. "Congenital anomalies induced in hamster embryos with ribavirin." *Science* 195:413-4.

50. Kleinberg, M.L. and Quinn, M.J. 1981. "Airborne drug levels in a laminar-flow hood."*Am. J. Hosp. Pharm.*38:1301-3.

51. Kolmodin-Hedman, B., Hartvig, P., Sorsa, M., and Falck, K. 1983. Occupational handling of cytostatic drugs. *Arch. Toxicol.*

52. Kyle, R. A. 1984. "Second malignancies associated with chemotherapy." *In Toxicity of Chemotherapy,* Perry, M.C., and J.W. Yarbro (eds.), 479-506. Orlando: Grune & Stratton.

53. Laidlaw, J.L., Connor, T.H., Theiss, J.C., et al. 1984. "Permeability of latex and polyvinyl chloride gloves to 20 antineoplastic drugs." *Am. J. Hosp. Pharm.* 41:2618-23.

54. Laidlaw, J.L., Connor, T.H., Theiss, J.C., et al. 1985. "Permeability of four disposable protective-clothing materials to seven antineoplastic drugs." *Am. J. Hosp. Pharm.* 42:2449-54.

55. Lee, S.B. 1988. "Ribavirin - Exposure to health care workers." *Am. Ind. Hyg. Assoc.* 49:A13-A14.

56. LeRoy, M.L., Roberts, M.J., and Theisen, J.A. 1983. "Procedures for handling antineoplastic injections in comprehensive cancer centers." *Am. Hosp Pharm.* 40:601-3.

57. Lunn, G. and Sansone, E.B. 1989. "Validated methods for handling spilled antineoplastic agents." *Am. J. Hosp. Pharm.* 46:1131.

58. Lunn, G., Sansone, E.B., Andrews, A.W., and Hellwig, L.C. 1989. "Degradation and disposal of some antineoplastic drugs." *J. Pharm. Sciences* 78:652-9.

59. Matthews, T. and Boehme, R. 1988. "Antiviral activity and mechanism of action of ganciclovir." *Rev. Infect. Diseases* 10(suppl 3):s490-94.

60. McDevitt, J.J., Lees, P.S.J., and McDiarmid, M.A. 1993. "Exposure of Hospital Pharmacists and Nurses to Antineoplastic Agents." *J. Occup. Med.* 35:57-60.

61. McDiarmid, M.A., Egan, T., Furio, M., et al. 1988. "Sampling for airborne fluorouracil in a hospital drug preparation area." *Am. J. Hosp. Pharm.* 43:1942-5.

62. McDiarmid, M.A. and Emmett, E.A. 1987. "Biological monitoring and medical surveillance of workers exposed to antineoplastic agents." *Seminars in Occup. Med.* 2:109-17.

63. McDiarmid, M.A., and Jacobson-Kram, D. 1989. "Aerosolized pentamidine and public health." *Lancet* ii:863-864.

64. McDiarmid, M.A. 1990. "Medical surveillance for antineoplastic drug handlers." *Am. J. Hosp. Pharm.* 47:1061-6.

65. McDiarmid, M.A., Gurley, H. T., and Arrington, D. 1991. "Pharmaceuticals as hospital hazards: Managing the risks." *J. Occup. Med.* 33:155-8.

66. McDiarmid, M.A., Kolodner, K., Humphrey, F., et al. 1992. "Baseline and phosphoramide mustard-induced sister-chromatid exchanges in pharmacists handling anticancer drugs." *Mutation Research* 279:199-204.

67. McDiarmid, M.A., Schaefer, J., Richard, C.L., Chaisson, R.E., and Tepper, B.S. 1992. "Efficacy of engineering controls in reducing occupational exposure to aerosolized pentamidine." *Chest* 102:1764-6.

68. McEvoy, G. K., ed. 1993. *American Hospital Formulary Service Drug Information. Bethesda*: American Society of Hospital Pharmacists.

69. McLendon, B.F. and Bron, A.F. 1978. "Corneal toxicity from vinblastine solution." *Br. J. Ophthalmol.* 62;97-9.

70. National Sanitation Foundation. 1990. *Standard No. 49 for Class II (Laminar Flow) Biohazard Cabinetry.* Ann Arbor: National Sanitation Foundation.

71. National Study Commission on Cytotoxic Exposure. 1983. *Recommendations for Handling Cytotoxic Agents.* Louis P. Jeffrey, Sc.D., Chairman, Rhode Island Hospital, Providence, Rhode Island.

72. National Study Commission on Cytotoxic Exposure. 1984. *Consensus Responses to Unresolved Questions Concerning Cytotoxic Agents*. Louis P. Jeffrey, Sc.D., Chairman, Rhode Island Hospital, Providence, Rhode Island.

73. Neal, A.D., Wadden, R.A., and Chiou, W. L. 1983. "Exposure of hospital workers to airborne antineoplastic agents." *Am. J. Hosp. Pharm.* 40:597-601.

74. Nikula E., Kiviniitty K., Leisti J., and Taskinen, P. "Chromosome aberrations in lymphocytes of nurses handling cytostatic agents." Scand. J. Work Environ. Health. 1984;10:71-4.

75. Norppa, H., Sorsa, M., Vainio, H., et al. 1980. "Increased sister chromatid exchange frequencies in lymphocytes of nurses handling cytostatic drugs." *Scand. J. Work Envir. Health* 6:299-301.

76. Nguyen, T.V., Theiss, J.C., and Matney, T.S. "Exposure of pharmacy personnel to mutagenic antineoplastic drugs." *Cancer Research* 42:4792-6.

77. Palmer, R.G., Dore, C.J., and Denman, A.M. 1984. "Chlorambucil-induced chromosome damage to human lymphocytes is dose-dependent and cumulative." *Lancet* i:246-9.

78. Perry, M.C. and Yarbro, J. W., eds. 1984. *Toxicity of Chemotherapy*. Orlando: Grune & Stratton.

79. Physician's Desk Reference. 1991. *Physician's Desk Reference*. 45th ed. Page 730. Barnhart, E.R., pub. Oradell, New Jersey: Medical Economics Data.

80. Pohlova, H., Cerna, M., and Rossner, P. 1986. "Chromosomal aberrations, SCE and urine mutagenicity in workers occupationally exposed to cytostatic drugs." *Mutation Res*. 174:213-7.

81. Pyy, L., Sorsa, M., and Hakala, E. 1988. "Ambient monitoring of cyclophosphamide in manufacture and hospitals." *Am. Ind. Hyg. Assoc*. J. 49:314-7.

82. Reich, S.D. and Bachur, N.R. 1975. "Contact dermatitis associated with adriamycin (NSC-123127) and daunorubicin (NSC-82151)." *Cancer Chemotherap. Reports* 59:677-8.

83. Reynolds, R.D., Ignoffo, R., Lawrence, J., et al. 1982. "Adverse reactions to AMSA in medical personnel." *Cancer Treat. Rep*. 66:1885.

84. Rogers, B. 1987. "Health hazards to personnel handling antineoplastic agents." Occupational Medicine: *State of the Art Reviews* 2:513-24.

85. Rogers, B. and Emmett, E.A. 1987. "Handling antineoplastic agents: Urine mutagenicity in nurses." *IMAGE Journal of Nursing Scholarship*. 19:108-113

86. Rosner, F. 1976. "Acute leukemia as a delayed consequence of cancer chemotherapy." *Cancer*37:1033-6.

87. Rudolph, R., Suzuki, M., and Luce, J. K. 1979. "Experimental skin necrosis produced by adriamycin." *Cancer Treat. Reports* 63:529-37.

88. Schafer, A.L. 1981. "Teratogenic effects of antileukemic therapy." *Arch. Int. Med*. 141:514-5.

89. Selevan, S.G., Lindbolm, M.L., Hornung, R.W., and Hemminki, K. 1985. "A study of occupational exposure to antineoplastic drugs and fetal loss in nurses." *New England J. Med*. 313:1173-8.

90. Sieber, S.M. 1975. "Cancer chemotherapeutic agents and carcinogenesis." *Cancer Chemotherap. Reports* 59:915-8.

91. Sieber, S.M., and Adamson, R.H. 1975. "Toxicity of antineoplastic agents in man: Chromosomal aberrations, antifertility effects, congenital malformations, and carcinogenic potential." *Adv. Cancer Res*. 22:57-155.

92. Siebert, D. and Simon, U. 1973. "Cyclophosphamide: Pilot study of genetically active metabolites in the urine of a treated human patient." *Mutat. Res*. 19:65-72.

93. Slevin, M.L., Ang, L.M., Johnston, A., and Turner, P. 1984. "The efficiency of protective gloves used in the handling of cytotoxic drugs." *Cancer Chemo. Pharmacol*. 12:151-3.

94. Smaldone, G.C., Vincicuerra, C., and Marchese, J. 1991. "Detection of inhaled pentamidine in health care workers." *New England J. Med*. 325:891-2.

95. Sorsa, M., Hemminki, K., and Vanio, H. 1985. "Occupational exposure to anticancer drugs -- Potential and real hazards." *Mut. Res*. 154:135-49.

96. Sotaniemi, E.A., Sutinen, S., Arranto, A.J., et al. 1983. "Liver damage in nurses handling cytostatic agents."*Acta Med. Scand*. 214:181-9.

97. Stellman, J.M. 1987. "The spread of chemotherapeutic agents at work: Assessment through stimulation." *Cancer Investigation* 5:75-81.

98. Stephens, J.D., Golbus, M.S., Miller, T. R., et al. 1980. "Multiple congenital abnormalities in a fetus exposed to 5-fluorouracil during the first trimester." *Am. J. Obstet. Gynecol.* 137:747-9.

99. Stiller, A., Obe, G., Bool, I., and Pribilla, W. 1983. "No elevation of the frequencies of chromosomal aberrations as a consequence of handling cytostatic drugs." *Mut. Res.* 121:253-9.

100. Stoikes, M.E., Carlson, J.D., Farris, F.F., and Walker, P.R. 1987. "Permeability of latex and polyvinyl chloride gloves to fluorouracil and methotrexate." *Am. J. Hosp. Pharm.* 44:1341-6.

101. Stucker, I., Hirsch, A., Doloy, T., et al. 1986. "Urine mutagenicity, chromosomal abnormalities and sister chromatid exchanges in lymphocytes of nurses handling cytostatic drugs." *Int. Arch. Occup. Environ. Health* 57:195-205.

102 Stucker, I., Caillard, J.F., Collin, R., et al. 1990. "Risk of spontaneous abortion among nurses handling antineoplastic drugs." *Scand. J. Work Environ. Health* 16:102-7.

103. U.S. Department of Health and Human Services. Public Health Service. National Institutes of Health. 1992. *Recommendations for the Safe Handling of Cytotoxic Drugs. NIH Publication* No. 92-2621.

104 U.S. Department of Health and Human Services. Public Health Service. Centers for Disease Control. National Institute for Occupational Safety and Health. 1988. *Guidelines for Protecting the Safety and Health of Health Care Workers.* DHHS (NIOSH) Publication No. 88-119.

105. U.S. Department of Labor, Occupational Safety and Health Administration. 1984. Respiratory Protection Standard. 29 CFR 1910.134.

106. U.S. Department of Labor, Occupational Safety and Health Administration. 1986. *Work practice guidelines for personnel dealing with cytotoxic (antineoplastic) drugs.* OSHA Publication #8-1.1.

107. U.S. Department of Labor, Occupational Safety and Health Administration. 1989. Hazard Communication Standard. 29 CFR 1910.1200, as amended February 9, 1994.

108. U.S. Department of Labor, Occupational Safety and Health Administration. 1990. Access to Employee and Medical Records Standard. 29 CFR 1910.1020.

109. U.S. Department of Labor, Occupational Safety and Health Administration. 1991. Occupational Exposure to Bloodborne Pathogens Standard. 29 CFR 1910.1030.

110. Vaccari, F.L., Tonat, K., DeChristoforo, R., et al. 1984. "Disposal of antineoplastic wastes at the NIH." *Am. J. Hosp. Pharm.* 41:87-92.

111. Valanis, B., Vollmer, W.M., Labuhn, K., Glass, A., and Corelle, C. 1992. "Antineoplastic drug handling protection after OSHA guidelines: Comparison by profession, handling activity, and worksite." *J. Occup. Med.* 34:149-55.

112. Venitt, S., Crofton-Sleigh, C., Hunt, J., et al. 1984. "Monitoring exposure of nursing and pharmacy personnel to cytotoxic drugs: Urinary mutation assays and urinary platinum as markers of absorption." *Lancet* i:74-6.

113. Waksvik, H., Klepp, O., and Brogger, A. 1981. "Chromosome analyses of nurses handling cytostatic agents." *Cancer Treat. Reports* 65:607-10.

114. Wall, R.L. and Clausen, K.P. 1975. "Carcinoma of the urinary bladder in patients receiving cyclophosphamide." *New England J. Med.* 293:271-3.

115. Weisburger, J.H., Griswold, D.P., Prejean J.D., et al. 1975. "Tumor induction by cytostatics. The carcinogenic properties of some of the principal drugs used in clinical cancer chemotherapy." *Recent Results Cancer Res.* 52:1-17.

116. Zimmerman, P.F., Larsen, R.K., Barkley, E.W., and Gallelli, J.F. 1981. "Recommendations for the safe handling of injectable antineoplastic drug products." *Am. J. Hosp. Pharm.* 38:1693-5.

APPENDIX VI: 2-1. SOME COMMON DRUGS THAT ARE CONSIDERED HAZARDOUS.

Appendix VI:2-1 is not all-inclusive, should not be construed as complete, and represents an assessment of some, but not all, marketed drugs at a fixed point in time. Appendix VI:2-1 was developed through consultation with institutions that have assembled teams of pharmacists and other health care personnel to determine which drugs should be handled with caution. These teams reviewed product literature and drug information when considering each product.

Sources for this appendix are the "Physicians Desk Reference," Section 10:00 in the American Hospital Formulary Service Drug Information,[68] IARC publications (particularly Volume 50),[43] the Johns Hopkins Hospital, and the National Institutes of Health, Clinical Center Nursing Department. No attempt to include investigational drugs was made, but they should be prudently handled as hazardous drugs until adequate information becomes available to exclude them.

Any determination of the hazard status of a drug should be periodically reviewed and updated as new information becomes available. Importantly, new drugs should routinely undergo a hazard assessment.

CHEMICAL/GENERIC NAME	SOURCE*
ALTRETAMINE	C
AMINOGLUTETHIMIDE	A
AZATHIOPRINE	ACE
L-ASPARAGINASE	ABC
BLEOMYCIN	ABC
BUSULFAN	ABC
CARBOPLATIN	ABC
CARMUSTINE	ABC
CHLORAMBUCIL	ABCE
CHLORAMPHENICOL	E
CHLOROTIANISENE	B
CHLOROZOTOCIN	E
CYCLOSPORIN	E
CISPLATIN	ABCE
CYCLOPHOSPHAMIDE	ABCE
CYTARRABINE	ABC
DACARBAZINE	ABC
DACTINOMYCIN	ABC

DAUNORUBICIN	ABC
DIETHYLSTILBESTROL	BE
DOXORUBICIN	ABCE
ESTRADIOL	B
ESTRAMUSTINE	AB
ETHINYL ESTRADIOL	B
ETOPOSIDE	ABC
FLOXURIDINE	AC
FLUOROURACIL	ABC
FLUTAMIDE	BC
GANCICLOVIR	AD
HYDROXYUREA	ABC
IDARUBICIN	AC
IFOSFAMIDE	ABC
INTERFERON-A	BC
ISOTRETINOIN	D
LEUPROLIDE	BC
LEVAMISOLE	C
LOMUSTINE	ABCD
MECHLORETHAMINE	BC
MEDROXYPROGESTERONE	B
MEGESTROL	BC
MELPHALAN	ABCE
MERCAPTOPURINE	ABC
METHOTREXATE	ABC
MITOMYCIN	ABC
MITOTANE	ABC
MITOXANTRONE	ABC
NAFARELIN	C
PIPOBROMAN	C
PLICAMYCIN	BC
PROCARBAZINE	ABCE
RIBAVIRIN	D
STREPTOZOCIN	AC
TAMOXIFEN	BC
TESTOLACTONE	BC
THIOGUANINE	ABC
THIOTEPA	ABC
URACIL MUSTARD	ACE
VIDARABINE	D
VINBLASTINE	ABC
VINCRISTINE	ABC
ZIDOVUDINE	D

*Sources

A The National Institutes of Health, Clinical Center Nursing Department
B Antineoplastic drugs in the Physicians' Desk Reference
C American Hospital Formulary, Antineoplastics
D Johns Hopkins Hospital
E International Agency for Research on Cancer

APPENDIX VI:2-2. SOME AEROSOLIZED DRUGS.

RIBAVIRIN

Ribavirin is a synthetic nucleoside with antiviral activity against respiratory syncytial virus (RSV). It appears to restrict synthesis of viral proteins by interfering with mRNA production, but the exact mechanism of action remains unknown.[34] It is reconstituted from a lyophilized powder for aerosol administration.

Ribavirin is usually administered in the aerosolized form via mask or oxygen tent for 12-18 hours per day for 3 to 7 days. A small particle aerosol generator (SPAG) creates respirable particles of 1.3 micrometer median diameter. Under current practice, excess drug is exhausted into the patient's room, causing additional exposures.

Studies have shown Ribavirin to be teratogenic in rodents and embryolethal in rabbits. Ribavirin induces cell transformation in an in-vitro mammalian system (Balb/C 3T3 cell line). In vivo carcinogenicity studies are incomplete.

Human studies on nurses who administer the drug by oxygen tent calculate that the absorbed dose of ribavirin per workshift is 13.5 mg/kg.[31] This estimated dose exceeded 1/100 (the safety factor) of the short term daily dose levels toxic in animal models described above. No symptoms were reported by any health care worker in this study. However, minor pulmonary function abnormalities have been seen among healthy adult volunteers in clinical studies.[12,18]

PENTAMIDINE

Aerosolized pentamidine isethionate (4,4'-diami-dinophenoxypentane) is FDA-approved for the treatment and prophylaxis of pneumonia caused by the protozoan Pneumocystis carinii. The exact mode of action is not fully understood; some studies indicate that pentamidine interferes with nuclear metabolism, inhibiting the synthesis of DNA, RNA, phospholipids, and proteins. It possesses two amidine groups and resembles other compounds called electrophiles which form DNA adducts. Pentamidine is administered as an aerosol after being reconstituted from a lyophilized powder.

No studies have been performed to evaluate the potential carcinogenic, mutagenic, or reproductive effects of pentamidine in animals or humans.[63]

Studies among health care workers have demonstrated pentamidine uptake by those personnel who administer the drug. Side effects include coughing, sneezing, mucous membrane irritation, headache, and bronchospasm. Pulmonary function tests have demonstrated transitory decreases in carbon monoxide diffusing capacity (DLCO). However, one respiratory therapist followed for 14 months has not returned to baseline after exposure.[29]

SECTION VII: CHAPTER 1

BACK DISORDERS AND INJURIES

Contents:

I. INTRODUCTION.

A. GENERAL. Back disorders can develop gradually as a result of microtrauma brought about by repetitive activity over time or can be the product of a single traumatic event. Because of the slow and progressive onset of this internal injury, the condition is often ignored until the symptoms become acute, often resulting in disabling injury. Acute back injuries can be the immediate result of improper lifting techniques and/or lifting loads that are too heavy for the back to support. While the acute injury may seem to be caused by a single well-defined incident, the real cause is often a combined interaction of the observed stressor coupled with years of weakening of the musculoskeletal support mechanism by repetitive micro-trauma. Injuries can arise in muscle, ligament, vertebrae, and discs, either singly or incombination.

B. INCIDENCE. Although back injuries account for no work-related deaths, they do account for a significant amount of human suffering, loss of productivity, and economic burden on compensation systems. Back disorders are one of the leading causes of disability for people in their working years and afflict over 600,000 employees each year with a cost of about $50 billion annually in 1991 according to NIOSH. The frequency and economic impact of back injuries and disorders on the work force are expected to increase over the next several decades as the average age of the work force increases and medical costs go up.

II. BACK DISORDERS.

A. FACTORS ASSOCIATED WITH BACK DISORDERS. Back disorders result from exceeding the capability of the muscles, tendons, discs, or the cumulative effect of several contributors:

- Reaching while lifting.
- Poor posture--how one sits or stands.
- Stressful living and working activities--staying in one position for too long.
- Bad body mechanics--how one lifts, pushes, pulls, or carries objects.
- Poor physical condition-losing the strength and endurance to perform physical tasks without strain.
- Poor design of job or work station.
- Repetitive lifting of awkward items, equipment, or (in health-care facilities) patients.
- Twisting while lifting.
- Bending while lifting.
- Maintaining bent postures.
- Heavy lifting.

- Fatigue.
- Poor footing such as slippery floors, or constrained posture.
- Lifting with forceful movement.
- Vibration, such as with lift truck drivers, delivery drivers, etc.

B. SIGNS AND SYMPTOMS. Signs and symptoms include pain when attempting to assume normal posture, decreased mobility, and pain when standing or rising from a seated position.

III. REPORTS OF BACK INJURIES.

A. CONTRIBUTING FACTORS. These factors usually account for very few work-related back injuries.

- Congenital defects of the spine.
- Increase in static standing or sitting tasks.
- An aging work force.
- Decreases in physical conditioning and exercise.
- Increased awareness of workplace hazards.
- Job dissatisfaction.

B. MANUAL MATERIALS HANDLING. Manual materials handling is the principal source of compensable injuries in the American work force, and four out of five of these injuries will affect the lower back.

IV. INVESTIGATION GUIDELINES.

A. RECORDS REVIEW: OSHA 200 LOG.

1. Note when back or other musculoskeletal disorders appear excessive from Lost Work Day Injury and Illness (LWDII) rate calculations. Understand that excessiveness is relative, since there is no firm figure established that delineates safe from unsafe. A better measure is to look for trends of escalating number of injuries or of increasing severity of injuries. Comparing your target population with BLS data, other company rates, other lines, departments, wings, or occupational titles can yield a meaningful measuring point to gauge excessiveness.

   Back injuries should be treated as an injury on the OSHA 200 log regardless of whether the injury was the result of an acute or chronic exposure.

2. To determine if trends exist, at least several years of the OSHA 200 log will be needed for review.

3. Record or copy information, including occupational titles, departments, dates of injury or illness, from the OSHA 200 log and pertinent OSHA 101 (or equivalent). This information can be used to calculate the LWDII and Severity rates (see Appendix VII:1-1).

4. If you determine that there is a need for a more in-depth analysis of the extent and magnitude of the back disorders, see Appendix VII:1-1.

B. EMPLOYER, EMPLOYEE INTERVIEWS.

1. Walkaround.

   a. Ask employees about their opinion on the difficulty of the task as well as personal experiences of back pain.

   b. Observe worker postures and lifting.

   c. Determine weight of objects lifted.

   d. Determine the frequency and duration of lifting tasks.

e. Measure the dimensions of the workplace and lift.

2. Evaluation.

a. Videotapes should be taken of the work task for later review and for evidence of recognized musculoskeletal hazards (see Appendix VII:1-3).

b. Manual lifting:

- Repetitive material handling increases the likelihood of a disorder.
- Principal variables in evaluating manual lifting tasks to determine how heavy a load can be lifted are: the horizontal distance from the load to the employee's spine, the vertical distance through which the load is handled, the amount of trunk twisting the employee utilized during the lifting, the ability of the hand to grasp the load, and the frequency with which the load is handled.
- Additional variables include floor and shoe traction, space constraints, two-handed lifts, size and stability of the load.
- The NIOSH Lifting Formula uses the principal variables to compute a theoretically safe lift.

V. PREVENTION AND CONTROL.

A. ENGINEERING CONTROLS.

1. General. Alter the task to eliminate the hazardous motion and/or change the position of the object in relation to the employee's body -- such as adjusting the height of a pallet or shelf.

2. Manual Handling Tasks.

a. Material handling tasks should be designed to minimize the weight, range of motion, and frequency of the activity.

b. Work methods and stations should be designed to minimize the distance between the person and the object being handled.

c. Platforms and conveyors should be built at about waist height to minimize awkward postures. Conveyors or carts should be used for horizontal motion whenever possible. Reduce the size or weight of the object(s) lifted.

d. High-strength push-pull requirements are undesirable, but pushing is better than pulling. Material handling equipment should be easy to move, with handles that can be easily grasped in an upright posture.

e. Workbench or workstation configurations can force people to bend over. Corrections should emphasize adjustments necessary for the employee to remain in a relaxed upright stance or fully supported, seated posture. Bending the upper body and spine to reach into a bin or container is highly undesirable. The bins should be elevated, tilted or equipped with collapsible sides to improve access.

f. Repetitive or sustained twisting, stretching, or leaning to one side are undesirable. Corrections could include repositioning bins and moving employees closer to parts and conveyors.

g. Store heavy objects at waist level.

h. Provide lift-assist devices, and lift tables.

B. CONTROLS AND WORK PRACTICES.

1. Engineering controls are preferred.

2. Worker training and education:

- Training should include general principles of ergonomics, recognition of hazards and injuries, procedures for reporting hazardous conditions, and methods and procedures for early reporting of injuries. Additionally, job specific training should be given on safe work practices, hazards, and controls.
- Strength and fitness training can reduce compensation costs.

3. Rotating of employees, providing a short break every hour, or using a two-person lift may be helpful.

4. Rotation is not simply a different job, but must be a job that utilizes a completely different muscle group from the ones that have been over-exerted.

C. OTHER.

1. Standing for extended periods places excessive stress on the back and legs. Solutions include a footrest or rail, resilient floor mats, height-adjustable chairs or stools, and opportunities for the employee to change position.

2. Where employees are seated the chairs or stools must be properly chosen.

3. Proper adjustable lumbar support may be provided.

4. Static seated postures with bending or reaching should be avoided.

VI. BIBLIOGRAPHY.

Cailliet, R. 1968. *Low Back Pain Syndrome*. 2nd. ed. Philadelphia: F.A. Davis Co.

Eastman Kodak Company. 1983. *Ergonomic Design for People at Work*. Vol. 1 & 2. Belmont California: Lifetime Learning Publications.

National Institute for Occupational Safety and Health (NIOSH), DHHS, *Work Practices Guide for Manual Lifting* 1981 & 1991.

Singleton, W.T. ed. 1982. *The Body at Work: Biological Ergonomics*. 1st. ed. New York: Cambridge University Press.

Chaffin and Andersson, 1984. *Occupational Biomechanics*. John Wiley & Sons, New York/Chichester/Brisbane/Toronto/Singapore

Mital, A.; Nicholson, A.S.; Ayoub, M.M., 1993. *A Guide to Manual Materials Handling* 1993. Taylor and Francis, London/Washington, D.C.

APPENDIX VII:1-1. IN-DEPTH ANALYSIS.

The usefulness of the information gained from the review of the OSHA 200 log is limited by internal practices of recording injuries and illnesses. Some plants record everything and some record only those cases that are sent to see a physician. With back disorders, these cases are not always recognized as being work-related and therefore are not recorded. Thus, the Compliance Officer must determine the internal procedure for recording on the OSHA 200, i.e., who records, what cases are recorded, and when cases are recorded (see OSHA Field Inspection Reference Manual).

The following is a systematic approach to identifying the extent and magnitude of a disease or injury and is based on epidemiological principles. The approach consists of records review, worker surveys, and job analysis respectively. This information may be used to determine which jobs pose a risk to workers.

WORKERS' COMPENSATION RECORDS.

Advantages:

Identify additional cases, departments, and job titles.

Limitations:

Does not include cases where treatment is paid for directly by the employee or comprehensive health insurance.

Describes only most severe and advanced problems. May fail to identity problems in early stage of development.

PLANT MEDICAL RECORDS.

In larger plants that employ health care providers, individual employee medical folders, or records, will be maintained and every visit to the health office may be recorded in the record. Such records should be accessed through a medical access order.

There may also be a first-aid log or health office sign-in log. Entries in these records often include:

- Date of visit;
- Department or location where employee works;
- Description of injury or illness;
- Treatment given, including medications; and
- Work restrictions recommended.

Monthly summaries of employee visits to the health office are often compiled by health office personnel. If there are too many records, review a random sample of records to identify cases of back disorders.

SAFETY AND ACCIDENT REPORTS.

Internal reports that may be available in the health, safety or personnel office. These cases may not be noted on the OSHA 200 Log or in worker compensation records. Employee may just want to report the injury or disorder and not seek treatment.

PAYROLL RECORDS (If available).

Used to obtain information on number of hours worked.

Serves as crude measure of exposure potential and can be used to compare jobs in terms of incidence rates of all forms of back disorders.

Useful in identifying job titles or departments with high absentee or turnover rates.

AVAILABLE INFORMATION FROM RECORDS REVIEW.

- Total number of back injuries reported to the company;
- Date each case reported;
- Department or specific job of those who are injured or ill; and
- Number of workers on the same job or in the same department.

WHAT YOU CAN DO WITH THE INFORMATION.

The incidence rate can be calculated for the entire establishment and for each department. This procedure allows comparison between and within the same departments from year to year.

Lost Work Day Injury and Illness rate (LWDII) = (Number of cases)*(200,000)/Total population at risk in a given period

- Numerator: Number of lost or restricted time incidents (cases) in specified group or department that experiences a disorder in a specified time period multiplied by 200,000. Multiplying the number of employees by 200,000 normalizes the observed work population to a standard work population of 100 employees working a 50-week year.

- Denominator: Total number of hours worked in a specified group or department within the same time period. If these numbers are not available an approximation can be made by multiplying the observed number of employees by 2000.

Severity Rate (SR) = This is the same calculation as was performed to produce the LWDII except that the days away from work or restricted days are substituted into the numerator for the number of incidents. This calculation provides a measure of the severity of the cases and is used in conjunction with the LWDII to determine the magnitude of the case.

NOTE: If counting system recognized only lost-time or Workers Compensation cases, relatively low incidence rates may be computed. If the company has instituted an ergonomics program the LWDII may rise dramatically, but there should be a corresponding drop in the SR.

SURVEY THE WORKERS.

Purpose:

Assist in identifying new or early cases of back injury and disorders in the work force. Useful in smaller facilities where data gathered from records review may be limited or in facilities where reporting disincentives limit the number of reported cases. The major reason for this is to collect data on the number of workers that may be experiencing some form of back injury or disorder. This is also a good method for identifying departments or jobs where potential back problems exist.

Factors to be considered in designing a questionnaire or symptom survey:

- Reading level and primary language of workers if the questionnaire or survey is self-administrated. Wording is very important and must be geared to particular respondents.
- Length of the questionnaire (usually should not exceed 20 minutes).
- Instructions: Are they clear?
- Important questions should be asked first.
- Sensitive or personal questions should be asked later in the survey.
- Multiple-choice questions are easier to evaluate but limit the potential responses of the person being questioned.

Mass Medical Screening:

Mass medical screening could be useful but must be performed by trained medical staff -- so the process can be costly in both personnel and resources. The Office of Occupational Medicine should be consulted prior to attempting mass medical screenings.

JOB ANALYSIS AND OBSERVATION.

Each job in which workers have a greater incidence of back disorders might be subject to a job analysis after an appropriate records review and worker survey.

Work-Methods Analysis:

- Observe employees at work: -Notice what employees are doing to make themselves more comfortable in the workplace. For example, look for improvised foot rests, padding, or homemade tools and devices; -Watch for repeated motions and the position of the arms, wrist and trunk (e.g. over-stretching or unusual posture).
- Record the movements used to perform the task. Use a video camera for subsequent slow-motion analysis. See Appendix VII:1-3 for instructions on videotaping techniques.
- Describe the positions seen. Ask questions to determine if staffing levels and production pace is truly representative of the normal operation. Obtain production data if at all possible.

- An ergonomic check-list can be helpful on inspections to record information which is not readily identifiable from the videotape. A sample checklist is shown in Appendix VII:1-4.
- Work station and tool evaluation may be necessary.

APPENDIX VII:1-2. EVALUATION OF LIFTING TASKS. [Completely Revised]

NIOSH WORK PRACTICE GUIDE FOR MANUAL LIFTING.

In 1981, NIOSH developed an equation to assess lifting conditions. In 1991, NIOSH issued a revised equation for the design and evaluation of manual lifting tasks. The 1991 equation uses six factors that have been determined to influence lifting difficulty the most, combining the factors into one equation. Two of the factors which are new to the revised equation include twisting (asymmetry) and the quality of the worker's grip on the load (coupling). Using the equation involves calculating values for the six factors in the equation for a particular lifting and lowering task, thereby generating a Recommended Weight Limit (RWL) for the task. The RWL is the load that nearly all healthy employees (90% of the adult population, 99% of the male and 75% of the female workforce) can lift over a substantial period of time (i.e., up to 8 hours) without placing an excessive load on the back.

The revised equation also incorporated a term called the Lifting Index, which is defined as a relative estimate of the level of physical stress associated with a particular manual lifting task. The estimate of the level of physical stress is defined by the relationship of the weight of the load lifted divided by the recommended weight limit. A level greater than one indicates that the lifted weight exceeded the RWL and should be addressed using either administrative or engineering controls. A level greater than three indicates that the lifted weight exceeds the capacity to safely lift for most of the population, is likely to cause injury, and should be modified by implementation of engineering controls.

The 1991 equation still maintains the 1981 biomechanical criteria for establishing the maximum lower back compression force of 770 lbs. For the revised equation, the load constant was reduced from 90 pounds to 51 pounds. This reduction was driven by the need to increase the minimum horizontal distance from 6 inches to 10 inches (which is believed to be the minimum attainable horizontal distance as measured from the spine during lifting) in the 1991 equation. Aside from this reduction the 1991 revised equation represents only a two-pound reduction from the 1981 version when adjusted for revised horizontal distance.

Application of the NIOSH lifting tasks assumes the following:

- Lifting task is two-handed, smooth, in front of the body, hands are at the same height or level, moderate-width loads (i.e., they do not substantially exceed the body width of the lifter), and the load is evenly distributed between both hands.
- Manual handling activities other than lifting are minimal and do not require significant energy expenditure, especially when repetitive lifting tasks are performed (i.e., holding, pushing, pulling, carrying, walking or climbing).
- Temperatures (66-79°F) or humidity (35-50%) outside of the ranges may increase the risk of injury.
- One-handed lifts, lifting while seated or kneeling, lifting in a constrained or restricted work space, lifting unstable loads, wheelbarrows and shovels are not tasks designed to be covered by the lifting equation.
- The shoe sole to floor surface coupling should provide for firm footing.
- Lifting and lowering assumes the same level of risk for low back injuries.
- Using the Guidelines in situations that do not conform to these ideal assumptions will typically underestimate the hazard of the lifting task under investigation.

The computed values of the Recommended Weight Limit are used by the CSHO as a guide to estimate risk. The numbers by themselves do not identify a hazardous activity. The employer's incidence of injuries and lack of programs for training, work practice controls, and engineering controls related to lifting are elements used to determine the seriousness of the hazard.

CALCULATIONS.

The revised lifting equation for calculating the Recommended Weight Limit (RWL) is based on a multiplicative model that provides a weighting for each of six variables:

RWL=LC × HM × VM × DM × AM × FM × CM

where:

LC=Load Constant (51 pounds)

HM=Horizontal Multiplier (10/H)

FIGURE VII: 1-1. HORIZONTAL MEASUREMENT.

Horizontal location of the hands (H): The horizontal location of the hands at both the start (origin) and end (destination) of the lift must be measured. The horizontal location is measured as the distance from the mid-point between the employee's ankles to a point projected on the floor directly below the mid-point of the hands grasping the object (the middle knuckle can be used to define the mid-point). The horizontal distance should be measured when the object is lifted (when the object leaves the surface).

VM=Vertical Multiplier (1 - (0.0075|V-30|))

Vertical location of the hands (V): The vertical location is measured from the floor to the vertical mid-point between the two hands (the middle knuckle can be used to define the mid-point).

DM=Distance Multiplier (0.82 + (1.8 / D)

Travel Distance of the load (D): The total vertical travel distance of the load during the lift is determined by subtracting the vertical location of the hands (V) at the start of the lift from the vertical location of the hands (V) at the end of the lift. For lowering, the total vertical travel distance of the load is determined by subtracting the vertical location of the hands (V) at the end of the lower from the vertical location of the hands (V) at the start of the lower.

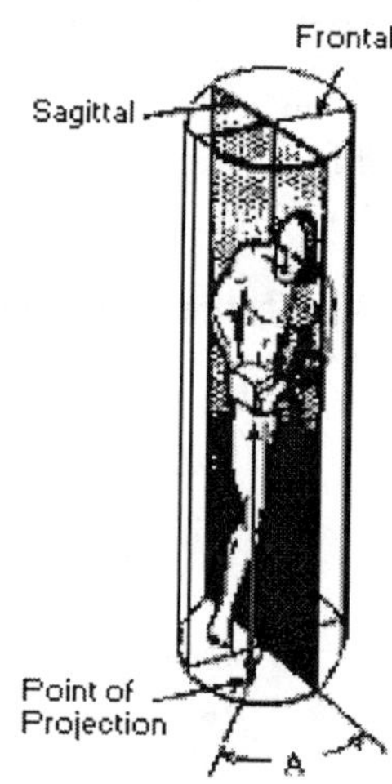

FIGURE VII: 1-2. MEASURE OF ASYMETRY ANGLE A.

AM=Asymmetric Multiplier (1 - (0.0032A))

Asymmetry Angle(A): The angular measure of the perpendicular line that intersects the horizontal line connecting the mid-point of the shoulders and the perpendicular line that intersects the horizontal line connecting the outer mid-point of the hips.

FM=Frequency Multiplier (See Frequency Table Below (Table VII:1-1))

Lifting Frequency (F): The average lifting frequency rate, expressed in terms of lifts per minute, must be determined. The frequency rate can be determined by observing a typical 15 minute work period, and documenting the number of lifts performed during this time frame. The number of lifts observed is divided by 15 to determine the average lifts per minute. Duration is measured using the following categories: Short (Less than one hour); Moderate (1 to 2 hours); Long (2 to 8 hours).

Help using the lifting formula is available through the Directorate of Technical Support.

LIFTING ANALYSIS WORKSHEET TABLE.

The actual worksheet can be found in Appendix VII:1-5. The lifting analysis should be performed using both the average and maximum weights.

APPENDIX VII:1-3. VIDEOTAPE GUIDELINES AND ANALYSIS. [Completely Revised]

VIDEO GUIDELINES FOR ERGONOMIC EVALUATIONS

TABLE VII: 1-1. FREQUENCY MULTIPLIER TABLE (FM)

Frequency Lifts/min (F) ‡	Work Duration					
	≤ 1 Hour		> 1 but ≤ 2 Hours		> 2 but ≤ 8 Hours	
	V < 30 †	V ≥ 30	V < 30	V ≥ 30	V < 30	V ≥ 30
≤ 0.2	1.00	1.00	.95	.95	.85	.85
0.5	.97	.97	.92	.92	.81	.81
1	.94	.94	.88	.88	.75	.75
2	.91	.91	.84	.84	.65	.65
3	.88	.88	.79	.79	.55	.55
4	.84	.84	.72	.72	.45	.45
5	.80	.80	.60	.60	.35	.35
6	.75	.75	.50	.50	.27	.27
7	.70	.70	.42	.42	.22	.22
8	.60	.60	.35	.35	.18	.18
9	.52	.52	.30	.30	.00	.15
10	.45	.45	.26	.26	.00	.13
11	.41	.41	.00	.23	.00	.00
12	.37	.37	.00	.21	.00	.00
13	.00	.34	.00	.00	.00	.00
14	.00	.31	.00	.00	.00	.00
15	.00	.28	.00	.00	.00	.00
> 15	.00	.00	.00	.00	.00	.00

†Values of V are in inches.
‡For lifting less frequently than once per 5 minutes, set F = 2 lifts/minute.

CM=Coupling Multiplier (See Coupling Table Below (Table VII:1-2))

Object coupling (C): The classification of the quality of the hand-to-object coupling (rated as Good, Fair, or Poor).

Obtaining good video documentation for ergonomic evaluations can be difficult -- as the tasks are often performed in inaccessible areas with poor lighting conditions and a lot of extraneous movement taking place. This guide presents suggestions for capturing effective video documentation of potential ergonomic hazards.

PREPARATION

Use the OSHA Form 200 logs and 101's, complaint information, and interviews to help prioritize areas for taping. It is desirable to have at least a two-person team when performing an evaluation. One person can operate the video camera while the other can record task and employee information.

The equipment needed for an ergonomic inspection will generally include:

- Video Camera with extra tapes and charged batteries
- Tape measure
- Small Notebook
- Fanny pack
- Small scale (Chattillon or fish scale that can measure pull forces)

- Bungee cord or small piece of rope
- Questionnaires for employee interviews concerning ergonomic factors.

TABLE VII: 1-2. COUPLING TABLE

GOOD	FAIR	POOR
CM = 1.00	***V < 30" then CM = 0.95***	***CM = 0.90***
	V > or = to 30" then CM = 1.00	
1. For containers of optimal design, such as some boxes, crates, etc., a "Good" hand-to-object coupling would be defined as handles or hand-hold cut-outs of optimal design.	1. For containers of optimal design, a "Fair" hand-to-object coupling would be defined as handles or hand-hold cut-outs of less than optimal design.	1. Containers of less than optimal design or loose parts or irregular objects that are bulky or hard to handle.
2. For loose parts or irregular objects, which are not usually containerized, such as castings, stock, supply materials, etc., a "Good" hand-to-object coupling would be defined as a comfortable grip in which the hand can be easily wrapped around the object.	2. For containers of optimal design with no handles or hand-hold cut-outs or for loose parts or irregular objects, a "Fair" hand-to-object coupling is defined as a grip in which the hand can be flexed about 90 degrees.	2. Lifting non-rigid bags (i.e., bags that sag in the middle).

Other useful items may include:

- Stop watch
- Lens cleaning paper
- Extra batteries for internal clock
- Skylight UV filter. This is a must in a dirty environment if you do not have a protective case.

The following are general suggestions on camera usage which, if reviewed prior to going on-site, will provide the best video documentation for the analyst and ensure that all pertinent information is obtained and documented.

- Become familiar with the camera and read the operators manual. Shoot some test footage so you are familiar with all the functions of the camera.
- Always activate the date and time mechanism on the camera so that this information is displayed on the video during the entire taping series. This will provide additional reference points with which to correlate written information with the videotape footage. Be aware of the position of the date and time printout on the video footage to make sure that it is not superimposed over the top of important features of the video documentation.
- For operations with extraneous movement it may be necessary to use the manual focus to avoid the camera refocusing on irrelevant moving objects. Determine where the focus point is for the camera you are using. It may not be in the center of the viewfinder. To make this determination place the camera on auto focus and try to focus on a small item such as a hanging pendant that has nothing else in the same plane. Hang the item from a doorway and try to focus by moving the item back and forth in the field of the viewfinder. You have found the focus point when the camera focuses on the item.

- If the camera has a high speed shutter, turn it off and use the auto shutter. high speed requires too much light for most industrial tasks. If you are taping a worker with dark clothes against a light background (such as a window, or a white wall), activate the "back lit" capability on the camera.
- Practice visual slating of information. This should be done by filming a piece of paper with information clearly written on it just prior to or directly after videotaping the task. Use a marker or dark pen that can be clearly seen. The macro-zoom on your camera will permit use of a small notebook or journal to be used as a slate. A small notebook is easy to carry and any pertinent notes can be recorded on the slate sheet for easy correlation and future reference. Macro-zoom is also helpful for documentation of small informational areas such as labels.
- If visual slating is absolutely not possible, cover the lens with your hand and record the information verbally before the actual job taping begins. Be aware that you will need to speak directly into the camera microphone to be clearly understood. Use of an external microphone can be helpful in audio slating.
- Hold the camera as still as possible or use a tripod if available. Don't walk with the camera unless absolutely necessary to record the task. When you change location, move slowly and minimize camera movement. Use the zoom instead of walking whenever possible. Use the manual focus whenever there is extraneous movement in the frame of action to ensure the focus will be on the items of interest.

VIDEOTAPING TASKS

The following items outline the procedures used for obtaining useful video documentation.

- If possible tape the operation in the order of production. Do the beginning of the production process first and proceed through all tasks of interest.
- Visually SLATE at least the name of the task just prior to or directly after videotaping the task.
- Tape 5-10 minutes for all jobs including approximately 10 cycles. A cycle is considered to be a set of repeated motions during which one part or assembly is processed. Jobs that have relatively long cycle times in excess of 30-60 seconds may require fewer than 10 cycles if all aspects of the job are recorded at least 3-4 times.
- Begin each task with a whole-body view of the worker from the side including the chair and/or the floor. Hold this view for 2-3 cycles and then zoom the camera in for a closer view of the area of principal interest. Tape from a variety of angles to allow a determination of wrist deviation, arm postures, back angles, etc. Tape from both sides and the front if possible. The total footage may be distributed between these different angles.
- Videotape the operation from a distance to give perspective to the analyst about workstation layout.
- Find an entity of known dimension in the frame of the picture and measure it for reference purposes. The employee's forearm from the wrist to the elbow is a convenient landmark since it is in most frames and is measurable on the television screen. If possible place a piece of contrasting tape on the reference points to provide a more distinct and identifiable location point. Record the reference dimensions either by visually slating the information or verbally recording the data. If using a ruler or tape measure as your reference point, ensure that the increments are clearly visible.
- Obtain video footage of tools or machinery that are used on the job. Videotape labels from hand tools, machinery, weight from boxes, etc.

ANALYSIS OF VIDEOTAPES

It is usually best to contact the Salt Lake Technical Center before you submit a videotape for analysis. This will allow us to provide you with a time frame for analysis completion and to make any pertinent inquiries concerning the nature of the request. Send a copy of the tape rather than the original, since we keep all materials for our files, and any written documentation which was obtained about the inspection. Allow plenty of time for analysis as there is generally significant backlog.

There is information that cannot be readily obtained by visual inspection of the video documentation. The following information should be recorded on the videotape slate at the beginning of the taping sequence or provided in the written "Request For Tape Evaluation" accompanying the tape (example attached):

- Name, SIC code, and location of the facility being inspected.
- Date of inspection.

- Name of Compliance Officer(s), and OSHA office performing the inspection.

Additionally, the following information should be visually slated at the beginning of each individual task or recorded in a written Supplemental Factors Checklist (example attached). Written information should be referenced to the video documentation and must accompany the videotape when submitted for analysis.

- The name of the task and employee.
- Anthropometry (height) of the employee.
- Ambient conditions when working in extreme areas (freezers, furnaces, etc.).
- Clothing and PPE (materials, etc.).
- The period of time in which the task is performed including work-rest schedules.
- The nature of injuries as determined from the 200's, 101's, or interviews.
- Weight and dimension of loads lifted.
- Dimensions of the work items seen in the shot (i.e., pallets, tables, shelving units, etc.).
- Vertical distance between origin and destination of lift. Horizontal distance the load is held from the body at the beginning and end of the lift. These distances can be estimated directly from the video documentation if measuring will significantly interfere with the operation. To do this there must be a clear view of the entire body and the work space, preferably in profile. Provide dimensional information on as many work items seen in the footage as possible.
- Distance loads must be carried.
- Production data to aid in determining if the video segment is representative of normal activity.
- Conditions that might affect grip or traction (ie., sand on the floor, ice on boxes being lifted, etc.).

APPENDIX VII:1-4. SUPPLEMENTAL FACTORS FOR ERGONOMIC TAPE EVALUATION [New]

Worker Information:

Worker's Name:
Worker's Height:
Distance from wrist to elbow or between contrast markers:
Past medical or CTD problems (prior to this task):
Current medical or CTD problems:
Non-occupational activities:

Job Information:

Job Name:
Location:
Time of videotape sequence:
Job Description (task frequency, cycle time, time on job; is the job always performed in this manner?):

Number of employees on job (now and normally):
Line speed; Self or machine-paced; (pieces/min):
Break schedule (micro-rest breaks possible):
Rotation schedule:
Jobs in rotation:

Job and Ergonomic Training

Given by Whom:
Hands-on or theoretical:
Time spent in training:
Updates:
How often:
Last update:
Update given by whom:

Workstation Information

*Make a sketch of the workstation layout on separate sheet of paper or reference to area of tape that shows the entire work layout.

Is the workstation adjustable, can it be tilted or rotated (mechanism and range of motion, heights, dimensions):
Table:
Chair:
What is the worker standing on (concrete, wood):
Is it slippery:
Can work positions be changed (sit/stand):
Reach distances:
Horizontal:
Vertical:

Tool Information

*Answer these questions and make a sketch or identify the tool on the video segment for each tool used. Use other sheets of paper if needed.

Name of tool:
Type of tool and Power Source:
Torque:
Reciprocating or vibrating:
Other:
Weight of tool:
Handle:
Span and Length:
Material:
If air-powered, is the exhaust away from the hand:

Miscellaneous Information:

What objects or materials are handled and their weight (for patients estimate the amount of help they can provide):
Name & Weight:
Temperature of work environment:
Personal protective equipment:
Are gloves worn, What material:
Estimate of task exertion level:
Estimate of hand and finger exertion level:
(1 = Low; hold coffee cup; to 5 = high; open large "Bulldog" paper clip)

Does the employee have an opinion concerning the nature of the problem, management commitment, and possible corrective measures?

Employee Signature (optional, but desirable)__________________________________

APPENDIX VII:1-5. LIFTING ANALYSIS WORKSHEET [New]

LIFTING ANALYSIS WORKSHEET

DEPARTMENT ____________ JOB DESCRIPTION

JOB TITLE ____________ ____________

ANALYST'S NAME ____________ ____________

DATE

STEP 1. Measure and record task variables

Object Weight (lbs)		Hand Location				Vertical Distance	Asymmetric Angle (deg.)		Frequency Rate	Duration	Object Coupling
		Origin		Dest			Origin	Destination	lifts/min	Hrs	
L(AVG)	L(MAX)	H	V	H	V	D	A	A	F		C

STEP 2. Determine the multipliers and compute the RWLs

RWL = LC x HM x VM x DM x AM x FM x CM

		LC		HM		VM		DM		AM		FM		CM		
ORIGIN	RWL =	51	x		x		x		x		x		x		=	
DEST.	RWL =	51	x		x		x		x		x		x		=	

STEP 3. Compute the LIFTING INDEX

ORIGIN LIFT INDEX $\frac{\text{OBJECT WEIGHT}}{\text{RWL}}$ = ____________ = ____________

DESTINATION LIFT INDEX $\frac{\text{OBJECT WEIGHT}}{\text{RWL}}$ = ____________ = ____________

SECTION VIII: CHAPTER 1

CHEMICAL PROTECTIVE CLOTHING

Contents:

I. INTRODUCTION.

A. The purpose of chemical protective clothing and equipment is to shield or isolate individuals from the chemical, physical, and biological hazards that may be encountered during hazardous materials operations. During chemical operations, it is not always apparent when exposure occurs. Many chemicals pose invisible hazards and offer no warning properties.

B. These guidelines describe the various types of clothing that are appropriate for use in various chemical operations, and provides recommendations in their selection and use. The final paragraph discusses heat stress and other key physiological factors that must be considered in connection with protective clothing use.

C. It is important that protective clothing users realize that no single combination of protective equipment and clothing is capable of protecting you against all hazards. Thus protective clothing should be used in conjunction with other protective methods. For example, engineering or administrative controls to limit chemical contact with personnel should always be considered as an alternative measure for preventing chemical exposure. The use of protective clothing can itself create significant wearer hazards, such as heat stress, physical and psychological stress, in addition to impaired vision, mobility, and communication. In general, the greater the level of chemical protective clothing, the greater the associated risks. For any given situation, equipment and clothing should be selected that provide an adequate level of protection. Overprotection as well as under-protection can be hazardous and should be avoided.

II. DESCRIPTIONS.

A. PROTECTIVE CLOTHING APPLICATIONS.

1. Protective clothing must be worn whenever the wearer faces potential hazards arising from chemical exposure. Some examples include:

- Emergency response;
- Chemical manufacturing and process industries;
- Hazardous waste site cleanup and disposal;
- Asbestos removal and other particulate operations; and
- Agricultural application of pesticides.

2. Within each application, there are several operations which require chemical protective clothing. For example, in emergency response, the following activities dictate chemical protective clothing use:

- Site Survey: The initial investigation of a hazardous materials incident; these situations are usually characterized by a large degree of uncertainty and mandate the highest levels of protection.
- Rescue: Entering a hazardous materials area for the purpose of removing an exposure victim; special considerations must be given to how the selected protective clothing may affect the ability of the wearer to carry out rescue and to the contamination of the victim.
- Spill Mitigation: Entering a hazardous materials area to prevent a potential spill or to reduce the hazards from an existing spill (i.e., applying a chlorine kit on railroad tank car). Protective clothing must accommodate the required tasks without sacrificing adequate protection.
- Emergency Monitoring: Outfitting personnel in protective clothing for the primary purpose of observing a hazardous materials incident without entry into the spill site. This may be applied to monitoring contract activity for spill cleanup.
- Decontamination: Applying decontamination procedures to personnel or equipment leaving the site; in general a lower level of protective clothing is used by personnel involved in decontamination.

B. THE CLOTHING ENSEMBLE. The approach in selecting personal protective clothing must encompass an "ensemble" of clothing and equipment items which are easily integrated to provide both an appropriate level of protection and still allow one to carry out activities involving chemicals. In many cases, simple protective clothing by itself may be sufficient to prevent chemical exposure, such as wearing gloves in combination with a splash apron and faceshield (or safety goggles).

1. The following is a checklist of components that may form the chemical protective ensemble:

- Protective clothing (suit, coveralls, hoods, gloves, boots);
- Respiratory equipment (SCBA, combination SCBA/SAR, air purifying respirators);
- Cooling system (ice vest, air circulation, water circulation);
- Communications device;
- Head protection;
- Eye protection;
- Ear protection;
- Inner garment; and
- Outer protection (overgloves, overboots, flashcover).

2. Factors that affect the selection of ensemble components include:

- How each item accommodates the integration of other ensemble components. Some ensemble components may be incompatible due to how they are worn (e.g., some SCBA's may not fit within a particular chemical protective suit or allow acceptable mobility when worn).
- The ease of interfacing ensemble components without sacrificing required performance (e.g. a poorly fitting overglove that greatly reduces wearer dexterity).
- Limiting the number of equipment items to reduce donning time and complexity (e.g. some communications devices are built into SCBA's which as a unit are NIOSH certified).

C. LEVEL OF PROTECTION.

1. Table VIII:1-1 lists ensemble components based on the widely used EPA Levels of Protection: Levels A, B, C, and D. These lists can be used as the starting point for ensemble creation; however, each ensemble must be tailored to the specific situation in order to provide the most appropriate level of protection. For example, if an emergency response activity involves a highly contaminated area or if the potential of contamination is high, it may be advisable to wear a disposable covering such as Tyvek coveralls or PVC splash suits, over the protective ensemble.

TABLE VIII:1-1. EPA LEVELS OF PROTECTION

LEVEL A:

Vapor protective suit (meets NFPA 1991)
Pressure-demand, full-face SCBA
Inner chemical-resistant gloves, chemical-resistant safety boots, two-way radio communication

OPTIONAL: Cooling system, outer gloves, hard hat

Protection Provided: Highest available level of respiratory, skin, and eye protection from solid, liquid and gaseous chemicals.

Used When: The chemical(s) have been identified and have high level of hazards to respiratory system, skin and eyes. Substances are present with known or suspected skin toxicity or carcinogenity. Operations must be conducted in confined or poorly ventilated areas.

Limitations: Protective clothing must resist permeation by the chemical or mixtures present. Ensemble items must allow integration without loss of performance.

LEVEL B:

Liquid splash-protective suit (meets NFPA 1992)
Pressure-demand, full-facepiece SCBA
Inner chemical-resistant gloves, chemical-resistant safety boots, two-way radio communications
Hard hat.

OPTIONAL: Cooling system, outer gloves

Protection Provided: Provides same level of respiratory protection as Level A, but less skin protection. Liquid splash protection, but no protection against chemical vapors or gases.

Used When: The chemical(s) have been identified but do not require a high level of skin protection. Initial site surveys are required until higher levels of hazards are identified. The primary hazards associated with site entry are from liquid and not vapor contact.

Limitations: Protective clothing items must resist penetration by the chemicals or mixtures present. Ensemble items must allow integration without loss of performance.

LEVEL C:

Support Function Protective Garment (meets NFPA 1993)
Full-facepiece, air-purifying, canister-equipped respirator
Chemical resistant gloves and safety boots
Two-way communications system, hard hat

OPTIONAL: Faceshield, escape SCBA

Protection Provided: The same level of skin protection as Level B, but a lower level of respiratory protection. Liquid splash protection but no protection to chemical vapors or gases.

Used When: Contact with site chemical(s) will not affect the skin. Air contaminants have been identified and concentrations measured. A canister is available which can remove the contaminant. The site and its hazards have been completely characterized.

Limitations: Protective clothing items must resist penetration by the chemical or mixtures present. Chemical airborne concentration must be less than IDLH levels. The atmosphere must contain at least 19.5% oxygen.

Not Acceptable for Chemical Emergency Response

LEVEL D:

Coveralls, safety boots/shoes, safety glasses or chemical splash goggles

OPTIONAL: Gloves, escape SCBA, face-shield

Protection Provided: No respiratory protection, minimal skin protection.

Used When: The atmosphere contains no known hazard. Work functions preclude splashes, immersion, potential for inhalation, or direct contact with hazard chemicals.

Limitations: This level should not be worn in the Hot Zone. The atmosphere must contain at least 19.5% oxygen.

Not Acceptable for Chemical Emergency Response

2. The type of equipment used and the overall level of protection should be reevaluated periodically as the amount of information about the chemical situation or process increases, and when workers are required to perform different tasks. Personnel should upgrade or downgrade their level of protection only with concurrence with the site supervisor, safety officer, or plant industrial hygienist.

3. The recommendations in Table VIII:1-1 serve only as guidelines. It is important for you to realize that selecting items by how they are designed or configured alone is not sufficient to ensure adequate protection. In other words, just having the right components to form an ensemble is not enough. The EPA levels of protection do not define what performance the selected clothing or equipment must offer. Many of these considerations are described in the "limiting criteria" column of Table VIII: 1-1. Additional factors relevant to the various clothing and equipment items are Described in subsequent Paragraphs.

D. ENSEMBLE SELECTION FACTORS.

1. Chemical Hazards. Chemicals present a variety of hazards such as toxicity, corrosiveness, flammability, reactivity, and oxygen deficiency. Depending on the chemicals present, any combination of hazards may exist.

2. Physical Environment. Chemical exposure can happen anywhere: in industrial settings, on the highways, or in residential areas. It may occur either indoors or outdoors; the environment may be extremely hot, cold, or moderate; the exposure site may be relatively uncluttered or rugged, presenting a number of physical hazards; chemical handling activities may involve entering confined spaces, heavy lifting, climbing a ladder, or crawling on the ground. The choice of ensemble components must account for these conditions.

3. Duration of Exposure. The protective qualities of ensemble components may be limited to certain exposure levels (e.g. material chemical resistance, air supply). The decision for ensemble use time must be made assuming the worst case exposure so that safety margins can be applied to increase the protection available to the worker.

4. Protective Clothing or Equipment Available. Hopefully, an array of different clothing or equipment is available to workers to meet all intended applications. Reliance on one particular clothing or equipment item may severely limit a facility's ability to handle a broad range of chemical exposures. In its acquisition of equipment and clothing, the safety department or other responsible authority should attempt to provide a high degree of flexibility while choosing protective clothing and equipment that is easily integrated and provides protection against each conceivable hazard.

E. CLASSIFICATION OF PROTECTIVE CLOTHING.

1. Personal protective clothing includes the following:

   - Fully encapsulating suits;
   - Nonencapsulating suits;
   - Gloves, boots, and hoods;
   - Firefighter's protective clothing;
   - Proximity, or approach clothing;
   - Blast or fragmentation suits; and
   - Radiation-protective suits.

2. Firefighter turnout clothing, proximity gear, blast suits, and radiation suits by themselves are not acceptable for providing adequate protection from hazardous chemicals.

3. Table VIII:1-2 describes various types of protection clothing available, details the type of protection they offer, and lists factors to consider in their selection and use.

TABLE VIII:1-2. TYPES OF PROTECTIVE CLOTHING FOR FULL BODY PROTECTION

Description	***Type of Protection***	***Use Considerations***
Fully encapsulating suit One-piece garment. Boots and gloves may be integral, attached and replaceable, or separate.	Protects against splashes, dust gases, and vapors.	Does not allow body heat to escape. May contribute to heat stress in wearer, particularly if worn in conjunction with a closed-circuit SCBA; a cooling garment may be needed. Impairs worker mobility, vision, and communication.
Nonencapsulating suit Jacket, hood, pants or bib overalls, and one-piece coveralls.	Protects against splashes, dust, and other materials but not against gases and vapors. Does not protect parts of head or neck.	Do not use where gas-tight or pervasive splashing protection is required. May contribute to heat stress in wearer. Tape-seal connections between pant cuffs and boots and between gloves and sleeves.
Aprons, leggings, and sleeve protectors Fully sleeved and gloved apron. Separate coverings for arms and legs. Commonly worn over nonencapsulating suit.	Provides additional splash protection of chest, forearms, and legs.	Whenever possible, should be used over a nonencapsulating suit to minimize potential heat stress. Useful for sampling, labeling, and analysis operations. Should be used only when there is a low probability of total body contact with contaminants.
Firefighters' protective clothing Gloves, helmet, running or bunker coat, running or bunker pants (NFPA No. 1971, 1972, 1973, and boots (1974).	Protects against heat, hot water, and some particles. Does not protect against gases and vapors, or chemical permeation or degradation. NFPA	Decontamination is difficult. Should not be worn in areas where protection against gases, vapors, chemical splashes or permeation is required.

	Standard No. 1971 specifies that a garment consists of an outer shell, an inner liner and a vapor barrier with a minimum water penetration of 25 lb/in^2 ($1.8\ kg/cm^2$) to prevent passage of hot water.	
Proximity garment (approach suit) One- or two-piece overgarment with boot covers, gloves, and hood of aluminized nylon or cotton fabric. Normally worn over other protective clothing, firefighters' bunker gear, or flame-retardant coveralls.	Protects against splashes, dust, gases, and vapors.	Does not allow body heat to escape. May contribute to heat stress in wearer, particularly if worn in conjunction with a closed-circuit SCBA; a cooling garment may be needed. Impairs worker mobility, vision, and communication.
Blast and fragmentation suit Blast and fragmentation vests and clothing, bomb blankets, and bomb carriers.	Provides some protection against very small detonations. Bomb blankets and baskets can help redirect a blast.	Does not provide for hearing protection.
Radiation-contamination protective suit Various types of protective clothing designed to prevent contamination of the body by radioactive particles.	Protects against alpha and beta particles. Does *not* protect against gamma radiation.	Designed to prevent skin contamination. If radiation is detected on site, consult an experienced radiation expert and evacuate personnel until the radiation hazard has been evaluated.
Flame/fire retardant coveralls Normally worn as an undergarment.	Provides protection from flash fires.	Adds bulk and may exacerbate heat stress problems and impair mobility

F. CLASSIFICATION OF CHEMICAL PROTECTIVE CLOTHING. Table VIII:1-3 provides a listing of clothing classifications. Clothing can be classified by design, performance, and service life.

TABLE VIII:1-3. CLASSIFICATION OF CHEMICAL PROTECTIVE CLOTHING

By Design	*By Performance*	*By Service Life*
gloves	particulate protection	single use
boots	liquid-splash protection	limited use
aprons, jackets, coveralls, full body suits	vapor protection	reusable

1. Design. Categorizing clothing by design is mainly a means for describing what areas of the body the clothing item is intended to protect.

   In emergency response, hazardous waste site cleanup, and dangerous chemical operations, the only acceptable types of protective clothing include fully or totally encapsulating suits and nonencapsulating or "splash" suits plus accessory clothing items such as chemically resistant gloves or boots. These descriptions apply to how the clothing is designed and not to its performance.

2. Performance. The National Fire Protection Association (NFPA) has classified suits by their performance as:

   a. Vapor-protective suits (NFPA Standard 1991) provide "gas-tight" integrity and are intended for response situations where no chemical contact is permissible. This type of suit would be equivalent to the clothing required in EPA's Level A.

   b. Liquid splash-protective suits (NFPA Standard 1992) offer protection against liquid chemicals in the form of splashes, but not against continuous liquid contact or chemical vapors or gases. Essentially, the type of clothing would meet the EPA Level B needs. It is important to note, however, that by wearing liquid splash-protective clothing, the wearer accepts exposure to chemical vapors or gases because this clothing does not offer gas-tight performance. The use of duct tape to seal clothing interfaces does not provide the type of wearer encapsulation necessary for protection against vapors or gases.

   c. Support function protective garments (NFPA Standard 1993) must also provide liquid splash protection but offer limited physical protection. These garments may comprise several separate protective clothing components (i.e., coveralls, hoods, gloves, and boots). They are intended for use in nonemergency, nonflammable situations where the chemical hazards have been completely characterized. Examples of support functions include proximity to chemical processes, decontamination, hazardous waste clean-up, and training. Support function protective garments should not be used in chemical emergency response or in situations where chemical hazards remain uncharacterized.

   d. These NFPA standards define minimum performance requirements for the manufacture of chemical protective suits. Each standard requires rigorous testing of the suit and the materials that comprise the suit in terms of overall protection, chemical resistance, and physical properties. Suits that are found compliant by an independent certification and testing organization may be labeled by the manufacturer as meeting the requirements of the respective NFPA standard. Manufacturers also have to supply documentation showing all test results and characteristics of their protective suits.

   e. Protective clothing should completely cover both the wearer and his or her breathing apparatus. In general, respiratory protective equipment is not designed to resist chemical contamination. Level A protection (vapor-protective suits) require this configuration. Level B ensembles may be configured either with the SCBA on the outside or inside. However, it is strongly recommended that the wearer's respiratory equipment be worn inside the ensemble to prevent its failure and to reduce decontamination problems. Level C ensembles use cartridge or canister type respirators which are generally worn outside the clothing.

3. Service Life.

   a. Clothing item service life is an end user decision depending on the costs and risks associated with clothing decontamination and reuse. For example, a Saranex/Tyvek garment may be designed to be a coverall (covering the wearer's torso, arms, and legs) intended for liquid splash protection, which is disposable after a single use.

   b. Protective clothing may be labeled as:

      - Reusable, for multiple wearings; or
      - Disposable, for one-time use.

   The distinctions between these types of clothing are both vague and complicated. Disposable clothing is generally lightweight and inexpensive. Reusable clothing is often more rugged and costly. Nevertheless, extensive contamination of any garment may render it disposable. The basis of this classification really depends on the costs involved in purchasing, maintaining, and reusing protective clothing versus the al ternative of disposal following exposure. If an end user can anticipate obtaining several uses out of a garment while still maintaining adequate protection from that garment at lower cost than its disposal, the suit becomes reusable. Yet, the key assumption in this determination is the viability of the garment following exposure. This issue is further discussed in the Paragraph on decontamination.

III. PROTECTIVE CLOTHING SELECTION FACTORS.

A. CLOTHING DESIGN. Manufacturers sell clothing in a variety of styles and configurations.

1. Design Considerations.

   - Clothing configuration;
   - Components and options;
   - Sizes;
   - Ease of donning and doffing;
   - Clothing construction;
   - Accommodation of other selected ensemble equipment;
   - Comfort; and
   - Restriction of mobility.

B. MATERIAL CHEMICAL RESISTANCE. Ideally, the chosen material(s) must resist permeation, degradation, and penetration by the respective chemicals.

1. Permeation is the process by which a chemical dissolves in or moves through a material on a molecular basis. In most cases, there will be no visible evidence of chemicals permeating a material.

   Permeation breakthrough time is the most common result used to assess material chemical compatibility. The rate of permeation is a function of several factors such as chemical concentration, material thickness, humidity, temperature, and pressure. Most material testing is done with 100% chemical over an extended exposure period. The time it takes chemical to permeate through the material is the breakthrough time. An acceptable material is one where the breakthrough time exceeds the expected period of garment use. However, temperature and pressure effects may enhance permeation and reduce the magnitude of this Safety factor. For example, small increases in ambient temperature can significantly reduce breakthrough time and the protective barrier properties of a protective clothing material.

2. Degradation involves physical changes in a material as the result of a chemical exposure, use, or ambient conditions (e.g. sunlight). The most common observations of material degradation are discoloration, swelling, loss of physical strength, or deterioration.

3. Penetration is the movement of chemicals through zippers, seams, or imperfections in a protective clothing material.

   It is important to note that no material protects against all chemicals and combinations of chemicals, and that no currently available material is an effective barrier to any prolonged chemical exposure.

4. Sources of information include:

   - Guidelines for the Selection of Chemical Protective Clothing, 3rd Edition. This reference provides a matrix of clothing material recommendations for approximately 500 chemicals based on an evaluation of chemical resistance test data, vendor literature, and raw material suppliers. The major limitation for these guidelines are their presentation of recommendations by generic material class. Numerous test results have shown that similar materials from different manufacturers may give widely different performance. That is to say manufacturer A's butyl rubber glove may protect against chemical X, but a butyl glove made by manufacturer B may not.
   - Quick Selection Guide to Chemical Protective Clothing. Pocket size guide that provides chemical resistance data and recommendations for 11 generic materials against over 400 chemicals. The guide is color-coded by material-chemical recommendation. As with the "Guidelines..." above, the major limitation of this reference is its dependence on generic data.
   - Vendor data or recommendations. The best source of current information on material compatibility should be available from the manufacturer of the selected clothing. Many vendors supply charts which show actual test data or their own recommendations for specific chemicals. However, unless vendor data or the recommendations are well documented, end users must approach this information with caution. Material recommendations must be based on data obtained from tests performed to standard ASTM methods. Simple ratings of "poor," "good," or "excellent" give no indication of how the material may perform against various chemicals.

5. Mixtures of chemicals can be significantly more aggressive towards protective clothing materials than any single chemical alone. One permeating chemical may pull another with it through the material. Very little data is available for chemical mixtures. Other situations may involve unidentified substances. In both the case of mixtures and unknowns, serious consideration must be given to deciding which protective clothing is selected. If clothing must be used without test data, garments with materials having the broadest chemical resistance should be worn, i.e. materials which demonstrate the best chemical resistance against the widest range of chemicals.

C. PHYSICAL PROPERTIES.

1. As with chemical resistance, manufacturer materials offer wide ranges of physical qualities in terms of strength, resistance to physical hazards, and operation in extreme environmental conditions. Comprehensive manufacturing standards such as the NFPA Standards set specific limits on these material properties, but only for limited applications, i.e. emergency response.

2. End users in other applications may assess material physical properties by posing the following questions:

   - Does the material have sufficient strength to withstand the physical strength of the tasks at hand?
   - Will the material resist tears, punctures, cuts, and abrasions?
   - Will the material withstand repeated use after contamination and decontamination?
   - Is the material flexible or pliable enough to allow end users to perform needed tasks?
   - Will the material maintain its protective integrity and flexibility under hot and cold extremes?
   - Is the material flame-resistant or self-extinguishing (if these hazards are present)?
   - Are garment seams in the clothing constructed so they provide the same physical integrity as the garment material?

D. EASE OF DECONTAMINATION. The degree of difficulty in decontaminating protective clothing may dictate whether disposable or reusable clothing is used, or a combination of both.

E. COST. Protective clothing end users must endeavor to obtain the broadest protective equipment they can buy with available resources to meet their specific application.

F. CHEMICAL PROTECTIVE CLOTHING STANDARDS. Protective clothing buyers may wish to specify clothing that meets specific standards, such as 1910.120 or the NFPA standards (see Paragraph on classification by performance). The NFPA Standards do not apply to all forms of protective clothing and applications.

IV. GENERAL GUIDELINES.

A. DECIDE IF THE CLOTHING ITEM IS INTENDED TO PROVIDE VAPOR, LIQUID-SPLASH, OR PARTICULATE PROTECTION.

1. Vapor protective suits also provide liquid splash and particulate protection. Liquid splash protective garments also provide particulate protection. Many garments may be labeled as totally encapsulating but do not provide gas-tight integrity due to inadequate seams or closures. Gas-tight integrity can only be determined by performing a pressure or inflation test and a leak detection test of the respective protective suit. This test involves:

   - Closing off suit exhalation valves;
   - Inflating the suit to a prespecified pressure; and
   - Observing whether the suit holds the above pressure for a designated period.

   ASTM Standard Practice F1052 (1987 Edition) offers a procedure for conducting this test.

2. Splash suits must still cover the entire body when combined with the respirator, gloves, and boots. Applying duct tape to a splash suit does not make it protect against vapors. Particulate protective suits may not need to cover the entire body, depending on the hazards posed by the particulate. In general, gloves, boots and some form of face protection are required. Clothing items may only be needed to cover a limited area of the body such as gloves on hands. The nature of the hazards and the expected exposure will determine if clothing should provide partial or full body protection.

B. DETERMINE IF THE CLOTHING ITEM PROVIDES FULL BODY PROTECTION.

1. Vapor-protective or totally encapsulating suit will meet this requirement by passing gas-tight integrity tests.

2. Liquid splash-protective suits are generally sold incomplete(i.e. fewer gloves and boots).

3. Missing clothing items must be obtained separately and match or exceed the performance of the garment.

4. Buying a PVC glove for a PVC splash suit does not mean that you obtain the same level of protection. This determination must be made by comparing chemical resistance data.

C. EVALUATE MANUFACTURER CHEMICAL RESISTANCE DATA PROVIDED WITH THE CLOTHING.

1. Manufacturers of vapor-protective suits should provide permeation resistance data for their products, while liquid and particulate penetration resistance data should accompany liquid splash and particulate protective garments respectively. Ideally data should be provided for every primary material in the suit or clothing item. For suits, this includes the garment, visor, gloves, boots, and seams.

2. Permeation data should include the following:

   - Chemical name;
   - Breakthrough time (shows how soon the chemical permeates);
   - Permeation rate (shows the rate that the chemical comes through);
   - System sensitivity (allows comparison of test results from different laboratories); and
   - A citation that the data was obtained in accordance with ASTM Standard Test Method F739-85.

3. If no data are provided or if the data lack any one of the above items, the manufacturer should be asked to supply the missing data. Manufacturers that provide only numerical or qualitative ratings must support their recommendations with complete test data.

4. Liquid penetration data should include a pass or fail determination for each chemical listed, and a citation that testing was conducted in accordance with ASTM Standard Test Method F903-86. Protective suits which are certified to NFPA 1991 or NFPA 1992 will meet all of the above requirements.

5. Particulate penetration data should show some measure of material efficiency in preventing particulate penetration in terms of particulate type or size and percentage held out. Unfortunately, no standard tests are available in this area and end users may have little basis for company products.

6. Suit materials which show no breakthrough or no penetration to a large number of chemicals are likely to have a broad range of chemical resistance. (Breakthrough times greater than one hour are usually considered to be an indication of acceptable performance.) Manufacturers should provide data on the ASTM Standard Guide F1001-86 chemicals. These 15 liquid and 6 gaseous chemicals listed in Table VIII:1-4 below represent a cross-section of different chemical classes and challenges for protective clothing materials. Manufacturers should also provide test data on other chemicals as well. If there are specific chemicals within your operating area that have not been tested, ask the manufacturer for test data on these chemicals.

TABLE VIII:1-4. RECOMMENDED CHEMICALS TO EVALUATE THE PERFORMANCE OF PROTECTIVE CLOTHING MATERIALS

Chemical	***Class***
Acetone	Ketone
Acetonitrile	Nitrile
Ammonia	Strong base (gas)
1,3-Butadiene	Olefin (gas)
Carbpm Dosi;fode	Sulfur-containing organic
Chlorine	Inorganic gas
Dichloromethane	Chlorinated hydrocarbon
Diethylamine	Amine
Dimethyl formamide	Amide
Ethyl Acetate	Ester
Ethyl Oxide	Oxygen heterocyclic gas
Hexane	Aliphatic hydrocarbon
Hydrogen Chloride	Acid gas
Methanol	Alcohol
Methyl Chloride	Chlorinated hydrocarbon (gas)
Nitrobenzene	Nitrogen-containing organic
Sodium Hydroxide	Inorganic base
Sulfuric Acid	Inorganic acid
Tetrachloroethylene	Chlorinated hydrocarbon
Tetrahydrofuran	Oxygen heterocyclic
Toluene	Aromatic hydrocarbon

D. OBTAIN AND EXAMINE THE MANUFACTURER'S INSTRUCTION OR TECHNICAL MANUAL.

1. This manual should document all the features of the clothing, particularly suits, and describe what material(s) are used in its construction. It should cite specific limitations for the clothing and what restrictions apply to its use. Procedures and recommendations should be supplied for at least the following:

    - Donning and doffing;
    - Inspection, maintenance, and storage;
    - Decontamination; and
    - Use.

    The manufacturer's instructions should be thorough enough to allow the end users to wear and use the clothing without a large number of questions.

E. OBTAIN AND INSPECT SAMPLE CLOTHING ITEM GARMENTS. Examine the quality of clothing construction and other features that will impact its wearing. The questions listed under "Protective Clothing Selection Factors, Clothing Design" should be considered. If possible, representative clothing items should be obtained in advance and inspected prior to purchase, and discussed with someone who has experience in their use. It is also helpful to try out representative garments prior to purchase by suiting personnel in the garment and having them run through exercises to simulate expected activities.

F. FIELD SELECTION OF CHEMICAL PROTECTIVE CLOTHING.

1. Even when end users have gone through a very careful selection process, a number of situations will arise when no information is available to judge whether their protective clothing will provide adequate protection. These situations include:

    - Chemicals that have not been tested with the garment materials;
    - Mixtures of two or more different chemicals;
    - Chemicals that cannot be readily identified;
    - Extreme environmental conditions (hot temperatures); and
    - Lack of data in all clothing components (e.g. seams, visors).

2. Testing material specimens using newly developed field test kits may offer one means for making an on-site clothing selection. A portable test kit has been developed by the EPA using a simple weight loss method that allows field qualification of protective clothing materials within one hour. Use of this kit may overcome the absence of data and provide additional criteria for clothing selection.

3. Selection of chemical protective clothing is a complex task and should be performed by personnel with both extensive training and experience.

    Under all conditions, clothing should be selected by evaluating its performance characteristics against the requirements and limitations imposed by the application.

V. MANAGEMENT PROGRAM.

A. WRITTEN MANAGEMENT PROGRAM.

1. A written Chemical Protective Clothing Management Program should be established by all end users who routinely select and use protective clothing. Reference should be made to 1910.120 for those covered.

    The written management program should include policy statements, procedures, and guidelines. Copies should be made available to all personnel who may use protective clothing in the course of their duties or job. Technical data on clothing, maintenance manuals, relevant regulations, and other essential information should also be made available.

2. The two basic objectives of any management program should be to protect the wearer from safety and health hazards, and to prevent injury to the wearer from incorrect use and/or malfunction of the chemical protective clothing. To accomplish these goals, a comprehensive management program should include: hazard

identification; medical monitoring; environmental surveillance; selection, use, maintenance, and decontamination of chemical protective clothing; and training.

B. PROGRAM REVIEW AND EVALUATION. The management program should be reviewed at least annually. Elements which should be considered in the review include:

- The number of person-hours that personnel wear various forms of chemical protective clothing and other equipment;
- Accident and illness experience;
- Levels of exposure;
- Adequacy of equipment selection;
- Adequacy of the operational guidelines;
- Adequacy of decontamination, cleaning, inspection, maintenance, and storage programs;
- Adequacy and effectiveness of training and fitting programs;
- Coordination with overall safety and health program;
- The degree of fulfillment of program objectives;
- The adequacy of program records;
- Recommendations for program improvement and modification; and
- Program costs.

The results of the program evaluation should be made available to all end users and presented to top management so that program changes may be implemented.

C. TYPES OF STANDARD OPERATING PROCEDURES. Personal protective clothing and equipment can offer a high degree of protection only if it is used properly. Standard Operating Procedures (SOP's) should be established for all workers involved in handling hazardous chemicals. Areas that should be addressed include:

- Selection of protective ensemble components;
- Protective clothing and equipment donning, doffing, and use;
- Decontamination procedures;
- Inspection, storage, and maintenance of protective clothing/equipment; and
- Training.

D. SELECTION OF PROTECTIVE CLOTHING COMPONENTS.

1. Protective clothing and equipment SOP's must take into consideration the factors presented in the Clothing Ensemble and Protective Clothing Applications Paragraphs of this chapter. All clothing and equipment selections should provide a decision tree that relates chemical hazards and information to levels of protection and performance needed.

2. Responsibility in selecting appropriate protective clothing should be vested in a specific individual who is trained in both chemical hazards and protective clothing use such as a safety officer or industrial hygienist. Only chemical protective suits labeled as compliant with the appropriate performance requirements should be used. In cases where the chemical hazards are known in advance or encountered routinely, clothing selection should be predetermined. That is, specific clothing items should be identified in specific chemical operations without the opportunity for individual selection of other clothing items.

VI. CLOTHING DONNING, DOFFING, AND USE.

The procedures below are given for vapor protective or liquid-splash protective suit ensembles and should be included in the training program.

A. DONNING THE ENSEMBLE.

1. A routine should be established and practiced periodically for donning the various ensemble configurations that a facility or team may use. Assistance should be provided for donning and doffing since these operations are difficult to perform alone, and solo efforts may increase the possibility of ensemble damage.

2. Table VIII:1-5 below lists sample procedures for donning a totally encapsulating suit/SCBA ensemble. These procedures should be modified depending on the suit and accessory equipment used. The procedures assume the wearer has previous training in respirator use and decontamination procedures.

3. Once the equipment has been donned, its fit should be evaluated. If the clothing is too small, it will restrict movement, increase the likelihood of tearing the suit material, and accelerate wearer fatigue. If the clothing is too large, the possibility of snagging the material is increased, and the dexterity and coordination of the wearer may be compromised. In either case, the wearer should be recalled and better-fitting clothing provided.

B. DOFFING AN ENSEMBLE.

1. Exact procedures for removing a totally encapsulating suit/SCBA ensemble must be established and followed in order to prevent contaminant migration from the response scene and transfer of contaminants to the wearer's body, the doffing assistant, and others.

2. Sample doffing procedures are provided in Table VIII:1-6 below. These procedures should be performed only after decontamination of the suited end user. They require a suitably attired assistance. Throughout the procedures, both wearer and assistant should avoid any direct contact with the outside surface of the suit.

C. USER MONITORING AND TRAINING.

1. The wearer must understand all aspects of clothing/equipment operation and their limitations; this is especially important for fully encapsulating ensembles where misuse could potentially result in suffocation. During protective clothing use, end users should be encouraged to report any perceived problems or difficulties to their supervisor. These malfunctions include, but are not limited to:

   - Degradation of the protection ensemble;
   - Perception of odors;
   - Skin irritation;
   - Unusual residues on clothing material;
   - Discomfort;
   - Resistance to breathing;
   - Fatigue due to respirator use;
   - Interference with vision or communication;
   - Restriction of movement; and
   - Physiological responses such as rapid pulse, nausea, or chest pain.

2. Before end users undertake any activity in their chemical protective ensembles, the anticipated duration of use should be established. Several factors limit the length of a mission, including:

   - Air supply consumption as affected by wearer work rate, fitness, body size, and breathing patterns;
   - Suit ensemble permeation, degradation, and penetration by chemical contaminants, including expected leakage through suit or respirator exhaust valves (ensemble protection factor);
   - Ambient temperature as it influences material chemical resistance and flexibility, suit and respirator exhaust valve performance, and wearer heat stress; and
   - Coolant supply (if necessary).

VII. DECONTAMINATION PROCEDURES.

TABLE VIII:1-5. SAMPLE DONNING PROCEDURES

1. Inspect clothing and respiratory equipment before donning (see Paragraph on Inspection).
2. Adjust hard hat or headpiece if worn, to fit user's head.
3. Open back closure used to change air tank (if suit has one) before donning suit.
4. Standing or sitting, step into the legs of the suit; ensure proper placement of the feet within the suit; then gather the suit around the waist.
5. Put on chemical-resistant safety boots over the feet of the suit. Tape the leg cuff over the tops of the boots.
   If additional chemical-resistant safety boots are required, put these on now.
   Some one-piece suits have heavy-soled protective feet. With these suits, wear short, chemical resistant safety boots inside the suit.
6. Put on air tank and harness assembly of the SCBA. Don the facepiece and adjust it to be secure, but comfortable. Do not connect the breathing hose. Open valve on air tank.
7. Perform negative and positive respirator facepiece seal test procedures.
   To conduct a negative-pressure test, close the inlet part with the palm of the hand or squeeze the breathing tube so it does not pass air, and gently inhale for about 10 seconds. Any inward rushing of air indicates a poor fit. Note that a leaking facepiece may be drawn tightly to the face to form a good seal, giving a false indication of adequate fit.
   To conduct a positive-pressure test, gently exhale while covering the exhalation valve to ensure that a positive pressure can be built up. Failure to build a positive pressure indicates a poor fit.
8. Depending on type of suit:
   Put on long-sleeved inner gloves (similar to surgical gloves). Secure gloves to sleeves, for suits with detachable gloves (if not done prior to entering the suit).
   Additional overgloves, worn over attached suit gloves, may be donned later.
9. Put sleeves of suit over arms as assistant pulls suit up and over the SCBA. Have assistant adjust suit around SCBA and shoulders to ensure unrestricted motion.
10. Put on hard hat, if needed.
11. Raise hood over head carefully so as not to disrupt face seal of SCBA mask. Adjust hood to give satisfactory comfort.
12. Begin to secure the suit by closing all fasteners on opening until there is only adequate room to connect the breathing hose. Secure all belts and/or adjustable leg, head, and waistbands.
13. Connect the breathing hose while opening the main valve.
14. Have assistant first ensure that wearer is breathing properly and then make final closure of the suit.
15. Have assistant check all closures.
16. Have assistant observe the wearer for a period of time to ensure that the wearer is comfortable, psychologically stable, and that the equipment is functioning properly.

TABLE VIII: 1-6. SAMPLE DOFFING PROCEDURES

If sufficient air supply is available to allow appropriate decontamination before removal:

1. Remove any extraneous or disposable clothing, boot covers, outer gloves, and tape.
2. Have assistant loosen and remove the wearer's safety shoes or boots.
3. Have assistant open the suit completely and lift the hood over the head of the wearer and rest it on top of the SCBA tank.
4. Remove arms, one at a time, from suit. Once arms are free, have assistant lift the suit up and away from the SCBA backpack--avoiding any contact between the outside surface of the suit and the wearer's body--and lay the suit out flat behind the wearer. Leave internal gloves on, if any.
5. Sitting, if possible, remove both legs from the suit.
6. Follow procedure for doffing SCBA.
7. After suit is removed, remove internal gloves by rolling them off the hand, inside out.
8. Remove internal clothing and thoroughly cleanse the body.

If the low-pressure warning alarm has sounded, signifying that approximately 5 minutes of air remain:

1. Remove disposable clothing.
2. Quickly scrub and hose off, especially around the entrance/exit zipper.
3. Open the zipper enough to allow access to the regulator and breathing hose.
4. Immediately attach an appropriate canister to the breathing hose (the type and fittings should be predetermined). Although this provides some protection against any contamination still present, it voids the certification of the unit.
5. Follow Steps 1 through 8 of the regular doffing procedure above. Take extra care to avoid contaminating the assistant and the wearer.

A. DEFINITION AND TYPES.

1. Decontamination is the process of removing or neutralizing contaminants that have accumulated on personnel and equipment. This process is critical to health and safety at hazardous material response sites. Decontamination protects end users from hazardous substances that may contaminate and eventually permeate the protective clothing, respiratory equipment, tools, vehicles, and other equipment used in the vicinity of the chemical hazard; it protects all plant or site personnel by minimizing the transfer of harmful materials into clean areas; it helps prevent mixing of incompatible chemicals; and it protects the community by preventing uncontrolled transportation of contaminants from the site.

2. There are two types of decontamination:

   - Gross decontamination: To allow end user to safely exit or doff the chemical protective clothing.

- Decontamination for reuse of chemical protective clothing.

B. PREVENTION OF CONTAMINATION. The first step in decontamination is to establish Standard Operating Procedures that minimize contact with chemicals and thus the potential for contamination. For example:

- Stress work practices that minimize contact with hazardous substances (e.g. do not walk through areas of obvious contamination, do not directly touch potentially hazardous substances).
- Use remote sampling, handling, and container-opening techniques (e.g. drum grapples, pneumatic impact wrenches).
- Protect monitoring and sampling instruments by bagging.
- Make openings in the bags for sample ports and sensors that must contact site materials.
- Wear disposable outer garments and use disposable equipment where appropriate.
- Cover equipment and tools with a strippable coating that can be removed during decontamination.
- Encase the source of contaminants, e.g. with plastic sheeting or overpacks.
- Ensure all closures and ensemble component interfaces are completely secured; and that no open pockets that could serve to collect contaminant are present.

C. TYPES OF CONTAMINATION.

1. Surface Contaminants. Surface contaminants may be easy to detect and remove.

2. Permeated Contaminants. Contaminants that have permeated a material are difficult or impossible to detect and remove. If contaminants that have permeated a material are not removed by decontamination, they may continue to permeate the material where they can cause an unexpected exposure.

   Four major factors affect the extent of permeation:

   - Contact time. The longer a contaminant is in contact with an object, the greater the probability and extent of permeation. For this reason, minimizing contact time is one of the most important objectives of a decontamination program.
   - Concentration. Molecules flow from areas of high concentration to areas of low concentration. As concentrations of chemicals increase, the potential for permeation of personal protective clothing increases.
   - Temperature. An increase in temperature generally increases the permeation rate of contaminants.
   - Physical state of chemicals. As a rule, gases, vapors, and low-viscosity liquids tend to permeate more readily than high-viscosity liquids or solids.

D. DECONTAMINATION METHODS.

1. Decontamination methods either (1) physically remove contaminants; (2) inactivate contaminants by chemical detoxification or disinfection/sterilization; or (3) remove contaminants by a combination of both physical and chemical means.

2. In general, gross decontamination is accomplished using detergents (surfactants) in water combined with a physical scrubbing action. This process will remove most forms of surface contamination including dusts, many inorganic chemicals, and some organic chemicals. Soapy water scrubbing of protective suits may not be effective in removing oily or tacky organic substances (e.g. PCB's in transformer oil). Furthermore, this form of decontamination is unlikely to remove any contamination that has permeated or penetrated the suit materials. Using organic solvents such as petroleum distillates may allow easier removal of heavy organic contamination but may result in other problems, including:

   - Permeation into clothing components, pulling the contaminant with it;
   - Spreading localized contaminant into other areas of the clothing; and
   - Generating large volumes of contaminated solvents that require disposal.

3. One promising method for removing internal or matrix contamination is the forced circulation of heated air over clothing items for extended periods of time. This allows many organic chemicals to migrate out of the materials and evaporate into the heated air. The process does require, however, that the contaminating chemicals be volatile. Additionally, low level heat may accelerate the removal of plasticizer from garment materials and affect the adhesives involved in garment seams.

4. Unfortunately, both manufacturers and protective clothing authorities provide few specific recommendations for decontamination. There is no definitive list with specific methods recommended for specific chemicals and materials. Much depends on the individual chemical-material combination involved.

E. TESTING THE EFFECTIVENESS OF DECONTAMINATION.

1. Protective clothing or equipment reuse depends on demonstrating that adequate decontamination has taken place. Decontamination methods vary in their effectiveness and unfortunately there are no completely accurate methods for nondestructively evaluating clothing or equipment contamination levels.

2. Methods which may assist in a determination include:

    - Visual examination of protective clothing for signs of discoloration, corrosive effects, or any degradation of external materials. However, many contaminants do not leave any visible evidence.
    - Wipe sampling of external surfaces for subsequent analysis; this may or may not be effective for determining levels of surface contamination and depends heavily on the material-chemical combination. These methods will not detect permeated contamination.
    - Evaluation of the cleaning solution. This method cannot quantify clean method effectiveness since the original contamination levels are unknown. The method can only show if chemical has been removed by the cleaning solution. If a number of garments have been contaminated, it may be advisable to sacrifice one garment for destructive testing by a qualified laboratory with analysis of contamination levels on and inside the garment.

F. DECONTAMINATION PLAN.

1. A decontamination plan should be developed and set up before any personnel or equipment are allowed to enter areas where the potential for exposure to hazardous substances exists. The decontamination plan should:

    - Determine the number and layout of decontamination stations;
    - Determine the decontamination equipment needed;
    - Determine appropriate decontamination methods;
    - Establish procedures to prevent contamination of clean areas;
    - Establish methods and procedures to minimize wearer contact with contaminants during removal of personal protective clothing; and
    - Establish methods for disposing of clothing and equipment that are not completely decontaminated.

2. The plan should be revised whenever the type of personal protective clothing or equipment changes, the use conditions change, or the chemical hazards are reassessed based on new information.

3. The decontamination process should consist of a series of procedures performed in a specific sequence. For chemical protective ensembles, outer, more heavily contaminated items (e.g. outer boots and gloves) should be decontaminated and removed first, followed by decontamination and removal of inner, less contaminated items (e.g. jackets and pants). Each procedure should be performed at a separate station in order to prevent cross contamination. The sequence of stations is called the decontamination line.

4. Stations should be separated physically to prevent cross contamination and should be arranged in order of decreasing contamination, preferably in a straight line. Separate flow patterns and stations should be provided to isolate workers from different contamination zones containing incompatible wastes. Entry and exit points to exposed areas should be conspicuously marked. Dressing stations for entry to the decontamination area should

be separate from redressing areas for exit from the decontamination area. Personnel who wish to enter clean areas of the decontamination facility, such as locker rooms, should be completely decontaminated.

5. All equipment used for decontamination must be decontaminated and/or disposed of properly. Buckets, brushes, clothing, tools, and other contaminated equipment should be collected, placed in containers, and labeled. Also, all spent solutions and wash water should be collected and disposed of properly. Clothing that is not completely decontaminated should be placed in plastic bags, pending further decontamination and/or disposal.

6. Decontamination of workers who initially come in contact with personnel and equipment leaving exposure or contamination areas will require more protection from contaminants than decontamination workers who are assigned to the last station in the decontamination line. In some cases, decontamination personnel should wear the same levels of protective clothing as workers in the exposure or contaminated areas. In other cases, decontamination personnel may be sufficiently protected by wearing one level lower protection (e.g. wearing Level B protection while decontaminating workers who are wearing Level A).

G. DECONTAMINATION FOR PROTECTIVE CLOTHING REUSE. Due to the difficulty in assessing contamination levels in chemical protective clothing before and after exposure, the responsible supervisor or safety professional must determine if the respective clothing can be reused. This decision involves considerable risk in determining clothing to be contaminant-free. Reuse can be considered if, in the estimate of the supervisor:

- No "significant" exposures have occurred.
- Decontamination methods have been successful in reducing contamination levels to safe or acceptable concentrations.

Contamination by known or suspected carcinogens should warrant automatic disposal. Use of disposable suits is highly recommended when extensive contamination is expected.

H. EMERGENCY DECONTAMINATION.

1. In addition to routine decontamination procedures, emergency decontamination procedures must be established. In an emergency, the primary concern is to prevent the loss of life or severe injury to personnel. If immediate medical treatment is required to save a life, decontamination should be delayed until the victim is stabilized. If decontamination can be performed without interfering with essential life-saving techniques or first aid, or if a worker has been contaminated with an extremely toxic or corrosive material that could cause severe injury or loss of life, decontamination should be continued.

2. If an emergency due to a heat-related illness develops, protective clothing should be removed from the victim as soon as possible to reduce the heat stress. During an emergency, provisions must also be made for protecting medical personnel and disposing of contaminated clothing and equipment.

VIII. INSPECTION, STORAGE, AND MAINTENANCE.

The end user in donning protective clothing and equipment must take all necessary steps to ensure that the protective ensemble will perform as expected. During emergencies is not the right time to discover discrepancies in the protective clothing. Teach end user care for his clothing and other protective equipment in the same manner as parachutists care for parachutes. Following a standard program for inspection, proper storage, and maintenance along with realizing protective clothing/equipment limitations is the best way to avoid chemical exposure during emergency response.

A. INSPECTION.

1. An effective chemical protective clothing inspection program should feature five different inspections:

- Inspection and operational testing of equipment received as new from the factory or distributor.
- Inspection of equipment as it is selected for a particular chemical operation.
- Inspection of equipment after use or training and prior to maintenance.
- Periodic inspection of stored equipment.

- Periodic inspection when a question arises concerning the appropriateness of selected equipment, or when problems with similar equipment are discovered.

2. Each inspection will cover different areas with varying degrees of depth. Those personnel responsible for clothing inspection should follow manufacturer directions; many vendors provide detailed inspection procedures. The generic inspection checklist provided in Table VIII:1-7 may serve as an initial guide for developing more extensive procedures.

3. Records must be kept of all inspection procedures. Individual identification numbers should be assigned to all reusable pieces of equipment (many clothing and equipment items may already have serial numbers), and records should be maintained by that number. At a minimum, each inspection should record:

    - Clothing/equipment item ID number;
    - Date of the inspection;
    - Person making the inspection;
    - Results of the inspection; and
    - Any unusual conditions noted.

   Periodic review of these records can provide an indication of protective clothing which requires excessive maintenance and can also serve to identify clothing that is susceptible to failure.

B. STORAGE.

1. Clothing must be stored properly to prevent damage or malfunction from exposure to dust, moisture, sunlight, damaging chemicals, extreme temperatures and impact. Procedures are needed for both initial receipt of equipment and after use or exposure of that equipment. Many manufacturers specify recommended procedures for storing their products. These should be followed to avoid equipment failure resulting from improper storage.

2. Some guidelines for general storage of chemical protective clothing include:

    - Potentially contaminated clothing should be stored in an area separate from street clothing or unused protective clothing.
    - Potentially contaminated clothing should be stored in a well-ventilated area, with good air flow around each item, if possible.
    - Different types and materials of clothing and gloves should be stored separately to prevent issuing the wrong material by mistake (e.g. many glove materials are black and cannot be identified by appearance alone).
    - Protective clothing should be folded or hung in accordance with manufacturer instructions.

C. MAINTENANCE.

1. Manufacturers frequently restrict the sale of certain protective suit parts to individuals or groups who are specially trained, equipped, or authorized by the manufacturer to purchase them. Explicit procedures should be adopted to ensure that the appropriate level of maintenance is performed only by those individuals who have this specialized training and equipment. In no case should you attempt to repair equipment without checking with the person in your facility who is responsible for chemical protective clothing maintenance.

2. The following classification scheme is recommended to divide the types of permissible or nonpermissible repairs:

    - Level 1: User or wearer maintenance, requiring a few common tools or no tools at all.
    - Level 2: Maintenance that can be performed by the response team's maintenance shop, if adequately equipped and trained.
    - Level 3 : Specialized maintenance that can be performed only by the factory or an authorized repair person.

TABLE VIII: 1-7. SAMPLE PPE INSPECTION CHECKLISTS

	Clothing
Before use:	Determine that the clothing material is correct for the specified task at hand.
Visually inspect for:	• Imperfect seams; • Nonuniform coatings; • Tears; and • Malfunctioning closures.
Hold up to light and check for pinholes	
Flex product:	• Observe for cracks. • Observe for other signs or shelf deterioration.
If the product has been used previously, inspect inside and out for signs of chemical attaack:	• Discoloration • Swelling • Stiffness
During the work task, periodically inspect for:	• Evidence of chemical attack such as discoloration, swelling, stiffening and softening. • Keep in mind, however, that chemical permeation can occur without any visible effects. • Closure failure • Tears • Punctures • Seam discontinuities
	Gloves
Before use:	Pressurize glove to check for pinholes. Either blow into glove, then roll gauntlet towards fingers or inflate glove and hold under water. In either case, no air should escape.
	Fully Encapsulating Suits

Before use:	• Check the operation of pressure relief valves
	• Inspect the fitting of wrists, ankles, and neck
	• Check faceshield, if so equipped, for:
	  - cracks
	  - crazing
	  - fogginess

3. Each facility should adopt the above scheme and list which repairs fall into each category for each type of protective clothing and equipment. Many manufacturers will also indicate which repairs, if performed in the

field, void the warranty of their products. All repairs made must be recorded on the records for the specific clothing along with appropriate inspection results.

IX. TRAINING.

A. BENEFITS. Training in the use of protective clothing:

- Allows the user to become familiar with the equipment in a nonhazardous, nonemergency condition.
- Instills confidence of the user in his/her equipment.
- Makes the user aware of the limitations and capabilities of the equipment.
- Increases worker efficiency in performing various tasks.
- Reduces the likelihood of accidents during chemical operations.

B. CONTENT. Training should be completed prior to actual clothing use in a non-hazardous environment and should be repeated at the frequency required by OSHA SARA III legislation. As a minimum the training should point out the user's responsibilities and explain the following, using both classroom and field training when necessary, as follows:

- The proper use and maintenance of selected protective clothing, including capabilities and limitations.
- The nature of the hazards and the consequences of not using the protective clothing.
- The human factors influencing protective clothing performance.
- Instructions in inspecting, donning, checking, fitting, and using protective clothing.
- Use of protective clothing in normal air for a long familiarity period.
- The user's responsibility (if any) for decontamination, cleaning, maintenance, and repair of protective clothing.
- Emergency procedures and self-rescue in the event of protective clothing/ equipment failure.
- The buddy system.

The discomfort and inconvenience of wearing chemical protective clothing and equipment can create a resistance to its conscientious use. One essential aspect of training is to make the user aware of the need for protective clothing and to instill motivation for the proper use and maintenance of that protective clothing.

X. RISKS.

A. HEAT STRESS. Wearing full body chemical protective clothing puts the wearer at considerable risk of developing heat stress. This can result in health effects ranging from transient heat fatigue to serious illness or death. Heat stress is caused by a number of interacting factors, including:

- Environmental conditions;
- Type of protective ensemble worn;
- The work activity required; and
- The individual characteristics of the responder.

When selecting chemical protective clothing and equipment, each item's benefit should be carefully evaluated for its potential for increasing the risk of heat stress. For example, if a lighter, less insulating suit can be worn without a sacrifice in protection, then it should be. Because the incidence of heat stress depends on a variety of factors, all workers wearing full body chemical protective ensembles should be monitored.

Review Paragraph III: Chapter 4, Heat Stress, in the OSHA Technical Manual. The following physiological factors should be monitored.

B. HEART RATE. Count the radial pulse during a 30-second period as early as possible in any rest period. If the heart rate exceeds 110 beats per minute at the beginning of the rest period, the next work cycle should be shortened by one-third.

C. ORAL TEMPERATURE.

1. Do not permit an end user to wear protective clothing and engage in work when his or her oral temperature exceeds 100.6°F (38.1°C).

2. Use a clinical thermometer (three minutes under the tongue) or similar device to measure oral temperature at the end of the work period (before drinking), as follows:

   - If the oral temperature exceeds 99.6°F (37.6°C), shorten the next work period by at least one-third.
   - If the oral temperature exceeds 99.6°F (37.6°C) at the beginning of a response period, shorten the mission time by one-third.

D. BODY WATER LOSS. Measure the end user's weight on a scale accurate to plus or minus 0.25 pounds prior to any response activity. Compare this weight with his or her normal body weight to determine if enough fluids have been consumed to prevent dehydration. Weights should be taken while the end user wears similar clothing, or ideally, in the nude. The body water loss should not exceed 1.5% of the total body weight loss from a response.

XI. BIBLIOGRAPHY.

Barker, R.L. and Coletta, G.C. 1986. *Performance of Protective Clothing.* American Society for Testing Materials: Philadelphia.

Forsberg, K. and Keith, L.H. 1989. *Chemical Protective Clothing Performance Index Book.* John Wiley & Sons: New York.

Forsberg, K. and Mansdorf, S.Z. 1989. *Quick Selection Guide to Chemical Protective Clothing.* Van Nostrand-Reinhold: New York.

Perkins, J.L. and Stull, J.O., ed. 1989. *Chemical Protective Clothing Performance in Chemical Emergency Response.* American Society for Testing Materials: Philadelphia.

Schwope, A.D., et al. 1987. *Guidelines for the Selection of Chemical Protective Clothing.* Third Ed. ACGIH: Cincinnati.

SECTION VIII: CHAPTER 2 [New addition to OTM]

RESPIRATORY PROTECTION

Contents:

I. INTRODUCTION.

A. Wearing respiratory protective devices to reduce exposure to airborne contaminants is widespread in industry. An estimated 5.0 million workers wear respirators, either occasionally or routinely. Although it is preferred industrial hygiene practice to use engineering controls to reduce contaminant emissions at their source, there are operations where this type of control is not technologically or economically feasible or is otherwise inappropriate.

B. Since respirators are not as consistently reliable as engineering and work practice controls, and may create additional problems, they are not the preferred method of reducing exposures below the occupational exposure levels. Accordingly, their use as a primary control is restricted to certain circumstances. In those circumstances where engineering and work practice controls cannot be used to reduce airborne contaminants below their occupational exposure levels (e.g., certain maintenance and repair operations, emergencies, or during periods when engineering controls are being installed), the use of respirators could be justified to reduce worker exposure. In other cases, where work practices and engineering controls alone cannot reduce exposure levels to below the occupational exposure level, the use of respirators would be essential for supplemental protection.

C. There are many variables that affect the degree of protection afforded by respiratory protective devices, and the misuse of respirators can be hazardous to employee safety and health. Selection of the wrong equipment, one of the most frequent errors made in respiratory protection, can result in the employee being exposed to increased concentrations of the harmful contaminant. This error may result in a broad range of health effects caused by the harmful contaminants, including silicosis, asbestosis, permanent lung damage, and cancer. Respirators that are not maintained and inspected can be less effective at reducing exposure to the harmful contaminants, and can place a greater burden on the respiratory system. Respirators that are not clean can cause dermatitis or skin irritation. Because respirator use may give the employee a false sense of security and presumed protection, an improper respirator program can actually present a high degree of hazard for the employee.

D. Respirators can only provide adequate protection if they are properly selected for the task; are fitted to the wearer and are consistently donned and worn properly; and are properly maintained so that they continue to provide the protection required for the work situation. These variables can only be controlled if a comprehensive respiratory protection program is developed and implemented in each workplace where respirators are used. When respirator use is augmented by an appropriate respiratory protection program, it can prevent fatalities and illnesses from both acute and chronic exposures to hazardous substances.

E. The primary aim of this chapter is to give detailed instruction in the selection of the proper respirator and its use and maintenance. The emphasis is on the implementation of a respiratory protection program developed in a logical progression of steps, outlined below:

- A clear definition of the hazards that will be encountered and the degree of protection required;
- The selection and fitting of the respirator;
- Medical evaluation for respirator selection and use;
- The required training in the correct use and care of the respirator; and
- The implementation of a maintenance program that will ensure that a high level of respiratory protection is maintained.

II. HISTORY OF THE DEVELOPMENT OF RESPIRATORY PROTECTION.

A. EARLY PRACTICES. The concept of using respiratory protective devices to reduce or eliminate hazardous exposures to airborne contaminants first came from Pliny (circa A.D. 23-79) who discussed the idea of using loose fitting animal bladders in Roman mines to protect workers from the inhalation of red oxide of lead. (See proposed respiratory protection standard, 59 Federal Register 58885.) Later, in the 1700's, the ancestors of modern atmosphere-supplying devices, such as the self-contained breathing apparatus or hose mask, were developed. Although the devices themselves have become more sophisticated in design and materials, respirators' performance is still based on one of two basic principles: purifying the air by removing contaminants before they reach the breathing zone of the worker, or providing clean air from an uncontaminated source.

B. DEVELOPMENT OF MODERN METHODS. In 1814, a particulate-removing filter encased in a rigid container was developed -- the predecessor of modern filters for air-purifying respirators. In 1854 it was recognized that activated charcoal could be used as a filtering medium for vapors. During World War I, with the use of chemical warfare, improvements in the design of respirators was necessary. In 1930 the development of the resin-impregnated dust filter made available efficient, inexpensive filters that have good dust-loading characteristics and low breathing resistance.

C. LATEST ADVANCES. A more recent development was the high efficiency particulate filter made with very fine glass fibers. These extremely efficient filters are used for very small airborne particles and produce little breathing resistance. Some features that are currently being incorporated into respirator design include a smaller facepiece, which translates into a better field of vision and a low profile that permits the respirator to fit under other protective gear such as a welder's helmet. Over the years there have been continuing major developments in the basic design of respirators. Modern design improvements have created products that are both more comfortable to wear and more protective than earlier respirators.

III. GENERAL INFORMATION.

A. PURPOSE. The purpose of a respirator is to prevent the inhalation of harmful airborne substances and/or an oxygen-deficient atmosphere. Functionally, a respirator is designed as an enclosure that covers the nose and mouth or the entire face or head. Respirators are of two general "fit" types, tight-fitting and loose-fitting.

1. The tight-fitting respirator (Figure VIII:2-1) is designed to form a seal with the face of the wearer. It is available in three types: quarter mask, half mask, and full facepiece. The quarter mask covers the nose and mouth, where the lower sealing surface rests between the chin and the mouth. The half mask covers the nose and mouth and fits under the chin. The full facepiece covers the entire face from below the chin to the hairline.

2. The loose-fitting respirator (Figure VIII:2-2) has a respiratory inlet covering that is designed to form a partial seal with the face. These include loose-fitting facepieces, as well as hoods, helmets, blouses, or full suits, all of which cover the head completely. The best known loose-fitting respirator is the supplied air hood used by the abrasive blaster. The hood covers the head, neck, and upper torso, and usually includes a neck cuff. Air is delivered by a compressor through a hose leading into the hood. Because the hood is not tight-fitting, it is important that sufficient air is provided to maintain a slight positive-pressure inside the hood relative to the

environment immediately outside the hood. In this way, an outward flow of air from the respirator will prevent contaminants from entering the hood.

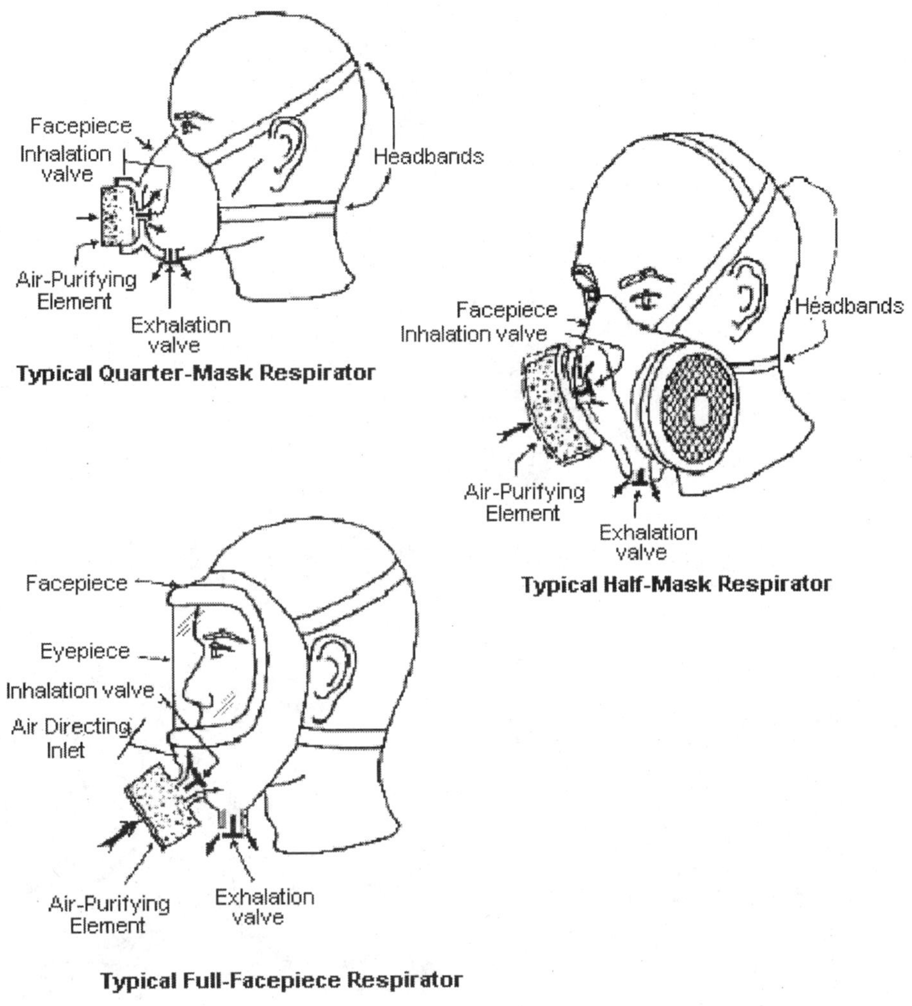

FIGURE VIII: 2-1. TIGHT-FITTING RESPIRATORS.

B. AIRBORNE (OR RESPIRATORY) HAZARDS may result from either an oxygen deficient atmosphere or breathing air contaminated with toxic particles, vapors, gases, fumes or mists. The proper selection and use of a respirator depend upon an initial determination of the concentration of the hazard or hazards present in the workplace, or the presence of an oxygen deficient atmosphere.

Airborne hazards generally fall into the following basic categories:

1. Dusts. Particles that are formed or generated from solid organic or inorganic materials by reducing their size through mechanical processes such as crushing, grinding, drilling, abrading, or blasting.

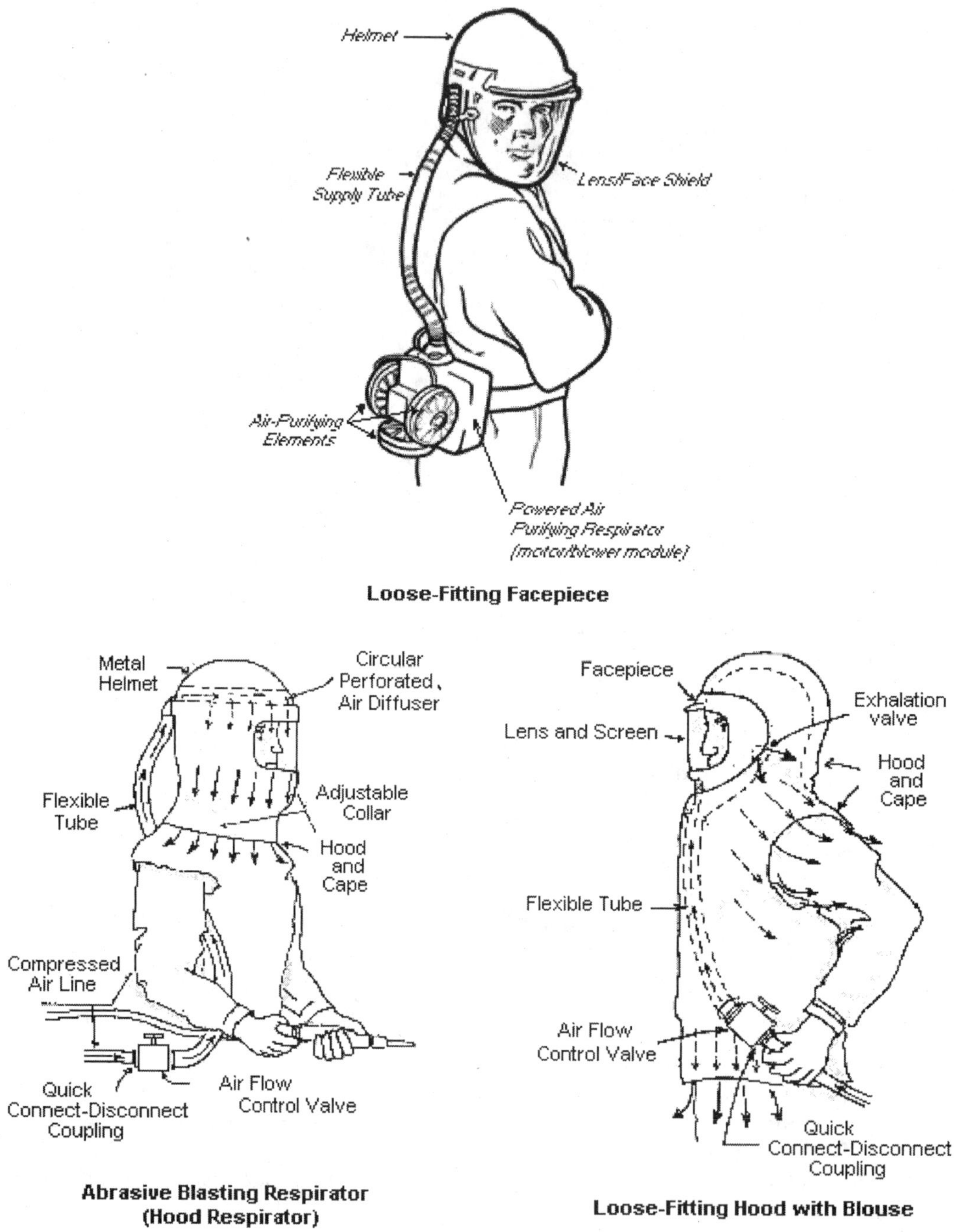

FIGURE VIII: 2-2. LOOSE-FITTING RESPIRATORS.

2. Fumes. Particles formed when a volatilized solid, such as a metal, condenses in cool air. This physical change is often accompanied by a chemical reaction, such as oxidation. Examples are lead oxide fumes from smelting, and iron oxide fumes from arc-welding. A fume can also be formed when a material such as magnesium metal is burned or when welding or gas cutting is done on galvanized metal.

3. Mists. A mist is formed when a finely divided liquid is suspended in the air. These suspended liquid droplets can be generated by condensation from the gaseous to the liquid state or by breaking up a liquid into a dispersed state, such as by splashing, foaming, or atomizing. Examples are the oil mist produced during cutting and grinding operations, acid mists from electroplating, acid or alkali mists from pickling operations, paint spray mist from spraying operations, and the condensation of water vapor to form a fog or rain.

4. Gases. Gases are formless fluids that occupy the space or enclosure and which can be changed to the liquid or solid state only by the combined effect of increased pressure and decreased temperature. Examples are welding gases such as acetylene, nitrogen, helium and argon; and carbon monoxide generated from the operation of internal combustion engines. Another example is hydrogen sulfide, which is formed wherever there is decomposition of materials containing sulfur under reducing conditions.

5. Vapors. Vapors are the gaseous form of substances that are normally in the solid or liquid state at room temperature and pressure. They are formed by evaporation from a liquid or solid, and can be found where parts cleaning and painting takes place and where solvents are used.

6. Smoke. Smoke consists of carbon or soot particles resulting from the incomplete combustion of carbonaceous materials such as coal or oil. Smoke generally contains droplets as well as dry particles.

7. Oxygen deficiency. An oxygen deficient atmosphere has an oxygen content below 19.5% by volume. Oxygen deficiency may occur in confined spaces, which include, but are not limited to, storage tanks, process vessels, towers, drums, tank cars, bins, sewers, septic tanks, underground utility tunnels, manholes, and pits.

C. RESPIRATOR CLASSIFICATIONS. Respirators provide protection either by removing contaminants from the air before they are inhaled or by supplying an independent source of respirable air. There are two major classifications of respirators

1. Air purifying respirators (devices that remove contaminants from the air); and

2. Atmosphere-supplying respirators (those devices that provide clean breathing air from an uncontaminated source).

   Each class of respirator may have tight-fitting and loose-fitting facepieces. An important aspect of respirator operation and classification is the air pressure within the facepiece. When the air pressure within the facepiece is negative during inhalation with respect to the ambient air pressure, the respirator is termed negative-pressure respirator. When the pressure is normally positive with respect to ambient air pressure throughout the breathing cycle, the respirator is termed a positive-pressure respirator. The concept of negative and positive pressure operation is important when considering potential contaminant leakage into the respirator.

D. AIR PURIFYING RESPIRATORS are grouped into three general types: particulate removing, vapor and gas removing, and combination. Elements that remove particulates are called filters, while vapor and gas removing elements are called either chemical cartridges or canisters. Filters and canisters/cartridges are the functional portion of air-purifying respirators, and they can generally be removed and replaced once their effective life has expired. The exception would be filtering facepiece respirators (commonly referred to as "disposable respirators," "dust masks," or "single-use respirators"), which cannot be cleaned, disinfected, or resupplied with an unused filter after use.

1. Particulate-removing respirators are designed to reduce inhaled concentrations of nuisance dusts, fumes, mists, toxic dusts, radon daughters, asbestos-containing dusts or fibers, or any combination of these substances, by filtering most of the contaminants from the inhaled air before they enter the breathing zone of the worker. They may have single-use or replaceable filters. These respirators may be non-powered or powered air-purifying. A

powered air-purifying respirator (PAPR) uses a blower to force the ambient atmosphere through air purifying elements to the inlet covering.

2. Vapor- and gas-removing respirators are designed with sorbent elements (canisters or cartridges) that adsorb and/or absorb the vapors or gases from the contaminated air before they can enter the breathing zone of the worker. Combination cartridges and canisters are available to protect against particulates, as well as vapors and gases.

E. ATMOSPHERE-SUPPLYING RESPIRATORS are respirators that provide air from a source independent of the surrounding atmosphere instead of removing contaminants from the atmosphere. These respirators are classified by the method that is used to supply air and the way in which the air supply is regulated. Basically, these methods are: self-contained breathing apparatus (air or oxygen is carried in a tank on the worker's back, similar to SCUBA gear); supplied-air respirators (compressed air from a stationary source is supplied through a high-pressure hose connected to the respirator); and combination self-contained and supplied-air respirators.

F. LIMITATIONS OF RESPIRATOR USE. Not all workers can wear respirators. Individuals with impaired lung function, due to asthma or emphysema for example, may be physically unable to wear a respirator. Individuals who cannot get a good facepiece fit, including those individuals whose beards or sideburns interfere with the facepiece seal, will be unable to wear tight-fitting respirators. An adequate fit is required for a respirator to be effective. In addition to these problems, respirators may also be associated with communication problems, vision problems, fatigue, and reduced work efficiency.

In principle, respirators usually are capable of providing adequate protection. However, problems associated with selection, fit, and use often render them less effective in actual application; these problems prevent the assurance of consistent and reliable protection, regardless of the theoretical capabilities of the respirator. Occupational safety and health experts have spent considerable effort over the years developing fit-testing procedures and methods of measuring respirator effectiveness, thereby improving protection for those employees required to wear them.

IV. RESPIRATOR PROTECTION PROGRAM.

A. THE STANDARD. Whenever respirators are required to be worn, a written respirator protection program must be developed and implemented in accordance with OSHA's respirator standard, 29 CFR 1910.134. (Additional program requirements may be found in the standards that regulate the hazards to which the employee is exposed.) Because workplaces differ substantially, each program must be tailored to the specific conditions of the workplace. The program must consist of worksite-specific procedures governing the selection, use, and care of respirators. The program must be updated as often as necessary to reflect changes in workplace conditions and respirator use.

B. THE WORKSITE-SPECIFIC PROCEDURES must contain all the information needed to maintain an effective respirator program to meet the user's individual requirements. These procedures are a set of step-by-step instructions written so that a task (i.e., respirator use, fit-testing procedures, cleaning and storage, etc.) can be performed by all personnel in a uniform and consistent way, while supplying the maximum protection for workers who use respirators in the workplace. The employer must anticipate both the routine and non-routine use of respirators, as well as any possible emergency use based on the conditions in the workplace in which they are to be used. Worksite-specific procedures must be written so as to be useful to those who are directly involved in the respirator program: the program administrator, those fitting the respirators and training the workers, respirator maintenance workers, and the supervisors responsible for overseeing respirator use on the job.

C. ADMINISTRATION. In addition, the respirator standard requires that the respiratory protection program be administered by one qualified individual to ensure that the integrity of the respiratory protection program is maintained through the continuous oversight of one responsible person. The program administrator must be qualified by appropriate training and/or experience in the proper selection, use, and maintenance of respirators, be responsible for implementing the respiratory protection program, and conduct regular evaluations of the program's effectiveness.

Although responsibility for respirator program oversight rests with the program administrator, he or she may delegate responsibilities to other qualified individuals. For instance, a large facility may find it practical and economical to have a staff of personnel involved in the respirator program, each with their own area of responsibility. However, each of these people must report to the one administrator who has overall responsibility for the program. This approach promotes coordination of all facets of the program. The administrator should have the full support of higher level management; without it, an effective respirator program is difficult to initiate and maintain.

D. ELEMENTS. The respiratory protection program must cover the following basic elements, as applicable:

- Procedures for selecting respirators for use in the workplace;
- Medical evaluations of employees required to use respirators;
- Fit testing procedures for tight-fitting respirators;
- Use of respirators in routine and reasonably foreseeable emergency situations;
- Procedures and schedules for cleaning, disinfecting, storing, inspecting, repairing, and otherwise maintaining respirators;
- Procedures to ensure adequate air quality, quantity and flow of breathing air for atmosphere-supplying respirators;
- Training of employees in the respiratory hazards to which they are potentially exposed;
- Training of employees in the proper use of respirators, including putting on and removing them, any limitations on their use, and maintenance procedures; and
- Procedures for regularly evaluating the effectiveness of the program.

V. RESPIRATOR SELECTION.

Respirator selection requires correctly matching the respirator with the hazard, the degree of hazard, and the user. The respirator selected must be adequate to effectively reduce the exposure of the respirator user under all conditions of use, including reasonably foreseeable emergency situations. Proper respirator selection involves choosing a device that fully protects the worker from the respiratory hazards to which he or she may be exposed and permits the worker to perform the job with the least amount of physical burden.

A. SELECTION FACTORS. Many factors must be considered carefully in respirator selection. In choosing the appropriate respirator, one must consider the nature and extent of the hazard, work requirements and conditions, and the characteristics and limitations of the respirators available. The following categories of information must be taken into account:

- Nature of the hazard, and the physical and chemical properties of the air contaminant;
- Concentrations of contaminants;
- Relevant permissible exposure limit or other occupational exposure limit;
- Nature of the work operation or process;
- Time period the respirator is worn;
- Work activities and physical/psychological stress;
- Fit testing; and
- Physical characteristics, functional capabilities and limitations of respirators.

1. Nature of the hazard, and the physical and chemical properties of the air contaminant. The nature of the hazard, whether it is in the form of a gas, dust, organic vapor, fume, mist, oxygen deficiency or any combination of hazards, needs to be taken into account. The physical and chemical properties of the contaminant that affect respirator selection, and the selection of respirator components such as cartridges, canisters, and filters must also be considered. Physical properties include such factors as particle size for dusts, and vapor pressure for gases and vapors. Chemical properties of the air contaminant that affect breakthrough times, and the ability of the filter material to remove, adsorb, or absorb the contaminant must also be considered.

2. Concentrations of contaminants. Sampling and analysis of the workplace air determines what degree of exposure is occurring, and thus what degree of protection is required. Where such sampling and analysis have been done, the results are to be used as a point of comparison with the occupational exposure level, i.e., to determine how much the concentration must be lowered by the respirator to reduce employee exposure to a safe level.

3. The relevant permissible exposure limit or other occupational exposure limit. Respirators selected must be capable of protecting against overexposure by reducing and maintaining exposure to or below the relevant exposure limit. In addition to the OSHA limits, employers should refer to the ACGIH (American Conference of Governmental Industrial Hygienists) recommended Threshold Limit Values (TLV's), the NIOSH (National Institute for Occupational Safety and Health) Recommended Exposure Limits (REL's), or other occupational exposure limits.

4. Nature of the work operation or process. The type of job operation, the equipment or tools that will be used, and any motion or travel the job requires can influence the type of respirator selected, particularly when supplied-air respirators, which require a connection to a clean air source, are used.

5. Time period respirator is worn. The employer must also consider the period of time during which the respirator will be used by employees during a work shift. Breakthrough times for different chemicals can vary greatly, and are dependent on the concentrations of contaminants in the workplace air, patterns of respirator use, and environmental factors including temperature and humidity. A respirator that provides adequate protection for one chemical may be inadequate for another chemical with a different breakthrough time. In addition, employees wearing respirators for longer periods of time may need respirators that impose the minimum possible physical burden.

6. Work activities and stress. The work activities of employees while wearing respirators are also a factor. Heavy work that is physically draining may affect an employee's capability of wearing certain types of respirators. Temperature and humidity conditions in the workplace may also affect the physical/psychological stress level associated with wearing a respirator, as well as the effectiveness of respirator filters and cartridges. These types of factors must be assessed in selecting the appropriate equipment for a particular work situation.

7. Fit testing. Some employees may be unable to achieve an adequate fit with certain respirator models or a particular type of respirator -- such as half-mask air-purifying respirators -- so an alternative respirator model with an adequate fit or other type of respirator that provides adequate protection must be used. Therefore, it is necessary for employers to provide a sufficient number of respirator models and sizes from which employees can choose an acceptable respirator that fits correctly.

8. Physical characteristics, functional capabilities, and limitations of respirators. The last category of information to be considered when selecting respiratory protection is the physical itself. Respirators selected must not impair the worker's vision, hearing, communication, and physical movement necessary to perform jobs safely. For example, airline respirators should not be used by mobile employees around moving machinery to avoid entanglement of the respirator in the equipment.

B. SELECTION. Once the above factors have been taken into account, the employer must select a NIOSH-certified respirator. Where NIOSH has not specifically certified any respirator for use against the particular contaminant present in the workplace, the employer must select a NIOSH-certified respirator that has no limitation prohibiting its use for that contaminant. The respirator must be appropriate for the contaminant's physical form and chemical properties and the conditions under which it will be used. All respirators must be chosen and used according to the limitations of the NIOSH certification, which appears on the NIOSH certification label.

C. ASSIGNED PROTECTION FACTORS. Until such time as OSHA addresses the issue of assigned protection factors (APF's), employers may rely on APF's published by NIOSH and ANSI. Where there are conflicts between the NIOSH and ANSI APF's, the employer should apply the more protective APF.

D. WARNING SYSTEM. When an air-purifying respirator is selected for protection against gases and vapors, a system must be in effect that will reliably warn respirator wearers of contaminant breakthrough. These systems are: a respirator equipped with an end-of-service life indicator (ESLI) certified by NIOSH for the contaminant, or an established and enforced cartridge/canister change schedule that is based on objective information or data that will ensure that canisters and cartridges are changed before the end of their service life.

E. ATMOSPHERES REQUIRING HIGHEST LEVEL OF PROTECTION. For atmospheres that are immediately dangerous to life and health (IDLH), the highest level of respiratory protection and reliability is required. These atmospheres, by definition, are the most dangerous environments in which respirators are used. In these atmospheres, there is no tolerance for respirator failure. Consequently, only the following respirators must be provided and used: full-facepiece pressure demand self-contained breathing apparatus (SCBA) certified for a minimum service life of thirty minutes, or a combination full-facepiece pressure demand supplied-air respirator (SAR) with an auxiliary self-contained air supply.

VI. MEDICAL EVALUATION.

A. OVERVIEW. Persons assigned to tasks that require the use of a respirator must be physically able to perform the work while using the respirator. Accordingly, employers have the responsibility of ensuring that employees are medically fit to tolerate the physical and psychological stress imposed by respirator use, as well as the physical stress originating from job and workplace conditions.

Employees must be medically evaluated and found eligible to wear the respirator selected for their use prior to fit testing or first-time use of the respirator in the workplace. Medical eligibility is to be determined by a physician or other licensed health care professional (referred to as a "PLHCP"). A variety of qualified health care providers, besides physicians, including occupational health nurses, nurse practitioners, and physician assistants, can perform the medical evaluations provided they are licensed to do so in the state in which they practice.

B. QUESTIONNAIRE. In assessing the employee's medical eligibility to use a respirator, the PLHCP must perform a medical evaluation using a medical questionnaire (Appendix C to 1910.134) or provide a medical examination that obtains the same information as the medical questionnaire. The medical evaluation must be administered confidentially and at a time and place, during working hours, that is convenient to the employee. Employers are free to provide respirator users with a medical examination in lieu of the medical questionnaire if they chose to do so, but they are not required by the standard to administer a medical examination unless the employee gives a positive response to specific questions on the questionnaire.

C. MEDICAL FACTORS AND CONDITIONS. The purpose of a medical evaluation program is to determine if employees can tolerate the physiological burden associated with respirator use, including: the burden imposed by the respirator itself (e.g., its weight and breathing resistance during both normal operation and under conditions of filter, canister, or cartridge overload); musculoskeletal stress (e.g., when the respirator to be worn is a SCBA); limitations on auditory, visual, and olfactory sensations; and isolation from the workplace environment. Since certain jobs and workplace conditions in which a respirator is used can also impose a physiological burden on the user, the medical evaluation must also consider the following factors: type and weight of the respirator to be worn; duration and frequency of respirator use; expected physical work effort; use of protective clothing and equipment to be worn; and temperature and humidity extremes that may be encountered. This information must be provided to the PLHCP before the PLHCP makes a recommendation regarding an employee's ability to use a respirator.

The medical evaluation is designed to identify general medical conditions that place employees who use respirators at risk of serious medical consequences. Medical conditions known to compromise an employee's ability to tolerate respirator-, job-, and workplace-related physiological stress include: cardiovascular and respiratory diseases (e.g., a history of high blood pressure, angina, heart attack, cardiac arrhythmias, stroke, asthma, chronic bronchitis, emphysema); reduced pulmonary function caused by other factors (e.g., smoking or prior exposure to respiratory hazards); neurological or musculoskeletal disorders (e.g., ringing in the ears, epilepsy, lower back pain); impaired sensory function (e.g., perforated ear drums, reduced or absent ability to smell); and psychological disorders (e.g., claustrophobia and severe anxiety).

D. STANDARD OF EVALUATION. The employer must obtain a written recommendation from the PLHCP on whether the employee is medically able to wear a respirator. The recommendation must identify any limitations on the employee's use of the respirator, as well as the need for follow-up medical evaluations that are needed to assist the PLHCP in making a recommendation. The employee must also receive a copy of the PLHCP's written recommendations. A powered air-purifying respirator (PAPR) must be provided to an employee if information from the medical evaluation indicates that the employee can use a PAPR but not a negative pressure respirator. If, subsequent to this evaluation, the PLHCP determines that the employee is able to wear a negative pressure respirator, the employer is no longer required to provide a PAPR to that employee.

In addition, the standard requires the employer to medically re-evaluate an employee when:

- That employee reports medical signs or symptoms that are related to the employee's ability to use a respirator;
- A PLHCP, supervisor, or the respirator program administrator observes that the employee is having a medical problem during respirator use and they inform the employer of their observation;
- Information from the respiratory protection program, including observations made during fit testing and program evaluation, indicates a need for employee re-evaluation; or
- A change occurs in workplace conditions (e.g., physical work effort, type of respirator used, protective clothing, temperature) that may result in a substantial increase in the physiological burden placed on an employee.

VII. FIT TESTING.

It has long been recognized that respirators must fit properly to provide protection. To obtain adequate respiratory protection, there must be a proper match between respirator and wearer. Respirators that don't seal properly around the face offer only the illusion of protection. To accommodate the variability of face size characteristics among individuals, a number of manufacturers offer facepieces in several sizes and models.

A. PURPOSE. The primary purpose of fit testing is to identify the specific make, model, style, and size of respirator best suited for each employee. In addition, fit testing also provides an opportunity to check on problems with respirator wear, and reinforces respirator training by having wearers review the proper methods of donning and wearing the respirator.

B. REQUIREMENT. Fit testing is required for all negative or positive pressure tight-fitting facepiece respirators. The OSHA respiratory protection standard requires that fit testing be performed before an employee first starts wearing a respirator in the work environment, whenever a different respirator facepiece is used, and at least annually thereafter.

C. METHOD. Prior to the actual fit test, the employee must be shown how to put on a respirator, position it on the face, set strap tension, and determine an acceptable fit. Next, the employee must be allowed to choose a respirator from a sufficient number of models and sizes so that the employee can find an acceptable and correctly fitting respirator. Once an acceptable respirator has been found -- which takes into account the position of the mask on the face, nose, and cheeks; room for eye protection; and room to talk -- a user seal check must be conducted (refer to on "Use of Respirators").

D. TYPES OF FIT TESTING. Fit testing may either be qualitative (QLFT) or quantitative (QNFT), and must be administered using an OSHA-accepted QLFT or QNFT protocol. These protocols are described in mandatory Appendix A to 1910.134. Prior to the commencement of the fit test, the employee must be given a description of the fit test and a description of the exercises that he or she will be performing during fit testing. The respirator to be tested must be worn for at least five minutes before the start of the fit test. The employee must be fit tested with the same make, model, style, and size of respirator that will be used in the workplace.

1. Qualitative fit testing (QLFT). Qualitative fit testing involves the introduction of a gas, vapor, or aerosol test agent into an area around the head of the respirator user. A determination is then made as to whether or not the wearer can detect the presence of the test agent through means such as odor, taste, or nasal irritation. If the presence of the test agent is detected inside the mask, the respirator fit is considered to be inadequate.

   There are four qualitative fit test protocols approved in OSHA's standard. The isoamyl acetate (IAA) test determines whether a respirator is protecting a user by questioning whether the user can smell the distinctive odor of IAA. Both the saccharin and Bitrex™ tests involve substances with distinctive tastes that should not be detected through an effective respirator. The irritant smoke (e.g., stannic chloride) test involves a substance that elicits an involuntary irritation response in those exposed to it.

   Before conducting a qualitative test, the worker must undergo a sensitivity test to determine if he or she can taste, smell or react to the substance. When performing the isoamyl acetate test, the protocol requires that separate rooms be used for the odor screening and fit tests, and that the rooms be sufficiently ventilated to ensure that there is no detectable odor of IAA prior to a test being conducted. This will prevent olfactory fatigue among workers being fit tested by preventing a buildup of IAA in the general room air.

2. Quantitative fit testing (QNFT). In a quantitative fit test, the adequacy of respirator fit is assessed by numerically measuring the amount of leakage into the respirator. This testing can be done by either generating a test aerosol as a test atmosphere, using ambient aerosol as the test agent, or using controlled negative pressure (CNP) to measure the volumetric leak rate. Appropriate instrumentation is required to quantify respirator fit.

E. FIT TEST EXERCISES. The following test exercises must be performed for all fit testing methods described in the OSHA standards, except the CNP method which has its own fit testing exercise regimen:

1. Normal breathing in a normal standing position, without talking;

2. Deep breathing in a normal standing position, breathing slowly and deeply, taking precaution not to hyperventilate;

3. Turning the head slowly from side to side, while standing in place, with the employee holding his/her head momentarily at each extreme so that the employee can inhale at each side;

4. Moving the head up and down slowly, while standing in place, inhaling in the up position when looking toward the ceiling;

5. Talking out loud slowly, reading from a prepared text such as the Rainbow Passage (see Appendix A of the standard), counting backward from 100, or reciting a memorized poem or song;

6. Grimacing by smiling or frowning (only for QNFT testing);

7. Bending at the waist as if to touch toes (jogging in place can be done when the fit test enclosure doesn't permit bending at the waist); and

8. Normal breathing (as described above).

Each test exercise must be performed for one minute, except for the grimace exercise which must be performed for 15 seconds. The respirator must not be adjusted once the fit test exercises begin. Any adjustment voids the test, and the fit test must be repeated.

The employee must perform exercises in the test environment while wearing any applicable safety equipment that may be worn during actual respirator use and that could interfere with respirator fit. If the employee exhibits breathing difficulty during the fit test, he or she must be referred to a physician or other licensed health care professional to determine whether the employee can wear a respirator while performing his or her duties.

F. RETESTING. If the employee finds the fit of the respirator unacceptable, he or she must be given a reasonable opportunity to select a different respirator and to be retested. In addition, retesting is required whenever an employee reports, or the employer, PLHCP, supervisor, or program administrator observe changes in an employee's physical condition that could affect respirator fit. Such conditions include, but are not limited to, facial scarring, dental changes (e.g., wearing new dentures), cosmetic surgery, or an obvious change in body weight.

VIII. USE OF RESPIRATORS.

A. CONDITIONS. Once the respirator has been properly selected and fitted, it is necessary to ensure that the respirator is used properly in the workplace. The following conditions may compromise the effective use of the respirator and jeopardize worker protection: facepiece seal leakage; removing the respirator at the wrong times in hazardous atmospheres; not properly performing user seal checks; or not properly repairing defective parts. In these circumstances, there is the danger that employees may have a false sense of security in feeling that they are protected when they are not.

The employer must also be aware of the conditions in the work areas where employees are using respirators. Employers are required to routinely evaluate workplace conditions, the degree of employee exposure, and physical stress so that they can provide additional or different respiratory protection when necessary. By observing respirator use under actual workplace conditions, employers can note problems such as changes in the fit of a respirator due to the use of other protective equipment, or conditions leading to skin irritation.

B. FACEPIECE SEAL PROTECTION.

1. Seal of Tight-Fitting Respirators and Valve Function. The employer must not permit respirators with tight-fitting facepieces to be worn by employees who have conditions that would compromise the facepiece-to-face seal. Examples of these conditions include facial hair that interferes with the facepiece seal or valve function, absence of normally worn dentures, facial deformities (e.g., scars, deep skin creases, prominent cheekbones), or the use of jewelry or headgear that projects under the facepiece seal.

2. Corrective Glasses or Goggles. Corrective glasses or goggles, or other personal protective equipment, must be worn in such a way that they do not interfere with the seal of the facepiece to the face. Since eye glasses or goggles may interfere with the seal of half-facepieces, it is strongly recommended that full-facepiece respirators be worn where either corrective glasses or eye protection is required, since corrective lenses can be mounted inside a full-facepiece respirator. In addition, the full-facepiece respirator may be more comfortable, and less cumbersome, than the combination of a half-mask and chemical goggles. OSHA's current standard on respiratory protection, unlike the previous one, allows the use of contact lenses with respirators where the wearer has successfully worn such lenses before.

3. User Seal Check. A user seal check (formerly known as a fit check) must be performed every time a tight-fitting respirator is put on or adjusted to ensure proper seating of the respirator to the face. The user seal check conducted must be either the positive and/or negative pressure checks described in Appendix VIII:2-2 of this chapter, or the manufacturer's recommended procedures (when equally protective). If the employee fails the user seal check test, another facepiece must be selected.

   The employee must not have any hair growth (e.g., beard stubble, sideburns, or beard) that comes between the sealing surface of the respirator facepiece and the face, as well as hair that interferes with valve function, or any other condition that might interfere with the face-to-facepiece seal such as jewelry or facial makeup. The user seal check must be used for all respirators on which such checks are possible. If a user seal check cannot be performed on a tight-fitting respirator, the OSHA standard prohibits that respirator from being used.

C. CONTINUING RESPIRATOR EFFECTIVENESS.

1. Skin or Eye Irritation. Skin or eye irritation can result from wearing a respirator in hot, humid conditions, as well as in contaminated environments. Such irritation can be distressing to workers, causing them to remove or adjust the respirator, or to refrain from wearing the respirator altogether. Therefore, to prevent skin or eye irritation associated with respirator use, employees must be permitted to leave the respirator use area to wash their faces and respirator facepieces as needed.

2. Filter, Canister, and Cartridge Elements for Air-Purifying Respirators. Whenever the respirator user can detect vapor or gas breakthrough (by odor, taste, and/or irritation effects), a change in breathing resistance or leakage of the facepiece, the worker must be allowed to leave the respirator use area to replace the respirator or the filter, cartridge, or canister elements. Similarly, employees must be permitted to leave the respirator use area if they are replacing cartridge or canister elements according to a change schedule, or when the end-of-service-life indicator shows that the canister or cartridge(s) must be changed.

3. Repair, Disposal, and Replacement of Respirators. Since respirators must be in good working condition to function, it is imperative that they not be used if they have been impaired in any way. Impairments include a broken strap, loss of respirator shape, and a face seal that can no longer be maintained. Therefore, respirators that are not properly functioning must be replaced, repaired, or discarded. The respirator manufacturers can supply replacement parts for damaged parts on elastomeric respirators. Only when the respirator has been replaced or repaired can the employee return to the respirator use area.

D. IMMEDIATELY DANGEROUS TO LIFE OR HEALTH (IDLH) ATMOSPHERES. Atmospheres are IDLH when they pose an immediate threat to life, would cause irreversible adverse health effects, or would interfere with an individual's ability to escape from a dangerous atmosphere. Care must be exercised in these situations since failure of the respirator to provide the appropriate protection may result in serious injury or death. Consequently, the employer must develop and implement specific procedures for the use of respirators in IDLH atmospheres that include the following provisions:

1. At least one employee (referred to as the "standby employee") is to be located outside the IDLH atmosphere and maintain visual, voice, or signal line communication with the employee(s) in the IDLH atmosphere;

2. The standby employee(s) located outside the IDLH atmosphere must be trained and equipped to provide effective emergency rescue;

3. The employer or authorized designee is to be notified before the standby employees(s) enter the IDLH atmosphere to provide emergency rescue;

4. The employer or authorized designee, once notified of such entry, must provide the necessary assistance appropriate to the situation;

5. Standby employee(s) must be equipped with pressure demand or other positive pressure SCBA, or a pressure demand or other positive pressure supplied-air respirator with auxiliary SCBA; and

6. Standby employee(s) must be equipped with appropriate retrieval equipment for lifting or removing the employee from the hazardous atmosphere, or, when such retrieval equipment cannot be used because it would increase the overall risk resulting from entry, ensure that equivalent provisions for rescue have been made.

E. INTERIOR STRUCTURAL FIREFIGHTING. In the ultra-hazardous situation of interior structural firefighting, firefighters must operate using a buddy system. Safeguards that may be adequate for well-controlled and well-characterized IDLH situations are not adequate in the uncontrolled and unpredictable situation characterized by a burning building. Therefore, in addition to the above safeguards for IDLH atmospheres, the following requirements apply to interior structural fire fighting:

1. Two or more firefighters must always be sent in together and remain in visual or voice contact with one another at all times;

2. At least two standby personnel are to be located outside the fire area; and

3. All personnel engaged in interior structural firefighting must use SCBA.

The "two-in/two-out" requirement does not take effect until firefighters begin to perform interior structural fire fighting. While the fire is in the incipient stage (as determined by the commander or other person in charge), or when emergency rescue operations are required before the entire team has assembled, the standard does not require two-member teams inside and outside the structure.

IX. MAINTENANCE AND CARE.

A. REQUIREMENTS. The OSHA standard requires that employers provide each respirator user with a respirator that is clean, sanitary, and in good working order. These requirements are a vital part of any successful respiratory protection program. To ensure that the respirator remains serviceable and delivers effective protection, a maintenance program must be in place prior to respirator use.

The OSHA respirator standard strongly emphasizes the importance of a good maintenance program, but permits its tailoring to the type of facilities, working conditions, and hazards involved. However, all programs are required to include at least:

- Cleaning and disinfecting procedures;
- Proper storage;
- Regular inspections for defects (including leak check); and
- Repair methods.

In addition to the OSHA requirements, the manufacturer's instructions for inspection, cleaning, and maintenance of respirators should be consulted to ensure that the respirator continues to function properly. A proper maintenance program ensures that the worker's respirator remains as effective as when it was new.

B. CLEANING AND DISINFECTING.

1. Cleaning and sanitizing respirators are necessary to prevent skin irritation, dermatitis, and to encourage worker acceptance. Where the contaminant is a dust, mist, or fume, build-up on the respirator face-to-facepiece seal or within the respirator will reduce the protection provided by the respirator because the contaminant is in the breathing zone or has compromised the seal. In addition, the build-up of contamination on the respirator can contribute to the deterioration of the respirator's materials, which can lead to reduced protection. Full facepieces must be cleaned to ensure that employees can see through the facepiece.

2. Respirators that are issued for the exclusive use of an employee must be cleaned and disinfected as often as necessary to be maintained in a sanitary condition. Respirators used by more than one employee must be cleaned and disinfected prior to being used by a different individual. Respirators maintained for emergency use as well as respirators used in fit testing and training, must be cleaned and disinfected after each use. The employer must use either the OSHA cleaning and disinfecting procedures recommended in Appendix VIII:2-3 of this chapter or the procedures recommended by the respirator manufacturer, as long as they are equivalent in effectiveness to the OSHA method.

C. STORAGE.

1. All respirators must be stored so that they are protected against damage, contamination, dust, sunlight, extreme temperatures, excessive moisture, and damaging chemicals. When respirators are packed or stored, the facepiece and exhalation valve must be stored in a manner that will prevent deformation. Each respirator should be positioned so that it retains its natural configuration. Synthetic materials and even rubber will warp if stored in an unnatural shape, thus affecting the fitting characteristics of the facepiece.

2. Respirators intended for emergency use must be kept accessible to the work area, but not in an area that might itself be involved in the emergency because such an area may become contaminated or inaccessible. Emergency-use respirators must be stored in compartments or covers that are clearly marked to indicate that they contain emergency respirators, and stored according to any applicable manufacturer instructions.

D. INSPECTION. To ensure the continued reliability of respiratory equipment, it must be inspected on a regular basis. The frequency of inspection and the procedures to be followed depend on whether the respirator is intended for non-emergency, emergency, or escape use only.

1. The OSHA standard requires that all respirators used in non-emergency situations be inspected before each use and during cleaning. Respirators designated for use in an emergency situation are to be inspected at least monthly and in accordance with the manufacturer's instructions, and checked for proper function before and after each use. Emergency escape-only respirators must be inspected before being carried into the workplace.

2. For all respirators, inspections must include a check of respirator function, tightness of connections, and the condition of the various parts including, but not limited to, the facepiece, head straps, valves, connecting tube, and cartridges, canisters, or filters. In addition, the elastomeric parts must be evaluated for pliability and signs of deterioration.

3. For SCBA's, which require monthly inspections, the air and oxygen cylinders must be maintained in a fully charged state and recharged when the pressure falls to 90% of the manufacturer's recommended pressure level. In addition, the regulator and warning devices must be inspected to ensure that they function properly.

4. For respirators that are maintained for use in emergencies, the OSHA standard requires certifying the respirator by documenting the date that the inspection was performed, the name or signature of the inspector, the findings of the inspection, any required remedial action, and a serial number or other means of identifying the inspected respirator. This information must be provided on a tag or label that is attached to the storage compartment for

the respirator, is kept with the respirator, or is stored in the form of inspection reports (paper or electronic). The information must be maintained until it is replaced following a subsequent certification.

E. REPAIR. Respirators that fail to pass inspection or are otherwise found to be defective, must be removed from service, and discarded, repaired, or adjusted. Repairs or adjustments to respirators must be done only by appropriately trained personnel, using only the respirator manufacturer's NIOSH-approved parts designed for that respirator. The repairs also must be made in accordance with the manufacturer's recommendations and specifications regarding the type and extent of repairs to be performed. Because components such as reducing and admission valves, regulators, and alarms are complex and essential to the safe functioning of the respirator, they are required to be adjusted and repaired only by the manufacturer or a technician trained by the manufacturer.

X. BREATHING AIR QUALITY AND USE.

A. STANDARDS AND SPECIFICATIONS.

1. Breathing air for atmosphere-supplying respirators must be of high purity, meet quality levels for content, and not exceed certain contaminant levels and moisture requirements. Compressed air, compressed oxygen, liquid air, and liquid oxygen used for respiration must be in accordance with the following requirements:

   - Compressed and liquid oxygen must meet the United States Pharmacopoeia for medical or breathing oxygen.
   - Compressed breathing air must meet at least the requirements for Grade D breathing air as described in the ANSI/Compressed Gas Association Commodity Specification for Air, G-7.1-1989.

2. Compressed oxygen must not be used in atmosphere-supplying respirators, including open circuit SCBA's, that have previously used compressed air. This prohibition is intended to prevent fires and explosions that could result if high-pressure oxygen comes into contact with oil or grease that has been introduced to the respirator or the air lines during compressed-air operations. In addition, oxygen in concentrations greater than 23.5% can only be used in equipment designed for oxygen service or distribution.

3. Breathing air may be supplied to respirators from cylinders or air compressors. Where cylinders are used, they must be tested and maintained as prescribed in the Shipping Container Specification Regulations of the Department of Transportation (49 CFR parts 173 and 178). Cylinders of purchased breathing air must have a certificate of analysis from the supplier stating that the air meets the requirements for Grade D breathing air. The moisture content of the compressed air in the cylinder cannot exceed a dew point of-50°F (-45.6°C) at 1 atmosphere pressure. This requirement will prevent respirator valves from freezing, which can occur when excess moisture accumulates on the valves. All breathing gas containers must be marked in accordance with the NIOSH respirator certification standard, 42 CFR part 84.

B. OTHER SPECIFIC REQUIREMENTS.

1. Where compressors are used for supplying air, the compressor must be constructed and situated so contaminated air cannot enter the air-supply system. The location of the air intake is very important, and must be in an uncontaminated area where exhaust gases from nearby vehicles, the internal combustion engine that is powering the compressor itself (if applicable), or other exhaust gases being ventilated from the plant will not be picked up by the compressor air intake.

2. In addition, compressors must be equipped with suitable in-line, air-purifying sorbent beds and filters to further ensure breathing air quality, and to minimize moisture content so that the dew point at 1 atmosphere pressure is 10°F (5.56°C) below the ambient temperature. Sorbent beds and filters must be maintained and replaced or refurbished periodically according to the manufacturer's recommendations, and a tag must be kept at the compressor indicating the most recent change date and the signature of the person authorized by the employer to perform the change.

3. For compressors that are not oil-lubricated, the employer must ensure that carbon monoxide levels do not exceed 10 ppm. This requirement can be met by several different methods, including the use of continuous

carbon monoxide alarms, carbon monoxide sorbent materials, proper air intake location in an area free of contaminants, frequent monitoring of air quality, or the use of high-temperature alarms and automatic shutoff devices, as appropriate. Employers have flexibility in selecting the method(s) most appropriate for conditions in their workplace. Since no single method will be appropriate in all situations, several methods may be needed. For example, it may be necessary to combine the use of a carbon monoxide alarm with a carbon monoxide sorbent bed where conditions are such that a reliable carbon monoxide-free area for air intake cannot be found.

4. Oil-lubricated compressors can produce carbon monoxide if the oil enters the combustion chamber and is ignited. This problem can be particularly severe in older compressors with worn piston rings and cylinders. Consequently, if an oil-lubricated compressor is used, it must have a high-temperature or carbon monoxide alarm, or both, to monitor carbon monoxide levels. If only a high-temperature alarm is used, the air from the compressor must be tested for carbon monoxide at intervals sufficient to prevent carbon monoxide in the breathing air from exceeding 10 ppm.

5. Breathing air couplings must be incompatible with outlets for non-respirable plant air or other gas systems to prevent accidental servicing of air line respirators with non-respirable gases or oxygen. Also, no asphyxiating substance must be allowed in the breathing air lines.

XI. PROGRAM LOGISTICS.

A. IDENTIFICATION OF FILTERS, CARTRIDGES, AND CANISTERS. The employer must ensure that all filters, cartridges, and canisters used in the workplace are labeled and color coded with the NIOSH approval label, and ensure that the label is not removed and remains legible.

B. TRAINING AND INFORMATION.

1. Employee training is an important part of the respiratory protection program and is essential for correct respirator use. The OSHA respiratory protection standard requires employers to provide training before the employee uses a respirator in the workplace. For the training to be effective, the training information must be comprehensive and presented in an understandable way.

2. Employers should develop training programs based upon the employees' educational level and language background. Such an approach will ensure that all employees receive training that enables them to maximize the effectiveness of the respirators they use. As a result of this training, the employee will be able to understand the operation of the respirator and demonstrate the ability to properly use the respirator.

3. Employee training must include a discussion of why the use of the respirator is necessary. Such training would address the identification of the hazards involved, the extent of employee exposures to those hazards, and the potential health effects of such exposures.

4. Information regarding the consequences of improper fit, usage, or maintenance on respirator effectiveness must also be provided to employees. Inadequate attention to any of these program elements would obviously defeat the effectiveness of the respirator. Proper fit, usage, and maintenance of respirators are critical to ensure employee protection.

5. Employees must also be provided with an explanation of the limitations and capabilities of the respirator selected for employee use. A discussion of the limitations and capabilities of the respirator must address how the respirator operates. This training would include, for example, an explanation of how the respirator provides protection by either filtering the air, absorbing the vapor or gas, or providing clean air from an uncontaminated source. Where appropriate, it should include limitations on the use of the equipment, such as prohibitions against using an air-purifying respirator in IDLH atmospheres and an explanation of why such a respirator should not be used in these situations.

6. Employees must also know how to use the respirator effectively in emergency situations, including those in which the respirator malfunctions. Comprehensive training is necessary where respirators are used in IDLH situations, including oxygen-deficient atmospheres such as those that occur in fire fighting, rescue operations, and confined-area entry.

7. Training must include the procedures for inspecting the respirator, donning and removing it, checking the fit and respirator seal, and actually wearing the respirator. Employees must also be capable of recognizing any problems that may threaten the continued protective capability of the respirator. The training must include the steps employees are to follow if they discover any problems during inspection, that is, who the problems are to be reported to and where they can obtain replacement equipment if necessary.

8. Instructions must be given to respirator users regarding the proper procedures for maintenance and storage of respirators. The extent of training may vary according to workplace conditions. In some cases, where employees are responsible for performing some or all respirator maintenance and for storing respirators while not in use, detailed training in maintenance and storage procedures may be necessary. In other facilities, where specific personnel or central repair facilities are assigned to perform these tasks, most employees may need to be informed only of the maintenance and storage procedures without having to learn detailed technical information. By providing this training, respirator users will be able to identify respirator deficiencies that can result from improper maintenance and storage of respirators so that they will not use improperly functioning respirators.

9. The training program must also provide employees with medical information that is sufficient for them to recognize the signs and symptoms of medical conditions (e.g., shortness of breath, dizziness) that may limit or prevent the effective use of respirators. Employee knowledge of this information is important to ensure implementation of a successful respirator program.

10. In addition to specific training requirements regarding the proper use of respirators, employees must be informed of the general requirements of the OSHA respiratory protection standard. This discussion could simply inform employees that employers are obligated to develop a written program, properly select respirators, evaluate respirator use and correct deficiencies in use, conduct medical evaluations, provide for the maintenance, storage, and cleaning of respirators, and retain and provide access to specific records. Thus, employees will know in general what the employer's obligations are under the standard with respect to employee protection.

11. At a minimum, annual training is required by the OSHA respiratory protection standard. With few exceptions, a new employee must be provided with respirator training prior to using a respirator in the workplace. OSHA believes that annual training is necessary and appropriate to ensure that employees know about the respiratory protection program and that they cooperate and actively participate in the program. Training and interaction with respirator instructors on at least an annual basis reinforces employee knowledge about the correct use of respirators and other pertinent elements of the respiratory protection program. It also builds employee confidence when using respirators.

12. Under some conditions, additional training will be required to supplement the annual training. Circumstances which require additional training include situations where changes in the workplace (e.g., process changes, increase in exposure, emergence of new hazards) or the type of respirator used by the employee render previous training obsolete. Additional training is also required when the employee has not retained the requisite understanding or skill to use the respirator properly, or when any other situation arises in which retraining appears necessary.

C. PROGRAM EVALUATION.

1. The employer must conduct evaluations of the workplace as necessary to ensure that the provisions of the current written respirator program are being properly implemented for all employees required to use respirators. In addition, evaluations must be conducted to ensure the continued effectiveness of the program. Evaluations of the workplace will determine whether the correct respirators are being used and worn properly, and will also serve to determine whether the training program is effective.

2. The employer must regularly consult with employees wearing respirators to ascertain the employees' views on program effectiveness and to identify any problems. This assessment must determine if the respirators are properly fitted. It must also evaluate whether: employees are able to wear the respirators without interfering with effective workplace performance; respirators are correctly selected for the hazards encountered; respirators

are being worn when necessary; and respirators are being maintained properly. The employer must correct any problems associated with wearing a respirator that are identified by employees, or that are revealed during any other part of this evaluation.

D. RECORDKEEPING. The OSHA respiratory protection standard requires the employer to establish and retain written information regarding medical evaluations, fit testing, and the respirator program. This information will promote employee involvement in the respirator program, assist the employer in auditing the adequacy of the program, and provide a record for compliance determinations by OSHA.

1. The employer must retain a medical evaluation record for each employee subject to medical evaluation. This record is to include the result of the medical questionnaire and, if applicable, a copy of the PLHCP's written opinion and recommendations, including the results of relevant medical examinations and tests. Records of medical evaluations must be retained and made available as required by 29 CFR 1910.1020, OSHA's Access to Employee Exposure and Medical Records rule.

2. Fit test records must be retained for respirator users until the next fit test is administered. These records consist of:

   - Name or identification of the employee tested;
   - Type of fit test performed (QLFT, QNFT -- irritant smoke, saccharin, etc.);
   - Make, model, and size of the respirator fitted;
   - Date of the fit test;
   - Pass/fail results if a QLFT is used; or
   - Fit factor and strip chart recording or other record of the test results if quantitative fit testing was performed.

3. If the employee's use of a respirator is discontinued (e.g., because of a change of duties or successful implementation of engineering controls), fit test records need not be retained for the employee. Fit test records must be maintained to determine whether annual fit testing has been done, and whether the employee who was tested passed the QLFT, or passed the QNFT with a fit factor that was appropriate for the type of respirator being used.

4. All written materials required to be maintained under the recordkeeping requirements must be made available, upon request, to the employee who is subject of the records and to the Assistant Secretary for OSHA or designee for examination and copying.

E. NIOSH GUIDELINES FOR THE SELECTION AND USE OF PARTICULATE RESPIRATORS. In June 1995, NIOSH updated and modernized the Federal Regulation for certifying air-purifying particulate respirators [42 CFR part 84]. As a consequence of this new regulation, NIOSH developed a User's Guide to familiarize respirator users with the new Part 84 certification regulations for particulate respirators, and to provide guidance for the selection and use of the new particulate respirators. The new regulation became effective on July 10, 1995, and replaces 30 CFR part 11 under which NIOSH and the Mine Safety and Health Administration (MSHA) jointly certified respirators before that date. The respirators certified under this new regulation are tested under much more demanding conditions than under the old regulation to provide increased worker protection. See Appendix VIII:2-4 of this chapter for a summary of the NIOSH Guide to the Selection and Use of Particulate Respirators Certified Under 42 CFR 84.

APPENDIX VIII: 2-1. GLOSSARY.

Air-purifying respirator a respirator with an air-purifying filter, cartridge, or canister that removes specific air contaminants by passing ambient air through the air-purifying element.

Assigned protection factor (APF) [reserved]

Atmosphere-supplying respirator a respirator that supplies the respirator user with breathing air from a source independent of the ambient atmosphere, and includes supplied-air respirators (SAR's) and self-contained breathing apparatus (SCBA) units.

Canister or cartridge a container with a filter, sorbent, or catalyst, or a combination of these items, that removes specific contaminants from the air passed through the container.

Demand respirator an atmosphere-supplying respirator that admits breathing air to the facepiece only when a negative pressure is created inside the facepiece by inhalation.

Emergency situation any occurrence such as, but not limited to, equipment failure, rupture of containers, or failure of control equipment that may or does result in an uncontrolled substantial release of an airborne contaminant.

Employee exposure an exposure to a concentration of an airborne contaminant that would occur if the employee were not using respiratory protection.

End-of-service-life indicator (ESLI) a system that warns the respirator user of the approach of the end of adequate respiratory protection; for example, that the sorbent is approaching saturation or is no longer effective.

Escape-only respirator a respirator intended to be used only for emergency exit.

Filtering facepiece (dust mask) a negative pressure particulate respirator with a filter as an integral part of the facepiece or with the entire facepiece composed of the filtering medium.

Filter or air purifying element a component used in respirators to remove solid or liquid aerosols from the inspired air.

Fit factor a quantitative estimate of the fit of a particular respirator to a specific individual, and typically estimates the ratio of the concentration of a substance in ambient air to its concentration inside the respirator when worn.

Fit test the use of a protocol to qualitatively or quantitatively evaluate the fit of a respirator on an individual. See also "Qualitative fit test (QLFT)" and "Quantitative fit test (QNFT)."

Helmet a rigid respiratory inlet covering that also provides head protection against impact and penetration.

High efficiency particulate air (HEPA) filter a filter that is at least 99.97% efficient in removing monodisperse particles of 0.3 micrometers in diameter and larger. The equivalent NIOSH 42 CFR part 84 particulate filters are the N100, R100, and P100 filters.

Hood a respiratory inlet covering that completely covers the head and neck, and may also cover portions of the shoulders and torso.

Immediately dangerous to life or health (IDLH) an atmosphere that poses an immediate threat to life, would cause irreversible adverse health effects, or would impair an individual's ability to escape from a dangerous atmosphere.

Interior structural firefighting the physical activity of fire suppression, rescue or both, inside of buildings or enclosed structures that are involved in a fire situation beyond the incipient stage.

Loose-fitting facepiece a respiratory inlet covering that is designed to form a partial seal with the face.

Maximum use concentration (MUC) [reserved]

Negative pressure respirator (tight fitting) a respirator in which the air pressure inside the facepiece is negative during inhalation with respect to the ambient air pressure outside the respirator.

Oxygen deficient atmosphere an atmosphere with an oxygen content below 19.5% by volume.

Physician or other licensed health care professional (PLHCP) an individual whose legally permitted scope of practice (i.e., license, registration, or certification) allows him or her to independently provide, or be delegated the responsibility to provide, some or all of the health care services required by 29 CFR 1910.134(e), "Medical evaluation."

Positive-pressure a respirator in which the pressure inside the respiratory inlet covering exceeds the ambient air pressure outside the respirator.

Powered air-purifying respirator (PAPR) an air-purifying respirator that uses a blower to force the ambient air through air-purifying elements to the inlet covering.

Pressure demand respirator a positive pressure atmosphere-supplying respirator that admits breathing air to the facepiece when the positive pressure is reduced inside the facepiece by inhalation.

Qualitative fit test (QLFT) a pass/fail fit test to assess the adequacy of respiratory fit that relies on the individual's response to the test agent.

Quantitative fit test (QNFT) an assessment of the adequacy of respirator fit by numerically measuring the amount of leakage into the respirator.

Respiratory inlet covering the portion of a respirator that forms the protective barrier between the user's respiratory tract and an air-purifying device or breathing air source, or both. It may be a facepiece, helmet, hood, suit, or a mouthpiece respirator with nose clamp.

Self-contained breathing apparatus (SCBA) an atmosphere-supplying respirator for which the breathing air source is designed to be carried by the user.

Service life the period of time that a respirator, filter or sorbent, or other respiratory equipment provides adequate protection to the wearer.

Supplied-air respirator (SAR) or airline respirator an atmosphere-supplying respirator for which the source of breathing air is not designed to be carried by the user.

Tight-fitting facepiece a respiratory inlet covering that forms a complete seal with the face.

User seal check an action conducted by the respirator user to determine if the respirator is properly seated to the face.

APPENDIX VIII:2-2. USER SEAL CHECK.

A. Facepiece Positive and/or Negative Pressure Checks

Positive Pressure Check

Close off the exhalation valve and exhale gently into the facepiece. The face fit is considered satisfactory if a slight positive pressure can be built up inside the facepiece without any evidence of outward leakage of air at the seal. For most respirators, this method of leak testing requires the wearer to first remove the exhalation valve cover before closing off the exhalation valve, and then carefully replacing it after the test.

Negative Pressure Check

Close off the inlet opening of the canister or cartridge(s) by covering it with the palm of the hand(s) or by replacing the filter seal(s). Inhale gently so that the facepiece collapses slightly, and hold your breath for ten seconds. The design of the inlet opening of some cartridges cannot be effectively covered with the palm of the hand, which requires that the test be performed by covering the inlet opening of the cartridge with a thin latex or nitrile glove. If the facepiece remains in its slightly collapsed condition, and no inward leakage of air is detected, the tightness of the respirator is considered satisfactory.

B. Manufacturer's Recommended User Seal Check Procedures

The respirator manufacturer's recommended procedures for performing a user seal check may be used instead of the positive and/or negative pressure check procedures, provided that the employer demonstrates that the manufacturer's procedures are equally effective in detecting seal leakage compared to the positive pressure and negative pressure checks described above.

APPENDIX VIII:2-3. RECOMMENDED PROCEDURES FOR CLEANING RESPIRATORS.

These procedures are provided for employer use when cleaning respirators. They are general in nature, and the employer, as an alternative, may use the cleaning recommendations provided by the manufacturer of the respirators used by their employees, provided such procedures are as effective as those listed here. Equivalent effectiveness simply means that the procedures used must accomplish the objectives set forth in this Appendix (i.e., must ensure that the respirator is properly cleaned and disinfected in a manner that prevents damage to the respirator and does not cause harm to the user).

A. Remove filters, cartridges, or canisters. Disassemble facepieces by removing speaking diaphragms, demand or pressure-demand valve assemblies, hoses, or any components recommended by the manufacturer. Discard or repair any defective parts.

B. Wash components in warm (43°C/110°F maximum) water with a mild detergent or with a cleaner recommended by the manufacturer. A stiff bristle (not wire) brush may be used to facilitate the removal of dirt.

C. Rinse components thoroughly in clean, warm (43°C/110°F maximum), preferably running, water. Drain the components.

D. When the cleaner used does not contain a disinfecting agent, respirator components should be immersed for two minutes in:

- Hypochlorite solution (50 ppm of chlorine) made by adding approximately one milliliter of laundry bleach to one liter of water at 43°C/110°F; or
- Aqueous solution of iodine (50 ppm iodine) made by adding approximately 0.8 milliliters of tincture of iodine (6-8 grams ammonium and/or potassium iodine/100 cc of 45% alcohol) to one liter of water at 43°C/110°F; or
- Other commercially available cleansers of equivalent disinfectant quality when used as directed, if their use is recommended or approved by the respirator manufacturer.

E. Rinse components thoroughly in clean, warm (43°C/110°F maximum), preferably running, water. Drain the components. The importance of thorough rinsing cannot be overemphasized. Detergents or disinfectants that dry on facepieces may result in dermatitis. In addition, some disinfectants may cause deterioration of rubber or corrosion of metal parts if not completely removed.

F. Components should be hand-dried with a clean, lint-free cloth, or air-dried.

G. Reassemble facepiece, replacing filters, cartridges, and canisters where necessary.

H. Test the respirator to ensure that all components work properly.

APPENDIX VIII:2-4. NIOSH GUIDE TO THE SELECTION AND USE OF PARTICULATE RESPIRATORS CERTIFIED UNDER 42 CFR 84.

Summary for Respirator Users

This summary presents a brief overview of what the respirator user needs to know about the new categories of particulate respirators certified by the National Institute for Occupational Safety and Health (NIOSH).

NIOSH has developed a new set of regulations in 42 CFR 84 (also referred to as "Part 84") for testing and certifying nonpowered, air-purifying, particulate-filter respirators. The new Part 84 respirators have passed a more demanding certification test than the old respirators (e.g., dust and mist [DM], dust, fume and mist [DFM], spray paint, pesticide, etc.) certified under 30 CFR 11 (also referred to as "Part 11").

Changes in the new regulations involve only nonpowered, air-purifying, particulate-filter respirators. Certification requirements for all other classes of respirators (e.g., chemical cartridges, self-contained breathing apparatus [SCBA], airlines, gas masks without a particulate filter, powered air-purifying respirators [PAPR's] equipped with high-particulate air [HEPA] filters, etc.) have been transferred to Part 84 without change. Until further notice, the Occupational Safety and Health Administration (OSHA) is allowing the continued use of Part 11 particulate-filter respirators. Under Part 84, NIOSH is allowing manufacturers to continue selling and shipping Part 11 particulate filters as NIOSH-certified until July 10, 1998.

The new Part 84 regulation provides for nine classes of filters (three levels of filter efficiency, each with three categories of resistance to filter efficiency degradation). The three levels of filter efficiency are 95%, 99%, and 99.97%. The three categories of resistance to filter efficiency degradation are labeled N, R, and P. The class of filter will be clearly marked on the filter, filter package, or respirator box. For example, a filter marked N95 would mean an N-series filter that is at least 95% efficient. Chemical cartridges that include particulate filter elements will carry a similar marking that pertains only to the particulate filter element.

Filter efficiency is the stated percentage of particles removed from the air. Filter efficiency degradation is defined as a lowering of filter efficiency or a reduction in the ability of the filter to remove particles as a result of workplace exposure.

The new classes of nonpowered particulate respirators require new decision logic for selection of the proper respirator. The selection process for using the new particulate classification is outlined as follows and is discussed in Section II of NIOSH Guide to the Selection and Use of Particulate Respirators Certified Under 42 CFR 84:

1. The selection of N-, R-, and P-series filters depends on the presence or absence of oil particles, as follows:

    - If no oil particles are present in the work environment, use a filter of any series (i.e., N-, R-, or P-series).
    - If oil particles (e.g., lubricants, cutting fluids, glycerine, etc.) are present, use an R-or P-series filter. Note: N-series filters cannot be used if oil particles are present.
    - If oil particles are present and the filter is to be used for more than work shift, use only a P-series filter.

Note: To help you remember the filter series, use the following guide:

N for Not resistant to oil
R for Resistant to oil
P for oil-Proof

2. Selection of filter efficiency (i.e., 95%, 99%, or 99.97%) depends on how much filter leakage can be accepted. Higher filter efficiency means lower filter leakage.

3. The choice of facepiece depends on the level of protection needed -- that is, the assigned protection factor (APF) needed.

    Call 1-800-35-NIOSH (1-800-356-4674) for additional information or for free single copies of the complete document NIOSH Guide to the Selection and Use of Particulate Respirators Certified Under 42 CFR 84 [DHHS (NIOSH) Publication No. 96-101].

    NIOSH is the National Institute for Occupational Safety and Health, Centers for Disease Control and Prevention, Public Health Service, U.S. Department of Health and Human Services.

SECTION X: CHAPTER 1

METRIC SYSTEM CONVERSION

Contents:

I. INTRODUCTION.

A. Use of metric measurement standards in the United States has been authorized by law since 1866. In 1988, Congress enacted legislation (Public Law 00-418) to establish the metric system as the preferred system of weights and measures for all domestic trade and commerce. This legislation also required the use of metric measurement standards in all Federal activities. On July 25, 1991, the President issued Executive Order 12770, which reiterated the order to implement the metric system "as the preferred system of weights and measures for United States trade and commerce." This executive order directed all Federal agencies to implement "metrification" to the extent economically feasible by September 30, 1992.

B. In 1988, when Congress enacted metrification legislation, the U.S. was the only major industrial nation not using the metric system. As a result, domestic industries were often at a competitive disadvantage and were sometimes even excluded from the international marketplace. In its deliberations, Congress also found that the inherent simplicity of the metric system could lead to major cost savings, further enhance our international competitive position, and provide substantial advantage to the Federal Government in its own operations. Commercially, metric measure is sometimes used as a household measure in the U.S., e.g. two-liter soft drinks. Most food labels list both metric and English measure. Metric measure labeling has opened markets for business products in areas which use metric measure daily.

C. OSHA's safety compliance operations and industrial hygiene efforts have an advantage in metrification because the biological, chemical, and physical sciences have long used the metric system. Students of these subjects have been using metric weights and measures along with daily use of the "English" system of measures for decades. Time-weighted averages (TWA), permissible exposure limits (PEL), and sampling and reporting forms all make use of the metric system. For example, the industrial hygiene Air Sampling Report, OSHA-91(S) (Rev. 1/84) requires data as metric measures and weights. Also examine OSHA reference Table Z-1-A (29 CFR 1910 Subpart Z).

II. THE METRIC SYSTEM.

A. BASE UNITS.

1. Originally, there were only two basic reference points, the meter for length and the gram for weight. The list of reference points has been expanded for the International System (SI) to include the kilogram (instead of the gram) for mass, the second for time, the ampere for electric current, the degree Kelvin for temperature, and the candela for light intensity (see Table X:1-1).

2. The reference objects for these International standards units are maintained for comparisons and checked periodically against other international references by the National Institute for Standards and Technology (NIST). The base units used in the metric system are given in Table X:1-1 below.

B. POWERS OF TEN.

1. The metric system entails the use of multiples or powers of ten to describe magnitudes greater or less than the basic units of meter, gram, ampere, and so forth. The prefixes for the multiples and submultiples of basic units are presented in Table X:1-2. For example, mass or weight can be presented as:

- the kilogram is 1000 grams; and
- the milligram is 1/1000 gram.

In addition, there are:

- 100 centimeters to one meter; and
- 1000 meters to one kilometer.

Also, volume can be presented as:

- 10 deciliters in one liter;
- 10 liters in one decaliter; and
- 10 decaliters in one hectoliter.

TABLE X:1-1. METRIC BASE UNITS

Unit Measurement	***Unit name***	***symbol***
Length	meter	m
Mass	kilogram	kg
Time	second	s
Electric current	ampere	A
Thermodynamic temperature	Kelvin	K
Amount of substance	mole	mol
Luminous intensity	candela	cd

2. Area and volume in both inch-pound and the metric system are calculated similarly. Area is length times width with the measure of each side in the same units for both length and width for square feet (feet2) or square meters (m2). Volume is likewise length times width times height: cubic yards (yard3) or cubic meters (m3). The metric measure of a liter is defined as a volume of 1000 cubic centimeters (cc) or a volume of 10 cm by 10 cm by 10 cm.

3. In physics, some measurements are defined by several metric units. These derived units are named, generally, for the scientists who first defined the measurement. The metric expression as a combination of metric units is presented in Table X:1-3 below.

III. CONVERSION.

A. THE CONVERSION PROCESS.

1. The conversion equivalent for an inch is 1 inch equals 2.54 centimeters or 25.4 millimeters. The measure and units to be converted are multiplied by the conversion factor to yield the new measure and units. Do include the units in the conversion calculation and cancel the units as part of the calculation process (Table X:1-4).

2. Each conversion factor has the metric units per inch-pound units and should be included in the calculation converting inch-pound units to metric units until the use of the conversion factor is ingrained in the user. During a calculation, like units divided by like units equal one and cancel themselves, as shown in Table X:1-4 below.

TABLE X:1-2. PREFIXES AND SYMBOLS FOR DECIMAL MULTIPLES AND SUBMULTIPLES OF UNITS1

Power of 10	***Prefix***	***Symbol***
10^{12}	tera	T
10^{9}	giga	G
10^{6}	mega	M
10^{3}	kilo	k
10^{2}	hecto	h
10^{1}	deca*	da
10^{-1}	deci	d
10^{-2}	centi	c
10^{-3}	milli	m
10^{-6}	micro	m
10^{-9}	nano	n
10^{-12}	pico	p
10^{-15}	femto	f
10^{-18}	atto	a

* Also "deka."

[1] Conférence générale des Poid et Mesures, *Comtes rendus de séances de la 11e Conférence générale des Poids et Mesures,* Paris 1960, Gauthier-Villars, Paris 1961, page 87; Conférence générale des Poids et Mesures, *Comtes rendus des séances de la 12e Conférence générale des Poids et Mesures,* Paris 1964, Gauthier-Villars, Paris, 1964, page 94.

3. Because of the number of conversion factors presented in Federal Standard 376A (May 5, 1983), include with the numeric conversion factor both metric and inch-pound unit labels during the calculations. Cancel like units during the calculation. Then only the new measurement unit labels will remain (see Table X:1-4).

B. CONVERSION PRECISION.

1. Some conversion factors have been developed to seven significant digits, e.g. Federal Standard 376A includes the conversion factor for the foot (survey) as used in the United States as:

   1 ft (survey) = 0.3048006 m

2. When using a conversion factor, round resulting calculations to the number of significant digits of the original inch-pound measurement or to fewer significant decimal positions.

3. Precision is not increased by converting from inch-pound to metric by a multiple-digit conversion factor.

TABLE X:1-3. DERIVED SI UNITS WITH SPECIAL NAMES

	Unit name	*Symbol*	*Formula*	*Expressions as metric base units*
Frequency	hertz	Hz	1/s or s^{-1}	s^{-1}
Force	newton	N	m • kg • s^{-2}	m • kg • s^{-2}
Energy, Work	joule	J	N • m	m^2 • kg • s^{-2}
Power	watt	W	J/s	m^2 • kg • s^{-3}
Electric charge	coulomb	C	A • s	A • s
Electric potential	volt	V	W/A	m^2 • kg • s^{-3} • A^{-1}
Electric resistance	ohm	W	V/A	m^2 • kg • s^{-3} • A^{-2}
Electric capacitance	farad	F	C/V	m^{-2} • kg^{-1} • s^4 • A^2
Magnetic flux	weber	Wb	V • s	m^2 • kg • s^{-2} • A^{-1}
Pressure, Stress	pascal	Pa	N/m^2	m^{-1} • kg • s^{-2}
Conductance	siemens	S	A/V	m^{-2} • kg^{-1} • s^3 • A^2
Magnetic flux density	tesla	T	Wb/m^2	kg • s^{-2} • A^{-1}
Inductance	henry	H	Wb/A	m^2 • kg • s^{-2} • A^{-2}
Luminous flux	lumen	lm	cd • sr	cd • sr
Illuminance	lux	lx	lm/m^2	m^{-2} • cd • sr

TABLE X:1-4. AN EXAMPLE OF AN INCH-POUND TO SI METRIC CONVERSION

Original inch-pound measure	*Conversion*	*Calculated metric measure*
1 inch	2.54 cm/inch 1 inch • 2.54 cm/inch 1 inch • 2.54 cm/inch	2.54 cm (Or 25.4 mm)

4. Calculations or conversion of a measurement from inch-pound to metric measure requires the application of three rules of calculation, often done by habit.

5. Significant number of digits. A distance shown as 61 miles has two significant digits, implying measurement accuracy to a whole number of miles. A distance written as 61.6 miles is three significant digits, two plus one decimal digit and implies accuracy to one-tenth of a mile. The number of digits presented usually reflects the accuracy of the actual measurement. The figures in 61 miles or 61.2 miles may represent an automobiles odometer measurement. A surveyors measurement may be represented as 61.243 miles from measurement stake point to stake point: five significant digits including three decimals.

6. Number of carried digits.

   a. If the conversion number will result in a larger number, 1.609 km/mi, carry only the total number of first number digits, e.g.:

      - 61 mi × 1.609 km/mi calculates to 98.149 km but is presented as 98 km.
      - 61.2 mi × 1.609 km/mi calculates to 98.4708 km but is presented as 98.5 km.
      - 61.243 mi × 1.609 km/mi calculates to 98.539987 km but is presented as 98.540 km.

   b. If the conversion number will result in a smaller number, carry only the total number of first number digits plus one extra digit, e.g.:

      - 4 lb × 0.453 lb/kg calculates to 1.812 kg but is presented as 1.8 kg (for measurement by a household scale).
      - 4.3 lb × 0.453 lb/kg calculates to 1.9479 kg but is rounded up and is presented as 1.95 kg (for measurement by a postal scale).
      - 4.33 lb × 0.453 lb/kg calculates to 1.96149 kg is presented with no rounding as 1.961 kg (for measurement by a laboratory scale).

7. "Rounding" of numbers. To "round off" calculated numbers to a minimum number of significant digits, follow these rules:

   a. When a number such as 13.34 is to be rounded to 13.3+, the digit 4 is less than 5 so the number is rounded to an unchanged 13.3.

   b. When a number as 13.37 is to be rounded to 13.3+, the digit 7 is equal to or greater than 5 so the number is rounded or increased to a changed 13.4.

C. CONVERSION EQUIVALENTS.

1. The document to be used as a primary reference for inch-pound to metric conversions is Federal Standard 376A, Preferred Metric Units for General Use by the Federal Government, May 5, 1983. A selected number of conversion factors are presented in Tables X:1-5 through X:1-15 below.

2. The basic units of the measurement system are by definition to man-made reference standards for mass or weight (the kilogram) and length (the meter). Some units of measure are mathematically derived, as plane and solid angle measures (from geometry), or derived from astronomical phenomena, time, or are not related to either mass or length measurement references. The units for these measures have not changed. A measure of frequency, cycles per second (cps), has been renamed Hertz to honor the German physicist H. R. Hertz.

3. Measures of light were defined in physics originally in terms of the metric system and have not changed. If the measure of light was converted to the inch-pound system, and includes inch-pound area measures as part of its definition, a conversion to metric equivalents is necessary. These exceptions to light measurements are presented in the appropriate table.

4. Metrification or establishment of the measure system as a base of ten is extended to temperature measures.

   a. G.D. Fahrenheit, the German physicist, defined a scale of measure for his alcohol thermometer in 1709 and used the same scale again in 1714 for his mercury thermometer. On this scale, water freezes at 32° and boils at 212°. The base of ten temperature scale definition of Anders Celsius, the Swedish astronomer, is used in the metric system for temperature measures. The point at which water freezes or becomes a solid is 0° Celsius and the point of transition of water from a liquid to a gas is at 100° Celsius.

      [These definitions assume that the atmospheric pressure is at a standard of 760 mm of mercury pressure.]

b. The relationship between the arbitrary Fahrenheit scale and the Celsius scale used within the metric system is:

$$\frac{degrees\,Fahrenheit - 32}{9}\,x5 = degrees\,Celsius$$

TABLE X:1-5. LENGTH

$\frac{\text{1.609347 kilometers (km)}}{\text{1 mile (mi) [statute]}}$
$\frac{\text{1.609344 kilometers (km)}}{\text{1 mile (mi) [international]}}$
$\frac{\text{1.828 meters (m)}}{\text{1 fathom (fm)}}$
$\frac{\text{0.9144 meter (m)}}{\text{1 yard (yd)}}$
$\frac{\text{0.3048006 meter (m)}}{\text{1 foot (ft) [survey]}}$
$\frac{\text{0.3048 meter (m)}}{\text{1 foot (ft) [international]}}$
$\frac{\text{2.54 centimeters (cm)}}{\text{1 inch (in)}}$
$\frac{\text{25.4 millimeters (mm)}}{\text{1 inch (in)}}$
$\frac{25.4\ \mu m}{\text{1 mil}}$
$\frac{1\ \mu m}{\text{1 micron}}$
$\frac{0.0254\ \mu m}{1\ \mu in}$
$\frac{\text{0.1 nm}}{\text{1 angstrom (Å)}}$

The equation must be applied for the conversion. For every nine-degree interval on the Fahrenheit temperature scale (°F), the Celsius temperature scale (°C) changes only five units. Please note that at the freezing point for water, the Celsius scale is at zero but the Fahrenheit scale is at 32°. It is convenient to remember that only at -40° are both scales equivalent, or : -40°C = -40°F

TABLE X:1-6. AREA

$\frac{\text{2.589998 square kilometers (km}^2\text{)}}{\text{1 square mile (m}^2\text{)}}$
$\frac{\text{0.4046873 hectare (ha)}}{\text{1 acre}}$
$\frac{\text{4046.873 square meters (m}^2\text{)}}{\text{1 acre}}$
$\frac{\text{0.8361274 square meter (m}^2\text{)}}{\text{1 square yard (yd}^2\text{)}}$
$\frac{\text{0.09290304 square meter (m}^2\text{)}}{\text{1 square foot (ft}^2\text{)}}$
$\frac{\text{6.4516 square centimeters (cm}^2\text{)}}{\text{1 square inch (in}^2\text{)}}$
$\frac{\text{645.16 square millimeters (cm}^2\text{)}}{\text{1 square inch (in}^2\text{)}}$

TABLE X:1-7. VELOCITY

0.3048 meter/second (m/s)
1 foot/second (ft/s)

1.609347 kilometers/hour (km/h)
1 mile/hour (mi/h)

TABLE X: 1-8. VOLUME

1233.489 cubic meters (m^3)
1 acre foot (acre ft)

0.7645549 cubic meter (m^3)
1 cubic yard (yd^3)

0.1589873 cubic meter (m^3)
1 barrel (bbl) *[42 gallons]*

0.02831685 cubit meter (m^3)
1 cubic foot (ft^3)

28.31685 liters (l)
1 cubic foot (ft^3)

0.002359737 cubic meter (m^3)
1 board foot (bd ft)

35.23907 liters (l)
1 bushel (bu)

3.785412 liters (l)
1 gallon (gal) *[liquid]*

0.9463529 liter (l)
1 quart (qt) *[liquid]*

0.4731765 liter (l)
1 pint (pt) *[liquid]*

29.57353 milliliters (ml or cc)
1 fluid ounce (fl oz)

16.38706 cm^3
1 in^3

TABLE X: 1-9. MASS WEIGHT

1.016047 metric tons (t)
1 long ton
[where 1 t = 1000 kg]

0.90718474 metric ton (t)
1 short ton
[where 1 short ton = 2000 pounds]

0.45359237 kilogram (kg)
1 pound (lb avdp)

31.10348 grams (g)
1 ounce (oz troy)

28.34952 grams (g)
1 ounce (oz avdp)

64.79891 milligrams (mg)
1 grain (gr)

TABLE X:1-10. DENSITY

1.186553 metric ton/m^3 / 1 short ton/yd^3

16.01846 kg/m^3 / 1 lb/ft^3

119.8264 g/l / 1 lb/gal

7.489152 g/l / 1 oz/gal

TABLE X:1-11. PRESSURE

101.325 kPa / 1 atmosphere (atm) *[standard]*

100 kPa / 1 bar

6.894757 kPa / 1 lbf/in^2 *[formerly psi]*

3.38638 kPa / 1 inch Hg *[at 32°F]*

2.98898 kPa / 1 foot H_2O *[at 39.2°F]*

0.24884 kPa / 1 inch H_2O *[at 60°F]*

0.133322 kPa / 1 mm Hg *[at 0°C]*

0.1 kPa / 1 millibar

TABLE X:1-12. STRESS

6.894757 MPa / 1 kip/in^2 (formerly ksi)

0.006894757 MPa / 1 lbf/in^2 (formerly psi)

6.894757 kPa / 1 lbf/in^2 (formerly psi)

0.04788026 kPa / 1 lbf/ft^2

TABLE X:1-13. WORK

3.6 MJ / 1 kWh
4.184 J / 1 cal th
1.055056 kJ / 1 Btu
1.355818 J / 1 ft • lbf
105.4808 MJ / 1 therm (U.S.)

TABLE X:1-14. POWER

3.516800 kW / 1 ton (refrigeration)
1.055056 kW / 1 Btu/s
0.7456999 kW / 1 hp (550 lbf/s)
0.746 kW / 1 hp (electric)
0.2930711 W / 1 Btu/h

TABLE X:1-15. MISCELLANEOUS

Quantity	Conversion
Heat flow rate	1.055056 kW / 1 Btu/s
Heat capacity	1.899108 kJ/°C / 1 Btu/°F
Magnetic field strength	79.57747 A/m / 1 oersted (Oe)
Conductivity	100 S/m / 1 mho/cm
Wavelength	0.1 nm / 1 angstrom (Å)
Luminance	0.3183099 cd/cm^2 / 1 lambert (L)
	0.1550003 cd/cm^2 / 1 cd/in^2
	3.426259 cd/cm^2 / 1 ft-lambert (ft-L)
	10.76391 lx / 1 foot-candle (fc)

IV. REFERENCES.

General Services Administration. Federal Standard 376A, *Preferred Metric Units for General Use by the Federal Government*, May 5, 1983.

Government Institutes Mini-Catalog

PC #	ENVIRONMENTAL TITLES	Pub Date	Price
629	ABCs of Environmental Regulation: Understanding the Fed Regs	1998	$49
627	ABCs of Environmental Science	1998	$39
585	Book of Lists for Regulated Hazardous Substances, 8th Edition	1997	$79
579	Brownfields Redevelopment	1998	$79
4088	CFR Chemical Lists on CD ROM, 1997 Edition	1997	$125
4089	Chemical Data for Workplace Sampling & Analysis, Single User Disk	1997	$125
512	Clean Water Handbook, 2nd Edition	1996	$89
581	EH&S Auditing Made Easy	1997	$79
587	E H & S CFR Training Requirements, 3rd Edition	1997	$89
4082	EMMI-Envl Monitoring Methods Index for Windows-Network	1997	$537
4082	EMMI-Envl Monitoring Methods Index for Windows-Single User	1997	$179
525	Environmental Audits, 7th Edition	1996	$79
548	Environmental Engineering and Science: An Introduction	1997	$79
643	Environmental Guide to the Internet, 4rd Edition	1998	$59
560	Environmental Law Handbook, 14th Edition	1997	$79
353	Environmental Regulatory Glossary, 6th Edition	1993	$79
625	Environmental Statutes, 1998 Edition	1998	$69
4098	Environmental Statutes Book/CD-ROM, 1998 Edition	1997	$208
4994	Environmental Statutes on Disk for Windows-Network	1997	$405
4994	Environmental Statutes on Disk for Windows-Single User	1997	$139
570	Environmentalism at the Crossroads	1995	$39
536	ESAs Made Easy	1996	$59
515	Industrial Environmental Management: A Practical Approach	1996	$79
510	ISO 14000: Understanding Environmental Standards	1996	$69
551	ISO 14001: An Executive Report	1996	$55
588	International Environmental Auditing	1998	$149
518	Lead Regulation Handbook	1996	$79
478	Principles of EH&S Management	1995	$69
554	Property Rights: Understanding Government Takings	1997	$79
582	Recycling & Waste Mgmt Guide to the Internet	1997	$49
603	Superfund Manual, 6th Edition	1997	$115
566	TSCA Handbook, 3rd Edition	1997	$95
534	Wetland Mitigation: Mitigation Banking and Other Strategies	1997	$75

PC #	SAFETY and HEALTH TITLES	Pub Date	Price
547	Construction Safety Handbook	1996	$79
553	Cumulative Trauma Disorders	1997	$59
559	Forklift Safety	1997	$65
539	Fundamentals of Occupational Safety & Health	1996	$49
612	HAZWOPER Incident Command	1998	$59
535	Making Sense of OSHA Compliance	1997	$59
589	Managing Fatigue in Transportation, *ATA Conference*	1997	$75
558	PPE Made Easy	1998	$79
598	Project Mgmt for E H & S Professionals	1997	$59
552	Safety & Health in Agriculture, Forestry and Fisheries	1997	$125
613	Safety & Health on the Internet, 2nd Edition	1998	$49
597	Safety Is A People Business	1997	$49
463	Safety Made Easy	1995	$49
590	Your Company Safety and Health Manual	1997	$79

GOVERNMENT INSTITUTES ORDER FORM

4 Research Place, Suite 200 • Rockville, MD 20850-3226
Tel (301) 921-2323 • Fax (301) 921-0264
Internet: http://www.govinst.com • E-mail: giinfo@govinst.com

3 EASY WAYS TO ORDER

1. Phone: **(301) 921-2323**
   Have your credit card ready when you call.
2. Fax: **(301) 921-0264**
   Fax this completed order form with your company purchase order or credit card information.
3. Mail: **Government Institutes**
   4 Research Place, Suite 200
   Rockville, MD 20850-3226 USA
   Mail this completed order form with a check, company purchase order, or credit card information.

PAYMENT OPTIONS

❏ **Check** *(payable to Government Institutes in US dollars)*

❏ **Purchase Order** *(This order form must be attached to your company P.O. Note: All International orders must be prepaid.)*

❏ **Credit Card** ☐ VISA ☐ MasterCard ☐ American Express

Exp.___/___

Credit Card No. ______________________________

Signature ______________________________

(Government Institutes' Federal I.D.# is 13-2695912)

CUSTOMER INFORMATION

Ship To: (Please attach your purchase order)

Name: ______________________________

GI Account # (*7 digits on mailing label*): ______________

Company/Institution: ______________________________

Address: ______________________________
(Please supply street address for UPS shipping)

____

City: ______________ State/Province: ______________

Zip/Postal Code: ______________ Country: ______________

Tel: (   ) ______________________________

Fax: (   ) ______________________________

Email Address: ______________________________

Bill To: (if different from ship-to address)

Name: ______________________________

Title/Position: ______________________________

Company/Institution: ______________________________

Address: ______________________________
(Please supply street address for UPS shipping)

____

City: ______________ State/Province: ______________

Zip/Postal Code: ______________ Country: ______________

Tel: (   ) ______________________________

Fax: (   ) ______________________________

Email Address: ______________________________

Qty.	Product Code	Title	Price

❏ **New Edition No Obligation Standing Order Program**

Please enroll me in this program for the products I have ordered. Government Institutes will notify me of new editions by sending me an invoice. I understand that there is no obligation to purchase the product. This invoice is simply my reminder that a new edition has been released.

Subtotal ________

MD Residents add 5% Sales Tax ________

Shipping and Handling (see box below) ________

Total Payment Enclosed ________

Within U.S:	**Outside U.S:**
1-4 products: $6/product	Add $15 for each item (Airmail)
5 or more: $3/product	Add $10 for each item (Surface)

15 DAY MONEY-BACK GUARANTEE

If you're not completely satisfied with any product, return it undamaged within 15 days for a full and immediate refund on the price of the product.

SOURCE CODE: BP01